Vol. 51 1997

Theilheimer's
Synthetic Methods
of Organic Chemistry

Editor **Alan F. Finch, Cambridge**

Assistant Editor Gillian Tozer, Derwent Information
Editorial Consultant William Theilheimer, Nutley, N.J.

Basel · Freiburg
Paris · London
New York · New Delhi
Bangkok · Singapore
Tokyo · Sydney

Deutsche Ausgaben

Vol. 1	1946	1. Auflage
	1948	2., unveränderte Auflage
	1950	3., unveränderte Auflage
Vol. 2	1948	
Vol. 3	1949	with English Index key
	1953	2., unveränderte Auflage
	1966	3., unveränderte Auflage
	1975	4., unveränderte Auflage
Vol. 4	1950	with English Index key
	1966	2., unveränderte Auflage

English Editions

Vol. 1	1948	Interscience Publishers
	1975	(Karger) Second Edition
Vol. 2	1949	Interscience Publishers
	1975	(Karger) Second Edition
Vol. 5	1951	with Reaction Titles Vol. 1-5 and Cumulative Index
	1966	Second Edition
Vol. 6	1952	
	1975	Second Edition
Vol. 7	1953	
	1975	Second Edition
Vol. 8	1954	
	1975	Second Edition
Vol. 9	1955	
Vol. 10	1956	with Reaction Titles Vol. 6-10 and Cumulative Index
	1975	Second Edition
Vol. 11	1957	
	1975	Second Edition
Vol. 12	1958	
	1975	Second Edition
Vol. 13	1959	
	1975	Second Edition
Vol. 14	1960	
	1975	Second Edition

Vol. 15	1961	with Reaction Titles Vol. 11-15 and Cumulative Index
Vol. 16	1962	
Vol. 17	1963	
Vol. 18	1964	
Vol. 19	1965	
Vol. 20	1966	with Reaction Titles Vol. 16-20 and Cumulative Index
Vol. 21	1967	
Vol. 22	1968	
Vol. 23	1969	
Vol. 24	1970	
Vol. 25	1971	with Reaction Titles Vol. 21-25 and Cumulative Index
Vol. 26	1972	
Vol. 27	1973	
Vol. 28	1974	
Vol. 29	1975	
Vol. 30	1976	with Reaction Titles Vol. 26-30 and Cumulative Index
Vol. 31	1977	
Vol. 32	1978	
Vol. 33	1979	
Vol. 34	1980	
Vol. 35	1981	with Reaction Titles Vol. 31-35 and Cumulative Index
Vol. 36	1982	
Vol. 37	1983	
Vol. 38	1984	
Vol. 39	1985	
Vol. 40	1986	with Reaction Titles Vol. 36-40 and Cumulative Index
Vol. 41	1987	
Vol. 42	1988	
Vol. 43	1989	
Vol. 44	1990	
Vol. 45	1991	with Reaction Titles Vol. 41-45 and Cumulative Index
Vol. 46	1992	
Vol. 47	1993	
Vol. 48	1994	
Vol. 49	1995	
Vol. 50	1996	

Library of Congress, Cataloging-in-Publication Data

Theilheimer's synthetic methods of organic chemistry = Synthetische Methoden der organischen Chemie. – Vol. 51 (1997) - Basel; New York: Karger, © 1982 -
v.
Continues: Synthetic methods of organic chemistry.
Editor: Alan F. Finch.
1. Chemistry, Organic – yearbooks   I. Finch, Alan F.   II. Theilheimer, William, 1914–
ISBN 3-8055-6529-1

All rights reserved.
No part of this publication may be translated into other languages, reproduced or utilized in any form or by any means, electronic or mechanical, including photocopying, recording, microfilming, or by any information storage and retrieval system, without permission in writing from the publisher.

© Copyright 1997 by S. Karger AG, Basel (Switzerland), and Derwent Information Ltd., London
Distributed by S. Karger AG, Allschwilerstrasse 10, P.O. Box, CH-4009 Basel (Switzerland)
Printed in Switzerland on acid-free paper by Schüler AG, Biel
ISBN 3-8055-6529-1

Theilheimer's
Synthetic Methods
of Organic Chemistry

Vol. 51

Contents

Preface to Volume 51	VI
Advice to the User	VII
General Remarks	VII
Methods of Classification	VIII
High-Coverage Searches	X
Trends and Developments in Synthetic Organic Chemistry 1997	XI
Systematic Survey	XXI
Abbreviations and Symbols	XXIII
Reactions	1
Reviews	238
Subject Index	246
Supplementary References	287

Preface

Volume *51* of the *Theilheimer* series represents an important departure from the norm: we shall now be publishing *two* volumes *per annum* so that the reader can appraise, even more quickly, the key developments in the field of synthetic organic chemistry. The style and format are the same as before, but the volume of data (and, of course, price!), are approximately half that of the last annual volume. Volume *52* will be published in the autumn, and subsequent volumes in the Spring and Autumn of each year.

This particular volume contains abstracts and supplementary data from papers published mainly in the latter half of 1995 and the first half of 1996. For browsing purposes, these are displayed according to the Systematic Classification (symbol notation) so that reactions of the same type and associated data appear together. For example, all deprotections appear in the early sections (under HO⇅C, HN⇅C, HS⇅C); reduction of oxo compds. and carbon-carbon multiple bonds under HC⇊OC and HC⇊CC, respectively; peptide coupling under NC⇅O; halogenation under HalC⇅H; syntheses involving C–C bond formation in the latter half of the book; and data on resolutions (Res) at the end. A list of reaction symbols and reference thereto is given in the Systematic Survey (p. XXI).

The displayed data are supported by the customary in-depth Subject Index, and access to supplementary data can be made in the usual manner via the Supplementary References Index, e.g. the reader interested in updates to asym. dihydroxylation (Synth. Meth. *47*, 114), will note from p. 293 that additional references can be found on p. 40 of this volume.

As usual, the volume contains a 'Reviews' section (p. 238, covering reviews published up to and including December 1996, and a 'Trends' section (p. XI) incorporating key developments in synthetic chemistry up to February 1997. These latter references will appear as abstracts in the next volume.

I would like to express my gratitude to Dr. Theilheimer for his encouragement in the preparation of these yearbooks, and to my colleagues at Derwent Information Ltd., London, whose *Journal of Synthetic Methods* provides the data for inclusion in these volumes. A special thank you goes to David Penn and Kath Ince of Derwent for developing and overseeing the new electronic processing system by which these volumes are now published.

April 1997 *A.F. Finch*, Editor

Advice to the User

General Remarks

New methods for the synthesis of organic compounds and improvements of known methods are being recorded continuously in this series.

Reactions are classified on a simple though purely formal basis by symbols, which can be arranged systematically. Thus searches can be performed without knowledge of the current trivial or author names (e.g. 'Oxidation' and 'Friedel-Crafts reaction').

Users accustomed to the common notations will find these in the subject index. By consulting this index, use of the classification system may be avoided. It is thought that the volumes should be kept close at hand. The books should provide a quick survey, and obviate the immediate need for an elaborate library search. Syntheses are therefore recorded in the index by starting materials and end products, along with the systematic arrangement for the methods. This makes possible a sub-classification within the reaction symbols by reagents, a further methodical criterion. Complex compounds are indexed with cross reference under the related simpler compounds. General terms, such as synthesis, replacement, heterocyclics, may also be brought to the attention of the reader.

A brief review, *Trends and Developments in Synthetic Organic Chemistry*, stresses highlights of general interest and calls attention to key methods too recent to be included in the body of the text.

The abstracts are limited to the information needed for an appraisal of the applicability of a desired synthesis. In order to carry out a particular synthesis it is therefore advisable to have recourse to the original papers or, at least, to an abstract journal. In order to avoid repetition, selections are made on the basis of most detailed description and best yields whenever the same method is used in similar cases. Continuations of papers already included will not be abstracted, unless they contain essentially new information. They may, however, be quoted at the place corresponding to the abstracted papers. These supplementary references (see page 287) make it possible to keep abstracts of previous volumes up-to-date.

Syntheses that are divided into their various steps and recorded in different places can be followed with the help of the notations such as *startg. m. f.* (starting material for the preparation of ...).

Method of Classification

Reaction Symbols. As summarized in the Systematic Survey (pXXI), reactions are classified firstly according to the bond formed in the synthesis, secondly according to the reaction type, and thirdly according to the bond broken or the element eliminated. This classification is summarized in the reaction symbol, e.g.

$$\underset{\text{Bond formed}}{\nearrow} \overset{OC \Uparrow N}{\underset{\text{Reaction type}}{\uparrow}} \underset{\text{Bond broken or element eliminated}}{\nwarrow}$$

The first part of the symbol refers to the chemical bond formed during the reaction, expressed as a combination of the symbols for the two elements bonded together, e.g. HN, NC, CC. The order of the elements is as follows:

H, O, N, Hal (Halogen), S, Rem (Remaining elements), and C.

Thus, for the formation of a hydrogen-nitrogen bond, the notation is HN, not NH.

If two or more bonds are formed in a reaction, the 'principle of the latest position' applies. Thus, for the reduction

RCH=O + H$_2$ ⟶ R–CH(H)-OH

in which both hydrogen-oxygen and hydrogen-carbon bonds are formed, the symbol is HC⇓OC and not HO⇓OC.

The second part of the symbol refers to the reaction type. Four types are distinguished: addition (⇓), rearrangement (∩), exchange (⇅), and elimination (⇑), e.g.

RCH=CH$_2$ + H$_2$O ⟶ R–CH(OH)-CH$_3$		OC⇓CC
(thiophene-allyl) ⟶ (cyclized product)		CC∩SC
R-Cl + CN⁻ ⟶ R-CN [+ Cl⁻]		CC⇅Hal
R-CH(Br)-CH$_3$ ⟶ RCH=CH$_2$ [+ HBr]		CC⇑Hal

Monomolecular reactions are either rearrangements (∩), where the molecular weight of the starting material and product are the same, or eliminations (⇑), where an organic or inorganic fragment is lost; bimolecular and multicomponent reactions are either additions (⇓), such as intermolecular

Diels-Alder reactions, Michael addition and 1,4-addition of organometallics, or exchanges (⇵), such as substitutions and condensations, where an organic or inorganic fragment is lost.

The last part of the symbol refers to the essential bond broken or, in the case of exchange reactions and eliminations, to a characteristic fragment which is lost. While the addition symbol is normally followed by the two elements denoting the bond broken, in the case of valency expansion, where no bonds are broken, the last part of the symbol indicates the atom at which the addition occurs, e.g.

R_2S ⟶ R_2SO		OS⇵S
RONO ⟶ $RONO_2$		ON⇵N

For addition, exchanges, and eliminations, the 'principle of the latest position' again applies if more than one bond is broken. However, for rearrangements, the most descriptive bond-breakage is used instead. Thus, for the thio-Claisen rearrangement depicted above, the symbol is CC∩SC, and not CC∩CC.

Deoxygenations, quaternizations, stable radical formations, and certain rare reaction types are included as the last few methods in the yearbook. The reaction symbols for these incorporate the special symbols El (electron pair), Het (heteropolar bond), Rad (radical), Res (resolutions), and Oth (other reaction types), e.g.

$R_2S=O$ ⟶ R_2S		ElS⇵O
R_3N + R'Cl ⟶ R_3N^+R' Cl⁻		Het⇵N

The following rules simplify the use of the reaction symbols:

1. The chemical bond is rigidly classified according to the structural formula without taking the reaction mechanism into consideration.

2. Double or triple bonds are treated as being equivalent to two or three single bonds, respectively.

3. Only stable organic compounds are usually considered: intermediates such as Grignard compounds and sodiomalonic esters, and inorganic reactants, such as nitric acid, are therefore not expressed in the reaction symbols.

Reagents. A further subdivision, not included in the reaction symbols, is based on the reagents used. The sequence of the reagents usually follows that of the periodic system. Reagents made up of several components are arranged according to the element significant for the reaction (e.g. $KMnO_4$ under Mn, NaClO under Cl). When a constituent of the reagent forms part of the product, the remainder of the reagent, which acts as a 'carrier' of this constituent, is the criterion for the classification; for example, phosphorus is the carrier in a chlorination with PCl_5 and sodium in a nitrosation with $NaNO_2$.

High-Coverage Searches

A search through *Synthetic Methods* provides a selection of key references from the journal literature. For greater coverage, as for bibliographies, a supplementary search through the following publications is suggested:

Derwent Reaction Service[1]. Designed for both current awareness and retrospective retrieval. Its monthly publication, the *Derwent Journal of Synthetic Methods*, covers the journal and patent literature, and provides 3,000 abstracts of recently published papers annually.

On-line REACCS and keyword access is available to over 80,000 reactions, including the data in all the abstracts in *Synthetic Methods*.

Science Citation Index[2]. For which *Synthetic Methods* serves as a source of starting references. This is particularly useful for accessing papers quoting details of a particular method which has been included in these volumes from a preliminary communication.

Chemical Abstracts Service[3]. References may not be included in *Synthetic Methods* (1) to reactions which are routinely performed by well known procedures; (2) to subjects which can be easily located in handbooks and indexes of abstracts journals, such as the ring system of heterocyclics or the metal in case of organometallic compounds, and (3) to inadequately described procedures, especially if yields are not indicated.

References to less accessible publications such as those in the Chinese or Japanese language are usually only included if the method in question is not described elsewhere.

---

[1] Derwent Information Ltd., 14 Great Queen Street, London WC2B 5DF, England.
[2] Institute for Scientific Information, Philadelphia, Pa., USA.
[3] Chemical Abstracts Service, Columbus, Ohio, USA.

Trends and Developments in Synthetic Organic Chemistry 1997

Method of the moment is surely ring closing metathesis! Engineered by Grubbs and Schrock with their respective Mo- and Ru-carbene complexes (Synth. Meth. *48*, 988 and *49*, 985), the principle has been extensively adapted for both carbo- and hetero-cyclic synthesis[1], with the added bonus of a polymer-based variant[2]. Watch the review literature, and note developments in the related ring opening metathesis[3].

The methodology of the moment is still combinatorial synthesis of large compound libraries[4]. Here, the emergence of a multitude of polymer-based reactions[5] and the advent of such features as 'traceless' supports[6] have increased the potential enormously, though the real future may well lie in *solution*-based chemistry. Hence the development of *fluorous* 2-phase media, specifically designed for the same rapid and clean separations as possible by polymer-supported routes[7]. One device is to 'label' a substrate with a fluorocarbon substituent, so that the product (still bearing the label) can be extracted from the fluorocarbon phase in which it alone is soluble. In reactions with singlet oxygen in the same medium, it is the sensitizer - a polyfluoroalkylated porphyrin - which is retrievable from the fluorous phase, which further facilitates such oxidations be enhancing the solubility of O_2[8]. The same effect is evident in catalytic epoxidation under catalysis with a polyfluorinated cobalt complex (at 0.1 mol%!)[9].

In the wider context of catalyst regeneration, especially relevant to processes requiring such expensive chiral reagents as Ru-BINAP and Mn-salen complexes, a particular reference should be made to catalysis within elastomeric polydimethylsiloxane membranes[10]. Not only is the reactivity and enantioselectivity of the catalyst retained, the membranes themselves can be simply regenerated with the guest complex by a washing procedure. The same Ru-BINAP catalyst is also active and recoverable in 'molten' ionic salts[11] which, in addition, extend the lifetime of Pd(0) in Heck reactions and facilitate work-up by direct distillation[12].

From the ecological perspective, particularly at the industrial level, non-toxic alternatives to chlorinated solvents are especially welcome, as are new procedures based on aqueous media[13]. Replacement of heavy metals by non-toxic reagents is also at a premium, as in a recent Hunsdiecker reaction for which LiOAc may be used efficiently in place of the conventional HgO or Pb(OAc)$_4$[14]. By the same token, PtCl$_4$ is more 'friendly' than Hg-salts for alkyne hydration to ketones[15], while handling of toxic by-products from tin hydride-mediated syntheses can now be effectively avoided by using a catalytic amount of the reagent in combination with a stoichiometric amount of non-toxic organosilicon hydride[16] or NaBH$_3$CN[17]. In a similar vein, non-toxic allylsilanes can now be used in place of allylstannanes in Hosomi-type asym. synthesis of homoallyl alcohols, reaction being catalyzed by a highly reactive chiral fluorotitanium(IV)-BINOLate as Lewis acid[18]. Note, also, the use of *neutral* pentacoordinate allylsilanes for the same conversion *without* Lewis acid,[19] as well as the same conversion with allylgermanes in aq. media[20].

Under catalysis, certain general aspects merit highlighting. Firstly, the concept of enhancing reactivity by encapsulation within self-assemblies in order to increase reactant concentration[21]; secondly, the notion of encapsulation of complex catalytic species within inorganic supports in order to increase stability, adjust selectivity, and improve work-up, exemplified by a recent alkene oxidation catalyzed by a Ru-porphyrin complex[22]; and thirdly, the advent of 'catalytic membranes' to avoid loss of expensive complexes (*q.v.*)[10]. Under general transition metal catalysis, there is the relatively new concept of bimetal catalysis wherein each component of a bimolecular reaction is independently, but simultaneously, activiated by a different metallic species[23]. A further bonus is utilization of the same catalyst in tandem fashion to combine two mechanistically distinct processes in one operation, as in dehydrogenative O-silylation-intramolecular hydrosilylation with a Rh(I) complex[24]; and lastly, catalysis of a 2-step process reliant on sequential and chemoselective activation of two reactive functionalities[25]. This is illustrated in a total synthesis of the estrane skeleton from an o,b-dibromostyrene, the β-function reacting preferentially in a Pd-catalyzed vinylation, followed by intramolecular arylation in the second step with a second Pd complex[26].

Under Lewis acid catalysis, the aspect of enhancing reactivity by *bidentate* coordination is illustrated by dual carbonyl activation with a bis(dimethylaluminum aroxide)[27], while Lewis acid-base complexation is responsible for the kinetically-controlled generation of the *more substituted* enolate on deprotonation of ketones with LDA in the presence of a bulky aluminum triaroxide[28]. Lewis acid mediation of radical processes is developing, and there are broad applications of water-tolerant lanthanide(III) and scandium(III) triflates, notably in processes requiring regeneration of the catalyst following aqueous work-up (or, indeed, reaction in aqueous media)[29]. Scandium(III) triflimide presents even greater reactivity in as little as 1 mol%[30], as does the parent Brønsted acid Tf$_2$NH[31]. Among supported Lewis acids, a water-tolerant polymer-based scandium(III) N-triflylamide is also available for rapid recovery on work-up[32]. Lewis and Brønsted acid sites on inorganic supports confer an even broader spectrum of reactivity[33], with the potential of adjusting ('tailoring') regioselectivity by surface or internal modification of the support itself. The latter is apparent in a dye-sensitized ene reaction with singlet oxygen within a zeolite support[34], while external surface characteristics determine the regioselectivity of zeolite-catalyzed photo-Fries and -Claisen rearrangement[35].

In the context of asymmetric synthesis, an outline was given in the *Trends* last year of the many and various elements of this key aspect which is central to synthetic organic chemistry. Many new features have surfaced, among which the *in situ*-generation of chiral catalysts from inexpensive racemic complexes by an *activation* process. This is recently reported in an asym. carbonyl ene reaction with a racemic Ti(IV)-BINOL complex in the presence of a chiral activator, which generates a catalytic species affording *greater* face-selectivity than the homochiral Ti(IV)-BINOL complex itself[36]. Interestingly, the same transformation with the latter catalyst in the presence of molecular sieves is considered to involve a μ_3-oxo-species formed by *donation* of an entrapped water molecule from the support[37]. The aspect of double asym. induction by the same chiral auxiliary at two mechanistically distinct stages of a one-pot process is illustrated in a tandem 1,2-migration from a higher-order zincate coupled with a homoaldol condensation[38], while a *two-directional* asym. induction is featured in a double Mukaiyama-type aldol condensation, again with a chiral Ti(IV)-BINOL complex[39]. Interestingly, by appending a salen residue to the latter, enantioselectivity in such condensations can be significantly enhanced at much lower catalyst loading, the preparation of the complex (from Ti(OPr-*i*)$_4$) being effected in the presence of Me$_3$SiCl *without* resorting to azeotropic removal of isopropanol[40].

The many successes with chiral transition metal BINOL complexes, much exalted by the development of Noyori-type asym. reduction of ketones[41], have been paralleled in many other areas of ligand design. The corresponding TADDOLates emerged more recently and are now available in heterogenized form[42], as are the related chiral tartrate ligands much publicized and boosted largely by the development of Sharpless epoxidation[43]. The incorporation of chiral spacer ligands in asym. intramolecular conversions is a relatively new concept, evident in asym. oxidative biaryl coupling[44] and asym. intramolecular [2+2]-cycloaddition[45]. Jacobsen-type Mn-salen complexes have added a new dimension to catalytic asym. oxygen transfer, and it is interesting to note that the handedness of the corresponding Cr-salen complexes is quite the opposite in asym. epoxidation[46]. Asym. nucleophilic addition to epoxides can also be effected by the same chromium complexes, typified by asym. ring opening with azide ion (Synth.Meth. *51*, 121). Ring opening with acetic acid, however, is far more efficient with the corresponding Co-salen complex[47]. The polycyclic chiral diamine, sparteine, has also emerged to excellent effect in the arena of asym. deprotonation[48], and is the control element in asym. electrocatalytic oxidation[49]. It is arguable, however, that chiral Δ^2-oxazolines are ligands of the moment. As both internal and external auxiliaries they feature in a multitude of disparate asym. conversions[50], which, doubtless, can now be optimized with a 'toolbox' of such ligands![51] Chiral ansa-metallocenes, exemplified by Brintzinger-type complexes, are still very much in evidence, as illustrated in a catalytic asym. Pauson-Khand reaction[52]. So too, of course, are the ultimate chiral inducers - enzymes - now routinely applied in asym. reduction and lysis, principally in the context of kinetic resolution of sec. alcohols. Note that the classical kinetic resolution by asym. transesterification with enol esters (Synth. Meth. *44*, 214) can now be performed with 1-ethoxyvinyl acetate *without* the liberation of a by-product which deactivates the enzyme[53]. Their days do, however, appear numbered if we are to judge by the recent entry of a range of non-enzymatic methods to achieve the same end[54]. Perhaps the future lies somewhere in-between with the evolution of such enzyme-like catalysts as polyribofuranosides[55] or, perhaps, low-molecular weight DNA?[56]

Under palladium catalysis, developments in Heck chemistry, Stille and Suzuki coupling[57], allylation, and the principal of atom economy (*à la* Trost) are as abundant today as they were last year. Renewed interest in inter- and intra-molecular Pd-catalyzed N-arylation (Synth. Meth. *51*, 171 and 190) has presented a platform to explore arylation in a broader context. In O-arylation, for example, arylpalladium alkoxides are invoked as intermediates in the conversion of ar. bromides to phenols via *tert*-butyl phenolethers[58], while intramolecular O-arylation of alcohols encompasses the same mechanistic features via a cyclic arylpalladium alkoxide[59]. In the unprecedented formation of benzophenones from ar. aldehydes, an *o*-hydroxy-function serves to deliver the aryl residue via oxidative addition to the hydrogen-carbon bond of the formyl group[60], a similar feature being evident in anion-accelerated intramolecular biaryl formation with a phenolate-activated aryl residue serving as nucleophile[61].

Developments of Stille-type coupling of unsatd. stannanes have perhaps overshadowed those with unsatd. silanes, which is surprising in view of the reduced toxicity of the latter. However, let us hope that the balance will be redressed with the report that silanes can be coupled efficiently with iodonium salts under much milder conditions[62]. Notwithstanding, one still recognises in the chemistry of organotin compds. the unique characteristics of such compounds as 1,2-bis(trialkylstannyl)ethylenes, which, in a recent two-directional approach, have been incorporated ('stapled') into the Taxol A ring by combining organocuprate chemistry at one terminus and Stille coupling at the other[63]. With regard to macrocyclization, an intramolecular Stille coupling with enestannanes has been applied in a (−)-rapamycin synthesis[64], while a Stille-type cyclodimerization can be effected under Cu(I) catalysis[65]. *in situ*-Variants of such coupling processes are welcome[66], as also are polymer-based versions[67], and their enhancement under microwave irradiation[68]. An asym. cascade ring closure via intramolecular cyclopalladation has also been reported[69]. Finally, brief mention of palladium cluster catalysis, which, as in enol lactone cycloisomerization, is more efficient than under conventional control with more familiar catalysts[70]. A Pd-cluster is also effective in catalytic hydrogenation of CFCs[71].

Radical chemistry has been highlighted in recent years, due in some measure to advances in the Giese reaction (radical 1,4-addition) and tin hydride-mediated radical ring closures. A new facet is asym. synthesis of cyclic ketones with a chiral acetal serving as acyl radical equivalent[72]. The generation of oxiranyl radicals by Barton fragmentation is also worthy of note[73], as is radical dehalogenation by H-atom transfer from cyclohexane (solvent) initiated by dilauroyl peroxide[74]. Intramolecular Giese-type addition, extensively applied with the agency of Bu_3SnH, can also be effected under more friendly conditions with SmI_2, as in a recent carbohydrate-carbocycle conversion[75]. This same, extraordinarily versatile, reagent is also implicated in the expanding field of 1,3-dipolar cycloaddition: a g-lactam synthesis, for example, is achievable via Sm(III)-azomethinium ylids[76], while non-stabilized carbonyl ylids can be generated cleanly from 1,1'-dichloroethers[77]. A novel SmI_2-mediated synthesis of γ-lactones has been reported[78], and the catalytic potential of the reagent is manifested in pinacol coupling[79], intramolecular ene reaction[80], and, in the generation of enoxysilanes from ketones under mild conditions[81].

Through the years, there have been sporadic references to the tandem coupling of radical chemistry with anionic processes via single electron transfer, e.g. Synth. Meth. *46,* 790. One feels the methodology would, perhaps, develop more quickly if a 'snappy' label (lets propose 'radion reactions') could be affixed for indexing purposes. In the meantime, let it suffice to feature a new mediator for such processes, $Mn/PbCl_2$, which has emerged in a new synthesis of (E)-γ,δ-ethylenecarboxylic acids via radical 1,4-addition-Claisen rearrangement[82].

Under developments with stoichiometric titanium reagents, 'Sato chemistry' with titanium(II) η^2-olefin complexes is expanding after the initial surge highlighted last year. A titanium version of the Pauson-Khand reaction[83] and a new pyrrole synthesis[84] come to mind. A new olefin synthesis from mercaptals via a titanocene carbene complex has also been reported, together with a route based on benzotriazole, incorporating a low-valent titanium reduction[85].

Syntheses with chromium carbene and carbonyl complexes are no less prevalent[86], and a chromium-mediated asym. Reformatski-type reaction with a chiral N-(α-bromoacetyl)-2-oxazolidone is significant in that diastereoselectivity is the reverse of that obtained with zinc ester enolates [87].

The stabilization of an incipient positive charge at the β-position via aziridinium or thiiranium ions is well documented, but the same effect with a cobalt π-cation is relatively new[88]. So, too, the remarkable 1,2-shift of cobalt carbonyl-complexed alkynyl groups which benefits from the same stabilizing influence[89]. The 'β-effect' of silyl substitution is also well established, and responsible for the enhanced reactivity of cuprates by such functionality[90]. A remarkable conformational effect of silyl substitution has also been registered *across* the acetylene bond in the asym. reduction of α,β-acetyleneketones[91]. Equally notable are effects of *remote* chirality, which, in acyclic systems, are normally considered irrelevant beyond a certain distance from the reaction centre. Magnus et al. now find that chiral information can be transmitted through as many as *nine* achiral connecting atoms, and that such effects are magnified at *alternate* carbon centres along the chain[92]. Long-range transmission of chiral information through the cycloalkene skeleton is thought responsible for the anomalous diastereoselectivity in asym. Michael addition of iminocyclohexanes[93], while a remote chiral spiroketal function dictates the formation of *all-trans*-tricyclic terpenoids in a photosensitized biomimetic polyene cyclization[94]. Lastly, in respect of conformational studies, an entirely new model has been developed to explain the face-selectivity of 1,3,2-oxazaborolidine-catalyzed asym. syntheses with aldehydes, the transition state involving hydrogen-bonding *from the formyl hydrogen* to the boron atom[95].

As chirality issues are so fundamental to synthetic organic chemistry, it is inevitable that one questions the handedness of life from time to time. The fact that circularly polarized light can induce asymmetry[96] lends a certain weight to the argument that cosmic processes were responsible for primeval induction of asymmetry on this planet. However, it is dubious to draw the same conclusion from a recent analysis of amino acids arriving on a meteorite[97]. Is the level of optical activity (e.e. <10%) again a consequence of some cosmic force? Or were these extra-terrestrial amino acids originally homochiral, as our own, and simply racemized (well, let's say ca. 45% racemized!) under the frenzied conditions of entry through our atmosphere?

[1] Recent examples; synthesis of β-lactams s. A.G.M. Barrett et al., Chem. Commun. *1997*, 155-6; of azasugars s. C.M. Huwe, S. Blechert, Synthesis *1997*, 61-7; of cyclic aminoacids s. F.P.J.T. Rutjes, H.E. Schoemaker, Tetrahedron Letters *38*, 677-80 (1997); of S-heterocyclics s. Y.-S. Shon, T.R. Lee, ibid. 1283-6; of brevetoxin ring sub-units s. J.S. Clark, J.G. Kettle, ibid. 123-6; 127-30.

[2] S. Blechert et al., Angew. Chem. Intern. Ed. *35*, 1979-80 (1996).

[3] Method s. *51*, 327; recent adaptation s. S. Blechert et al., Angew. Chem. Intern. Ed. *36*, 257-9 (1997).

[4] Recent reviews s. Synth. Meth. *50*, 555s*51*

[5] Recent 3-component polymer-based synthesis of 3-aminoalcohols s. S. Kobayashi et al., Tetrahedron Letters *37*, 7783-6 (1996); of 5- and 6-membered lactams s. K.M. Short, A.M.M. Mjalli, ibid. *38*, 359-62 (1997); of oligosaccharides s. K.C. Nicolaou et al., J. Am. Chem. Soc. *119*, 449-50 (1997); s.a. J.A. Hunt, W.R. Roush, ibid. *118*, 9998-9 (1996).

[6] Reductive cleavage of a 'traceless' arylthio support s. X. Zhao et al., Tetrahedron Letters *38*, 977-80 (1997); substitutive cleavage of a sulfur-linked support cf. ibid. 211-4; cleavage of phosphonium supports s. I. Hughes, ibid. *37*, 7595-8 (1996).

[7] Overview s. D.P. Curran et al., Science *275*, 823-6 (1997); s.a. Synth. Meth. *50*, 555s*51*.

[8] S.G. DiMagno et al., J. Am. Chem. Soc. *118*, 5312-3 (1996).

[9] G. Pozzi et al., Chem. Commun. *1997*, 69-70; hydroboration in perfluoroalkanes with work-up by molecular O_2 s. I. Klement, P. Knochel, Synlett *1996*, 1004-6.

[10] I.F.J. Vankelecom et al., Angew. Chem. Intern. Ed. *35*, 1346-8 (1996).

[11] A.L. Monteiro et al., Tetrahedron:Asym. *8*, 177-9 (1997).

[12] D.E. Kaufmann et al., Synlett *1996*, 1091-2.

[13] Benzotrifluoride as alternative to methylene chloride s. A. Ogawa, D.P. Curran, J. Org. Chem. *62*, 450-1 (1997); recent procedures in aq. media include: Diels-Alder reaction s. T.-P. Loh et al., Chem. Commun. *1996*, 2315-6; Michael addition s. R. Ballini, G. Bosica, Tetrahedron Letters *37*, 8027-30 (1996); and radical bromination s. H. Shaw et al., J. Org. Chem. *62*, 236-7 (1997).

[14] S. Chowdhury, S. Roy, J. Org. Chem. *62*, 199-200 (1997).

[15] W. Baidossi et al., J. Org. Chem. *62*, 669-72 (1997).

[16] D.S. Hays et al., J. Org. Chem. *61*, 6751-2 (1996).

[17] D. Crich, S. Sun, J. Org. Chem. *61*, 7200 (1996).

[18] D.R. Gauthier, Jr., E.M. Carreira, Angew. Chem. Intern. Ed. *35*, 2363-5 (1996).

[19] M. Kira et al., Organometallics *15*, 5335-41 (1996).

[20] T. Akiyama, J. Iwai, Tetrahedron Letters *38*, 853-6 (1997).

[21] J. Kang, J. Rebek, Jr., Nature *385*, 50-2 (1997).

[22] C.-J. Liu et al., Chem. Commun. *1997*, 65-6; s.a. ref. 34.

[23] M. Sawamura et al., J. Am. Chem. Soc. *118*, 3309-10 (1996).

[24] X. Wang et al., Chem. Commun. *1996*, 2561-2.

[25] Review of sequential Pd-catalyzed reactions s. Synth. Meth. 45, 555s51.
[26] L.F. Tietze et al., Angew. Chem. Intern. Ed. 35, 2259-61 (1996).
[27] T. Ooi et al., J. Am. Chem. Soc. 118, 11307-8 (1996).
[28] S. Saito et al., J. Am. Chem. Soc. 119, 611-2 (1997); Lewis acid-catalyzed asym. Baylis-Hillman reaction s. V.K. Aggarwal et al., Chem. Commun. 1996, 2713-4.
[29] Recent syntheses with La(OTf)$_3$ or Sc(OTf)$_3$ s. A.G.M. Barrett, D.C. Braddock, Chem. Commun. 1997, 351-2; L.-B. Yu et al., J. Org. Chem. 62, 208-11 (1997); S. Kobayashi, S. Nagayama, ibid. 232-3; P.K. Jadhav, H.-W. Man, J. Am. Chem. Soc. 119, 846-7 (1997).
[30] K. Ishihara et al., Synlett 1996, 839-41.
[31] Example s. K. Ishihara et al., Synlett 1996, 1045-6.
[32] S. Kobayashi et al., Tetrahedron Letters 37, 9221-4 (1996); J. Am. Chem. Soc. 118, 8977-8 (1996).
[33] Recent example of deprotection with clay supports s. J. Asakura et al., J. Org. Chem. 61, 9026-7 (1996).
[34] X. Li, V. Ramamurthy, J. Am. Chem. Soc. 118, 10666-7 (1996); reviews of reactions on inorganic supports s. 49, 372s51.
[35] K. Pitchumani et al., J. Am. Chem. Soc. 118, 9428-9 (1996).
[36] K. Mikami, S. Matsukawa, Nature 385, 613-5 (1997); review of chiral metal BINOL catalysts s. 47, 45s51; of chiral auziliaries and reagents s. 47, 646s51.
[37] M. Terada et al., Chem. Commun. 1997, 281-2.
[38] J.C. McWilliams et al., J. Am. Chem. Soc. 118, 11970-1 (1996).
[39] K. Mikami et al., Tetrahedron Letters 38, 579-82 (1997).
[40] R.A. Singer, E.M. Carreira, Tetrahedron Letters 38, 927-30 (1997).
[41] Asym. Ru-catalyzed transfer-hydrogenation of dialkyl ketones with tridentate phosphinobis(Δ^2-oxazolines) cf. Y. Jiang et al., Tetrahedron Letters 38, 215-8 (1997).
[42] D. Seebach et al., Helv. Chim. Acta 79, 1710-40 (1996).
[43] L. Canali et al., Chem. Commun. 1997, 123-4.
[44] B.H. Lipshutz et al., Tetrahedron Letters 38, 753-6 (1997).
[45] S. Faure et al., Tetrahedron Letters 38, 1045-8 (1997).
[46] H. Imanishi, T. Katsuki, Tetrahedron Letters 38, 251-4 (1997); asym α-hydroxylation with a chiral Mn-salen complex s. W. Adam et al., ibid. 37, 6531-4 (1996); review of catalytic oxidation with Mn complexes s. 48, 134s51.
[47] E.N. Jacobsen et al., Tetrahedron Letters 38, 773-6 (1997).
[48] Recent reference s. P. Beak et al., J. Am. Chem. Soc. 118, 12218-9 (1996).
[49] Y. Kashiwagi et al., Chem. Commun. 1996, 2745-6; s.a. Y. Yanagisawa et al., Chem. Lett. 1996, 1043-4.
[50] Asym. homologization with Yb(OTf)$_3$/chiral bis(oxazolines) s. P.K. Jadhav, H.-W. Man, J. Am. Chem. Soc. 119, 846-7 (1997); asym. Diels-Alder reaction with cationic chiral copper(II)-bis(oxazolines) s. D.A. Evans, D.M. Barnes, Tetrahedron Letters 38, 57-8 (1997); asym. synthesis of homoallyl alcohols with Zn(OTf)$_2$/chiral bis(oxazolines) s. P.G. Cozzi et al., ibid. 145-8; asym. induction with a chiral phosphinooxazoline as ligand s. L. Ripa, A. Hallberg, J. Org. Chem. 62, 595-602 (1997); with a chiral phosphinobis(oxazoline) s. ref. 41.
[51] I.W. Davies et al., Tetrahedron Letters 38, 1145-8 (1997).
[52] F.A. Hicks, S.L. Buchwald, J. Am. Chem. Soc. 118, 11688-9 (1996); review of Brintzinger-type complexes s. 50, 15s51.
[53] M. Schudok, G. Kretzschmar, Tetrahedron Letters 38, 387-8 (1997).

54 Recent example s. J.C. Ruble et al., J. Am. Chem. Soc. *119*, 1492-3 (1997); s.a. Y. Kashiwagi et al., Chem. Commun. *1996*, 2745-6; T. Oriyama et al., Tetrahedron Letters *37*, 8543-6 (1996); S. Yamada, T. Ohe, ibid. 6777-80.
55 M.-J. Han et al., Chem. Commun. *1997*, 163-4.
56 R. Rawls, Chem. Eng. News *75*, 33-5 (1997).
57 Nolte: Suzuki coupling with cyclopropaneboronic acids s. X.-Z. Wang, M.-Z. Deng, J. Chem. Soc. Perkin Trans. I *1996*, 2663-4; cf. J.P. Hildebrand, S.P. Marsden, Synlett *1996*, 893-4.
58 G. Mann, J.F. Hartwig, J. Am. Chem. Soc. *118*, 13109-10 (1996); polymer-based N-arylation s. Y.D. Ward, V. Farina, Tetrahedron Letters *37*, 6993-6 (1996); s.a. C.A. Willoughby, K.T. Chapman, ibid. 7181-4.
59 M. Palucki et al., J. Am. Chem. Soc. *118*, 10333-4 (1996).
60 T. Satoh et al., Chem. Lett. *1996*, 823-4.
61 D.D. Hennings et al., J. Org. Chem. *62*, 2-3 (1997).
62 S.-K. Kang et al., Tetrahedron *53*, 3027-44 (1997); coupling with halides using NaOH in place of fluoride ion cf. E. Hagiwara et al., Tetrahedron Letters *38*, 439-42 (1997); copper(I)-catalyzed Stille and Suzuki coupling with iodonium salts cf. S.-K. Kang et al., J. Org. Chem. *61*, 9082-3 (1996).
63 F. Delalogue et al., Tetrahedron Letters *38*, 237-40 (1997).
64 A.B. Smith, III et al., J. Am. Chem. Soc. *119*, 962-73 (1997).
65 I. Paterson, J. Man, Tetrahedron Letters *38*, 695-8 (1997).
66 Example of *in situ*-Suzuki coupling s. Synth. Meth. *51*, 394; one-pot hydrostannylation-Stille coupling s. C.D.J. Boden et al., J. Chem. Soc. Perkin Trans. I *1996*, 2417-9.
67 Polymer-based Suzuki coupling s. S.R. Piettre, S. Baltzer, Tetrahedron Letters *38*, 1197-200 (1997); cf. J.W. Guiles et al., J. Org. Chem. *61*, 5169-71 (1996).
68 Heck arylation, Stille and Suzuki coupling s. M. Larhed, A. Hallberg, J. Org. Chem. *61*, 9582-4 (1996); polymer-based variants s. Tetrahedron Letters *37*, 8219-22 (1996).
69 S.P. Maddaford et al., J. Am. Chem. Soc. *118*, 10766-73 (1996).
70 M. Hidai et al., Angew. Chem. Intern. Ed. *35*, 2123-4 (1996).
71 R. Vilar et al., Chem. Commun. *1997*, 285-6.
72 D. Stien et al., J. Org. Chem. *62*, 275-86 (1997).
73 F.E. Ziegler, Y. Wang, Tetrahedron Letters *37*, 6299-302 (1996).
74 J. Boivin et al., Chem. Commun. *1997*, 353-4; radical addition s. M. Denieul et al., ibid. *1996*, 2511-2.
75 Z. Zhou, S.M. Bennett, Tetrahedron Letters *38*, 1153-6 (1997).
76 C. Alvarez-Ibarra et al., J. Org. Chem. *62*, 479-84 (1997).
77 M. Hojo et al., Tetrahedron Letters *37*, 9241-4 (1996).
78 S. Fukuzawa et al., J. Am. Chem. Soc. *119*, 1482-3 (1997).
79 R. Nomura et al., J. Am. Chem. Soc. *118*, 11666-7 (1996); with stoichiometric SmI_2 cf. T. Honda, M. Katoh, Chem. Commun. *1997*, 369-70.
80 T.K. Sarkar, S.K. Nandy, Tetrahedron Letters *37*, 5195-8 (1996).
81 J. Hydrio et al., Synthesis *1997*, 68-72.
82 K. Takai et al., J. Org. Chem. *61*, 8728-9 (1996); 1,4-addition-aldol condensation s. ibid. 7990-1; review of radical-ion ring closure s. *48*, 818s*51*.
83 F. Sato et al., J. Am. Chem. Soc. *118*, 8729-30 (1996); with $Cp_2Ti(CO)_2$ cf. F.A. Hicks et al., ibid. 9450-1.

[84] F. Sato et al., Tetrahedron Letters *37*, 7787-90 (1996); synthesis of 1,3-dienes s. ibid. 8865-8; condensed cyclopropanes s. J. Am. Chem. Soc. *118*, 8729-30 (1996).
[85] Y. Horikawa et al., J. Am. Chem. Soc. *119*, 1127-8 (1997); benzotriazole route s. A.R. Katritzky, J. Li, J. Org. Chem. *62*, 238-9 (1997).
[86] Asym. 1,4-addition-alkylation of chromium aryl(alkoxy)carbene complexes s. J. Barluenga et al., J. Am. Chem. Soc. *118*, 13099-100 (1996).
[87] T. Gabriel, L. Wessjohann, Tetrahedron Letters *38*, 1363-6 (1997).
[88] G.B. Gill et al., Tetrahedron Letters *37*, 9369-72 (1996); stereochemistry s. L.M. Grubb, B.P. Branchaud, J. Org. Chem. *62*, 242-3 (1997).
[89] T. Nagasawa et al., J. Am. Chem. Soc. *118*, 8949-50 (1996).
[90] S.H. Bertz et al., J. Am. Chem. Soc. *118*, 10906-7 (1996).
[91] E.J. Corey et al., J. Am. Chem. Soc. *118*, 10938-9 (1996).
[92] P. Linnane et al., Nature *385*, 799-801 (1997).
[93] M.J. Lucero, K.N. Houk, J. Am. Chem. Soc. *119*, 826-7 (1997).
[94] C. Heinemann, M. Demuth, J. Am. Chem. Soc. *119*, 1129-30 (1997).
[95] E.J. Corey, J.J. Rohde, Tetrahedron Letters *38*, 37-40 (1997).
[96] Y. Inoue et al., Chem. Commun. *1996*, 2627; with chiral sensitizers cf. J. Am. Chem. Soc. *119*, 472-8 (1997).
[97] J.R. Cronin, S. Pizzarello, Science *275*, 951-5 (1997).

Systematic Survey

Reaction symbol	Page
HO∩OC	1
HO⇅Rem	1
HO⇅C	2
HO⇑O	5
HN⇅O	5
HN⇅N	7
HN⇅S	9
HN⇅C	9
HS⇅C	11
HC⇓OC	11
HC⇓NC	17
HC⇓CC	18
HC∩HO	20
HC⇅O	20
HC⇅N	22
HC⇅Hal	22
HC⇅S	24
HC⇅Rem	25
HC⇅C	25
HC⇑O	26
ON⇓N	26
ON⇅H	26
OS⇓HO	27
OS⇓S	27
OS⇅C	29
ORem⇓HO	29
ORem⇓OC	30
ORem⇓Rem	30
ORem⇅O	30
ORem⇅Hal	31
ORem⇅C	31
OC⇓HC	31
OC⇓OO	32
OC⇓OC	32
OC⇓NC	34
OC⇓CC	35
OC∩HC	40
OC∩ON	41
OC∩OS	41
OC∩NC	41
OC∩CC	41
OC⇅H	42
OC⇅O	44
OC⇅N	48
OC⇅Hal	50
OC⇅S	52
OC⇅Rem	53
OC⇅C	55
OC⇑H	58
OC⇑O	60
OC⇑Hal	60
OC⇑S	61
OC⇑Rem	61
OC⇑C	61
NN⇓N	63
NN⇅H	63
NN⇅O	63
NN⇑H	63
NS⇅Rem	64
NRem⇅H	64
NRem⇅N	64
NC⇓ON	65
NC⇓OC	65
NC⇓NN	67
NC⇓NC	67
NC⇓CC	68
NC∩OC	70
NC∩CC	70
NC⇅H	70
NC⇅O	73
NC⇅N	81
NC⇅Hal	84
NC⇅S	87
NC⇅Rem	87
NC⇅C	89
NC⇑H	90
NC⇑O	90
NC⇑N	93
NC⇑Hal	93
NC⇑Rem	94
NC⇑C	94
HalRem⇅O	95
HalC⇓OC	95
HalC⇓SC	95
HalC⇓CC	96
HalC⇅H	97
HalC⇅O	101
HalC⇅N	102
HalC⇅Hal	103
HalC⇅S	103
HalC⇅Rem	104
HalC⇅C	104
SS⇅H	105
SS⇅O	105

Reaction Symbol	Page				
SS↕N	105	RemC⇓OC	115	CC∩CC	169
SS↕Hal	106	RemC⇓NC	115	CC↕H	171
SS↕C	106	RemC⇓CC	116	CC↕O	172
SRem↕C	106	RemC↕O	118	CC↕N	182
SC⇓OC	106	RemC↕N	119	CC↕Hal	186
SC⇓SS	107	RemC↕Hal	119	CC↕S	197
SC⇓CC	107	RemC↕Rem	122	CC↕Rem	200
SC∩OS	107	RemC↕C	124	CC↕C	216
SC∩OC	108	RemC⇑O	125	CC⇑H	219
SC↕H	108	CC⇓HC	125	CC⇑O	220
SC↕O	108	CC⇓OC	125	CC⇑N	223
SC↕N	111	CC⇓NC	139	CC⇑Hal	224
SC↕Hal	111	CC⇓SC	142	CC⇑S	229
SC↕S	112	CC⇓RemC	142	CC⇑Rem	231
SC↕Rem	113	CC⇓CC	142	CC⇑C	232
SC↕C	113	CC∩HO	161	ElN⇑O	234
SC⇑O	114	CC∩HN	161	ElN⇑O	234
SC⇑Hal	114	CC∩HC	162	ElS⇑C	234
RemRem∩ORem	114	CC∩OC	164	ElRem⇑Hal	235
RemRem↕H	115	CC∩NC	167	Het⇓N	235
RemRem↕C	115	CC∩SC	168	Het⇓S	235
		CC∩RemC	168	Res	235

Abbreviations and Symbols

abs.	absolute
alc.	alcoholic
aq.	aqueous
ar.	aromatic
atm.	atmosphere(s)
compd(s).	compound(s)
deriv(s).	derivative(s)
e.e.	enantiomeric excess
eq(s).	equivalent(s)
E.	Example
F.e.s.	Further example(s) see
M	molar
prepn.	preparation
prim.	primary
s*51*	supplementary reference in Volume *51*
sec.	secondary
startg. m.f.	starting material for (the preparation of ...)
subst.	substituted
sym.	symmetrical
tert.	tertiary
v.i.	via intermediates
w.a.r.	without additional reagents
Y *	Yield
↯	Electrolysis
⚡	Irradiation
○	Ring closure
◔	Ring contraction
◑	Ring expansion
↶	Ring opening
⊕	Ring hydrogenation
←	'see title on the left half of the page'

* Yields in parentheses refer to the immediately preceeding step of a multi-step reaction

Formation of H-O Bond

Rearrangement ↻

Oxygen/Carbon Type   HO ↻ OC

Via intermediates   v.i.
β-Hydroxyketones from 2,3-epoxyalcohols   C(OH)CHCO
Stereospecific conversion via N-(2,3-epoxyalkyl)pyridinium salts

1.

3-Deoxy-4-ketosugars. Benzyl 2,3-anhydro-β-L-ribopyranoside treated with 1 eq. Tf₂O in pyridine at room temp. for 3 h (or in py/CH₂Cl₂ at 0°), the resulting soln. of pyridinium salt (Y ca. 100%) diluted with an equal volume of water, heated at reflux (100°) for 3 h, solvent removed under reduced pressure, the residue taken up in water, treated with NaBH₄ at 0° for 2 h, and acidified with 0.1 N HCl → product. Y 82%. The pyridinio group facilitates ring opening of the epoxide. F.e. incl. cyclohexan-3-ol-1-one s. W. Voelter et al., Angew. Chem. Intern. Ed. **35**, 523-4 (1996).

Exchange ⇅

Remaining Elements ↑   HO ⇅ Rem

Potassium carbonate s. under MeOH/CCl₄   K₂CO₃

Sodium sulfide   Na₂S
Preferential O-de-*tert*-butoxy(diphenyl)silylation s. *46*, 3s*51*   OSi(OBu-*t*)Ph₂ → OH

Ammonium ceric nitrate   (NH₄)₂Ce(NO₃)₆
Tris[trinitratocerium(IV)] paraperiodate   [(NO₃)₃Ce]₂·H₂IO₆
Oxidative cleavage of silyl ethers under mild conditions   OSi≤ → OH

2.

Selective O-de-*tert*-butyldimethylsilylation. A soln. of startg. silyl ether in methanol treated with 1.2 eqs. CAN at 0°, and stirred for 15 min → product. Y 95%. Cleavage was also effective with 10 mol% CAN over a longer period, and ketal groups remained unaffected. F.e. and **cleavage of tetrahydropyran-2-yl ethers**, also selective cleavage of prim. *tert*-butyldimethylsilyl over sec. tetrahydro-2-furyl ethers, s. A. DattaGupta et al., Synlett *1996*, 69-71; cleavage of trimethylsilyl ethers with [(NO₃)₃Ce]₂·H₂IO₆ s. H. Firouzabadi, F. Shiriny, Synth. Commun. **26**, 423-32 (1996).

3.
Methanol/carbon tetrachloride or Potassium carbonate   *MeOH/CCl$_4$ or K$_2$CO$_3$*
*Tetra-*n-*butylammonium fluoride·boron fluoride*   *Bu$_4$NF·BF$_3$*
Preferential O-de-*tert*-butyldimethylsilylation   OSiMe$_2$Bu-*t* → OH

A soln. of startg. silyl ether in 1:1 methanol/carbon tetrachloride sonicated in a commercial ultrasonic bath (Crest 575-D, 39 kHz) at 40-50° for 1.5 h → product. Y 93%. The solvent mixture (pH 2) acts both as solvent and reagent under these conditions, and work-up is easy. *tert*-Butyldimethylsilyl derivatives of phenols, sec. and tert. alcohols were stable, and functional groups such as esters, ethers, amines, chlorides, ketones, aldehydes and amides were tolerated. F.e.s. A.S.-Y. Lee et al., Tetrahedron Letters *36*, 6891-4 (1995); with K$_2$CO$_3$, 3'-protected nucleosides, s. L. Le Hir de Fallois et al., ibid. 9479-80; with Bu$_4$NF·BF$_3$, also retention of triisopropylsilyl and *tert*-butyldiphenylsilyl ethers, s. S. Kawahara et al., ibid. *37*, 509-12 (1996).

Bis(acetonitrile)dichloropalladium(II)   *PdCl$_2$(MeCN)$_2$*
Preferential O-desilylation   OSi≤ → OH
with PdO/cyclohexene cf. *46*, 3; cleavage of aryl tri(m)ethyl- and *tert*-butyldimethyl-silyl ethers with PdCl$_2$(MeCN)$_2$ s. N.S. Wilson, B.A. Keay, Tetrahedron Letters *37*, 153-6 (1996); cleavage of Me$_3$Si- and Et$_3$Si-ethers, also selective and preferential cleavage of *t*-BuMe$_2$Si-ethers, with added *water* s. J. Org. Chem. *61*, 2918-9 (1996); preferential O-de-*tert*-butoxy(diphenyl)silylation with Na$_2$S cf. T. Schmittberger, D. Uguen, Tetrahedron Letters *36*, 7445-8 (1995).

Carbon ↑   HO ⇅ C

Electrolysis/tris(2,2'-bipyridyl)nickel bis(fluoroborate)   *⇵/Ni(II)*
Electrochemical O-deallylation   OCH$_2$CH=CH$_2$ → OH
with SmCl$_3$ cf. *48*, 6; with tris(2,2'-bipyridyl)nickel bis(fluoroborate), phenols and alcohols, s. S. Olivero, E. Duñach, Chem. Commun. *1995*, 2497-8.

Sodium/liq. ammonia   *Na/NH$_3$*
Regiospecific cleavage of 3-ethyleneethers s. *51* 438   OR → OH

Potassium carbonate   *K$_2$CO$_3$*
Cleavage of carboxylic acid 2-hydroxyethyl esters s. *7*, 246s*51*   COOCH$_2$CH$_2$OH → COOH

Cupric sulfate-silica gel   *CuSO$_4$-SiO$_2$*
Cleavage of cyclic acetals   C(OR)$_2$ → CO
with CuSO$_4$ cf. *20*, 20s*42*; with CuSO$_4$-SiO$_2$, selectivity, s. G.M. Caballero, E.G. Gros, Synth. Commun. *25*, 395-404 (1995).

Cadmium-lead   *Cd-Pb*
Cleavage of 2,2,2-trichloroethyl carbo(n,xyl)ates s. *24*, 9s*51*   ←

Magnesium/methanol or Magnesium methoxide   *Mg/MeOH or Mg(OMe)$_2$*
O-Deacylation   OAc → OH
with MgO/MeOH cf. *42*, 3; with Mg(OMe)$_2$, selective and preferential conversion, s. Y.-C. Xu et al., Tetrahedron Letters *37*, 455-8 (1996); with Mg/MeOH cf. ibid. *35*, 6207-10 (1994).

Ammonium ceric nitrate or 2,3-Dichloro-5,6-dicyanoquinone   *(NH$_4$)$_2$Ce(NO$_3$)$_6$ or DDQ*
Cleavage of tetrahydropyran-2-yl ethers   OThp → OH
s. *51*, 2; with DDQ in wet acetonitrile (cf. *48*, 10) s. S. Raina, V.K. Singh, Synth. Commun. *25*, 2395-400 (1995).

O-Decarbo-*tert*-butoxylation s. *51*, 15   OCOOBu-*t* → OH

Samarium/iodine/alcohol $Sm/I_2/ROH$
Deacylation under mild, neutral conditions   OCOOR → OH

4.

Selective O-deacylation. An equimolar mixture of startg. ester, Sm, and I_2 in ethanol stirred at room temp. for 20 min → product. Y 97%. The method is fast, simple and high-yielding, and leaves lactones and tert. esters unaffected. F.e. and catalytic procedure, **also preferential O-deacylation, N-deacylation** of lactam and uracil derivs., and N-decarbalkoxylation of imidodicarbonates, s. R. Yanada et al., Synlett *1995*, 1261-3.

Aminoesterase or Chymotrypsin   ←
Asym. hydrolysis of α-aminocarboxylic acid esters   COOR → COOH
of α-subst. aminoesters s. *48*, 8; on a large scale with an enzyme from *Humicola langinosa* cf. W. Liu et al., J. Chem. Soc. Perkin Trans. I *1995*, 553-9; of α-(benzylideneamino)carboxylic acid esters with chymotrypsin in aq. organic media cf. V.S. Parmar et al., J. Org. Chem. *61*, 1223-7 (1996).

Butyrylcholine esterase or Papain   ←
Protection of carboxylic groups as solubilizing choline esters $OCH_2CH_2N^+Me_3$ → OH
Enzymatic removal of the protective group under mild condition

5.

Choline esters are *solubility-enhancing* protective groups (cf. *21*, 426s27) for the enzymatic synthesis of **peptides** and **lipopeptides** (incl. S-palmitoylated and S-farnesylated derivs.), and may be cleaved under extremely mild, nearly neutral conditions. E: A soln. of Aloc-Leu-Pro-OCho and a little butyrylcholine esterase (from horse serum) in 0.05 M Na_3PO_4 buffer (pH 6.5) shaken at 37° until reaction complete by TLC (24-36 h) → product. Y 95%. F.e., prepn. of substrates, and [more slowly] with acetylcholine esterase from the electric eel s. H.-D. Jakubke et al., Angew. Chem. Intern. Ed. *35*, 106-9 (1996); cleavage of solubilizing 2-(2-methoxyethoxy)ethyl esters with papain, protected α-amino-β-glycosyloxycarboxylic acids, s. H. Kunz et al., J. Org. Chem. *61*, 2638-46 (1996).

Lipase   ←
Regiospecific O-deacylation   OAc → OH
s. *42*, 6; of (E)-2-ene-1,4-diol esters s. T. Itoh et al., Tetrahedron Letters *37*, 91-2 (1996); of acoxybenzoic acid esters s. A. Cipiciani et al., Tetrahedron *52*, 9869-76 (1996).

Preparation of chiral carboxylic acids and alcohols   COOR → COOH + HOR
by asym. hydrolysis of carboxylic acid esters
s. *28*, 13s*46*-9; factors influencing enantioselectivity s. M. Kinoshita, A. Ohno, Tetrahedron *52*, 5397-406 (1996); regulation by crown ethers s. T. Itoh et al., J. Org. Chem. *61*, 2158-63 (1996); asym. O-deacylation of acetoxy(alkylthio)acetic acid esters s. C.M. Rayner et al., Tetrahedron:Asym. *6*, 1903-6 (1995); of 5-acoxy-2(5H)-furanones s. H. van der Deen et al., Tetrahedron Letters *35*, 8441-4 (1994); of *o*-acoxysulfoxides s. A.N. Serreqi et al., Can. J. Chem. *73*, 1357-67 (1995); of *o*-acoxy-phosphines and -phosphine oxides s. J. Org. Chem. *59*, 7609-15 (1994); of α-acoxyphosphonates s. M. Drescher et al., Synthesis *1995*, 1267-72; of N-protected 2-aminocyclopropylcarbinol esters s. R. Csuk, Y. von Scholz, Tetrahedron *52*, 6383-96 (1996).

Cross-linked lipase
Asym. hydrolysis with cross-linked crystalline enzyme preparations   COOR→COOH
Enhancement of enantioselectivity

6.

Compared with the crude, commercial lipase preparation from *Candida rugosa*, which contains several competing enzymes as contaminants, the cross-linked crystalline form of the major hydrolase affords enhanced enantioselectivity in asym. hydrolysis and, being insoluble, is more easily recovered. It is also 2-3 orders of magnitude *more stable* than the soluble, crystalline (and expensive) protein from which it is prepared by cross-linking with glutaraldehyde. **E:** A suspension of (R,S)-ibuprofen methyl ester and the crystalline, cross-linked hydrolase from *C. rugosa* (Chiro CLEC-CR) in distilled water heated to 40° with vigorous stirring for 20 h, cooled to room temp., the enzyme removed by filtration for re-use, and worked up → (R)-ibuprofen methyl ester (Y 95.3%; e.e. 55.8%) and (S)-ibuprofen (Y 87%; e.e. 93% [>99% after crystallization of the Na-salt]). F.e.s. A.L. Margolin et al., J. Am. Chem. Soc. *117*, 6845-52 (1995); in *organic* media s. ibid. *118*, 5494-5 (1996).

Amberlite IR-120   ←
Cleavage of 4,4′-dimethoxytrityl ethers s. *12*, 16s*51*   OCAr₃ → OH
Dowex 50W-X8   ←
Cleavage of O,O-isopropylidene derivs.   C
with Amberlite cf. *15*, 146; preferential and selective cleavage of carbohydrate derivs. with Dowex 50W-X8 (cf. *12*, 15) s. K.H. Park et al., Tetrahedron Letters *35*, 9737-40 (1994).
Trifluoroacetic acid   CF₃COOH
Cleavage of *p*-methoxybenzyl ethers s. *37*, 7s*51*   OCH₂Ar → OH
2,3-Dichloro-5,6-dicyanoquinone s. under CAN   DDQ
1,1,1,3,3,3-Hexafluoro-2-propanol   CF₃CHOHCF₃
Triethylsilane/dichloroacetic acid   Et₃SiH/Cl₂CHCOOH
Cleavage of 4,4′-dimethoxytrityl ethers under mild, weakly acidic conditions   OCAr₃ → OH

7.

A soln. of 5-O-(4,4′-dimethoxytrityl)adenosine in 1,1,1,3,3,3-hexafluoro-2-propanol (pK$_a$ 9.3) left for 1 h at room temp. → adenosine. Conversion 87%. The method is generally applicable to nucleoside and deoxynucleoside derivs., as well as more simple systems. There was no N-glycosyl cleavage (as may take place with more acidic reagents), and a number of other protective groups were unaffected (N-acyl, O-*tert*-butyldimethylsilyl, N-allyloxycarbonyl). F.e.s. N.J. Leonard, Neelima, Tetrahedron Letters *36*, 7833-6 (1995); selective cleavage of 4,4′-dimethoxytrityl ethers with Et₃SiH/Cl₂CHCOOH (cf. *12*, 16s*51*) s. V.T. Ravikumar et al., ibid. 6587-90.

Silica/nitrogen dioxide   SiO₂/NO₂
Cleavage of acetals   C(OR)₂ → CO
with SiO₂/TsOH cf. *33*, 16s*39*; of cyclic ketals with SiO₂/NO₂ s. T. Nishiguchi et al., Chem. Commun. *1995*, 1121-2.

Trimethylsilyl chloride/sodium iodide $\qquad$ Me_3SiCl/NaI
Phosphoric acid mono- from tri-esters s. *51*, 51 $\qquad$ P—OMe → P—OH
Silica-supported guanidinium chloride/acetyl chloride $\qquad$ ←
Molybdenyl acetoacetonate $\qquad$ $MoO_2(acac)_2$
Cleavage of acetals under neutral conditions $\qquad$ $C(OR)_2$ → CO
with Ru complexes cf. *49*, 16; with $MoO_2(acac)_2$ s. M.L. Kantam et al., Synth. Commun. *25*, 2529-32 (1995); with silica-supported guanidinium chloride/acetyl chloride s. P. Gros et al., J. Chem. Res. (S) *1995*, 196-7.
Chlorine $\qquad$ Cl_2
Cleavage of trityl ethers $\qquad$ $OCAr_3$ → OH
with Amberlyst 15 cf. *12*, 16s*43*; with Cl_2 cf. J. Fuentes et al., Synth. Commun. *24*, 2237-45 (1994); of 4,4′-dimethoxytrityl ethers with Amberlite IR-120 s. S.V. Patil et al., ibid. 2423-8.
Iodine $\qquad$ I_2
Cleavage of *p*-methoxybenzyl ethers $\qquad$ OCH_2Ar → OH
with DDQ cf. *37*, 7; with I_2 s. A.R. Vaino, W.A. Szarek, Synlett *1995*, 1157-8; with trifluoroacetic acid cf. L. Yan, D. Kahne, ibid. 523-4.
*Hydrogen bromide/tetra-*n*-butylammonium bromide* $\qquad$ HBr/Bu_4NBr
O-Debenzylation $\qquad$ OBn → OH
with 48% HBr cf. *21*, 937; of phenolethers with added Bu_4NBr in a 2-phase medium s. U.T. Bhalerao et al., Synth. Commun. *25*, 1433-9 (1995).
Tris(2,2′-bipyridyl)nickel bis(fluoroborate) s. under ↯ $\qquad$ $Ni(bipy)_3(BF_4)_2$
Palladium-carbon/ammonia $\qquad$ $Pd-C/NH_3$
Cleavage of carboxylic acid benzyl esters s. *51*, 32 $\qquad$ COOBn → COOH
*Palladous acetate/tris(*m*-sulfophenyl)phosphine trisodium salt/sodium azide* $\qquad$ ←
O-Decarballyloxylation in aq. media $\qquad$ $OCOOCH_2CH=CH_2$ → OH
with Et_3N cf. *29*, 28s*48*; in neutral media with NaN_3 as allyl scavenger s. S. Sigismondi, D. Sinou, J. Chem. Res. (S) *1996*, 46-7.

Elimination ⇧

Oxygen ↑ $\qquad$ HO ⇧ O

Horseradish peroxidase $\qquad$ ←
α-Hydroxy- from α-hydroperoxy-carboxylic acid esters $\qquad$ OOH → OH
with kinetic resolution s. *51*, 498

Formation of H-N Bond

Exchange ⇅

Oxygen ↑ $\qquad$ HN ⇅ O

Sodium hydrogen selenide $\qquad$ NaSeH
Amines from nitro compds. $\qquad$ NO_2 → NH_2
with $Se/CO/Et_3N$ cf. *30*, 10as*36*; with $Se/NaBH_4$ [NaSeH], **also from nitroso compds.**, s. D.K. Dutta et al., J. Chem. Res. (S) *1994*, 388-9.

Cuprous bromide-dimethyl sulfide s. under NaBH₄ *CuBr-Me₂S*
Aluminum amalgam *Al,Hg*
Amines from nitro compds. NO₂ → NH₂
s. *15*, 19; 2-aminoalcohols under ultrasonication s. R.W. Fitch, F.A. Luzzio, Tetrahedron Letters *35*, 6013-6 (1994).
Borane-dimethyl sulfide/chiral β-hydroxysulfoximines ←
Prim. amines from alkoximes - Asym. reduction C=NOR → CHNH₂
with chiral borane-2-aminoalcohol complexes cf. *25*, 15s*42*; with chiral β-hydroxysulfoximine complexes s. C. Bolm, M. Felder, Synlett *1994*, 655-6.
Lithium tetrahydridoborate/trimethylsilyl chloride *LiBH₄ /Me₃SiCl*
Prim. amines from alkoximes
with LAH cf. *34*, 53; with LiBH₄/Me₃SiCl, 2-amino-2-deoxyglycosides, s. A. Banaszek, W. Karpiesiuk, Carbohyd. Res. *251*, 233-42 (1994).
Sodium tetrahydridoborate/cuprous bromide-dimethyl sulfide *NaBH₄/CuBr-Me₂S*
Ar. amines from nitro compds. NO₂ → NH₂
with Cu(OAc)₂ cf. *46*, 60s*49*; with CuBr-Me₂S s. H.V. Patel et al., Org. Prep. Proc. Intern. *27*, 81-3 (1995).
Sodium tetrahydridoborate/antimony trichloride *NaBH₄/SbCl₃*
with NaBH₄/BiCl₃ cf. *50*, 9; with less expensive SbCl₃ s. P.-D. Ren et al., Synth. Commun. *25*, 3799-803 (1995).
Sodium tetrahydridoborate/iodine *NaBH₄/I₂*
Anion exchanger-supported tetrahydridoborate/nickel acetate *[BH₄⁻]/Ni(OAc)₂*
Amines from oxime acetates C=NOAc or C=NOH → CHNH₂
with NaBH₄/MoO₃ cf. *24*, 23s*44*; with NaBH₄/I₂ s. D. Barbry, P. Champagne, Synth. Commun. *25*, 3503-7 (1995); **from oximes** with anion-exchanger-supported BH₄⁻/Ni(OAc)₂ s. B.P. Bandgar et al., ibid. 863-9.
Chiral 1,3,2-oxazaborolidines/borane-dimethyl sulfide ←
Cyclic *cis*-2-aminoalcohols from α-diketone monosiloximes CH(NH₂)CH(OH)
Asym. reduction

8.

(1S,2R)-2-Amino-1-tetralols. A soln. of startg. ketosiloxime in toluene added to a cold (-20°), stirred soln. of 10 mol% 1,3,2-oxazaborolidine-borane complex and 2.5 eqs. BH₃.SMe₂ in the same solvent over 4 h, stirred at -20° for 1 h, at 25° for 1 h, then heated at 80° overnight, methanol added, and refluxed for 30 min → product. Y 91% (*cis:trans* 6:1; 91% e.e. *cis*; 75% e.e. *trans*). This method creates two chiral centres in a single step with high stereoselectivity. F.e.s. R.D. Tillyer et al., Tetrahedron Letters *36*, 4337-40 (1995).

Samarium diiodide *SmI₂*
Cleavage of nitrogen-oxygen bonds under mild conditions N—O— → NH

9.

A stirred soln. of startg. alkoxylamine in dry THF treated dropwise at room temp. under N₂ with ca. 2.2 eqs. SmI₂ in THF, diluted with 3 volumes of methylene chloride after 5 h, then quenched with 2 volumes of 10% aq. Na-thiosulfate → product. Y 69%. The method is mild and general, compatible with trifluoroacetamides, and succeeds in certain instances where established methods fail. F.e. and isolation of product as the corresponding acylamine, also 3,6-dihydro-2*H*-1,2-oxazine ring opening, s. G.E. Keck et al., Tetrahedron Letters *36*, 7419-22 (1995); cleavage of free or N-acylated O-benzylhydroxylamines s. J. Marco-Contelles et al., J. Org. Chem. *61*, 359-60 (1996).

Antimony trichloride s. under NaBH₄ $SbCl_3$

Sodium dithionite/viologens/potassium carbonate ←
Reduction of nitro compds. ←
ar. amines cf. *49*, 21; aliphatic **hydroxylamines** s. J. Org. Chem. *60*, 6202-4 (1995).

Iodine s. under NaBH₄ I_2

Iron/acetic acid *Fe/AcOH*
Amines from N-oxide radicals $\geq NO^{\cdot} \rightarrow \geq NH$
with Na₂S cf. *28*, 17; with Fe/AcOH, selectivity, **and from nitrones** (cf. *41*, 991), s. K. Hideg et al., Synth. Commun. *25*, 2929-40 (1995).

Ferric chloride/N,N-dimethylhydrazine $FeCl_3 /Me_2NNH_2$
Cobaltous chloride/zinc $CoCl_2/Zn$
Amines from nitro compds. $NO_2 \rightarrow NH_2$
with FeCl₃/N₂H₄/C cf. *27*, 15s*31*; with FeCl₃/Me₂NNH₂, ar. amines, s. S.R. Boothroyd, M.A. Kerr, Tetrahedron Letters *36*, 2411-4 (1995); with CoCl₂/Zn cf. R.N. Baruah, Indian J. Chem. *33B*, 758, 1994.

Nickel acetate s. under Anion exchanger-supported BH₄⁻ $Ni(OAc)_2$

Chlorotris(triphenylphosphine)rhodium(I)/triethylsilane $RhCl(PPh_3)_3 /Et_3SiH$
Amines from nitro compds.
with RhCl(CO)(PPh₃)₂/sec.alcohols cf. *5*, 23s*38*; ar. amines with RhCl(PPh₃)₃/Et₃SiH s. H.R. Brinkman et al., Synth. Commun. *26*, 973-80 (1996).

Palladium-carbon *Pd-C*
2-Aminoalcohols from α-isonitrosoketones $CH(OH)CHNH_2$
s. *9*, 74; under ultrasonication cf. E. Dominguez et al., Tetrahedron *51*, 5361-8 (1995).

Palladous acetate/1,1'-bis(diphenylphosphino)ferrocene/triethylammonium formate ←
Prim. ar. amines from *o*-nitrosulfonates ←

10.

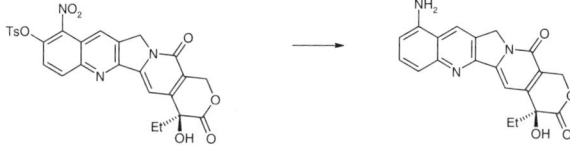

5.5 Mol% dppf and 5 mol% Pd(OAc)₂ added sequentially to a stirred soln. of startg. nitrosulfonate in dioxane at room temp. under argon, heated to 90°, *3 eqs.* 1.8 *M* HCOOH·Et₃N in dioxane added dropwise over 1.15 h, and cooled to room temp. → product. Y 88%. High yields were obtained independent of the sulfonate group, ligand and solvent. F.e.s. A. Bedeschi et al., Tetrahedron Letters *36*, 9197-200 (1995).

Water-soluble palladium phosphine complex/carbon monoxide/sodium hydroxide ←
Ar. amines from nitro compds. in aq. organic media $NO_2 \rightarrow NH_2$
with Ru₃(CO)₁₂/CO/*i*-Pr₂NH cf. *47*, 22; with a water-soluble Pd-phosphine complex/CO/NaOH s. A.M. Tafesh, M. Beller, Tetrahedron Letters *36*, 9305-8 (1995).

Nitrogen ↑ HN ⇅ N

Irradiation s. under N-Benzyl-1,4-dihydronicotinamide ⫻

Zinc bis(tetrahydridoborate) $Zn(BH_4)_2$
Tetra-n-butylammonium tetrahydridoborate Bu_4NBH_4
Dichloroborane/dimethyl sulfide $BHCl_2 \cdot SMe_2$
Amines from azides $N_3 \rightarrow NH_2$

with $NaBH_4$ cf. *23, 27*; with $Zn(BH_4)_2$, reduction of azides as well as carboxylic and sulfonic acid azides, s. B.C. Ranu et al., J. Org. Chem. *59*, 4114-6 (1994); with Bu_4NBH_4 or N-benzyl-4-aza-1-azoniabicyclo[2.2.2]octane tetrahydridoborate cf. H. Firouzabadi, G.R. Afsharifar, Bull. Chem. Soc. Japan. *68*, 2595-602 (1995); selective reduction with $BHCl_2 \cdot SMe_2$ s. A.M. Salunkhe, H.C. Brown, Tetrahedron Letters *36*, 7987-90 (1995).

Samarium diiodide SmI_2
Amines from azides under mild conditions

11.

Ar. amines. A carefully degassed soln. of 2-azidobiphenyl in anhydrous THF treated under N_2 with at least 3 eqs. commercial 0.1 M SmI_2 in THF, the mixture stirred at room temp. for ca. 30 min, then hydrolyzed with water and 5% aq. Na_2CO_3 → 2-aminobiphenyl. Y 96%. The method complements existing ones using a variety of reductants, with the advantage of SmI_2 being commercially available, inexpensive and non-toxic. Noteworthy, is the reduction of *o*-azido- to *o*-amino-oxo compds., and retention of ar. chlorine and nitro groups. F.e., also sulfonic acid amides from their azides s. L. Benati et al., Tetrahedron Letters *36*, 7313-4 (1995); arylcarboxylic acid amides from azides s. C. Goulaouic-Dubois, M. Hesse, ibid. 7427-30.

N-Benzyl-1,4-dihydronicotinamide/irradiation ←
Tri-n-butyltin hydride/azodiisobutyronitrile $Bu_3SnH/AIBN$
Amines from N-nitramines $NNO_2 \rightarrow NH$

12.

Cyclic sec. amines. An equimolar amount of Bu_3SnH and a little AIBN added sequentially to a soln. of N-nitromorpholine in anhydrous benzene, and heated under reflux for 6 h → morpholine. Y 75%. The procedure highlights a new method of generating **aminyl radicals**. (cf. *19, 21*). F.e., and N-nitrosamines with $(Me_3Si)_3SiH$, s. C. Imrie, J. Chem. Res. (S) *1995*, 328-9; with N-benzyl-1,4-dihydronicotinamide (or polymer-based variant) under irradiation cf. R.D. Chapman et al., Tetrahedron *52*, 9655-64 (1996).

Benzyltriethylammonium tetrathiomolybdate $(BnNEt_3)_2MoS_4$
Amines from azides s. *51*, 163 $N_3 \rightarrow NH_2$

Palladium-carbon Pd-C
Cytosines from tetrazolo[1,5-c]pyrimidin-5(6H)-ones s. *51*, 155 C.

Palladium-carbon/ammonia $Pd-C/NH_3$
Amines from azides s. *51*, 32 $N_3 \rightarrow NH_2$

Via intermediates v.i.
Amines from azides via phosphine imines
s. *12, 42*; with polymer-based triarylphosphines/NH_3 cf. T. Holletz, D. Cech, Synthesis *1994*, 789-91.

Sulfur ↑   HN ⇅ S

Samarium diiodide/N,N′-dimethyl-N,N′-propyleneurea SmI_2/DMPU
Tri-n-butyltin hydride/azodiisobutyronitrile Bu_3SnH/AIBN
N-Desulfonylation $NSO_2R \rightarrow NH$

13. $(PhCH_2)_2NSO_2Ph$ ⟶ $(PhCH_2)_2NH$

A mixture of startg. sulfonamide, DMPU, and 6 eqs. 0.1 *M* SmI_2 in THF refluxed under N_2 for 4.5 h → product. Y 92%. The method is mild and does not require any special apparatus. F.e. incl. amino acid and aziridine derivs., also from tosylamines, and with SmI_2/Sm s. E. Vedejs, S. Lin, J. Org. Chem. *59*, 1602-3 (1994); N-depyrid-2-ylsulfonation s. C. Goulaouic-Dubois et al., ibid. *60*, 5969-72 (1995); radical N-desulfonylation of N-aroylsulfonic acid amides with Bu_3SnH/AIBN cf. A.F. Parsons, R.M. Pettifer, Tetrahedron Letters *37*, 1667-70 (1996).

Carbon ↑   HN ⇅ C

Methylamine, Ethylenediamine or Polymer-based diamines   ←
Protection of prim. amino groups as tetrachlorophthalimides   C
Selective removal of the protective group under mild conditions

14.

An equimolar mixture of startg. tetrachlorophthalimide and the corresponding phthalimide in ethanol heated with 4 eqs. ethylenediamine at 50° for 8 h → N-unsubst. aminoglycoside (isolated as the N-acetyl deriv. in 84% yield) and unchanged N-phthalimide deriv. Significantly, the O^4-acetyl group and N-(pent-4-enoyl)-derivs. were also unaffected, while the presence of the tetrachlorophthalimide group in 2-amino-2-deoxyglycosyl donors does not inhibit neighbouring group participation essential for β-glycoside formation in disaccharide synthesis. F.e.s. B. Fraser-Reid et al., J. Am. Chem. Soc. *117*, 3302-3 (1995); with polymer-based diamines, **also cleavage of phthalimides**, s. P. Stangier, O. Hindsgaul, Synlett *1996*, 179-81; cleavage of N-allylphthalimides with $MeNH_2$ s. S.E. Sen, S.L. Roach, Synthesis *1995*, 756-8.

Cadmium-lead   Cd-Pb
Cleavage of 2,2,2-trichloroethyl esters $NCO_2CH_2CCl_3 \rightarrow NH$
decarbo-2,2,2-trichloroethoxylation with Cd cf. *24*, 9s37; with Cd-Pb, also cleavage of simple 2,2,2-trichloroethyl carboxylates, s. Q. Dong et al., Tetrahedron Letters *36*, 5681-2 (1995).

Mercuric acetate/sodium tetrahydridoborate $Hg(OAc)_2$/$NaBH_4$
Prim. amines from isocyanates $NCO \rightarrow NH_2$
with KOH cf. *11*, 399; aliphatic amines with $Hg(OAc)_2$/$NaBH_4$ s. C. Malanga et al., Tetrahedron Letters *36*, 8859-60 (1995).

Samarium/iodine/alcohol   ←
N-Deacylation under mild conditions s. *51*, 4 $NCOR \rightarrow NH$

15.

Ammonium ceric nitrate $(NH_4)_2Ce(NO_3)_6$
Catalytic decarbo-*tert*-butoxylation $NCO_2Bu\text{-}t \rightarrow NH$
under mild, neutral conditions

Chiral α-aminocarboxylic acid esters. Startg. *tert*-butoxycarbonylamine in acetonitrile treated with 0.2 eq. CAN, and the mixture refluxed for 2 h → product. Y 93%. The method is fast, simple, high-yielding, and requiring only a catalytic amount of the reagent. Benzyl esters and phthalimides remained unaffected, and there was no racemization of adjacent chiral centres. F.e. incl. O- and S-de-*tert*-butoxycarbonylation, and cleavage of *tert*-butyl esters, s. G.H. Hakimelahi et al., Tetrahedron Letters *37*, 2035-8 (1996); **catalytic N- and O-detritylation**, incl. cleavage of 4-methoxytrityl ethers, s. Chem. Commun. *1996*, 545-6; cleavage of bis(trimethylsilyl)methyl N-protective groups s. J.M. Aizpurua et al., Angew. Chem. Intern. Ed. *35*, 1239-41 (1996).

16.

Lipase or Dithiothreitol/triethylamine   ←
Protection of amino groups as carbo-*p*-acetoxybenzoxyamines $NCO_2CH_2Ar \rightarrow NH$
Selective removal of the protective group

A soln. of startg. protected peptide in 20 vol% methanol treated with the lipase from *Mucor miehei* at 30° in the presence of 0.2 M KI buffer (pH 5) → product. Y 65%. Terminal methyl and allyl esters, as well as S-farnesyl groups, were unaffected. KI (or NaHS) is requires to trap the liberated *p*-quinone methid which otherwise might react with amino functions. F.e. and enzymes (acetyl esterase and lipase from *Rhizopus arrhizus*) s. H. Waldmann, E. Nägele, Angew. Chem. Intern. Ed. *34*, 2259-62 (1995); **reductive cleavage of carbo-*p*-azidobenzoxyamines** with dithiothreitol/Et₃N cf. R.J. Griffin et al., J. Chem. Soc. Perkin Trans. I *1996*, 1205-11.

Trifluoroacetic acid CF_3COOH
N-Decarbo-*tert*-butoxylation $NCO_2Bu\text{-}t \rightarrow NH$
with retention of carbo-9-fluorenylmethoxyamines s. *51*, 166

Silica gel or Silicon tetrachloride/phenol SiO_2 or $SiCl_4/PhOH$
Preferential and selective N-decarbo-*tert*-butoxylation

17.

        Boc
        |
PhCH₂NAc   ⟶   PhCH₂NHAc

Acylamines. A soln. of startg. urethan in methylene chloride evaporated onto 10 times its weight of silica gel (35-70 mesh), and kept at 50°/0.2 mmHg for 15 h → product. Y 92%. Boc-derivs. of aliphatic amines were unaffected, whereas those attached to N-hetarenes, ar. amines or conjugated to carbonyl groups (as in N-acylurethans or imidodicarbonates) were removed. *tert*-Butyl esters were also unaffected. F.e.s. T. Apelqvist, D. Wensbo, Tetrahedron Letters *37*, 1471-2 (1996); pyrroles s. Tetrahedron *51*, 10323-42 (1995); cleavage of N-*tert*-butoxycarbonyl groups with SiCl₄/PhOH in solid phase peptide synthesis cf. K.M. Sivanandaiah et al., Tetrahedron Letters *37*, 5989-90 (1996).

Bromine Br_2
Prim. amines from sec. 1-phenylethylamines s. *51*, 240 $NHCH(Me)Ph \rightarrow NH_2$

Iodine I_2
Protection of amino groups as pent-4-enoylamines   NCOR → NH
Selective removal of the protective group under mild conditions

18.

A soln. of startg. N-(pent-4-enoyl)-deriv. (obtained in highly crystalline form from the corresponding prim. amine and pent-4-enoic anhydride) in 1:1 THF/water treated with 3 eqs. I_2, stirred until reaction complete, and quenched with ammonium thiosulfate → product. Y 84%. Deprotection of both prim. and sec. amines occurs rapidly and efficiently under *neutral* conditions without affecting methionine sulfur, *p*-methoxybenzyl and pent-4-enyl glycosidic linkages, acetals, O-acetyl groups, and the tetrachlorophthalimide residue. F.e.s. B. Fraser-Reid et al., J. Am. Chem. Soc. *117*, 3302-3 (1995).

Hydrogen chloride   *HCl*
Cleavage of N-(tetrahydropyran-2-yl) groups s. *51*, 352   NThp → NH
Protection of prim. amino groups   C
as perhydro-1,3,5-triazin-2-ones cf. *46*, 23; **as dihydro-1,3,5-dioxazines** and tetrahydro-1,3,5-oxadiazines, removal of the protective group with HCl, s. Y. Katsura, M. Aratani, Tetrahedron Letters *35*, 9601-4 (1994).

Palladium-carbon/ammonia   *Pd-C/NH₃*
N-Decarbobenzoxylation s. *51*, 32   NCOOBn → NH

Formation of H-S Bond

Exchange   ⇅

Carbon ↑   HS ⇅ C

Palladous acetate $Pd(OAc)_2$
S-Deacylation s. *51*, 229; *51*, 234   SCOR → SH

Formation of H-C Bond

Uptake   ⇓

Addition to Oxygen and Carbon   HC ⇓ OC

Cupric chloride/magnesium $CuCl_2/Mg$
2-Ethylenealcohols from α,β-ethylenealdehydes   CHO → CH_2OH
with Cu,Cd cf. *19*, 45; with $CuCl_2$/Mg, also benzylalcohols, s. N.B. Das et al., Tetrahedron Letters *36*, 7119-22 (1995).

Magnesium s. under $CuCl_2$ *Mg*
Zinc s.a. under $CoCl_2$ *Zn*

Zinc/aluminum chloride or acetic acid $\qquad$ $Zn/AlCl_3$ or $AcOH$
Alcohols from oxo compds. $\qquad$ CO → CHOH
with Zn cf. *4*, 48; *4*, 49; reduction of aryl or α,β-unsatd. oxo compds. and aliphatic aldehydes with Zn/AlCl₃ s. P.K. Chowdhury, P. Borah, J. Chem. Res (S) *1994*, 230-1; benzylalcohols (as acetate derivs.) and diarylcarbinols s. B.R. Rani et al., Bull. Chem. Soc. Japan *68*, 282-4 (1995).

Zinc/acetic acid $\qquad$ Zn/AcOH
Carboxylic acids from lactones $\qquad$ C̄
s. *13*, 59; from γ-lactones under ultrasonication s. J.R. Pedro et al., Tetrahedron Letters *36*, 8469-72 (1995); regioselectivity with Na/HMPA cf. J.K. Mukhopadhyaya et al., Indian J. Chem. *33B*, 132-6 (1994).

Zinc chloride s. under LiBH(s-Bu)₃ $\qquad$ $ZnCl_2$

Sodium tetrahydridoborate $\qquad$ $NaBH_4$
Alcohols from oxo compds. $\qquad$ CO → CHOH
s. *9*, 61; *threo*-2-hydroxyselenides, and f. reductants, s. S. Uemura et al., Bull. Chem. Soc. Japan *68*, 337-40 (1995).

Sodium tetrahydridoborate/chiral cobalt(II) Schiff base complex $\qquad$ ←
Asym. reduction of cyclic aryl ketones

19.

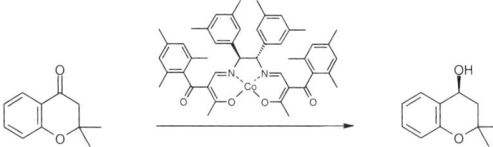

Ethanol (*500 mL*/mmol ketone) added to a suspension of 1.5 eqs. NaBH₄ in chloroform under argon, after stirring for 1 h at room temp. a soln. of 5 mol% of (S,S)-N,N'-bis(2-aroyl-3-oxobut-2-enylidene)ethylenediaminatocobalt(II) in chloroform added, cooled to -20°, 2,2-dimethylchroman-4-one in chloroform added, and the mixture stirred for 120 h before quenching with 1 M HCl → product. Y 98% (e.e. 92%). Since both enantiomers of the chiral diarylethylenediamine are commercially available or can be readily prepared by literature methods, either alcohol enantiomer is obtainable. Other borohydrides may be used, but NaBH₄ or KBH₄ gave better results. F.e. incl. reduction of acetophenone s. T. Mukaiyama et al., Angew. Chem. Intern. Ed. *34*, 2145-7 (1995).

Lithium tetrahydridoaluminate/chiral aminoalcohols $\qquad$ ←
Asym. reduction of ketones
with Darvon alcohol cf. *33*, 43; with (1R,2S,3S,5R)-(+)-10-anilino-3-ethoxy-2-hydroxypinane s. Y.-J. Cherng et al., Tetrahedron:Asym. *6*, 89-92 (1995); f. pinane-based modifiers (alkoxy- or amino-diol analogs) s. J. Chinese Chem. Soc. *41*, 205-8; 467-71 (1994); with NaBH₄ and monosaccharide derivs. cf. L. Sharma, S. Singh, Indian J. Chem. *33B*, 1183-6 (1994).

Diisopropoxytitanium(III) or Titanocene tetrahydridoborate s. under Titanium reagents $\qquad$ ←

Lithium hydridotri-sec-butylborate $\qquad$ $LiBH(s-Bu)_3$
Stereospecific 1,2-addition to α,β-ethylene-γ-(organothio)ketones $\qquad$ ←
Transmission of chiral information through conjugation

20.

In the absence of chelation control and adjacent stereogenic centres, the observed 1,4-diastereoselection on 1,2-addition to γ-sulfenylated enones is ascribable solely to a *stereo-electronic effect* transmitted through the conjugation. **E:** 2 eqs. 1 M Li-*sec*-Bu₃BH in THF added at -78° to a soln. of (E)-2,2,7,7-tetramethyl-6-phenylthio-4-octen-3-one in THF, stirred for 30 min, mixed with methanol, 1 N NaOH, and 30% H_2O_2, allowed to warm to room temp., and stirred for 30 min → (E)-3-hydroxy-2,2,7,7-tetramethyl-6-phenylthio-4-octene. Y 96% (3S*,6S*:3R*,6S* 97:3). The diastereofacial discrimination of the sulfenyl group is attributed to an effective s-p* interaction which fixes the C-S bond perpendicular to one side of the enone, thereby favouring backside attack of hydride ion. F.e. and carbophilic addition s. T. Sato et al., Angew. Chem. Intern. Ed. *34*, 2254-6 (1995).

Lithium hydridotri-sec-butylborate/zinc chloride	$LiBH(s-Bu)_3/ZnCl_2$
α-Hydroxy- from α-keto-carboxylic acid esters	CO → CHOH

Asym. reduction
with $LiAlH(OR)_3$ cf. *48*, 31; with $LiBH(s-Bu)_3/ZnCl_2$, asym. induction with chiral *cis*-1-arylsulfonamido-2-indenyl derivs., s. A.K. Ghosh, Y. Chen, Tetrahedron Letters *36*, 6811-4 (1995).

Diisobutylaluminum hydride	i-Bu_2AlH
Reduction of 2,3-epoxytosylates	╲o╱ → C(OH)CH

chiral sec. alcohols cf. *47*, 32; chiral glycol monotosylates *at -78°* s. Tetrahedron Letters *35*, 7197-200 (1994).

Tetra-n-butylammonium tetrahydridoborate/boron chloride	Bu_4NBH_4/BCl_3
Sodium hydridotriacetoxoborate	$NaBH(OAc)_3$
1-Hydroxy-6,8-diphenyl-3,4-dihydro-2-oxa-1-boranaphthalene/sodium tetrahydridoborate	←
1,3-Diols from β-hydroxyketones	CO → CHOH
Stereospecific reduction	

with $NaBH_4/R_2BOMe$ cf. *42*, 24; with Bu_4NBH_4/BCl_3 or $TiCl_4$ s. M. DiMare et al., J. Org. Chem. *61*, 868-73 (1996); cyclic 1,3-diols with $NaBH(OAc)_3$ s. T. Gallagher et al., J. Chem. Soc. Perkin Trans. I *1995*, 379-83; with 1-hydroxy-6,8-diphenyl-3,4-dihydro-2-oxa-1-boranaphthalene/$NaBH_4$, and cyclic 1,3-diols with $LiEt_3BH$ or *i*-Bu_2AlH cf. H. Yamashita, K. Narasaka, Chem. Lett. *1996*, 539-40.

Chiral 1,3,2-oxazaborolidines/borane-dimethyl sulfide or catecholborane ←
Chiral polymer-based 1,3,2-oxazaborolidines/borane-dimethyl sulfide ←
Asym. reduction of ketones
s. *43*, 45s*46-50*; of acylphosphonic acid esters s. C. Meier, W.H.G. Laux, Tetrahedron:Asym. *6*, 1089-92 (1995); also of β- and γ-ketophosphonic acid esters s. Liebigs Ann. *1995*, 1963-79; of ferrocenyl ketones s. J. Wright et al., J. Organometal. Chem. *476*, 215-7 (1994); with a conformationally rigid 1,3,2-oxazaborolidine s. M. Masui, T. Shioiri, Synlett *1996*, 49-50; with a polymer-based reagent s. P. Hodge et al., J. Chem. Soc. Perkin Trans. I *1995*, 345-9; stereoelectronic effect of remote substituents on face-selectivity, and asym. reduction of cobalt-complexed α,β-acetyleneketones, s. E.J. Corey, C.J. Helal, Tetrahedron Letters *36*, 9153-6 (1995); with chiral 1,3,2-oxazaphospholidines/BH_3-Me_2S cf. O. Chiodi et al., ibid. *37*, 39-42 (1996).

Chiral 1,3,2-oxazaborolidines/borane-dimethyl sulfide	←
B-Chlorodiisopinocampheylborane	R_2BCl
Diols from diketones by asym. reduction	

21.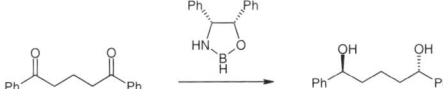

C_2-Symmetric diols. *1.4 eqs.* BH_3-Me_2S in THF added in one portion to a soln. of *1 eq.* (1S,2R)-2-amino-1,2-diphenylethanol in THF under N_2, stirred for 18 h, 1 eq. startg. dione in THF added dropwise over 1 h, stirred for 1 h, and cautiously quenched with methanol at 0° → product. Y 98% (99% S,S). The proportion of *meso*-diol by-product increases as the relative amount of oxazaborolidine decreases. F.e. incl. chiral glycols, 1,4- and 1,6-diols s. G.J. Quallich et al., Tetrahedron Letters *36*, 4729-32 (1995); with Ph_2SiH_2/chiral Rh-phosphine complex s. Y. Ito et al., ibid. 5239-42; C_2-symmetric 1,1'-ferrocenyldiols s. L. Schwink, P. Knochel, ibid. 37, 25-8 (1996); chiral hydrobenzoins with (S)-diphenylpyrrolidin-2-ylmethanol s. K.R.K. Prasad, N.N. Joshi, J. Org. Chem. *61*, 3888-9 (1996); chiral C_2-symmetric ar. diols with B-chlorodiisopinocampheylborane cf. H.C. Brown et al., Tetrahedron Letters *37*, 3795-8 (1996).

Asym. reduction of α-diketones monosiloximes s. *51*, 8   CO → CHOH
Catecholborane/chiral titanium(IV) alkoxides   ←
Asym. reduction of ketones   CO → CHOH

22.

Asym. borane-reduction of ketones in the presence of a chiral Ti(IV)-alkoxide has been reported for the first time. E: 10 Mol% Ti(OPr-*i*)$_4$ added to a soln. of 11.5 mol% chiral diol in *n*-hexane under inert gas, heated to reflux for 1-2 h, ca. 10% of the solvent removed by distillation, cooled to -30°, 1 eq. acetophenone and 1.1 eqs. catecholborane added sequentially, and treated with 1 *M* HCl after 1 h → (S)-1-phenylethanol. Y 78% (conversion >95%; purity >99%; e.e. 82%). Interestingly, the enantioselectivity is reversed with BH$_3$-Me$_2$S. F.e. and chiral diols, also with BH$_3$-THF and solvent effects, s. C. Wandrey et al., Angew. Chem. Intern. Ed. *34*, 2005-6 (1995).

Aluminum isopropoxide/trifluoroacetic acid $Al(OPr\text{-}i)_3/CF_3COOH$
Meerwein-Pondorf-Verley reduction
s. *8*, 48; modified procedure with 1 eq. TFA in an organic solvent s. K.G. Akamanchi, N.R. Varalakshmy, Tetrahedron Letters *36*, 3571-2 (1995); with a little Zr(OBu-*n*)$_4$ and 1-(4-dimethylaminophenyl)ethanol, optimization, s. B. Knauer, K. Krohn, Liebigs Ann. *1995*, 677-83; also with 1-tetralol, diastereoselectivity, s. ibid. 1347-51; **asym. reduction** with chiral diols s. Rec. Trav. Chim. Pays-Bas *115*, 140-4 (1996).

Diisobutylaluminum chloride   i-Bu$_2$AlCl
Selective and preferential reduction of oxo compds.

23.   PhCHO ⟶ PhCH$_2$OH

An equimolar mixture of benzaldehyde and cyclohexanone stirred with 1 eq. *i*-Bu$_2$AlCl in ether at 25° for 30 min, and quenched with 3 *N* HCl → benzyl alcohol. Y ≥99% (and cyclohexanone 100%). Discriminative reduction of aldehydes also occurs, with butanal being preferentially reduced in the presence of hexanal, and with benzaldehyde in the presence of either butanal or hexanal. Acyl chlorides, nitrile, ester and ethylene groups remained unaffected. Preferential reduction of structurally diverse ketones also takes place (≥99%), e.g. of cyclohexanone in the presence of cyclopentanone, 2-heptanone and acetophenone. F.e. incl. 1,2-reduction of α,β-ethyleneoxo compds. s. J.S. Cha et al., Synlett *1995*, 1055-6.

Yeast   ←
Asym. reduction of ketones
s. *29*, 36s*46-9*; of β-chloro-γ-ketocarboxylic acid esters s. O. Cabon et al., Tetrahedron:Asym. *6*, 2199-210 (1995); of γ-acoxy-β-diketones, regioselectivity, s. M. Utaka et al., ibid. 685-6; of *p*-subst. α,α,α-trifluoroacetophenones s. T. Fujisawa et al., ibid. *5*, 1095-8 (1994); of α-arylidenecycloalkanones s. G. Fogliato et al., Tetrahedron *51*, 10231 (1995); of 2-tetralones s. G. Speranza et al., ibid. 11531-46; of β-ketophosphonates s. B. Lejczak et al., ibid. 11809-14; of γ-nitroketones s. A. Guarna et al., ibid. 1775-88; of α,α,α-trifluoro-α'-sulfenylketones with different *Candida* species s. G. Resnati et al., Tetrahedron Letters *37*, 3903-6 (1996); of α-acoxyketones s. K. Ishihara et al., Bull. Chem. Soc. Japan *67*, 3314-8 (1994); of functionalized enones s. S. Koul et al., J. Chem. Soc. Perkin Trans. I *1995*, 2969-88; of α,β-epoxyketones and conversion to chiral glycols and 1,3-diols s. M. Takeshita et al., ibid. *1993*, 2901-5; structure-enantioselectivity relationships s. D. Zakarya et al., Tetrahedron Letters *35*, 4985-8 (1994); **reversal of enantioselectivity** with yeast from *Yarrowia lipolytica*, anti-Prelog sec. alcohols, s. A. Medici et al., Tetrahedron *52*, 3547-52 (1996); reversal with wet cells of *Pichia farinosa*, chiral diols from diketones (cf. *29*, 36s*49*; *51*, 21) s. H. Ohta et al., ibid. 8113-22.

Diphenylsilane/chiral rhodium phosphine complex
Sym. diols from diketones by asym. reduction s. *51*, 21
Diphenylsilane/chloro(cyclooctadiene)iridium(I) dimer/(R,R,R)-[2-(4,5-diphenyl-
  Δ²-oxazolin-2-yl)ferrocenyl]diphenylphosphine
Sec. alcohols from ketones via asym. hydrosilylation
Reversal of enantioselectivity

24.

Chiral sec. benzylalcohols. Acetophenone and 1.5 eqs. diphenylsilane added slowly and successively at 0° to a mixture of 0.25 mol% [Ir(cod)Cl]₂ and 5 mol% (R)-DIPOF in ether (pre-stirred at 25° for 1 h), the mixture stirred at 0° for 15 h, then quenched with methanol, followed by hydrolysis with 1 N HCl → (R)-1-phenylethyl alcohol. Y 100% (by GLC; e.e. 96%). In contrast, reaction with [Rh(cod)Cl]₂ [at 25°] gave the (S)-enantiomer [s. idem., Organometallics *14*, 5486-7 (1995)]. The method was effective for aryl methyl ketones, heterocyclic methyl ketones, and α,β-ethyleneketones, but a dialkyl ketone gave low enantioselectivity. F.e.s. S. Uemura et al., Chem. Commun. *1996*, 847-8.

Trimethoxysilane/lithium methoxide (MeO)₃SiH/LiOMe
Trichlorosilane/dimethylformamide Cl₃SiH/DMF
Alcohols from oxo compds.
with (EtO)₃SiH/KF cf. *36*, 670; 2,3-epoxyalcohols with (MeO)₃SiH/LiOMe, solvent effect on stereoselectivity, s. A. Hosomi et al., Tetrahedron Letters *36*, 571-4 (1995); with Cl₃SiH/DMF (cf. *25*, 470) and reduction of azomethines s. S. Kobayashi et al., Chem. Lett. *1996*, 407-8.

Polymethylhydrosiloxane s. under Cp₂Ti(OAr)₂

Titanocene bis(p-chlorophenoxide)/polymethylhydrosiloxane/tetra-n-butylammonium
  fluoride-alumina
Lactols from lactones via hydrosilylation
Catalytic reduction under mild conditions

25.

5 eqs. polymethylhydrosiloxane added via syringe to a slurry of 2 mol% titanocene bis(*p*-chlorophenoxide) and 1 mol% TBAF-on-alumina (15%) in toluene under argon, stirred at room temp. for 5-10 min, cooled with a cold water bath, δ-decanolactone added dropwise via syringe, stirred at room temp. until TLC indicated completion of reaction (0.5-5 h), the catalyst deactivated by exposure to air, dil. with THF, treated with 1 M NaOH (CAUTION: vigorous bubbling), and stirred for 1 h at room temp. → product. Y 94%. This catalytic procedure is more efficient than the standard route with a stoichiometric amount of air- and moisture-sensitive *i*-Bu₂AlH. The method is limited to 5- and 6-membered lactols, incl. α-subst. derivs., and tolerates ar. halides. F.e. and with retention of α-chirality s. S.L. Buchwald et al., J. Am. Chem. Soc. *117*, 12641-2 (1995).

Diisopropoxytitanium(III) or Titanocene tetrahydridoborate (i-PrO)₂TiBH₄ or Cp₂TiBH₄
Polymer-based zirconium(IV) tetrahydridoborate
Alcohols from oxo compds. s. *51*, 35 CO → CHOH

Chiral titanium(IV) alkoxides s. under Catecholborane Ti(OR)₄

Zirconium tetra-n-butoxide/1-(4-dimethylaminophenyl)ethanol or chiral diols
Catalyzed Meerwein-Pondorf-Verley reduction and asym. variant s. *8*, 48s*51*

Tri-n-butyltin hydride/tetra-n-butylammonium cyanide Bu₃SnH/Bu₄NCN
2,3-Epoxyalcohols from α,β-epoxyketones *anti*-isomers with Bu₂SnFH cf. *49*, 45; *syn*-isomers

with Bu₃SnH/Bu₄NCN s. I. Shibata et al., Tetrahedron Letters *35*, 8625-6 (1994).
*Tri-*n-*butyltin hydride/tri-*n-*butyltin iodide/triphenylphosphine oxide*   ←
Alcohols from epoxides $\overset{\triangledown}{\underset{\mathrm{O}}{\diagup}}$ → C(OH)CH
with Bu₃SnH/AlBN/NaI cf. *43*, 49; with Bu₃SnH/Bu₃SnI/Ph₃PO, selectivity, s. A. Baba et al., Tetrahedron Letters *36*, 9357-60 (1995).
Chiral 1,3,2-oxazaphospholidines/borane-dimethyl sulfide   ←
Asym. reduction of ketones s. *43*, 45s*51*
Chiral cobalt(II) Schiff base complex s. under NaBH₄   *Co(II)*
Cobalt(II) chloride/zinc   *CoCl₂/Zn*
Prim. alcohols from aldehydes   CHO → CH₂OH
with CoCl₂/(MeO)₂BH cf. *44*, 42; with CoCl₂/Zn in aq. DMF, also 1-deuterioalcohols with D₂O, s. R.N. Baruah, Indian J. Chem. *33B*, 182-3 (1994).

(η⁶-Arene)dichlororuthenium(II) dimer/chiral 2-aminoalcohol or (1S,2S)-N-tosyl-1,2-diphenylethylenediamine or (1R,1R′)-2,6-bis[1-(diphenylphosphino)ethyl]pyridine/ isopropanol/potassium hydroxide   ←
(η⁶-Arene)dichlororuthenium(II) dimer/(1S,2S)-N-tosyl-1,2-diphenylethylenediamine/formic acid/triethylamine   ←
Dichlorotris(triphenylphosphine)ruthenium(II)/2-[o-(diarylphosphino)phenyl]-Δ²-oxazolines/ isopropanol/sodium hydroxide   ←
Asym. transfer-hydrogenation of ketones under mild conditions   CO → CHOH

26.

By using formic acid/Et₃N in place of *i*-PrOH, Ru(II)-catalyzed asym. transfer-hydrogenation of aryl ketones can be achieved at much higher substrate concentration of 2-10 *M* (cf. <0.1 *M* with isopropanol). E: A 2 *M* soln. of acetophenone in 5:2 formic acid-triethylamine azeotrope treated with 0.5 mol% (R)-RuCl[(1S,2S)-*p*-TsNCHPhCHPhNH₂](η⁶-mesitylene) at 28° for 20 h → (S)-product. Y >99%; e.e. 98%. Unlike reduction with isopropanol, reaction is *irreversible* and proceeds to completion **under kinetic control**. Ar. chlorine, ethylene groups and carboxylic acid esters were unaffected. F.e.s. R. Noyori et al., J. Am. Chem. Soc. *118*, 2521-2 (1996); with *i*-PrOH/KOH cf. ibid. *117*, 7562-3 (1995); with chiral 2-aminoalcohols/*i*-PrOH/KOH cf. Chem. Commun. *1996*, 233-4; with RuCl₂(PPh₃)₃/2-[*o*-(diarylphosphino)phenyl]-Δ²-oxazolines/*i*-PrOH/NaOH cf. T. Langer, G. Helmchen, Tetrahedron Letters *37*, 1381-4 (1996); with (1R,1R′)-2,6-bis[1-(diphenylphosphino)ethyl]pyridine as ligand cf. X. Zhang et al., ibid. 797-800.

Dichlorotris(triphenylphosphine)ruthenium(II)/ethylenediamine/potassium hydroxide   ←
Dichloro[(R)-2,2′-bis(diphenylphosphino)-1,1′-binaphthyl]ruthenium(II)/chiral diamine/ potassium hydroxide   ←
Homogeneous hydrogenation of unsatd. oxo compds.

27.

A highly efficient system is now available for *transition metal*-catalyzed homogeneous hydrogenation of aldehydo and keto groups in the presence of ethylenic or acetylenic unsaturations. E: 0.01 Mol% RuCl₂(PPh₃)₃, 0.01 mol% ethylenediamine and 0.02 mol% KOH in degassed isopropanol (preliminarily sonicated for 30 min) containing benzalacetone pressurized in a glass autoclave with 4 atm H₂, and vigorously stirred at 25° for 18 h → (E)-4-phenyl-3-buten-2-ol. Y 97%. The unsaturation may be isolated or conjugated, and reaction is generally applicable to aromatic or aliphatic oxo compds. The procedure is mild, low-cost, simple, environmentally friendly and a valuable alternative to reduction with NaBH₄, notably for large-scale hydrogenation. F.e. and **asym. hydrogenation** with RuCl₂[(R)-binap](dmf)ₙ/chiral diamine/KOH s. R. Noyori et al., J. Am. Chem. Soc. *117*, 10417-8 (1995).

Dichloro[1-(diphenylphosphino)-2-ethoxy-1-(pyrid-2-yl)ethane]-
(triphenylphosphine)ruthenium(II)/isopropanol/sodium hydroxide ←
Transfer hydrogenation of ketones $CO \rightarrow CHOH$
with RuCl$_2$(PPh$_3$)$_3$ cf. *33*, 47s*47*; enhanced turnover with dichloro[1-(diphenylphosphino)-2-ethoxy-1-(pyrid-2-yl)ethane](triphenylphosphine)ruthenium(II) s. R. Mathieu et al., Chem. Commun. *1995*, 1721-2.

Chiral rhodium phosphine or phosphinite complexes (s.a. under Ph$_2$SiH$_2$) ←
Asym. homogeneous hydrogenation
s. *27*, 57s*46,47,49*; of α-acylamino-α,β-ethylenecarboxylic acids by electronic-tuning and placement of phosphinite ligands s. T.V. RajanBabu et al., J. Am. Chem. Soc. *116*, 4101-2 (1994); with surface active chiral bis(sulfophosphines) (cf. *27*, 57s*49*) s. H. Ding et al., Angew. Chem. Intern. Ed. *34*, 1645-7 (1995); with chiral rhodium(I) aminophosphine-phosphinite complexes s. A. Mortreux et al., Organometallics *14*, 2480-9 (1995); with disugar phosphine ligands s. S.R. Gilbertson et al., J. Org. Chem. *60*, 6226-8 (1995).

1,1'-Bis(diisopropylphosphino)ferrocene(cyclooctadiene)rhodium(I) triflate ←
Homogeneous hydrogenation
with cationic Rh(I) phosphine complexes s. *23*, 51s*37*; with crystalline and air-stable 1,1'-bis-(diisopropylphosphino)ferrocene(cyclooctadiene)rhodium(I) triflate, also hydrogenation of olefins and azomethines, s. M.J. Burk et al., Tetrahedron Letters *35*, 4963-6 (1994).

Palladium-carbon/ammonium formate $Pd\text{-}C/HCOONH_4$
Transfer hydrogenation of epoxides $\overset{\triangle}{O} \rightarrow C(OH)CH$
with Pd(OAc)$_2$/HCOOH/Et$_3$N cf. *42*, 30s*47*; with Pd-C/HCOONH$_4$, regioselectivity, s. P.S. Dragovich et al., J. Org. Chem. *60*, 4922-4 (1995); s.a. J.P. Varghese et al., Synth. Commun. *25*, 2267-73 (1995).

Via intermediates v.i.
3-Cycloalkenols from [n+5]-oxabicyclo[n.2.1]alkenes ⊂
Regiospecific ring opening via hydrostannylation-elimination

28.

3-Cyclohexenols. A soln. of 2.2 eqs. Bu$_3$SnH in dry toluene slowly added over 2 h via syringe pump to a soln. of startg. oxabicycloheptene, 2 mol% Pd$_2$dba$_3$, and 9 mol% Ph$_3$P in toluene → intermediate stannane (Y 95%), in dry THF treated at room temp. with 2.3 eqs. MeLi in ether, and after 10 min quenched with NH$_4$Cl → product (Y 89%). The regioselectivity complements that obtained via Ni-catalyzed hydroalumination-elimination (cf. *50*, 12), the tributylstannyl moiety being delivered preferentially to the less crowded carbon so that elimination affords the tert. alcohol. F.e. incl 3-cycloheptenols, also 4-acylaminocycloheptenes from the aza-analog, s. M. Lautens, W. Klute, Angew. Chem. Intern. Ed. *35*, 442-5 (1996).

Addition to Nitrogen and Carbon HC ⇓ NC

Triethylsilane/trifluoroacetic acid Et_3SiH/CF_3COOH
Acylhydrazines from acylhydrazones $C{=}N\text{-}NAc \rightarrow CHNH\text{-}NAc$
with NaBH$_4$ cf. *25*, 37; with Et$_3$SiH/TFA s. P.-L. Wu et al., Synthesis *1995*, 435-8; **tosylhydrazines** from tosylhydrazones s. ibid. *1996*, 249-52.

Trichlorosilane/dimethylformamide Cl_3SiH/DMF
Sec. amines from azomethines s. *36*, 670s*51* $C{=}N \rightarrow CHNH$

Di-n-butyl(chloro)stannane-hexamethylphosphoramide $Bu_2Sn(Cl)H\text{-}(Me_2N)_3PO$
Sec. amines from azomethines s. *51*, 126

Chiral (η⁶-arene)(N-arylsulfonyl-1,2-diamine)chlororuthenium(II) complexes/formic acid-triethylamine ←
Sec. amines from azomethines
Asym. transfer hydrogenation under mild conditions   C=N → CHNH

29.

Chiral cyclic sec. amines. 6 eqs. formic acid-Et₃N azeotrope (5:2) added to a soln. of startg. azomethine (0.5 M) and 0.5 mol% (S,S)-(η⁶-arene)(N-arylsulfonyl-1,2-diamine)chlororuthenium(II) complex in acetonitrile, and the mixture stirred at 28° for 3 h → (R)-product. Y >99% (e.e. 95%). The method is notably applicable to isoquinoline derivs. and cyclic ketimines, but enantioselectivity was lower (77%) with acyclic azomethines due to the inaccessibility of pure geoisometric (*syn/anti*) forms. Ketones and styryl groups were unaffected. F.e., catalysts, and solvent effects s. R. Noyori et al., J. Am. Chem. Soc. *118*, 4916-7 (1996).

Zwitterionic rhodium phosphine complex ←
1,1'-Bis(diisopropylphosphino)ferrocene(cyclooctadiene)rhodium(I) triflate ←
Homogeneous hydrogenation of azomethines
s. *23*, 51s*51*; with a zwitterionic rhodium phosphine complex cf. H. Alper et al., Organometallics *14*, 4209-12 (1995).

Addition to Carbon and Carbon   HC ⇓ CC

Lithium hydride s. under Ni(OAc)₂   LiH
Zinc s. under NiBr₂   Zn
Samarium/iodine/alcohols   Sm/I₂/ROH
Carboxylic from α,β-ethylenecarboxylic acid derivs.   C=C → CHCH
with SmI₂/HMPA cf. *47*, 46; acids, amides, esters and nitriles with less expensive Sm/I₂/ROH s. R. Yanada et al., Synlett *1995*, 443-4; β-deuterio-γ,δ-ethylene- from β-allene-boronates s. Synth. Commun. *26*, 393-405 (1996).
Chlorobis(cyclopentadienyl)hydridozirconium   Cp₂Zr(Cl)H
Regio- and stereo-specific hydrozirconation of functionalized acetylenes ←
(E)-enestannanes s. *48*, 48; (Z)-α,β-ethyleneboronates s. L. Deloux, M. Srebnik, J. Org. Chem. *59*, 6871-3 (1994).
Tricarbonyl(cyclopentadienyl)hydridotungsten/triflic acid   HW(CO)₃Cp/CF₃SO₃H
Ionic hydrogenation   C≡C → CH₂CH₂
of alkenes cf. *50*, 23; of alkynes s. J. Org. Chem. *60*, 7170-6 (1995).
Nickel boride   Ni₂B
Hydrogenation of carbon-carbon double bonds   C=C → CHCH
s. *25*, 46; oxo compds. or alcohols from α,β-ethyleneoxo compds., selectivity, s. C.M. Belisle et al., Tetrahedron Letters *35*, 5595-8 (1994).
Nickel acetate/lithium hydride/tert-butanol   ←
Selective hydrogenation of carbon-carbon double bonds s. *51*, 40
Nickel bromide/zinc/ethylenediamine   ←
Hydrogenation with active nickel(0)   ←
with Ni(II) halides/Mg cf. *16*, 72; (Z)-ethylene from acetylene derivs. with NiBr₂/Zn/ethylenediamine s. M. Sakai et al., Bull. Chem. Soc. Japan *67*, 1984-6 (1994).

Chiral ruthenium phosphine complexes
Noyori-type asym. hydrogenation   C=C → CHCH
s. *42*, 45s*50*; of 3-alkylidene-2-piperidones s. J.Y.L. Chung et al., Tetrahedron Letters 36, 7379-82 (1995); with diacetatobis[(S)-2,2'-bis(diphenylphosphino)-5,5',6,6',7,7',8,8'-octahydro-1,1'-binaphthyl]ruthenium(II) cf. H. Takaya et al., J. Chem. Soc. Perkin Trans. I *1994*, 2309-32.

Ruthenium trichloride/trioctylamine $RuCl_3/R_3N$
Hydrogenation of benzene rings
with Ru-Al$_2$O$_3$ cf. *11*, 95; with RuCl$_3$/trioctylamine, selectivity, s. F. Fache et al., Tetrahedron Letters *36*, 885-8 (1995).

Supramolecular rhodium(I) hydride complexes
Supramolecular catalysis

30.

Supramolecular Rh-catalysts exert *enzyme-like* selectivity by binding substrates in their molecular cavity prior to reaction. **E: Hydrogenation.** A soln. of 5-allylresorcinol in chloroform hydrogenated under 0.4 atm. H$_2$ at 25° for ca. 1 h in the presence of 10 mol% supramolecular Rh(I)-hydride complex and 1 eq. triphenyl phosphite → product. Y 56%. The initial rate of hydrogenation was ca. 5 times greater than the same reaction with HRh[P(OPh)$_3$]$_4$: evidence for specific binding of the allyl-subst. dihydroxyarene to receptor residues in the substrate. F.e. and catalyzed isomerization s. J.M. Nolte et al., J. Am. Chem. Soc. *117*, 11906-13 (1995).

Chiral cationic rhodium bis(phospholane) complexes
Asym. homogeneous hydrogenation   C=C → CHCH

31.

in supercritical carbon dioxide. Supercritical CO$_2$ is an excellent, environmentally friendly alternative to conventional organic solvents for asym. hydrogenation, with the added bonus that enantioselectivity can be considerably higher. **E:** Startg. enamide and 0.2 mol% chiral Rh-phosphine catalyst pressurized in a cylindrical stainless steel reactor with 200 psig H$_2$ and 3000 psig CO$_2$, warmed to 40° (when a homogeneous supercritical phase appeared at 5000 psig), and worked up after 24 h → product. Y 100% (e.e. 96.8%). This is the highest enantioselectivity reported to date for asym. hydrogenation of a β,β-disubst. α-enamidoester. Significantly, the density, polarity, viscosity, diffusivity and overall solvent strength of supercritical carbon dioxide can be dramatically varied by relatively small changes in the pressure and/or temperature. F.e.s. W. Tumas et al., J. Am. Chem. Soc. *117*, 8277-8 (1995); asym. hydrogenation of β-*branched* α-acylamino-α,β-ethylenecarboxylic acids and esters *in benzene* (cf. *27*, 57s*46,47,49,50*) s. ibid. 9375-6; with chiral Rh-2,2''-bis[1-(dialkylphosphino)ethyl]-1,1''-biferrocene complexes cf. ibid. 9602-3; asym. hydrogenation **of 1-arylenacylamines** in methanol s. ibid. *118*, 5142-3 (1996).

1,1'-Bis(diisopropylphosphino)ferrocene(cyclooctadiene)rhodium(I) triflate
Homogeneous hydrogenation s. *23*, 51s*51*   C=C → CHCH

Palladium-carbon/ammonia
Hydrogenation with retention of benzyl ethers

Pd-C/NH$_3$
C≡C → CHCH

32. Ph⌢⌢⌢OBn ⟶ Ph⌢⌢⌢OBn

A mixture of 3-benzyloxy-1-phenyl-1-propene, 0.5 eq. 2 *M* methanolic ammonia, and a little 5% Pd-C in methanol stirred under H$_2$ for 16 h at ambient temp. and pressure → 1-benzyloxy-3-phenylpropane. Y 98%. Ammonia effectively inhibits hydrogenolysis of benzyl ethers. The method is highly selective, mild, and sensibly complete within *30 min*. F.e. and inhibitors (NH$_4$OAc, pyridine), also reductive cleavage of azides, benzyl esters and N-Cbz-groups, s. H. Sajiki, Tetrahedron Letters *36*, 3465-8 (1995).

Rearrangement

Hydrogen/Oxygen Type

HC ∩ HO

Chloro(η^5-indenyl)bis(triphenylphosphine)ruthenium(II)/indium(III) chloride/
triethylammonium and ammonium hexafluorophosphate
(E)-α,β-Ethyleneoxo compds. from 2-acetylenealcohols
Two-metal catalytic redox isomerization

CH≡CHCO ←

33. ⟶

A 0.25 *M* soln. of startg. 2-acetylenealcohol in freshly distilled THF added to a mixture of 20-40 mol% InCl$_3$, 5 mol% chloro(η^5-indenyl)bis(triphenylphosphine)ruthenium(II), 5 mol% triethylammonium hexafluorophosphate, and 5 mol% ammonium hexafluorophosphate, the homogeneous red soln. stirred for several min at room temp., and heated to reflux until reaction complete (1.5 h) → product. Y 67%. InCl$_3$ possibly functions as a chloride scavenger, generating a reactive cationic species from the Ru-catalyst. Isolated keto, acetoxy, hydroxyl, alkyne and alkene groups remained unaffected. Enones were formed from internal 2-acetylenealcohols, but somewhat more slowly. F.e.s. B.M. Trost, R.C. Livingston, J. Am. Chem. Soc. *117*, 9586-7 (1995).

Exchange

Oxygen ↑

HC ↕ O

Zinc bis(tetrahydridoborate)
Prim. alcohols from carboxylic acids
with Zn(BH$_4$)$_2$/(CF$_3$CO)$_2$O cf. *48*, 61; with Zn(BH$_4$)$_2$ in THF, selectivity, s. S. Narasimhan et al., J. Org. Chem. *60*, 5314-5 (1995).

Zn(BH$_4$)$_2$
COOH → CH$_2$OH

Sodium tetrahydridoaluminate
Reductions with sodium tetrahydridoaluminate
s. *11*, 120; in stoichiometric amount s. J.S. Cha, H.C. Brown, Org. Prep. Proc. Intern. *26*, 459-64 (1994).

NaAlH$_4$ ←

Sodium hydridotris(diethylamino)aluminate
Aldehydes from carboxylic acid esters
with LiAlH$_4$/Et$_2$NH cf. *43*, 63s*45*; with NaAlH$_4$/Et$_2$NH under milder conditions s. J.S. Cha et al., Org. Prep. Proc. Intern. *27*, 95-8 (1995).

NaAlH(NEt$_2$)$_3$
COOR → CHO

Samarium/hydrogen chloride
Prim. alcohols from carboxylic acids or esters
with SmI$_2$/water cf. *49*, 62; with lanthanide metals and HCl, e.g. Sm/HCl, and f. reductions s. Y. Kamochi, T. Kudo, Chem. Pharm. Bull. *42*, 402-4 (1994).

Sm/HCl
COO(H,R) → CH$_2$OH

Samarium diiodide/trifluoroacetic acid SmI_2/CF_3COOH
Benzyl ethers from ar. acetals s. *34*, 723s*51* $C(OR)_2 \rightarrow CH(OR)$

1,3-Propanedithiol/tetra-n-*butylammonium fluoride/N-methylmorpholine*   ←
Ketones from α-acoxyketones under mild conditions $OAc \rightarrow H$

34.

A soln. of benzoin acetate, 5 eqs. 1,3-propanedithiol, and 3 eqs. N-methylmorpholine in THF treated with 3 eqs. Bu$_4$NF·H$_2$O in THF at room temp. under argon, stirred for 3 h, and quenched with CuSO$_4$ soln. → deoxybenzoin. Y 99%. This is a convenient method for cleaving prim. and sec. acoxy groups based on commercially available reagents. F.e.s. M. Ueki et al., Tetrahedron Letters *36*, 7467-70 (1995).

Polymethylhydrosiloxane s. *under Ti(OPr-*i*)$_4$*   ←

Diisopropoxytitanium(III) tetrahydridoborate $(i\text{-}PrO)_2TiBH_4$
Selective reductions with diisopropoxytitanium(III) tetrahydridoborate under mild conditions   ←

35.

Prim. alcohols from carboxylic acids. A soln. of Boc-L-Ala-OH in dry dichloromethane added to a soln. of 2 eqs. (*i*-PrO)$_2$TiBH$_4$ (prepared from BnEt$_3$NBH$_4$ and (*i*-PrO)$_2$TiCl$_2$ in 2:1 proportion) in the same solvent at -78°, allowed to warm to room temp. (25°), stirred for 4 h, a satd. soln. of K$_2$CO$_3$ added, and stirred for a further 15 min → Boc-L-Ala-ol. Y 70%. The reagent is easily prepared, versatile, and selective, and the method mild, fast, and free from racemization. Functional groups such as ethers, chlorides, carbamates, azides, esters, amides, nitriles, double bonds and aldoximes remained unaffected. F.e., **also from carboxylic acid chlorides** or aldehydes (cf. *49, 44*) and cyclic sec. alcohols from cyclic ketones, s. K.S. Ravikumar, S. Chandrasekaran, J. Org. Chem. *61*, 826-30 (1996); reduction of ketones with Cp$_2$TiBH$_4$ cf. M.C. Barden, J. Schwartz, ibid. *60*, 5963-5 (1995); *erythro*-2,3-epoxyalcohols from α,β-epoxyketones s. Tetrahedron *52*, 9137-42 (1996); reductions with polymer-based zirconium(IV) tetrahydridoborate cf. B. Tamami, N. Goudarzian, Chem. Commun. *1994*, 1079.

Titanium tetraisopropoxide/polymethylhydrosiloxane   ←
Prim. alcohols from carboxylic acid esters $CO_2R \rightarrow CH_2OH$
with Ti(OPr-*i*)$_4$/(EtO)$_3$SiH cf. *47*, 62s*48*; with polymethylhydrosiloxane, also from carboxylic acids and with Zr(OEt)$_4$, s. S.W. Breeden, N.J. Lawrence, Synlett *1994*, 833-5.

Hydrous tin(IV) oxide/isopropanol $SnO_2/i\text{-}PrOH$
Catalyzed gas-phase reductions with isopropanol   ←
with ZrO$_2$ cf. *45*, 28; with hydrous SnO$_2$, methylarenes, s. K. Takahashi et al., Bull. Chem. Soc. Japan *67*, 1107-12 (1994).

Acetyltetracarbonyl(triphenylphosphine)manganese/phenylsilane   ←
Ethers from carboxylic acid esters $CO_2R \rightarrow CH_2OR$
Catalytic hydrosilylation

36. PhCH$_2$COOMe $\xrightarrow{PhSiH_3}$ PhCH$_2$CHMe(OSiPh$_3$) $\longrightarrow$ PhCH$_2$CH$_2$OMe

A soln. of methyl phenylacetate (2 mmole scale) in benzene containing 1-1.2 eqs. phenylsilane treated with 1.5-3 mol% (PPh$_3$)(CO)$_4$MnC(O)Me for 25 min at room temp. (exothermic) → product. Y 83%. The active catalyst is a coordinatively unsatd. manganese silyl, (L)(CO)$_3$MnSiPh$_3$, which can also be formed [where L = CO] *in situ* by photolysis of (CO)$_5$MnSiMe$_2$Ph in the presence of excess of PhSiH$_3$. Reaction was slower with (CO)$_5$MnC(O)Me as pre-catalyst, while Mn(CO)$_5$Me and Mn(CO)$_5$Br were less effective. F.e. incl. **cyclic ethers** from lactones, also isolation of the intermediate **alkyl silyl acetals,** s. Z. Mao et al., J. Am. Chem. Soc. *117*, 10139-40 (1995).

Palladous acetate/1,3-bis(diphenylphosphino)propane/triethylsilane   *Pd(OAc)$_2$/dppp/Et$_3$SiH*
Replacement of triflyloxy groups by hydrogen   OSO$_2$CF$_3$ → H
with Pd(PPh$_3$)$_4$/NaBH$_4$ cf. *42*, 58s*43*; arenes and ethylene derivs. with Pd(OAc)$_2$/dppp/Et$_3$SiH under mild conditions s. H. Kotsuki et al., Synthesis *1995*, 1348-50.

Tris(dibenzylideneacetone)dipalladium/(R)-2-(diphenylphosphino)-2′-methoxy-1,1′-
 biphenanthryl/1,8-bis(dimethylamino)naphthalene/formic acid
Ethylene derivs. from 2-ethylenecarbonates - Asym. reduction   OCOOR → H
s. *49*, 64; chiral 2-ethylenesilanes s. Tetrahedron Letters *35*, 4813-6 (1994).

Nitrogen ↑   HC ⇅ N

Sodium/n-propanol   *Na/PrOH*
Prim. alcohols from carboxylic acid amides   CON< → CH$_2$OH
with Na/liq.NH$_3$ cf. *39*, 59; with Na/*n*-PrOH, 2-aminoalcohols, s. H.M. Moody et al., Tetrahedron Letters *35*, 1777-80 (1994); B. Kaptein et al., J. Chem. Soc. Perkin Trans. I *1994*, 1495-8.

Sodium tetrahydridoborate   *NaBH$_4$*
Prim. alcohols from N-acyl-N-heterocyclics
from 3-acylthiazolidine-2-thiones cf. *37*, 57; **from N-acylimidazoles**, selective reduction in aq. media, s. T.V. Ovaska et al., Synlett *1995*, 839-40.

Lithium tetrahydridoaluminate   *LiAlH$_4$*
Aldehydes from hydroxamic acid esters   CON(OMe)R → CHO
s. *45*, 510; peptide aldehydes from polymer-based hydroxamate derivs. s. J. Martinez et al., Tetrahedron Letters *36*, 7871-4 (1995).

Dimethylformamide/ferrous sulfate   *HCONMe$_2$/FeSO$_4$*
Hydrocarbons from diazonium fluoroborates   ArN$_2^+$ → ArH
with H$_3$PO$_2$/Cu$_2$O cf. *33*, 71; with DMF or DMA and FeSO$_4$ s. F.W. Wassmundt, W.F. Kiesman, J. Org. Chem. *60*, 1713-9 (1995).

Tri-n-butyltin hydride/di-tert-butyl hyponitrite   ←
Methyl ketones from diazomethyl ketones   C=N$_2$ → CH$_2$

37.   n-C$_5$H$_{11}$COCH=N$_2$ ⟶ n-C$_5$H$_{11}$COCH$_3$

A soln. of 1-diazoheptan-2-one and 1.025 eqs. Bu$_3$SnH in benzene stirred under N$_2$ for 30 min at 60° in the presence of a little di-*tert*-butyl hyponitrite as initiator, cooled, diluted with ether, and washed with satd. aq. KF then satd. brine → methyl pentyl ketone. Y 90%. A radical-chain mechanism was confirmed. F.e. and conditions s. H.-S. Dang, B.P. Roberts, J. Chem. Soc. Perkin Trans. I *1996*, 769-75.

Ferrous sulfate s. under HCONMe$_2$ *FeSO$_4$*

Halogen ↑   HC ⇅ Hal

Lithium/tert-butanol   *Li/t-BuOH*
Replacement of chlorine by hydrogen   Cl → H
s. *18*, 109; under ultrasonication cf. V.E. Uberti Costa et al., Synth. Commun. *25*, 2091-7 (1995).

Lithium hydride s. under Ni(OAc)₂   LiH
Sodium tetrahydridoborate/(2,12-dimethyl-3,7,11,17-tetraazabicyclo[11.3.1]heptadeca-1(17),2,11,13,15-pentaene)nickel(II) bis(tetrafluoroborate) or nickel chloride ←
Anion exchanger-supported tetrahydridoborate/nickel acetate ←
Replacement of halogen by hydrogen   Hal → H
with NaBH₄/NiCl₂ cf. *9*, 108s*36*; dechlorination of ar. and vinyl chlorides in isopropanol s. M. Cooke et al., Appl. Organometal. Chem. *9*, 297-303 (1995); with (2,12-dimethyl-3,7,11,17-tetraazabicyclo[11.3.1]heptadeca-1(17),2,11,13,15-pentaene)nickel(II) bis(tetrafluoroborate) in aq. ethanol s. M. Stiles, J. Org. Chem. *59*, 5381-5 (1994); with anion exchanger-supported BH₄⁻/ nickel acetate s. N.M. Yoon et al., ibid. 4687-8.

Diisopropoxytitanium(III) tetrahydridoborate   (i-PrO)₂TiBH₄
Prim. alcohols from carboxylic acid chlorides s. *51*, 35   COCl → CH₂OH

Tris(trimethylsilyl)germane/azodiisobutyronitrile   (Me₃Si)₃GeH/AIBN
Tris(trimethylsilyl)germane as alternative to tri-*n*-butyltin hydride ←

38.

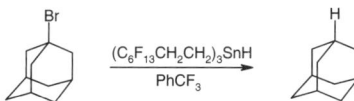

A soln. (0.2 M) of startg. halide, 2 eqs. tris(trimethylsilyl)germane and 10 mol% AIBN in toluene heated at 82° for 20-60 min → product. Y 99% (by GC). The reagent is effective for the reduction of chlorides, bromides, and iodides, deoxygenation of sec. alcohols via thiono-esters, deamination of prim. amines via isocyanides, and cleavage of phenylseleno and *tert*-nitro groups. It is notably effective for H-atom transfer to prim. alkyl radicals. F.e.s. C. Chatgilialoglu, M. Ballestri, Organometallics *14*, 5017-8 (1995).

Tri-n-butyltin hydride/sodium hydrogen carbonate   Bu₃SnH/NaHCO₃
Replacement of halogen by hydrogen in aq. media   Hal → H
s. *46*, 76; with water-soluble and [with added detergent] water-insoluble substrates s. U. Maitra, K.D. Sarma, Tetrahedron Letters *35*, 7861-2 (1994).

Trineophyltin deuteride/azodiisobutyronitrile   R₃SnD/RN=NR
Replacement of halogen by deuterium   Hal → D
s. *24*, 91; with trineophyltin deuteride, selectivity, s. J.C. Podestá et al., J. Org. Chem. *59*, 3747-8 (1994).

Tris[2-(perfluorohexyl)ethyl]tin hydride/sodium trihydridocyanoborate ←
Syntheses with tris[2-(perfluorohexyl)ethyl]tin hydride in a fluorous medium ←

39.

Routine syntheses conducted with toxic Bu₃SnH can be performed with tris[2-(perfluorohexyl)ethyl]tin hydride in a *partially*-fluorinated hydrocarbon solvent so that tin by-products can be removed more easily by liquid-liquid extraction. A catalytic variant with as little as 1 mol% of the reagent is equally effective and the hydride may be reused 5 times following extraction. **E**: A soln. of 1-bromoadamantane, 1.3 eqs. NaCNBH₃, and 10 mol% tris[2-(perfluorohexyl)ethyl]tin hydride in 1:1 benzotrifluoride/*tert*-butanol refluxed for 3 h → adamantane. Y 92%. Work-up involves aqueous extraction of inorganic salts, extraction of product with methylene chloride, and removal of stannane in the fluorous phase. Ionic reductions with the same reagent and Stille coupling with fluorinated stannanes have been performed in the same media. F.e. and **parallel liquid-phase combinatorial syntheses** s. D.P. Curran, S. Hadida, J. Am. Chem. Soc. *118*, 2531-2 (1996).

Nickel acetate/lithium hydride/tert-butanol
Lithium hydride complexes as reductant $\quad$ Ni(OAc)$_2$/LiH/t-BuOH ←

40.

Commercial LiH can be activated **as a hydride source** by metal complexation in the same manner as NaH (review s. *31*, 60s*38*). E: *tert*-Butanol in dry DME added dropwise to a suspension of 3 eqs. LiH and 0.5 eq. Ni(OAc)$_2$ in the same solvent at 65°, stirred for 5 h, 1-bromonaphthalene in the same solvent added, and worked up after 3.5 h → naphthalene. Y 99%. The reducing system [LiH-NiCRA] exhibited attenuated reducing properties by comparison with NaH-NiCRA. F.e. and reductions, incl. dehalogenation of alkyl bromides and chlorides, hydrogenation of alkenes and enones, reductive ring opening of epoxides, coupling of ar. halides, and desulfurization (the latter two processes with LiH-NiCRA-PPh$_3$ or LiH-NiCRA-bpy), s. Y. Fort, Tetrahedron Letters *36*, 6051-4 (1995).

Nickel(II) s. under NaBH$_4$ $\quad$ Ni(II)

Sulfur ↑ $\qquad\qquad$ HC ↕ S

tert-*Butyllithium* $\quad$ t-BuLi
Halides from α-halogenosulfoxides $\quad$ S(O)R → H
with EtMgBr cf. *47*, 75; with *t*-BuLi via 1,1-halogenoorganolithium compds., also 1,1-deuteriochlorides, s. Tetrahedron *52*, 2349-58 (1996).

Dilithium dimethyl(cyano)cuprate $\quad$ Li$_2$Cu(CN)Me$_2$
Zinc/zinc chloride-N,N,N′,N′-tetramethylethylenediamine $\quad$ Zn/ZnCl$_2$-TMEDA
Stepwise reduction of α-ketoketene mercaptals under mild conditions ←

41.

By increasing the proportion of reagent (Zn/ZnCl$_2$-TMEDA) and reaction time, reduction of α-ketoketene mercaptals may yield **3-ketothioenolethers, 3-ketothioethers or ketones**. E: 3 eqs. Zn dust and *1.5 eqs*. ZnCl$_2$-TMEDA added to a well-stirred soln. of startg. mercaptal in ethanol, and refluxed for *4 h* → product. Y 85%. The corresponding 3-ketothioethers were obtained with 3 eqs. ZnCl$_2$-TMEDA (7-13 h) and the corresponding ketones with 5 eqs. ZnCl$_2$-TMEDA (16-25 h). F.e.s. H. Junjappa et al., Tetrahedron *52*, 4679-86 (1996); (Z)-β-alkylthio-α,β-ethylenenitriles with Li$_2$Cu(CN)Me$_2$ s. A. Hosomi et al., Bull. Chem. Soc. Japan *67*, 1495-8 (1994).

Zinc/acetic acid $\quad$ Zn/AcOH
Cleavage of arylthio groups $\quad$ SAr → H
with Zn/Me$_3$SiCl cf. *5*, 78s*33*; of 2-pyridylthio groups with Zn/AcOH s. A. Schmitt et al., Tetrahedron Letters *36*, 7243-6 (1995).

Samarium diiodide/N,N′-dimethyl-N,N′-propyleneurea/methanol $\quad$ SmI$_2$/DMPU/MeOH
Replacement of sulfonyl groups by hydrogen $\quad$ SO$_2$R → H
with SmI$_2$/HMPA cf. *41*, 63s*46*; (E)-ethylene derivs. with SmI$_2$/DMPU/MeOH s. G.E. Keck et al., J. Org. Chem. *60*, 3194-204 (1995).

Triphenyltin hydride/azodiisobutyronitrile $\quad$ Ph$_3$SnH/AIBN
Ethers or cyclic ethers from thiono-carboxylic acid esters or -lactones $\quad$ C(S)OR → CH$_2$OR

42.

A soln. of startg. thionolactone and 5 eqs. Ph$_3$SnH in toluene at 110° treated at small intervals over 1 h with 0.15 eq. AIBN, and allowed to react to completion by TLC → product. Y 95%. No ring-opened by-products were observed. The method is applicable to prim. or sec. alcohol-derived thionoesters as well as thionolactones. Less reactive Bu₃SnH was less effective. F.e.s. K.C. Nicolaou et al., Chem. Commun. *1995*, 1583-5.

Tri-n-butyltin chloride/sodium trihydridocyanoborate/azodiisobutyronitrile ←
Replacement of sulfonyl groups by hydrogen $SO_2R \rightarrow H$
ketones from β-ketosulfones with Bu₃SnH cf. *45*, 41; cleavage of 2-hetarylsulfonyl groups, α-fluoroesters, s. S.F. Wnuk, M.J. Robins, J. Am. Chem. Soc. *118*, 2519-20 (1996); with Bu₃SnCl/NaBH₃CN, selectivity, s. R. Giovannini, M. Petrini, Synlett *1995*, 973-4.

Remaining Elements ↑   HC ⇅ Rem

Sodium/naphthalene $Na/C_{10}H_8$
Demercuration $Hg \rightarrow H$
with Li cf. *29*, 172; arenes from diarylmercury compds. with Na/C₁₀H₈ s. S. Rok Do, H.J. Shine, J. Org. Chem. *60*, 5414-8 (1995).

n-Butyllithium or Cesium fluoride   *n-BuLi or CsF*
Silver nitrite/potassium cyanide   *AgNO₂/KCN*
Trimethylsilyl chloride/potassium iodide   *Me₃SiCl/KI*
Ammonium fluoride/tetra-n-butylammonium fluoride   *NH₄F/Bu₄NF*
Replacement of silyl groups by hydrogen $Si \Leftarrow \rightarrow H$
with CsF s. *31*, 65; cleavage of arylsilane-linked polymer supports s. B. Chenera et al., J. Am. Chem. Soc. *117*, 11999-2000 (1995); ar. desilylation with Me₃SiCl/KI cf. F. Radner, L.-G. Wistrand, Tetrahedron Letters *36*, 5093-4 (1995); preferential cleavage of silylacetylenes with AgNO₂/KCN s. J. Alzeer, A. Vasella, Helv. Chim. Acta *78*, 177-93 (1995); cleavage of (1,1-dimethyl-3-hydroxypropyl)dimethylsilylacetylenes with *n*-BuLi s. ibid. 732-57; cleavage of chiral α-(2,3-dimethylbut-2-yldimethylsilyl)ketones s. *51*, 440.

Carbon ↑   HC ⇅ C

Samarium diiodide/hexamethylphosphoramide $SmI_2/(Me_2N)_3PO$
Replacement of cyano groups by hydrogen under mild conditions $CN \rightarrow H$

43.   Ph–CH(CN)–CN ⟶ Ph–CH₂–CN

Nitriles from malononitriles. 3 eqs. SmI₂ in 10:1 THF/HMPA added to a soln. of 3-phenyl-2-cyanopropanenitrile in THF *at 0°*, stirred for 1 h, and quenched with aq. NH₄Cl → 3-phenylpropanenitrile. Y 85%. The procedure is milder than the tributyltin hydride method (cf. *46*, 84), requiring refluxing in benzene, and can be performed in the presence of a remote ethylenic unsaturation (*without* radical ring closure) and a protected hydroxyl group. It also serves (at room temp.) for the preparation of **carboxylic from α-cyanocarboxylic acid esters**. F.e.s. Y.S. Cho et al., Tetrahedron Letters *36*, 7661-4 (1995).

Dihydridotetrakis(triphenylphosphine)ruthenium(II)/ammonium formate   ←
*Bis(dibenzylideneacetone)palladium(0)/1,3-bis(diphenylphosphino)propane/triethylamine/
   formic acid*   ←
Tetrakis(triphenylphosphine)palladium(0)/ammonium formate $Pd(PPh_3)_4/HCOONH_4$
3-Ethylenealcohols from 4-alkynyl-1,3-dioxolan-2-ones   C
(Z)-isomers with Pd(acac)₂ cf. *49*, 81; 2-allenealcohols with Pd(dba)₂/dppp/HCOOH/Et₃N and 3-acetylenealcohols with PBu₃ as ligand s. C. Darcel et al., Synlett 1994, 457-8; chiral 2-allenealcohols with RuH₂(PPh₃)₄, also chiral (E)-2-ethylenealcohols from 4-vinyl-derivs., s. S.-K. Kang et al., Synth. Commun. *26*, 1485-92 (1996); chiral 3-ethylenealcohols with Pd(PPh₃)₄ cf. ibid. *25*, 203-14 (1995).

Elimination

Oxygen ↑   HC ⇑ O

Chlorobis(cyclopentadienyl)hydridozirconium/lithium bis(trimethylsilyl)amide
α,β-Ethylene- from β-keto-carboxylic acid esters via zirconium(IV) enolates

44.

A soln. of startg. lithium enolate (generated by treatment of the corresponding acid ester with 1.2 eqs. LiN(SiMe$_3$)$_2$ in DME) added to a vigorously stirred suspension of 1.2 eqs. Cp$_2$ZrHCl in the same solvent at 0°, allowed to warm slowly to room temp., and worked up after 1-2 h → product. Y 65%. F.e.s. A.G. Godfrey, B. Ganem, Tetrahedron Letters *33*, 7461-4 (1992).

Formation of O-N Bond

Uptake   ⇓

Addition to Nitrogen   ON ⇓ N

m-Chloroperoxybenzoic acid/hydrogen fluoride/methanol
Methylrhenium oxide/hydrogen peroxide   MeReO$_3$ /H$_2$O$_2$
Cyclic N-oxides from N-heterocyclics   ≩N → ≩N→O
with peroxyacetic acid cf. *16*, 140; cyclic amino-N-oxides with *m*-CPBA/HF/MeOH s. S.H. Rhie, E.K. Ryu, Heterocycles *41*, 323-8 (1995); with MeReO$_3$/H$_2$O$_2$ s. *51*, 45.

Exchange   ⇅

Hydrogen ↑   ON⇅ H

Titanium silicate/hydrogen peroxide
Chromium silicate/tert-butyl hydroperoxide
Silicate-catalyzed oxidation of amines
oximes cf. *21*, 125s49; **nitrones** from sec. amines s. A. Sudalai et al., Synlett *1995*, 1177-8; **nitro compds.** from prim. amines with chromium silicate/*t*-BuOOH s. Tetrahedron *51*, 11305-18 (1995).

Urea-hydrogen peroxide/sodium tungstate
Nitrones from sec. amines   CHNHR → C=N(O)R
with Na$_2$WO$_4$/H$_2$O$_2$ cf. *47*, 80; more safely with urea-H$_2$O$_2$ s. E. Marcantoni et al., Tetrahedron Letters *36*, 3561-2 (1995).

Potassium peroxymonosulfate/acetone/sodium hydroxide/sodium hydrogen carbonate
Ar. nitro compds. from amines   NH$_2$ → NO$_2$
under phase transfer catalysis cf. *44*, 85; without catalyst, hydroxy- and carboxy-derivs., s. K.S. Webb, V. Seneviratne, Tetrahedron Letters *36*, 2377-8 (1995).

Methylrhenium oxide/hydrogen peroxide $MeReO_3/H_2O_2$
N-Oxidation ←

45.   t-BuNH$_2$ ⟶ t-BuNO$_2$

Nitro compds. from prim. amines. 0.08 eq. MeReO$_3$ added to a soln. of H$_2$O$_2$ in ethanol, the mixture added to *tert*-butylamine, and stirred at room temp. for 2 h → product. Y 100%. The method is simple, general (for aromatic and aliphatic amines as well as adamantylamine) and efficient. F.e. and from nitroso compds., also azoxy from azo compds., hydroxylamines from sec. amines, and cyclic N-oxides from cyclic amines, s. R.W. Murray et al., Tetrahedron Letters *37*, 805-8 (1996).

Formation of O-S Bond

Uptake ⇓

Addition to Hydrogen and Oxygen   OS ⇓ HO

Sulfotransferase ←
Via intermediates   *v.i.*
Regiospecific formation of sulfuric acid monoesters   OH → OSO$_3$H
with NaOH cf. *31*, 81; glycol monosulfates via cyclic dialkoxystannanes s. B. Guilbert et al., Tetrahedron Letters *35*, 6563-6 (1994); enzymatic sulfation of oligosaccharides with a sulfotransferase s. C.-H. Wong et al., J. Am. Chem. Soc. *117*, 8031-2 (1995).

Addition to Sulfur   OS ⇓ S

Montmorillonite s. under Mg-monoperoxyphthalate ←

Chiral 2-[N-(salicylidene)amino]alcohols s. under VO(acac)$_2$ ←

Pivaldehyde s. under Chiral β-oxoaldiminatomanganese(III) complex   RCHO

2,2,6,6-Tetramethylpiperidine nitroxyl s. under NaOCl   TEMPO

Magnesium monoperoxyphthalate/bentonite or montmorillonite ←
Sulfoxides from thioethers   >S → >SO
with Aliquat 336 as catalyst cf. *43*, 111; with bentonite or K10 montmorillonite as support s. M. Hirano et al., Synth. Commun. *25*, 3125-34 (1995).

Diphenyl sulfoxide s. under ReOCl$_3$(PPh$_3$)$_2$ Ph$_2$SO

1,2-Benziososelenazol-3(2H)-ones s. under H$_2$O$_2$ ←

Iodosobenzene/p-toluenesulfonic acid   PhIO/TsOH
Sulfoxides from thioethers
with *o*-iodosobenzoic acid/H$_2$SO$_4$ cf. *6*, 147s*48*; with PhIO/TsOH s. R.-Y. Yang, L.-X. Dai, Synth. Commun. *24*, 2229-36 (1994).

Perfluoro-cis-2,3-dialkyloxaziridines ←
with (E)-2-sulfonyloxaziridines cf. *35*, 57; with perfluoro-*cis*-2,3-dialkyloxaziridines, **also sulfones**, s. D.D. DesMarteau et al., J. Org. Chem. *59*, 2762-5 (1994); α,β-ethylene-α-(trifluoromethyl)-sulfoxides and -sulfones s. J.-P. Bégué et al., Synthesis *1996*, 399-402.

*Titanium tetraisopropoxide/diethyl tartrate/isopropanol/cumene hydroperoxide/
molecular sieves* ←
Vanadyl acetoacetonate/chiral 2-[N-(salicylidene)amino]alcohols/hydrogen peroxide ←
Sulfoxides from thioethers by asym. oxidation $>S \rightarrow >SO$

46.

Thioanisole added to a soln. of *1 mol%* VO(acac)$_2$ and *1.5 mol%* (S)-2-[N-(3'-*tert*-butyl-5'-nitrosalicylidene)amino]-3,3-dimethylbutan-1-ol in methylene chloride (pre-stirred for 5 min), followed by dropwise addition of 1.1 eqs. 30% H$_2$O$_2$, and the mixture stirred at room temp. for 16 h → (S)-product. Y 94% (e.e. 70%). The method proceeds simply (without the need to exclude air or moisture), and the oxidant is cheap and safe. The catalytic system is extremely efficient, *0.01 mol%* being sufficient to catalyze sulfoxide formation. F.e. and ligands, also oxidation of alkyl benzyl thioethers, cyclic mercaptals and α-(arylthio)ketones, s. C. Bolm, F. Bienewald, Angew. Chem. Intern. Ed. *34*, 2640-2 (1995); with Ti(OPr-*i*)$_4$/(R,R)-diethyl tartrate/*i*-PrOH/cumene hydroperoxide/molecular sieves (cf. *39,* 83s*49*) s. J.M. Brunel, H.B. Kagan, Synlett *1996*, 404-6.

Hydrogen peroxide s.a. under VO(acac)$_2$ H$_2$O$_2$
Hydrogen peroxide/1,2-benzisoselenazol-3(2H)-ones ←
Sulfoxides from thioethers
with 2-aryl-1,2-benzisoselenazol-3(2*H*)-one oxides cf. *42,* 498; catalytic procedure with 1,2-benzisoselenazol-3(2*H*)-ones/H$_2$O$_2$ and f. catalysts s. J. Mlochowski et al., Synth. Commun. *26,* 291-300 (1996).

Tetra-n-butylammonium persulfate (Bu$_4$N)$_2$S$_2$O$_8$
Sulfoxides from thioethers under mild, neutral conditions

47. PhSMe ⟶ PhS(O)Me

A soln. of thioanisole in methylene chloride treated with a soln. of 1 eq. tetra-*n*-butylammonium persulfate in the same solvent, stirred at room temp. for 1.5 h under N$_2$, poured into water, and extracted with dichloromethane → methyl phenyl sulfoxide. Y 98%. The reaction is generally applicable to dialkyl, alkyl aryl and diaryl sulfides, including those with strong electron-withdrawing substituents, affording the corresponding sulfoxides in excellent yield. Over-oxidation to sulfones was not observed. F.e.s. F. Chen et al., Synth. Commun. *26,* 253-60 (1996).

tert-*Butyl hypochlorite* t-*BuOCl*
Sulfoxides from thioethers
s. *23,* 196; *anti*-aryl benzyl sulfoxides s. T. Sato, J. Otera, Synlett *1995,* 365-6; 3-(sulfinylaryl)-Δ^1-diazirines s. C.W.G. Fishwick et al., Synthesis *1995,* 553-6.

*Sodium hypochlorite/2,2,6,6-tetramethylpiperidine nitroxyl/tetra-n-butylammonium chloride/
potassium bromide/sodium hydrogen carbonate* ←
Sulfoxides from thioethers in a 2-phase medium
with Cl$_2$/KHCO$_3$ cf. *34,* 79; with NaOCl/NaHCO$_3$/Bu$_4$NCl/KBr/2,2,6,6-tetramethylpiperidine nitroxyl s. R. Siedlecka, J. Skarzewski, Synthesis *1994,* 401-4.

Chiral β-oxoaldiminatomanganese(III) complexes/pivaldehyde ←
Sulfoxides from thioethers
Asym. catalyzed aerobic oxidation

48.

A soln. of 2-chlorophenyl methyl sulfide and 3 eqs. pivaldehyde in *m*-xylene added to a mixture of 18 mol% (S,S)-β-oxoaldiminatomanganese(III) complex in the same solvent, and stirred overnight at room temp. under O_2 → product. Y 72% (e.e. 72%). The catalyst was more effective than Mn(III)-salen complexes, and the abs. configuration was the reverse of that obtained with the latter catalysts by using PhIO or H_2O_2 as re-oxidant (cf. *46*, 106s*48,49*). F.e. and reversal of enantioselectivity in the presence of N-methylimidazole s. T. Mukaiyama et al., Chem. Lett. *1995*, 335-6; details s. Bull. Chem. Soc. Japan *68*, 3241-6 (1995).

Trichlorobis(triphenylphosphine)oxorhenium(III)/diphenyl sulfoxide $ReOCl_3(PPh_3)_2/Ph_2SO$
Sulfoxides from thioethers   >S → >SO
Catalytic oxygen atom transfer

49.   EtSCH$_2$CH$_2$OH   ⟶   EtS(O)CH$_2$CH$_2$OH

Startg. sulfide in deuteriochloroform treated with 0.05 mol% ReOCl$_3$(PPh$_3$)$_2$ and 1.3 eqs. Ph$_2$SO at 25° for 1.5 h → product. Y 98%. Oxygen atom transfer from sulfoxides provides a mild, efficient, and rapid route for sulfide oxidation, without production of sulfone by-products. Alcohols, amines, and ester groups remained unaffected, and the procedure is suitable on the preparative scale (in chloroform). F.e. and with Me$_2$SO-d$_6$ s. J.B. Arterburn, S.L. Nelson, J. Org. Chem. *61*, 2260-1 (1996).

Ferric bromide-dimethyl sulfoxide complex/nitric acid $FeBr_3·DMSO/HNO_3$
Sulfoxides from thioethers
with HNO$_3$/Bu$_4$NAuBr$_4$ cf. *29*, 84s*48*; with HNO$_3$/FeBr$_3$·DMSO s. A.R. Suárez et al., Tetrahedron Letters *36*, 1201-4 (1995).

Ruthenium trichloride/sodium periodate $RuCl_3/NaIO_4$
Sulfones from thioethers   >S → >SO$_2$
with RuO$_4$ cf. *9*, 174; with RuCl$_3$(0.05 mol%)/NaIO$_4$ for oxidation of electron-deficient thioethers s. W. Su, Tetrahedron Letters *35*, 4955-8 (1994).

Exchange   ⇅

Carbon ↑   OS ⇅ C

N-Bromosuccinimide   NBS
Sulfinic acid esters from methoxymethyl thioethers and alcohols   RS(O)OR′

50.   PhSCH$_2$OMe   —MeOH→   PhS(O)OMe

A stirred soln. of methoxymethyl phenyl sulfide in dry methanol at -40° treated with 2.1 eqs. NBS, allowed to warm to -10°, and stirring continued until reaction complete by TLC → product. Y 93%. Tert. alcohols failed to react under the reaction conditions. F.e. and with retention of halide, ester and ketone functional groups, s. D.-W. Kim et al., Synth. Commun. *25*, 2871-6 (1995).

Formation of O-Rem Bond

Uptake   ⇓

Addition to Hydrogen and Oxygen   ORem ⇓ HO

Pyridine/ammonium hydroxide C_5H_5N/NH_4OH
Phosphorous acid monoesters from alcohols   OH → OPO(H)OH
with HPO(OH)$_2$ cf. *50*, 43; with diphenyl phosphite/py/NH$_4$OH s. A. Kers et al., Synthesis *1995*, 427-30

Via intermediates *v.i.*
O-Phosphorylation of alcohols OH → OPO(OH)$_2$
via mixed phosphoric acid esters with di-*tert*-butyl N,N-diethylphosphoramidite cf. *43*, 83; with bis[2-(methyldiphenylsilyl)ethyl] N,N-diethylphosphoramidite s. S. Freeman et al., J. Chem. Soc. Perkin Trans. I *1995*, 421-6; solid-supported 5′-phosphorylation of oligonucleoside methylphosphonates with 2-cyanoethyl N,N-diisopropylchlorophosphoramidite, s. P. Bhan, Tetrahedron Letters *35*, 4895-8 (1994); with P(OMe)$_3$ s. *51*, 51; improved procedure with POCl$_3$ (cf. *15*, 131) s. A.M. Modro, T.A. Modro, Org. Prep. Proc. Intern. *24*, 57-9 (1992).

Addition to Oxygen and Carbon ORem ⇓ OC

Chlorobis(1,5-cyclooctadiene)rhodium(I) dimer/(R)-2,2′-bis(dicyclohexylphosphino)-1,1′-binaphthyl ←
Asym. hydrosilylation of ketones CO → CH(OSi≤)
with chiral phosphinites cf. *29*, 107s*44*; with (R)-Cybinap s. H. Takaya et al., Chem. Commun. *1994*, 2525-6; with chiral diferrocenyl dichalcogenides as ligand s. S. Uemura et al., Organometallics *15*, 370-9 (1996).

Addition to Remaining Elements ORem ⇓ Rem

Bromine Br$_2$
α-Amino-phosphonic from -phosphonous acids s. *51*, 240 PH(O)OH → PO(OH)$_2$
Iodine I$_2$
Oligonucleotide synthesis ←
s. *17*, 169s*46-9*; improved **phosphoramidite method** with diethyl N,N-diisopropylphosphoramidite s. S. Agrawal et al., Tetrahedron Letters *35*, 8565-8 (1994); with a polyfunctionalized phosphoramidite s. M. Komiyama et al., ibid. 5879-82; incorporation of ^{13}C-ribonucleosides s. ibid. 6649-52; P-chiral dinucleoside phosphorothioates by the phosphoramidite method s. G. Just, ibid. *37*, 973-6 (1996); **solid-phase RNA or DNA synthesis** with a novel nucleoside phosphoramidite adapter s. G.R. Gough et al., ibid. *36*, 27-30 (1995); with a universal allyl linker, 9-O-(4,4′-dimethoxytrityl)-10-undecenoic acid, s. X. Zhang, R.A. Jones, ibid. *37*, 3789-90 (1996); automated synthesis of branched oligodeoxynucleotides s. J. Wengel et al., Tetrahedron *51*, 8491-506 (1995); amide-linked oligodeoxynucleotides s. A. De Mesmaeker et al., Tetrahedron Letters *35*, 5225-8 (1994); 5′-5′-bridged oligodeoxyribonucleotides s. ibid. 5221-4; acridine- and/or lipid-containing oligodeoxynucleotides s. C.J. Marasco, Jr. et al., ibid. 3029-32; nucleic acid synthesis on a Fractogel support s. M.P. Reddy et al., ibid. 5771-4; **rapid cleavage of oligonucleotides** from the support and deprotection with MeNH$_2$/NH$_3$ s. ibid. 4311-4; **facile recycling of nucleosides** during solid-phase oligonucleotide synthesis s. W.K.-D. Brill, ibid. 3041-4 (1994).

Exchange ⇅

Oxygen ↑ ORem ⇅ O

Triphenylphosphine/diethyl azodicarboxylate Ph$_3$P/ROOCN=NCOOR
Benzotriazolyloxytris(dialkylamino)phosphonium hexafluorophosphate ←
Phosphorus acid esters from acids P—OH → P—OR
phosphonates s. *48*, 104; nucleoside 5′-phosphonates and 5′-phosphates s. M. Saady et al., Tetrahedron Letters *36*, 2239-42 (1995); **phosphorus(III) acid esters** s. I.D. Jenkins et al., ibid. *37*, 1087-90 (1996); mixed phosphonate di- from mono-esters with benzotriazolyloxytris(dialkylamino)phosphonium hexafluorophosphate s. J.-M. Campagne et al., J. Org. Chem. *60*, 5214-23 (1995).

Halogen ↑   ORem ⇅ Hal

Potassium hydroxide   *KOH*
Silanols from fluorosilanes s. *51*, 354   ≥SiF → ≥SiOH
2-Pyrrolidone magnesium salt or Triethylamine   ←
(Z)-Enoxysilanes from ketones   C=C(OSi≤)
with NaI/Et₃N cf. *43*, 89; with electrogenerated 2-pyrrolidone Mg-salt as mild, controllable base s. M. Bordeau et al., J. Organometal. Chem. *493*, 27-32 (1995); (E)-2-(arylseleno)enoxysilanes with Et₃N s. C. Paulmier et al., Tetrahedron *51*, 9569-80 (1995).

Carbon ↑   ORem ⇅ C

1,8-Diazabicyclo[5.4.0]undec-7-ene   *DBU*
Dithiophosphonic acid O-monoesters from alcohols s. *51*, 247   RPS(SH)OR
Carbon tetrabromide/pyridine CBr_4/C_5H_5N
O-Phosphorylation under mild conditions   ROPO(ORε)₂

51.   PhCH₂CH₂OH + P(OMe)₃ ⟶ PhCH₂CH₂OPO(OMe)₂ ⟶ PhCH₂CH₂OPO(OH)₂

Pyridine added to 2-phenylethanol and 1.1 eqs. carbon tetrabromide under N₂, cooled to 0°, 1.25 eqs. trimethyl phosphite added dropwise, and stirred at room temp. for 2.5 h → 2-phenylethyl dimethyl phosphate (Y 98%), dissolved in acetonitrile with 2.4 eqs. NaI, treated dropwise with 2.4 eqs. Me₃SiCl, and stirred in the dark for 2 h → 2-phenylethyl phosphate (Y 79%). The procedure is simple, inexpensive, and applicable to a wide range of prim. and sec. alcohols (incl. allyl alcohols) and phenol. Significantly, **preferential O-phosphorylation** of prim. alcohols takes place in the presence of sec. alcohols, and there was no N-phosphorylation of aminoalcohols (e.g. 2′,3′-isopropylideneadenosine). Carbon tetrachloride was ineffective. F.e.s. V.B. Oza, R.C. Corcoran, J. Org. Chem. *60*, 3680-4 (1995).

Formation of O-C Bond

Uptake   ⇓

Addition to Hydrogen and Carbon   OC ⇓ HC

Dimethyldioxirane/nickel(II) acetoacetonate   ←
α-Hydroxylation of β-dicarbonyl compds.   H → OH

52.

Startg. β-dicarbonyl compd. and 1 eq. 0.1 *M* dimethyldioxirane in acetone added successively to 0.1 eq. Ni(acac)₂ in water at room temp. (ca. 20°), and the mixture stirred for 3.5 h → 3-hydroxy-3-(1-oxoethyl)tetrahydrofuran-2-one. Y 99%. Reaction was sluggish with ketoesters or diesters in the absence of Ni(II)-salts. The method avoids the use of base. F.e. and with Ni(OAc)₂ s. W. Adam, A.K. Smerz, Tetrahedron *52*, 5799-804 (1996).

Sodium chlorite $NaClO_2$
Carboxylic acids from aldehydes   CHO → COOH
with NaClO₂/sulfamic acid cf. *30*, 66; without Cl₂ scavenger s. B.R. Babu, K.K. Balasubramaniam, Org. Prep. Proc. Intern. *26*, 123-5 (1994).

Ruthenium-substituted polyoxometalates
Tert. alcohols from hydrocarbons $\quad$ H → OH
with dioxiranes cf. *41*, 115s*46,48,49*; catalyzed oxidation **with molecular oxygen** in the presence of Ru-subst. polyoxometalates s. R. Neumann et al., Angew. Chem. Intern. Ed. *34*, 1587-9 (1995).

Addition to Oxygen-Oxygen Bonds $\quad$ OC ⇓ OO

5,10,15,20-Tetrakis(heptafluoropropyl)porphyrin/irradiation
Sensitized reactions with singlet oxygen in a fluorous 2-phase medium

53.

A 3 M soln. of cyclohexene in acetonitrile-d_3 layered on top of a 0.0002 M soln. of 5,10,15,20-tetrakis(heptafluoropropyl)porphyrin in perfluorohexanes, and the mixture photo-oxygenated with a 200 W illuminator (Dolan-Jenner Industries) with vigorous stirring at 0° for 46 h → product. Y 96%. The sensitizer is considerably more stable than the conventional tetraphenylporphyrin, due largely to its physical separation [with dissolved oxygen in the fluorous phase] from the reactant and product [in the organic phase]. Chromatographic separation is also much easier given that the highly nonpolar sensitizer is readily soluble in hexane. A further bonus is that singlet oxygen is more stable in fluorous media, and that preparative-scale syntheses can be carried out on reluctant substrates over a longer period of time without loss of sensitizer. F.e.s. S.G. DiMagno et al., J. Am. Chem. Soc. *118*, 5312-3 (1996).

Addition to Oxygen and Carbon $\quad$ OC ⇓ OC

Electrolysis
Glycol monoethers from epoxides
Regio- and stereo-specific ring opening under electrogenerated acid catalysis

54.

A mixture of styrene oxide in methanol containing 0.2 M NaClO$_4$ as electrolyte electrolyzed at room temp. in a divided cell fitted with Pt-electrodes for 5 min at a constant potential of 1.1 V until 0.0026 F/mol consumed, and stirred for a further 10 min without electrolysis → 2-phenyl-2-methoxyethanol. Y 98%. Both linear and branched alcohols participate in the reaction, and cycloalkene oxides afford the *trans*-isomers exclusively. Work-up is easy. F.e. and selectivity, **also glycols** in aq. acetonitrile, s. A. Safavi et al., Bull. Chem. Soc. Japan *68*, 2591-4 (1995).

Electrolysis/dibromo(cyclam)nickel(II)
1,3-Dioxolan-2-ones from epoxides.
with alkali metal halides cf. *42*, 108s*49*; with dibromo(cyclam)nickel(II) under electrolysis s. P. Tascedda, E. Duñach, Chem. Commun. *1995*, 43-4.

Pyridine/4-dimethylaminopyridine $\quad C_5H_5N/DMAP$
Dicarboxylic acid monoesters from anhydrides
with py cf. *2*, 147; with added 4-DMAP **under high pressure** for addition of *hindered* alcohols s. T. Nakata et al., Synlett *1995*, 650-2.

Sodium perborate/acetic anhydride/sodium carbonate $\quad NaBO_3/Ac_2O/Na_2CO_3$
Glycol monoesters from cyclic acetals
with O$_3$ cf. *39*, 107s*46*; with NaBO$_3$·4H$_2$O/Ac$_2$O/Na$_2$CO$_3$ s. S. Bhat et al., Synlett *1995*, 329-30.

Envirocat EPZG, Zeolite or Sulfated zirconia
Acylals from aldehydes $\quad$ CHO → CH(OAc)$_2$
Heterogeneous catalysis

55. $\quad$ o-NO$_2$C$_6$H$_4$CHO + Ac$_2$O ⟶ o-NO$_2$C$_6$H$_4$CH(OAc)$_2$

A mixture of startg. aldehyde, 1.1 eqs. acetic anhydride, and azeotropically dried Envirocat EPZG^R

(a solid-supported acid catalyst exhibiting both Brønsted and Lewis acidity) heated at 60-5° with stirring for 1.5 h → product. Y 99%. Work-up is easy, and the catalyst may be recycled. F.e.s. B.P. Bandgar et al., J. Chem. Res. (S) *1995*, 470-1; also ketone derivs. **on a solid support** (sulfated zirconia) s. S.V.N. Raju, ibid. *1996*, 68; on HY-zeolite cf. B. Gigante et al., Synthesis *1995*, 1077-8.

Ammonium ceric nitrate/ammonium nitrate $(NH_4)_2Ce(NO_3)_6/NH_4NO_3$
Glycol mononitrates from epoxides ▽O▽ → $C(OH)C(ONO_2)$
with $Tl(NO_3)_3$ cf. *36*, 108; with CAN/NH_4NO_3, regioselectivity, s. N. Iranpoor, P. Salehi, Tetrahedron *51*, 909-12 (1995).

2,4,5-Triphenyloxazole s. under O_3

Amberlyst-15 ←
Hydroxycarboxylic acid esters from lactones C
with TsOH cf. *34*, 190; with Amberlyst-15 s. R.C. Anand, N. Selvapalam, Synth. Commun. *24*, 2743-7 (1994).

Trichloroisocyanuric acid ←
Carboxylic acid esters from acetals $CH(OR)_2 → COOR$
electrochemically cf. *41*, 117; α-acoxycarboxylic acid esters with trichloroisocyanuric acid s. F. Ghelfi et al., Synth. Commun. *25*, 3463-70 (1995).

Titanium tetraisopropoxide/chiral α,α,α′,α′-tetraaryl-1,3-dioxolane-4,5-dimethanols ←
Dicarboxylic acid monoisopropyl esters from their anhydrides C
Desymmetrization

56.

Ar = 2-$C_{10}H_7$

1.2 eqs. Ti(OPr-*i*)$_4$ added dropwise to a soln. of 1.25 eqs. β-naphthyl-TADDOL in ether under argon, stirred at room temp. for 3 h, solvent removed under vacuum, the residue dried for 0.5 h, taken up in dry THF, cooled to -30°, a cold (ca. -30°) soln. of startg. anhydride in THF added, the soln. sealed, and stored for 7 days in a freezer (-30°) → product. Y 88% (99:1 mixture of enantiomers). The method is applicable to mono-, bi- and tri-cyclic γ-anhydrides; however, a β-subst. glutaric anhydride was opened with moderate (3:1) selectivity. F.e.s. D. Seebach et al., Angew. Chem. Intern. Ed. *34*, 2395-6 (1995).

Vanadyl acetate/tert-butyl hydroperoxide $VO(OAc)_2$/*t-BuOOH*
Diol monoesters from cyclic acetals
with $RuCl_3$ as catalyst cf. *39*, 107s48; with $VO(OAc)_2$ s. B.M. Choudary, P.M. Reddy, Synlett *1995*, 959-60.

Ozone/2,4,5-triphenyloxazole ←
Azole-ozone complexes as source of singlet oxygen ←

57.

In the presence of a *dilute* soln. of ozone *at -78°*, pyrroles, oxazoles and imidazoles appear to form O_3-complexes which serve as a source of singlet oxygen. The latter may then react with the heterocycle if in excess, or with a substrate added subsequently. E: Methylene chloride saturated with O_3 at -78° until a dark blue colour persisted, purged with N_2 until a pale blue colour remained (*0.006 M* in O_3), N_2 bubbled through the soln. thereby transferring a *low concentration* of O_3 to a connected flask containing 2,4,5-triphenyloxazole in the same solvent at -78°, the resulting complex swept free of excess ozone, 1,3-diphenylisobenzofuran added, and allowed to warm to room temp. → o-dibenzoylbenzene. Y 70%. F.e.s. H.H. Wasserman et al., J. Am. Chem. Soc. *117*, 9772-3 (1995).

Potassium dicarbonyl(cyclopentadienyl)ferrate   *K[CpFe(CO)$_2$]*
Carboxylic acid esters from two aldehyde molecules   2 RCHO → RCOOCH$_2$R
with K$_2$Fe(CO)$_4$/crown ether cf. *4*, 154s*49*; benzyl arylcarboxylates with the more active K[CpFe(CO)$_2$] s. T. Ohishi et al., Organometallics *13*, 4641-2 (1994).

Dibromo(cyclam)nickel(II) s. under ↙   ←

Addition to Nitrogen and Carbon   OC ⇓ NC

Potassium hydroxide/tetra-n-butylammonium hydrogen sulfate   *KOH/Bu$_4$NHSO$_4$*
Iminoesters from nitriles   CN → C(OR)=NH
with NaOR cf. *14*, 181; allyl, benzyl, and glycosyl trichloroacetimidates with KOH/Bu$_4$NHSO$_4$ s. V.J. Patil, Tetrahedron Letters *37*, 1481-4 (1996).

Sodium azide   *NaN$_3$*
Solvolysis of N-carbalkoxylactams   ↶
with KCN cf. *39*, 109s*47*; β-(carbalkoxyamino)esters with NaN$_3$ s. C. Palomo et al., Tetrahedron Letters *36*, 9027-30 (1995); opening of N-carbomethoxy- and N-sulfonyl-lactams with TsOH cf. A.N. Dixit et al., ibid. *35*, 6133-4 (1994).

Alumina   *Al$_2$O$_3$*
Carboxylic acid amides from nitriles under neutral conditions   CN → CONH$_2$
PhCN   ⟶   PhCONH$_2$

A mixture of benzonitrile and *unactivated* alumina (Brockmann, neutral, activity 1) allowed to react at 60° for 72 h → benzamide. Y 90%. The layer of hydroxyl groups on the surface of alumina acts as the source of water. The uncatalyzed reaction is far superior to hydrolysis on neutral Al$_2$O$_3$ promoted by CF$_3$SO$_3$H. F.e. incl. hydrolysis of aliphatic nitriles s. G.W. Kabalka et al., Tetrahedron Letters *36*, 3469-72 (1995).

Hydantoinases   ←
α-Ureidocarboxylic acids from hydantoins   ↶
Asym. hydrolysis

DL-5-*n*-Butylhydantoin added to buffer soln. [0.1 *M* glycine/NaOH, pH 8.5, 1mM Mn(II)] thermo statted at 50° under N$_2$, followed by addition of a little hydantoinase soln. [D-HYD-1], the pH maintained at 8.5 by continuous addition of 1 *N* NaOH, and allowed to react for 23.3 h → D-product. Y 67% (e.e. >99% after N-decarbamylation). Due to spontaneous racemization of the hydantoins under the reaction conditions, quantitative conversion can be achieved. The hydantoinases are thermally stable and commercially available. 5,5-Disubst. hydantoins were unreactive. F.e.s. O. Keil et al., Tetrahedron:Asym. *6*, 1257-60 (1995).

Distannoxanes   *(R$_3$Sn)$_2$O*
Diorganotin(IV) dicarboxylates   *R$_2$Sn(OCOR')$_2$*
Urethans from isocyanates   N=C=(O,S) → NHCO(OR,SR)
s. *49*, 105; chiral urethans (with distannoxanes) s. J. Otera et al., Synlett *1995*, 433-4; **thionourethans from isothiocyanates** (with Bn$_2$Sn(OAc)$_2$) s. G. Purnima, S. Roy, Indian J. Chem. *33B*, 291-2 (1994).

p-Toluenesulfonic acid   *TsOH*
Alcoholysis of N-protected lactams s. *39*, 109s*51*   ↶

Platinum phosphinite complex
Carboxylic acid amides from nitriles   CN → CONH$_2$
with PdCl$_2$ cf. *21*, 167s*29*; with a Pt-phosphinite complex (Pt(PPh$_3$)$_4$/Me$_2$HP(O)) s. T. Ghaffar, A.W. Parkins, Tetrahedron Letters *36*, 8657-60 (1995).

Addition to Carbon-Carbon Bonds   OC ⇓ CC

Electrolysis/potassium osmate/potassium hexacyanoferrate/1,4-bis(9-O-dihydroquinidine)phthalazine/potassium carbonate ←
Asym. electrochemical dihydroxylation   C=C → C(OH)C(OH)
with dihydroquinidine as ligand cf. *49*, 108; with (DHQD)$_2$PHAL s. S. Torri et al., Chem. Lett. *1995*, 319-20; effect of added I$_2$ cf. J. Org. Chem. *61*, 3055-60 (1996).

Diethylzinc/nickel(II) acetoacetonate/1,5-cyclooctadiene ←
Alcohols from ethylene derivs. via organozinc compds.   C=C → C(OH)CH
Regiospecific conversion

60.

via hydrozincation. 2.2 eqs. Et$_2$Zn added to a mixture of 1-phenyl-3-buten-1-ol, 5 mol% Ni(acac)$_2$, and 10 mol% 1,5-cyclooctadiene at -20°, stirred at 50° for 3.5 h, excess of Et$_2$Zn removed *in vacuo* (0.1 mm Hg) over 2 h at 50°, the residue dissolved in THF, cooled to 0°, and O$_2$ bubbled through the soln. at 0° for 1 h → 1-phenyl-1,4-butanediol. Y 67%. F.e. and with perfluorohexanes as solvent, also generation of the organozinc compds. via hydroboration and carbozincation, s. P. Knochel et al., Tetrahedron Letters *36*, 3161-4 (1995).

Catecholborane/borane-diethylaniline or dicyclohexylborane ←
Aldehydes from terminal acetylene derivs.   C≡C → CH$_2$CO
with dithexylborane cf. *38*, 111s*46*; with *in situ*-generated catecholborane and borane-diethylaniline (cat.) s. Y. Suseela, M. Periasamy, J. Organometal. Chem. *450*, 47-52 (1993); **also ketones** with a little dicyclohexylborane or 9-BBN s. A. Arase et al., Synth. Commun. *25*, 1957-62 (1995).

Envirocat EPZG ←
Protection of hydroxyl groups as tetrahydropyran-2-yl ethers   OH → OThp
with Mg-silicate cf. *39*, 128s*49*; with Envirocat EPZG s. B.P. Bandgar et al., Synth. Commun. *25*, 2211-5 (1995).

Bentonite s. under Mg-monoperoxyphthalate ←
Hydrotalcite s. under t-BuOOH or H$_2$O$_2$ ←
Chiral ketones s. under K-peroxymonosulfate ←
1-Dodecyl-1-methyl-4-oxopiperidinium triflate s. under K-peroxymonosulfate ←

Dicyanoketene ethylene acetal ←
Protection of alcohols as tetrahydropyran-2-yl ethers
under mild, neutral conditions

61.

3 eqs. 3,4-dihydro-2*H*-pyran added to a soln. of benzyl alcohol and 0.2 eq. dicyanoketene ethylene acetal in dry DMF, and stirred at 60° for 7 h → product. Y 88%. Prim. and [unhindered] sec. alcohols reacted smoothly, but tert. alcohols failed to react, and phenols reacted more slowly. Acid-sensitive acetals, allylic alcohols and 1,5-dienes were unaffected. F.e., incl. selective tetrahydropyran-2-ylation of 4-hydroxybenzyl alcohol (at the prim. alcohol group), s. T. Miura, Y. Masaki, Synth. Commun. *25*, 1981-7 (1995).

Chloroperoxidase ←
Asym. enzymatic epoxidation   C=C → epoxide
s. *44*, 118s*49*; of 1,1-disubst. ethylene derivs. s. L.P. Hager et al., J. Am. Chem. Soc. *117*, 6412-3 (1995).

Monooxygenase or Immobilized lipase or Yeast ←
Asym. enzymatic Baeyer-Villiger oxidation ←
s. *44*, 114s*46-9*; of mono- and bi-cyclic ketones s. R. Gagnon et al., J. Chem. Soc. Perkin Trans. I *1994*, 2537-43; of bicyclo[2.2.1]heptan-2-ones s. ibid. *1995*, 1505-11; oxidation of 2-subst. cycloalkanones s. Chem. Commun. *1995*, 1563-4; chiral caprolactones, comparison of microbiological (whole cell) and enzymatic methods, s. R. Furstoss et al., J. Chem. Soc. Perkin Trans. I *1996*, 1867-72; with an engineered whole-cell yeast culture s. J.D. Stewart et al., ibid. 755-7; chiral γ-lactones with *whole cells* of *Acinetobacter calcoaceticus* cf. R. Furstoss et al., ibid. *1995*, 2527-8; with peroxymyristic acid generated *in situ* with an immobilized lipase and myristic acid/H$_2$O$_2$ (with modest enantioselectivity) cf. S.C. Lemoult et al., ibid. 89-91.

tert-*Butyl hydroperoxide/1,8-diazabicyclo[5.4.0]undec-7-ene*   t-*BuOOH/DBU*
Epoxides from electron-deficient ethylene derivs.   C=C → \\o/

62.

The system, *t*-BuOOH/DBU, is more economical and safer than alkaline H$_2$O$_2$ or aq. organic procedures for epoxidation of electron-deficient alkenes. **E: α,β-Epoxy-δ-lactones.** A soln. of 6-(2-phenylethyl)-5,6-dihydropyran-2-one in dichloroethane added to a soln. of 1.3 eqs. DBU and 2 eqs. anhydrous *t*-BuOOH in the same solvent at 0°, and stirred at room temp. for 12 h → 3,4-epoxy-6-(2-phenylethyl)tetrahydropyran-2-one. Y 75%. Aq. base-sensitive functions and non-carbonyl-conjugated alkene groups (notably in 2,4-dienones) remained unaffected. The reaction is also applicable to cyclic and acyclic enones, but simple acyclic enoates were unreactive. There was no stereoselectivity in most instances. F.e. incl. epoxidation of alkylidenemalonic and α,β-ethylene-α-sulfonylcarboxylic acid ester. V.K. Yadav, K.K. Kapoor, Tetrahedron *51*, 8573-84 (1995).

tert-*Butyl hydroperoxide/hydrotalcite or chromium silicate*   ←
Epoxidation on inorganic supports
2,3-epoxyalcohols with *t*-BuOOH/molecular sieves cf. *36*, 235s*48*; with *t*-BuOOH/chromium silicate s. R. Kumar et al., Chem. Commun. *1995*, 1341-2; with titanium silicate/H$_2$O$_2$ stereoselectivity, s. ibid. 1315-6; epoxidation of acyclic enones with *t*-BuOOH/hydrotalcite and of cyclic enones with H$_2$O$_2$/hydrotalcite s. J.A. Mayoral et al., Tetrahedron Letters *36*, 4125-8 (1995); *37*, 5995-6 (1996); with metallosilicate xerogels s. R. Neumann et al., Chem. Commun. *1993*, 1685-7; with titanium silicate/urea-hydrogen peroxide, *threo*-2,3-epoxyalcohols, s. W. Adam et al., Angew. Chem. Intern. Ed. *35*, 880-2 (1996).

tert-*Butyl hydroperoxide/titanium tetraisopropoxide/*L-*diethyl tartrate*   ←
Sharpless epoxidation
s. *36*, 117; of (E)-2-cyano-2-ethylenealcohols s. M. Aiai et al., Tetrahedron:Asym. *6*, 2249-52 (1995); of unsym. divinylmethanols, regioselectivity, s. T. Honda et al., J. Chem. Soc. Perkin Trans. I *1996*, 1729-39.

Dimethyldioxirane or Perfluoro-cis-2,3-dialkyloxaziridines   ←
Epoxidation
s. *44*, 117s*49*; effect of solvent on stereoselectivity s. R.W. Murray et al., Tetrahedron Letters *36*, 2437-4 (1995); epoxidation of ethylene-*tert*-amines (as their BF$_3$ complexes) s. A. Messeguer et al., Chem. Commun. *1995*, 293-4; of enesilanes s. W. Adam et al., Tetrahedron Letters *36*, 4991-4 (1995); of cyclohexenes, diastereoselectivity, s. R.W. Murray et al., J. Org. Chem. *61*, 1830-41 (1996); of chalcone and isoflavone glycosides s. W. Adam et al., Liebigs Ann. *1995*, 1547-9; of chiral allyl alcohols **with asym. induction** s. Tetrahedron *51*, 13039-44 (1995); of chiral 5-alkylidene-1,3-dioxan-4-ones s. J. Liebscher et al., Tetrahedron:Asym. *6*, 1539-42 (1995); of glycals with perfluoro-*cis*-2,3-dialkyloxaziridines s. M. Cavicchioli et al., Chem. Commun. *1995*, 901-2.

Dowex 50W
Partial protection of sym. diols as tetrahydropyran-2-yl ethers   OH → OThp

63.   HO~~~~~OH + [dihydropyran] → HO~~~~~O-Othp

A mixture of hexane-1,6-diol and *undried* Dowex 50WX2 (50-100 mesh; 0.2 g/mmol) in 1:19 3,4-dihydro-2*H*-pyran (DHP)/toluene stirred at 30° until reaction complete by GC (210 min) → monotetrahydropyranyl ether. Y 95% (and 3% diether). The method is simple and practical, and applicable to symmetrical prim. and sec. diols. The reaction rate of the monoether is much lower than that of the diol, and does not increase greatly when most of the diol has been consumed (cf. *39*, 128a*48*), so that the timing of the termination of reaction is not so important. Each diol required a particular DHP/hydrocarbon ratio for optimum reaction. F.e. and resins s. T. Nishiguchi et al., Chem. Commun. *1995*, 2491-2.

m-Chloroperoxybenzoic acid/potassium fluoride   m-$ClC_6H_4COO_2H$/KF
Epoxidation   C=C → \o/
with *m*-CPBA s. *20*, 112s*49*; of allyl carbamates, effect of δ-hydroxy- and δ-acoxy-groups on stereoselectivity, s. K. Luthman et al., J. Org. Chem. *60*, 1026-32 (1995); epoxidation of α,β-ethylenelactams s. B. Li, M.B. Smith, Synth. Commun. *25*, 1265-75 (1995); of ethylene-*prim*-amines (as their arenesulfonate salts) s. G. Asensio et al., J. Org. Chem. *60*, 3692-9 (1995); of glycals with added KF, stereoselectivity, s. C. Chiappe et al., Tetrahedron Letters *35*, 8433-6 (1994); of α,β-ethylenefluorides s. D. Michel, M. Schlosser, Tetrahedron *52*, 2429-34 (1996).

Magnesium monoperoxyphthalate   ←
2,5-Dialkoxy-2,5-dihydrofurans from furans   ←
with Br_2/KOAc cf. *16*, 203; with magnesium monoperoxyphthalate under mild conditions s. A. D'Annibale, A. Scettri, Tetrahedron Letters *36*, 4659-60 (1995).

Magnesium monoperoxyphthalate/bentonite   ←
Baeyer-Villiger oxidation   ←
with Mg-monoperoxyphthalate in homogeneous media cf. *43*, 111; phthalides from benzo cyclobutenones s. T. Hosoya et al., Synlett 1995, 635-8; on a moist bentonite support, lactones, s. M. Hirano et al., Synth. Commun. *25*, 3765-75 (1995).

Disodium 3,3'-thiodipropionate s. under O_3   ←

Phenyl iodosoacetate/magnesium perchlorate $PhI(OAc)_2$/$Mg(ClO_4)_2$
6-Hydroxy-2,6-dihydro-3-pyrones from 2-furylcarbinols   ◯
with pyridinium chlorochromate cf. *33*, 138; with $PhI(OAc)_2$/$Mg(ClO_4)_2$ s. A. De Mico et al., Tetrahedron Letters *36*, 3553-6 (1995); with a polymer-based oxovanadium(V) complex/*tert*-butyl hydroperoxide s. S. Ponrathnam et al., J. Chem. Res. (S) *1996*, 202-3.

Titanium tetraisopropoxide s. under t-BuOOH   Ti(OPr-i)$_4$

Titanium silicate s. under H_2O_2   ←
Diphenylphosphinic anhydride s. under H_2O_2 $(Ph_2P(O))_2O$

Ozone/disodium 3,3'-thiodipropionate   ←
Ozonolysis of ethylene derivs. with an improved work-up   ←

64. EtO_2C-[cyclopentene] $\xrightarrow{O_3, (NaO_2CCH_2CH_2)_2S}$ EtO_2C-(CHO)(CHO)

Reductive quenching after ozonolysis of alkenes can be carried out more efficiently with 3,3'-thiodipropionic acid mono- or di-sodium salt than with the conventional dimethyl sulfide, which is normally used in large excess and may lead to undesirable acetal formation. **E:** Ethyl 3-cyclopentenecarboxylate and a little Sudan Red in 2:3 methanol/water treated with ozone/air at -10° until the pink colour faded, excess of O_3 removed by purging with air for 15 min, 2 eqs. disodium 3,3'-thiodipropionate added, and stirred at room temp. for <1 h → 3-carbethoxyglutaraldehyde. Y 93% (60% with 2 eqs. Me$_2$S after 1.5 h). The new quenching agents are also more friendly than Me$_2$S, and work-up is easy. F.e. and with a resin-supported thiodiacid s. R.B. Appell et al., Synth. Commun. 25, 3589-95 (1995).

Hydrogen peroxide/hydrotalcite, titanium silicate or other metallosilicates ←
Urea-hydrogen peroxide/titanium silicate ←
Epoxidation s. 36, 235s51 C=C → \o/
Hydrogen peroxide/diphenylphosphinic anhydride H$_2$O$_2$/(Ph$_2$P(O))$_2$O
Epoxidation
with added (EtO)$_2$P(O)CN cf. 16, 199s36; with other organophosphorus compds. (e.g. (Ph$_2$P(O))$_2$O, phosphoryl halides) s. A.S. Kende et al., Tetrahedron Letters 35, 8123-6 (1994).

Potassium peroxymonosulfate KHSO$_5$
Potassium peroxymonosulfate/α,α,α-trifluoroacetone/sodium hydrogen carbonate ←
Epoxidation
with Oxone/Bu$_4$N(H)SO$_4$ cf. 37, 128s40; in phosphate buffer (pH 6.8) and f. oxidations s. T.-C. Zheng, D.E. Richardson, Tetrahedron Letters 36, 833-6 (1995); with *in situ*-generated methyl(trifluoromethyl)dioxirane in homogeneous aq. solvent cf. D. Yang et al., J. Org. Chem. 60, 3887-9 (1995).

Potassium peroxymonosulfate/chiral [C$_2$-symmetric] ketones ←
Asym. epoxidation

65.

of *trans*-ethylene derivs. A mixture of startg. olefin, 1 eq. C$_2$-symmetric (R)-ketone, 5 eqs. Oxone, and 15.5 eqs. NaHCO$_3$ in 2:1.7 acetonitrile/aq. Na$_2$EDTA allowed to react at room temp. for 480 min → (S,S)-product. Y 82% (e.e. 87%). The procedure is notably applicable to the asym. epoxidation of unfunctionalized *trans*-olefins and trisubst. olefins, but failed with terminal and *cis*-olefins. This is the highest enantioselectivity for asym. epoxidation of a *trans*-olefin mediated by a chiral ketone (here generating a chiral dioxirane as the effective oxidant). F.e.s. D. Yang et al., J. Am. Chem. Soc. 118, 491-2 (1996); with chiral cyclic azomethinium salts cf. V.K. Aggarwal, M.F. Wang, Chem. Commun. 1996, 191-2.

Potassium peroxymonosulfate/1-dodecyl-1-methyl-4-oxopiperidinium triflate ←
Catalytic epoxidation with dioxiranes

66.

A *catalytic* version of Synth. Meth 37, 128s40 has been reported wherein a quaternary oxoammonium salt serves the dual function of phase transfer catalyst and substrate for *in situ*-generation of dioxirane in catalytic amount. E: A stirred mixture of startg. alkene and 0.1 eq. 1-dodecyl-1-methyl-4-oxopiperidinium triflate in 23:20 phosphate buffer (pH 7.8)/methylene chloride treated via syringe pump with 10 eqs. 0.472 M Oxone soln. at 0° during 8 h (the pH being maintained at 8 by automatic addition of 2 N aq. KOH), and kept at 0° for 24 h → *rel*-(2R,3R)-2-methyl-3-[3-(phenylmethoxy)propyl]oxirane. Y 91%. The structure of the ketone, its lipophilicity, and the counterion have a critical effect, while pH, reagent stoichiometry and rate of addition are of secondary importance. F.e.s. S.E. Denmark et al., J. Org. Chem. 60, 1391-407 (1995).

Chromium silicate s. under t-BuOOH ←
Hypofluorous acid-acetonitrile HOF MeCN
Epoxidation
s. 42, 125; epoxy-carboxylic acids and -alcohols s. Tetrahedron Letters 37, 531-4 (1996).

Magnesium perchlorate s. under PhI(OAc)$_2$ *Mg(ClO$_4$)$_2$*
Potassium permanganate/cupric sulfate   *KMnO$_4$/CuSO$_4$*
Oxidation of ethylene derivs.   C=C → ▽○∕
α-hydroxyketones cf. *45*, 72; β-epoxidation of unsatd. steroids s. E.J. Parish et al., Synth. Commun. *25*, 927-40 (1995).
Potassium permanganate/magnesium sulfate/sodium hydrogen carbonate   ←
α-Ketocarboxylic acid esters from alkoxyacetylenes   C≡C(OR) → COCO$_2$R

67.

A soln. of 0.6 eq. NaHCO$_3$ and 2 eqs. MgSO$_4$ in water added to a soln. of startg. ether in acetone, stirred vigorously at 23°, treated with 3 eqs. KMnO$_4$, and stirring continued for 2-5 min → 3-hydroxy-3,7-dimethyl-2-oxooct-6-enoic acid ethyl ester. Y 80%. Olefin functionality and sec. and tert. alcohols were unaffected. The procedure is rapid and high-yielding. F.e.s. J.H. Tatlock, J. Org. Chem. *60*, 6221-3 (1995).

Chiral manganese(III) Schiff base or salen complexes/dimethyldioxirane or sodium hypochlorite   ←
Asym. epoxidation   C=C → ▽○∕
s. *46*, 106s*46-9*; of benzocycloalkenes, enhancement of enantioselectivity by secondary kinetic resolution on subsequent benzylic hydroxylation s. J.F. Larrow, E.N. Jacobsen, J. Am. Chem. Soc. *116*, 12129-30 (1994); of tetrasubst. olefins s. B.D. Brandes, E.N. Jacobsen, Tetrahedron Letters *36*, 5123-6 (1995); of 2,2-dimethyl-2*H*-chromenes with *dimethyldioxirane* as oxygen donor s. W. Adam et al., ibid. 3669-72; of indenes with variously subst. salen ligands s. T. Katsuki et al., ibid. *37*, 4533-6 (1996); with *4-(3-phenylpropyl)pyridine N-oxide* as additive s. C.H. Senanayake et al., ibid. 3271-4; asym. epoxidation of chromenes with *isoquinoline N-oxide* as additive s. G.R. Green et al., ibid. 3895-8; regiospecific asym. epoxidation of 1,3-dienes s. K.A. Jørgensen et al., J. Chem. Soc. Perkin Trans. I *1995*, 2009-16; of unfunctionalized alkenes with *polymer-based* chiral Mn(III)-salen catalysts cf. P. Salvadori et al., Tetrahedron Letters *37*, 3375-8 (1996).

Manganese(III) porphyrin complexes/peroxyacetic acid or sodium hypochlorite   ←
Manganese(II) salt/1,4,7-trimethyl-1,4,7-triazacyclononane/hydrogen peroxide   ←
Epoxidation
s. *39*, 124s*47-9*; with tetrakis(polyfluoroacylaminoaryl)porphyrin ligands for enhancement of solubility s. G. Pozzi et al., Tetrahedron *52*, 11879-88 (1996); epoxidation of *cis*-alkenes with Mn(OAc)$_2$ (or MnSO$_4$) and *cis*-coordinating 1,4,7-trimethyl-1,4,7-triazacyclononane as ligand in the presence of H$_2$O$_2$ cf. D. De Vos, T. Bein, Chem. Commun. *1996*, 917-8; with Mn(III)-[aza]porphyrin complexes and peroxyacetic acid as oxygen donor s. S. Banfi et al., Tetrahedron Letters *36*, 2317-20 (1995); with competitive cleavage of methylenedioxy groups s. M. Ricci et al., ibid. *37*, 1091-4 (1996).

Methylrhenium oxide/hydrogen peroxide   *MeReO$_3$/H$_2$O$_2$*
α-Diketones from acetylene derivs. s. *51*, 107   C≡C → COCO
Methylrhenium oxide/urea-hydrogen peroxide   *MeReO$_3$/(NH$_2$)$_2$CO·H$_2$O$_2$*
Epoxidation   C=C → ▽○∕
with H$_2$O$_2$ cf. *47*, 111; with milder urea-H$_2$O$_2$ in non-aq. media s. W. Adam, C.M. Mitchell, Angew. Chem. Intern. Ed. *35*, 533-5 (1996).

Acylperrhenates/2,6-lutidine   ←
2-α-Hydroxytetrahydrofurans from 4-ethylenealcohols   ○
syn-oxidation with Re$_2$O$_7$/H$_5$IO$_6$ cf. *48*, 135; bis(tetrahydrofurans) with *less acidic* acylperrhenates, (RCOO)$_2$ReO$_3$, s. F.E. McDonald, T.B. Towne, J. Org. Chem. *60*, 5750-1 (1995).

Nickel(II) acetoacetonate s.a. under Et$_2$Zn   *Ni(acac)*
Nickel acetoacetonate/isobutyraldehyde   *Ni(acac)$_2$/RCHO*
Baeyer-Villiger oxidation   ←
s. *46*, 128; aryl formates, also with Ni(acac)$_2$/montmorillonite, s. C. Arnaud et al., Tetrahedron Letters *36*, 6679-80 (1995).

Tetracarbonyldi(μ-formato)bis(triphenylphosphine)diruthenium $[Ru(OCHO)(CO)_2PPh_3]_2$
Palladium tetrathiomolybdate cubane-type cluster complex/triethylamine ←
Enolesters from terminal acetylene derivs. $C\equiv CH \to C(OAc)=CH_2$
with a Ru(dppb) complex cf. *40*, 81s*49*; details s. J. Org. Chem. *60*, 7247-55 (1995); with $[Ru(OCHO)(CO)_2PPh_3]_2$ for addition of halogenaromatic acids s. P.H. Dixneuf et al., Tetrahedron *51*, 10901-12 (1995); (Z)-enolesters from electron-deficient acetylenes (cf. *40*, 81s*49*) with a Pd-tetrathiomolybdate cubane-type cluster complex cf. M. Hidai et al., Tetrahedron Letters *36*, 5585-8 (1995).

Osmium tetroxide/N-methylmorpholine N-oxide ←
Osmium trichloride/potassium hexacyanoferrate/quinuclidine/potassium carbonate/
   tert-*butanol/methanesulfonamide* ←
Dihydroxylation $C=C \to C(OH)C(OH)$
s. *21*, 858s*35*; carbocyclic nucleosides s. R. McCague et al., Tetrahedron Letters *37*, 4601-4 (1996); diastereoselective dihydroxylation of cyclic allyl alcohols s. T.J. Donohoe et al., ibid. 3407-10; β,β'-dihydroxylation of *meso*-tetraphenylchlorins and metallochlorins s. C. Brückner, D. Dolphin, ibid. *36*, 3295-8, 9425-8 (1995); preferential *cis*-dihydroxylation of (η^4-1,3,n-triene)tricarbonyliron(0) complexes s. W.A. Donaldson, L. Shang, ibid. *37*, 423-4 (1996); diastereoselective dihydroxylation of 5-(*p*-toluenesulfonamido)-3-hexen-2-ol derivs. s. J.-E. Bäckvall et al., J. Org. Chem. *60*, 1848-51 (1995); **Sharpless-type racemic dihydroxylation** with $OsCl_3$ and quinuclidine as accelerating ligand s. J. Eames et al., Tetrahedron Letters *36*, 1719-22 (1995).

Potassium osmate/potassium hexacyanoferrate/1,4-bis(9-O-dihydroquinidine)phthalazines/
   *potassium carbonate/*tert-*butanol/methanesulfonamide* ←
Potassium osmate/potassium hexacyanoferrate/3-(9-O-dihydroquinidine)- or 3,6-bis(9-O-
   *dihydroquinidine)-pyridazine/potassium carbonate/*tert-*butanol/methanesulfonamide* ←
Potassium osmate/potassium hexacyanoferrate/1,4-bis(9-O-dihydroquinidine)anthraquinone/
   potassium carbonate ←
Asym. dihydroxylation
s. *47*, 114s*48,49,50*; *48*, 142s*49*; of sym. dienes (with $(DHQD)_2PHAL$) s. M.L. Belley et al., Synlett *1996*, 92-4; effect of added K_2CO_3/I_2 or K_3PO_4-K_2HPO_4/I_2 s. S. Torri et al., J. Org. Chem. *61*, 3055-60 (1996); anomalous dihydroxylation of *o*-allylbenzamides s. P. Salvadori et al., ibid. 4190-1; of *cis*-alkenes with ligands based on 6,7-diphenylphthalazine and the 5,8-diaza-analog s. K.B. Sharpless et al., ibid. *60*, 3940-1 (1995); regiospecific asym. dihydroxylation of oligoprenyl derivs. with a naphth[2,3-*d*]phthalazine-based ligand s. E.J. Corey et al., Tetrahedron Letters *36*, 8741-4 (1995); with a 1,4-bis(9-O-dihydroquinidine)anthraquinone as ligand cf. H. Becker, K.B. Sharpless, Angew. Chem. Intern. Ed. *35*, 448-51 (1996); comparison of *polymer-based* ligands (cf. *43*, 121s*46*) s. P. Salvadori et al., Tetrahedron Letters *36*, 1549-52 (1994); s.a. B.B. Lohray et al., ibid. *35*, 6559-62 (1994); f. comparison of ligands s. K.B. Sharpless et al., ibid. 543-6; with a monoquinidine ligand based on 3-(9-O-dihydroquinidine)pyridazine cf. E.J. Corey et al., ibid. 6427-30; a caveat s. A.B. Smith, III et al., ibid. *36*, 2199-202 (1995).
Asym. dihydroxylation of *p*-methoxybenzoyloxyethylenes
of O-allyl derivs. s. *50*, 72; of homoallyl derivs. s. Tetrahedron Letters *36*, 3481-4 (1995); of bishomoallyl derivs. s. J. Am. Chem. Soc. *117*, 10805-16 (1995).

Rearrangement   ∩

Hydrogen/Carbon Type   OC ∩ HC

Tris(pentafluorophenyl)boron or Lithium perchlorate-etherate $(C_6F_5)_3B$ or $LiClO_4\cdot Et_2O$
Ketones from epoxides
with $(C_6F_5)_3B$ s. *26*, 149s*51*; from *internal* epoxides with $LiClO_4$-etherate cf. S. Sankararaman et al., J. Org. Chem. *61*, 1877-9 (1996).

Ammonium polysulfide/microwaves ←
Willgerodt reaction ←
s. *2*, 180; enhancement under microwave irradiation s. C.R. Strauss, R.W. Trainor, Org. Prep. Proc. Intern. *27*, 552-5 (1995).

Oxygen/Nitrogen Type OC ∩ ON

Without additional reagents w.a.r.
N-Heterocyclic carbinol esters from N-oxides ←
with Ac$_2$O cf. *13*, 851; with (CF$_3$CO)$_2$O at room temp., 2-α-hydroxypyridine derivs., s. C. Fontenas et al., Synth. Commun. *25*, 629-33 (1995).

Oxygen/Sulfur Type OC ∩ OS

Without additional reagents w.a.r.
Pummerer-type rearrangement of *p*-hydroxysulfoxides s. *51*, 95 ←

Nitrogen/Carbon Type OC ∩ NC

Sodium hydride NaH
2,3-Epoxysulfonylamines from N-sulfonylaziridin-2-ylcarbinols OC
Aza-Payne rearrangement

68.

(2S,3S)-3-Methyl-N-[(4-methylphenyl)sulfonyl]-2-aziridinemethanol in THF added to a stirred suspension of 4 eqs. NaH in 6:1 THF/HMPA at -40° under argon, allowed to warm to room temp., stirred for 2 h, and quenched with 5% citric acid at -78° with continued stirring → (2R,3S)-3-amino-1,2-epoxy-N-[(4-methylphenyl)sulfonyl]butane. Y 92%. Aq. base, LDA, BuLi, and DBU were ineffective for this rearrangement, which requires the strongly electron-withdrawing sulfonyl group on nitrogen. F.e. and with *t*-BuOK s. N. Fujii et al., J. Org. Chem. *60*, 2044-58 (1995).

Carbon/Carbon Type OC ∩ CC

Sodium hydride or Potassium tert-*butoxide* NaH or KOBu-t
Furans from γ,δ-acetyleneketones O
with TsOH/Ac$_2$O cf. *43*, 131; 2-carbalkoxymethylfurans with NaH s. R. Vieser, W. Eberbach, Tetrahedron Letters *36*, 4405-8 (1995); with KOBu-*t* cf. E. Rossi et al., ibid. *37*, 3387-90 (1996).
Silver or Silver(I) salts Ag or Ag(I)
Cycloisomerization of acetylenecarboxylic acids
s. *35*, 85s*48*; (Z)-alkylidenephthalides, isocoumarins and 5-alkylidene-2(5*H*)-furanones with Ag or Ag(I) salts s. Y. Ogawa et al., Heterocycles *41*, 2587-99 (1995).
Sulfuric acid H$_2$SO$_4$
Cyclopent-3-enol from bicyclo[2.1.0]pentan-2-ol ring 4

69.

Bicyclo[3.3.0]oct-1-en-4-ols. A mixture of startg. tricyclic alcohol and 20% H$_2$SO$_4$ in ether allowed to react at 20° → product. Y 84% (1:1.3 mixture of diastereomers). This is part of a multistep route **from the cyclobutene ring**. F.e.s. M. Franck-Neumann et al., Tetrahedron *51*, 4969-84 (1995).

Iodine
γ-Lactones from γ,δ-ethylenecarboxylic acids I_2

70.

α-Carboxy-γ-lactones. 0.2 eq. I_2 added to a stirred soln. of startg. ethylenecarboxylic acid in dry methylene chloride, stirred for 1 h, and quenched with 5% aq. Na-thiosulfate soln. → product. Y 95% (100% selectivity). The reaction is notably applicable to γ,γ-disubst. derivs. and assumed to involve HI generation. F.e.s. K.M. Kim, E.K. Ryu, Tetrahedron Letters 37, 1441-4 (1996).

Dichlorotris(triphenylphosphine)ruthenium(II)
2-Ethylenealcohols from 3-ethylenealcohols with position shift $RuCl_2(PPh_3)_3$ ←

71.

Cinnamyl alcohols. A mixture of startg. homoallyl alcohol in water treated with 2-4 mol% $RuCl_2(PPh_3)_3$ at 90-100° for 2 h with exposure to air → product. Y 65%. Reaction is thought to proceed **via a ruthenium π-allyl complex** which adds water to give the final product with elimination of active ruthenium hydride catalyst. Evidence for initial double bond shift to produce an allyl alcohol was provided by isomerization of the latter under the same conditions. The aryl group appears essential. F.e.s. C.-J. Li et al., J. Am. Chem. Soc. *117*, 12867-8 (1995).

Exchange ⇅

Hydrogen ↑ OC ⇅ H

Potassium tert-butoxide
2-Pyridone from pyridinium ring KOBu-t ←
with NaOH cf. *31*, 162; (iso)carbostyrils with KOBu-*t* under ultrasonication s. M. Grignon-Dubois, A. Meola, Synth. Commun. *25*, 2999-3006 (1995).

Sodium percarbonate s. under $(Bu_3Sn)_2CrO_4$ ←

Cuprous trifluoromethanesulfonate/chiral bis- or tris-(Δ²-oxazolines) ←
Asym. Kharasch acoxylation H → OCOR

72.

tert-Butyl perbenzoate added slowly via a polypropylene tube to a soln. of 5 eqs. cyclopentene and 5 mol% Cu(I)-triflate-chiral bis(Δ²-oxazoline) complex in acetonitrile (degassed with N_2) at 0°, and the mixture stirred at -20° for 5 days → product. Y 49% (based on perester; e.e. 81%). F.e., solvents, and ligands s. M.B. Andrus et al., Tetrahedron Letters *36*, 2945-8 (1995); with chiral bis(Δ²-oxazolin-2-yl)pyridines as ligand s. A. DattaGupta, V.K. Singh, ibid. *37*, 2633-6 (1996); with chiral tris[(4-phenyl-Δ²-oxazolin-2-yl)methyl]amine cf. K. Kawasaki et al., Synlett *1995*, 1245-6.

Cuprous chloride/4,4'-dihydroxy-2,2',3,3',5,5'-hexaphenyl-1,1'-biphenyl ←
Cupric chloride/18-crown-6 polyether/acetaldehyde $CuCl_2/18$-*crown-6/RCHO*
Ketones from hydrocarbons $CH_2 \rightarrow CO$
Catalyzed oxidation with molecular oxygen

73.

2.5 × 10^{-4} Mol% each of $CuCl_2$ and 18-crown-6 in methylene chloride stirred for 20 min, cyclohexane and 10 mol% acetaldehyde added, and allowed to react under 1 atm. O_2 at 70° for 24 h → cyclohexanone (Y 61%) and cyclohexanol (Y 10%) (yields based on acetaldehyde). The $CuCl_2$/18-crown-6 catalytic system is very efficient with extremely high turnover numbers: an important factor for industrial application. F.e.s. N. Komiya et al., Tetrahedron Letters *37*, 1633-6 (1996); diaryl ketones with CuCl/4,4'-dihydroxy-2,2',3,3',5,5'-hexaphenyl-1,1'-biphenyl cf. G. Barbiero et al., ibid. *35*, 5833-6 (1994).

Ammonium ceric nitrate/potassium bromate $(NH_4)_2Ce(NO_3)_6/KBrO_3$
Aryloxo compds. from alkylarenes
with $Ce(OTf)_4$ cf. *46*, 236; with $CAN/KBrO_3$ s. E. Ganin, I. Amer, Synth. Commun. *25*, 3149-54 (1995); pyrrole-2-carboxaldehydes and di-2-pyrryl ketones s. T. Thyrann, D.A. Lightner, Tetrahedron Letters *36*, 4345-8 (1995); J. Heterocyc. Chem. *33*, 221-2 (1996).

18-Crown-6 polyether/acetaldehyde s. under $CuCl_2$ *18-crown-6/RCHO*
Chiral Δ^2-oxazolines s. under CuOTf   ←
Enzymes   ←
Ar. aldehydes from methylarenes under mild conditions $CH_3 \rightarrow CHO$

74. p-$NO_2C_6H_4CH_3$ ⟶ p-$NO_2C_6H_4CHO$

A mixture of *p*-nitrotoluene, a little 2,2'-azinobis(3-ethylbenzothiazoline-6-sulfonic acid) diammonium salt, laccase soln., and acetate buffer (pH 4.5) stirred vigorously at room temp. under O_2 until the initial blue colour of the soln. disappeared → *p*-nitrobenzaldehyde. Y 98%. The method is widely applicable to methylarenes bearing electron-withdrawing or -donating substituents; while phenolic and benzylic hydroxyl groups and ar. and benzylic amino groups require protection, other functions such as aliphatic hydroxyl and amino groups, ar. chlorine, ar. alkoxy, multiple bonds, and alkyl groups other than ar. methyl are unaffected. F.e.s. C.-L. Chen et al., J. Org. Chem. *60*, 4320-1 (1995).

Bis(trifluoromethyl)dioxirane   ←
Regiospecific trifluoroacetoxylation of hydrocarbons $H \rightarrow OCOCF_3$

75. $PhCH_2CH_3$ + $(CF_3CO)_2O$ ⟶ $PhCH(OCOCF_3)CH_3$

Unactivated prim. and sec. carbon-hydrogen bonds are selectively oxidized by bis(trifluoromethyl)dioxirane in the presence of trifluoroacetic anhydride. E: A 1.2-fold excess of 0.37 M bis(trifluoromethyl)dioxirane in ketone-free methylene chloride added to a soln. of startg. alkylbenzene in the same solvent containing a 10-fold excess of trifluoroacetic anhydride at 0°, allowed to react for 20 min, and the stirred mixture treated with solid Na_2CO_3 at 0° for 1 h → product. Y 75% by GC (100% conversion). There was no over-oxidation. F.e.s. G. Asensio et al., Angew. Chem. Intern. Ed. *35*, 217-20 (1996).

Phenyl iodosoacetate $PhI(OAc)_2$
N,N'-Diacylhydrazines from aldehyde hydrazones $CH{=}NNH \rightarrow CONHNAc$
with $Pb(OAc)_4$ cf. *22*, 174; with $PhI(OAc)_2$ s. D. Kumar et al., J. Chem. Res. (S) *1993*, 244-5.

N-Hydroxyphthalimide ←
N-Hydroxyphthalimide/cobalt(III) acetoacetonate ←
N-Hydroxyphthalimide-catalyzed aerobic oxidation ←

76.

Benzylic oxidation. A soln. of fluorene and 10 mol% N-hydroxyphthalimide in benzonitrile stirred at 100° for 20 h under an O_2 balloon → fluorenone. Y 80%. F.e., incl. arylcarboxylic acid esters from benzyl ethers, benzolactones, and oxidation of cycloalkanes and aliphatic alcohols, s. Y. Ishii et al., J. Org. Chem. *60*, 3934-5 (1995); oxo compds. from alcohols and carboxylic acids from prim. alcohols with added Co(acac)$_3$ (0.5 mol%) cf. Tetrahedron Letters *36*, 6923-6 (1995); oxidation of cycloalkanes and alkylbenzenes s. J. Org. Chem. *61*, 4520-6 (1996).

p-Toluenesulfonyl chloride/sodium hydride   TsCl/NaH
Protection of alcohols as tetrahydrofuran-2-yl ethers ←
with $(Bu_4N)_2S_2O_8$ cf. *42*, 146s*49*; with TsCl/NaH s. B. Yu, Y. Hui, Synth. Commun. *25*, 2037-42 (1995).

Bis(tri-n-butylstannyl) chromate/sodium percarbonate/Adogen 464/p-toluenesulfonic acid ←
Keto from methylene groups $CH_2 \rightarrow CO$
with $(Ph_3Si)_2CrO_4$/*t*-BuOOH cf. *43*, 137s*49*; aryl ketones with $(Bu_3Sn)_2CrO_4$/Na-percarbonate/ Adogen 464/TsOH s. J. Muzart, S. Aït-Mohand, Tetrahedron Letters *36*, 5735-6 (1995).

Pyridinium chlorochromate $C_5H_5NH\cdot CrO_3Cl$
2-Ene-1,4-diones from 3-ethylenealcohols   COC=CCO
with pyridinium dichromate cf. *41*, 160; with pyridinium chlorochromate, 3,6-diketo-Δ^4-steroids (cf. *18*, 304), s. A. Nangia, A. Anthony, Synth. Commun. *26*, 225-30 (1996); K. Blaszczyk et al., ibid. *24*, 3255-9 (1994).

Potassium bromate s. under CAN $KBrO_3$

Potassium permanganate-alumina $KMnO_4$-Al_2O_3
Keto from methylene groups $CH_2 \rightarrow CO$
s. *6*, 202; aryl ketones with $KMnO_4$-Al_2O_3 cf. D. Zhao, D.G. Lee, Synthesis *1994*, 915-6.

Cobalt(III) acetoacetonate s. under N-Hydroxyphthalimide $Co(acac)_3$

Dichlorotris(triphenylphosphine)ruthenium(II)/tert-butyl hyperperoxide ←
4-Peroxy-2,5-cyclohexadienones from phenols s. *51*, 472 ←

Ruthenium trichloride/peroxyacetic acid $RuCl_3/MeCO_3H$
Carboxylic acids from prim. alcohols s. *15*, 261s*51* $CH_2OH \rightarrow COOH$

Oxygen ↑   OC ⇅ O

Without additional reagents   w.a.r.
O-Alkylation with (triphenylphosphoranylidene)acetic acid esters s. *51*, 137   OH → OR

Microwaves s. under Lewis acids and p-TsOH ←

Triethylamine Et_3N
α-Functionalized carboxylic acid amides via sulfonyl hydroxamates ←
s. *48*, 457; α-alkoxy- and α-hydrazino-derivs. s. J. Org. Chem. *60*, 7043-6 (1995).

4-Dimethylaminopyridine   DMAP
Carboxylic acid esters   COOH → COOR
from carboxylic acids and dialkyl carbonates
s. *15*, 190; acceleration by 4-DMAP or the polymer-based reagent s. K. Takeda et al., Synthesis *1994*, 1063-6.

Zinc acetate Zn(OAc)₂
Stereospecific ring expansion of 2-α-mesyloxy-O-heterocyclics

77.

A mixture of startg. cyclic ether in 1:1 acetic acid/water treated with 4 eqs. Zn(OAc)₂, and refluxed for 8 h → product. Y 95% (as a ca 5:4 mixture of 2-acetoxymethyl- and 2-hydroxymethyl-deriv.). Reaction is generally applicable to formation of 6- and 7-membered O-heterocyclics and affords a *single* stereoisomer. F.e. and ring expansion of 2-α-bromo-O-heterocyclics (with AgOAc) s. T. Nakata et al., Tetrahedron Letters *37*, 213-6 (1996); application to the synthesis of hemibrevetoxin B via double rearrangement s. ibid. 217-20.

Aluminum phosphate/alumina AlPO₄/Al₂O₃
Cyclic acetals from oxo compds.
with Al₂O₃ cf. *41*, 172; 1,3-dioxolanes with AlPO₄-Al₂O₃ and reverse reaction s. F.M. Bautista et al., J. Prakt. Chem. *336*, 620-2 (1994); with Al₂O₃ (or ZnCl₂, AlCl₃, TiCl₄, FeCl₃, TsOH) under microwave irradiation *without solvent* cf. F.M. Moghaddam, A. Sharifi, Synth. Commun. *25*, 2457-61 (1995).

Montmorillonite ←
2,3-Unsatd. glycosides by Ferrier reaction ←
with BF₃ s. *24*, 240; 2-deoxy-C-2-methyleneglycosides s. C. Booma et al., Chem. Commun. *1993*, 1394-5; with montmorillonite K-10 s. K. Toshima et al., Synlett *1995*, 306-8.

Zeolite ←
Protection of carbonyl groups as cyclic *o*-xylene acetals
with montmorillonite KSF cf. *28*, 141s*31*,*48*; with H-Y zeolite s. T.P. Kumar et al., J. Chem. Res. (S) *1994*, 394-5; carbohydrate O,O-alkylidene derivs. with zeolite HY, selectivity, s. A.P. Rauter et al., Tetrahedron *51*, 6259-40 (1995).

Lewis acids/microwaves ←
Cyclic acetals from oxo compds. without solvent s. *41*, 172s*51* CO → C(OR)₂
Boron fluoride or Ferric chloride BF₃ or FeCl₃
Triphenylcarbonium tetrakis(pentafluorophenyl)borate/lithium triflimide ←
Glycosides from aldoses ←
β-glycosides with AgClO₄/Lawesson's reagent cf. *49*, 153; β-D-ribofuranosides with TrB(C₆H₅)₄ s. H. Uchiro, T. Mukaiyama, Chem. Lett. *1996*, 79-80; α-glycosides with added LiNTf₂ s. ibid. 271-2; from *totally O-unprotected* glycosyl donors with FeCl₃ or BF₃ (under ultrasonication) cf. V. Ferrières et al., Tetrahedron Letters *36*, 2749-52 (1995); stereospecific formation of unprotected uronosides with BF₃ (cf. *28*, 146s*29*) s. J.-N. Bertho et al., Chem. Commun. *1995*, 1391-3.

Cation-exchanged mesoporous molecular sieves ←
Cerium(III)-exchanged montmorillonite ←
Acetals from oxo compds. CO → C(OR)₂
with montmorillonite cf. *28*, 141s*31*; with cation-exchanged mesoporous molecular sieves s. K.R. Kloetstra, H. van Bekkum, Chem. Commun. 1995, 1005-6; with Ce(III)-exchanged montmorillonite, chemoselectivity, s. J. Tateiwa et al., J. Org. Chem. *60*, 4039-43 (1995).

Scandium(III) triflimide/p-nitrobenzoic anhydride Sc(NTf₂)₃/(ArCO)₂O
Carboxylic acid esters from carboxylic acids COOH → COOR

78.

A mixture of menthol, 1.5 eqs. 2,4,6-trimethylbenzoic acid, 1.5-2 eqs. *p*-nitrobenzoic anhydride, and 2 mol% Sc(NTf$_2$)$_3$ in nitromethane allowed to react at 25° for 3 h → product. Y 99%. Sc(NTf$_2$)$_3$ is a powerful, air-stable, Lewis acid (better than Sc(OTf)$_3$), which is readily soluble in organic solvents. The method is high-yielding and fast, even with hindered alcohols (or phenols) and ar. carboxylic acids. F.e. and **O-acylation** with Ac$_2$O (cf. *50*, 87) s. K. Ishihara et al., Synlett *1996*, 265-6; O-acylation using Yb(NTf$_2$)$_3$ or Al(NTf$_2$)$_3$ cf. *51*, 81.

N,N,N',N'-Tetramethylazodicarboxamide s. under Bu$_3$P ←
2-Ethoxy-N-ethoxycarbonyl-1,2-dihydroquinoline ←
Carboxylic acid esters from carboxylic acids under neutral conditions   COOH → COOR

79.

1.2 eqs. EEDQ added to a soln. of Fmoc-Ala-OH in ethanol, and refluxed overnight → Fmoc-Ala-OEt. Y 88%. The alcohol must be in excess (as solvent for inexpensive alcohols or 5-600% excess for expensive alcohols in an inert solvent, e.g. chloroform). Reaction is generally applicable to esterification of aliphatic, ar. and α,β-unsatd. acids (incl. hindered ones) with prim., sec. and tert. alcohols or phenols. Acid- and base-sensitive protecting groups were tolerated (Boc, DiBoc, Fmoc), and there was no racemization of α-aminoacids. Furthermore, neither an inert atmosphere nor anhydrous media was required, and reaction is amenable to scale-up. F.e.s. B. Zacharie et al., J. Org. Chem. *60*, 7072-4 (1995).

Glycosyl transferase ←
Enzymatic glycosidation ←
s. *50*, 89; oligosaccharides from a soluble polymer-based sugar acceptor s. S.-I. Nishimura et al., Tetrahedron Letters *35*, 5657-60 (1994).

p-Nitrobenzoic anhydride s. under Sc(NTf$_2$)$_3$ (ArCO)$_2$O
Dimethylchloroformiminium chloride/potassium salt ←
Acoxy compds. from alcohols with inversion of configuration   OH → OCOPh

80.

Chiral sec. benzoyloxy compds. (R)-2-Octanol and ca. 5 eqs. K-benzoate added sequentially to a suspension of ca. 1 eq. dimethylchloroformiminium chloride (prepared from oxalyl chloride and DMF) in dry THF at 0° under N$_2$, and the mixture refluxed for 3 days → (S)-2-octyl benzoate. Y 91% (e.e. 98%). The reaction proceeded with clean inversion of configuration, except with α-methylbenzyl alcohol (e.e. 51%); however, reaction of axial alcohols gave lower yields due to competing elimination. This **alternative to the Mitsunobu reaction** (*29*, 188) should be particularly useful on the large scale (the side-products, DMF and KCl, being innocuous). F.e.s. A.G.M. Barrett et al., Chem. Commun. *1995*, 1403-4.

Titanium tetralkoxides/titanium tetrachloride   Ti(OR)$_4$/TiCl$_4$
Acetals from aldehydes   CHO → CH(OR)$_2$
with TiCl$_4$ cf. *18*, 234; with Ti(OR)$_4$/TiCl$_4$ s. R. Mahrwald, J. Prakt. Chem. *336*, 361-2 (1994).

Diisopropoxytitanium bis(triflimide)   (i-PrO)$_2$Ti(NTf$_2$)$_2$
O-Acylation under mild super-Lewis acid catalysis   OH → OAc

81.

Hindered compds. 2,6-Di-*tert*-butyl-*p*-cresol and 2 eqs. acetic anhydride added to a soln. of *1 mol%* (i-PrO)$_2$Ti(NTf$_2$)$_2$ in acetonitrile at room temp., and the mixture stirred for 1 min (1 h in

CH_2Cl_2) → 2,6-di-*tert*-butyl-4-methylphenyl acetate. Y 99%. The method is fast, simple, high-yielding, and notably applicable to the acylation of hindered phenols and sec. alcohols. F.e. and metal triflimides (e.g. $Al(NTf_2)_3$ and $Yb(NTf_2)_3$) s. K. Mikami et al., Synlett *1996*, 171-2.

Tri-n-butylphosphine/N,N,N',N'-tetramethylazodicarboxamide
Triphenylphosphine/diethyl azodicarboxylate $PPh_3/RN=NR$
Polymer-based Mitsunobu etherification   OH → OR

82.

Phenolethers. Startg. resin-supported phenol swelled in 1:1 THF/methylene chloride under N_2, 5 eqs. each of N,N,N',N'-tetramethylazodicarboxamide and *p*-bromobenzyl alcohol added sequentially, mixed until dissolved, treated with 5 eqs. Bu_3P via gas-tight syringe, mixed for 45-60 min, the resin washed several times, and cleaved with 90% TFA/water (twice for 20 min periods) → product. Y 87% (92% purity). The procedure is clean, facile, rapid and suitable for generating a vast **phenolether library**. cf. *31*, 170. F.e. incl. Mitsunobu O-benzylation of phenols using a resin-based benzyl alcohol s. T.A. Rano, K.T. Chapman, Tetrahedron Letters *36*, 3789-92 (1995); with PPh_3/DEAD cf. R.M. Valerio et al., ibid. *37*, 3019-22 (1996); enhancement by addition of tert. amine as base s. L.S. Richter, T.R. Gadek, ibid. *35*, 4705-6 (1994).

Mitsunobu etherification
s. *31*, 170s49; of N-protected serinates s. R.J. Cherney, L. Wang, J. Org. Chem. *61*, 2544-6 (1996); perfluoro-*tert*-butyl ethers s. D.P. Sebesta et al., ibid. 361-2; hexafluoro-2-phenylisopropyl ethers s. H.-S. Cho et al., J. Am. Chem. Soc. *116*, 8354-5 (1994); α-aryloxycarboxamides s. R.C. Desai et al., Synth. Commun. *25*, 2099-104 (1995); chromophoric azo-functionalized phenolethers s. R.D. Miller et al., Tetrahedron Letters *36*, 4393-6 (1995).

Chiral tert. phosphines R_3P
Asym. O-acylation with carboxylic acid anhydrides   OH → OCOR

83.

Desymmetrization of glycols. *cis*-Cyclohexane-1,2-diol treated with 5-8 mol% chiral phospholane and acetic anhydride in methylene chloride at 0-20° → product. Conversion 66% (e.e. 62-7%). F.e., phosphine catalysts, and details (in Supporting Information) s. E. Vedejs et al., J. Org. Chem. *61*, 430-1 (1996).

Benzotriazolyloxytris(dimethylamino)phosphonium hexafluorophosphate   *BOP*
Carboxylic acid esters from carboxylic acids   COOH → COOR
aryl esters s. *33*, 174; N-protected (Fom, Cbz) α-aminoesters s. M.H. Kim, D.V. Patel, Tetrahedron Letters *35*, 5603-6 (1994).

Tri-n-butylphosphine oxide/trifluoromethanesulfonic anhydride/ethyldiisopropylamine   ←
Phosphonium anhydrides as dehydrating agent   OH → OR
s. *44*, 166; 1,2-*cis*-ribofuranosides from riboses s. T. Mukaiyama, S. Suda, Chem. Lett. *1990*, 1143-6.

p-Toluenesulfonic acid   *TsOH*
Carboxylic acid esters from carboxylic acids   COOH → COOR
s. *7*, 246; 2-hydroxyethyl esters s. H.M. Puntambekar et al., Indian J. Chem. *32B*, 684-7 (1993).

p-Toluenesulfonic acid/microwaves ←
Syntheses using a continuous microwave reactor ←

84.

Routine laboratory-scale syntheses can now be conducted *quickly, and safely* in a range of solvents by using a *continuous* microwave reactor operating at pressures up to 1400 kPa and temperatures up to 200°. **E**: A soln. of glycerol and TsOH in acetone passed through a continuous microwave reactor (featuring a microwave-transparent coil held in a microwave cavity) during 1.2 min at 132-5° (1175 kPa mean pressure) → 2,2-dimethyl-1,3-dioxolane-4-methanol. Y 84%. For more sluggish reactions, multiple passes can be made to achieve complete conversion (as in the esterification of 2,4,6-trimethylbenzoic acid). F.e., reactions, and comparison with conventional procedures s. T. Cablewski et al., J. Org. Chem. *59*, 3408-12 (1994).

Molybdenum hexacarbonyl/benzyltriethylammonium chloride $Mo(CO)_6/BnEt_3NCl$
Acoxy-2-ethylenes from molybdenum π-allyl complexes s. *42*, 750s*51* OH → OC—C=C
Ferric chloride $FeCl_3$
Glycosides from unprotected aldoses s. *49*, 153s*51* OH → OR

Nitrogen ↑   OC ⇅ N

Without additional reagents w.a.r.
O-Formylation with 1-formylbenzotriazole s. *51*, 157 OH → OCHO

Irradiation s. under Methylene blue ⫽

Sodium hydride NaH
O-Acylation with 3-acylthiazolidine-2-thiones OH → OCOR
s. *43*, 162; selective O-acylation of aminophenols s. W.-M. Dai et al., Tetrahedron *51*, 12263-76 (1995).

Sodium/alcohol NaOR
Reactions of 1-imidoylbenzotriazoles as iminochloride equivalents ←

85.

1-Imidoylbenzotriazoles are readily prepared and storable substitutes for iminochlorides (which are extremely labile towards hydrolysis). **E**: Iminoesters. 2 eqs. NaOMe added to a soln. of startg. imidoylbenzotriazole in methanol, and refluxed for 18 h → product. Y 85%. F.e. **and thioiminoesters** s. A.R. Katritzky et al., Heterocycles *40*, 231-40 (1995).

Sodium nitrite s. under Me$_3$SiCl $NaNO_2$

Cupric nitrate/bentonite or Silver carbonate/bentonite ←
Cleavage of oximes C=NOH → CO
with CuCl$_2$ cf. *47*, 146s*50*; with Cu(NO$_3$)$_2$-on-bentonite s. R. Sanabria et al., Org. Prep. Proc. Intern. *27*, 480-2 (1995); of ketoximes with Ag$_2$CO$_3$-on-bentonite s. Synth. Commun. *24*, 2805-8 (1994).

Magnesium s. under SnCl$_2$ Mg
Zinc/trifluoroacetic acid Zn/CF_3COOH
Ketones from 1,1-nitroethylene derivs. C=C(NO$_2$) → CHCO
with Zn/AcOH cf. *10*, 182; more mildly and quickly with Zn/CF$_3$COOH s. N.C. Barua et al., J. Chem. Res. (S) *1996*, 124-5.

Sodium tetrahydridoborate $NaBH_4$
Ketones from 1-vinylpyridinium salts s. *51*, 1 ←
Sodium hydridotriaminoaluminates $NaAlH(NR_2)_3$
Aldehydes from nitriles CN → CHO
with Na-hydridodiethylpiperidinoaluminate cf. *48*, 65; ar. aldehydes with Na-hydrido-

triaminoaluminates s. J.S. Cha et al., Org. Prep. Proc. Intern. *26*, 583-8 (1994); with Li-analogs s. ibid. *24*, 331-4 (1992); s.a. Bull. Korean Chem. Soc. *13*, 451, 670 (1992).

Catecholalane/hydrogen chloride
Aldehydes from nitriles via aldimines CN → CHO

86. PhCN ⎯⎯→ [PhCH=NH] ⎯⎯→ PhCHO

One-pot procedure. 2 eqs. 0.9 *M* catecholalane (prepared in THF from catechol and 1 eq. AlH$_3$ at 0°) added slowly under N$_2$ to a vigorously stirred soln. of benzonitrile in the same solvent maintained at 25°, the mixture stirred for 48 h, hydrolyzed with 1 *N* HCl, then saturated with NaCl → benzaldehyde. Y 84%. The method is simple, general (for aliphatic and aromatic nitriles) and high-yielding. Catecholalane is a mild, readily prepared, reducing agent, leaving ar. chlorides, phenolethers and olefins unaffected. F.e.s. J.S. Cha et al., Synlett *1996*, 165-6.

Bentonite s. under Cu(NO$_3$)$_2$ ←
Formaldehyde/hydrogen chloride CH$_2$O/HCl
Oxo compds. from O-benzyloximes s. *51*, 424 C=NOBn → CO
Methylene blue/irradiation ←
Cleavage of acylhydrazones C=N-NAc → CO
with H$_2$SO$_4$ cf. *9*, 431; photo-cleavage of benzophenone aroylhydrazones with methylene blue as sensitizer s. A.J. Maroulis, E.A. Kalambokis, J. Chem. Res. (S) *1993*, 278-9.

o,o'-Bis(carbamyl) diselenides s. under H$_2$O$_2$ (RSe)$_2$
Phenyl iodosoacetate PhI(OAc)$_2$
Carboxylic acid esters from carboxylic acid hydrazides CONHNH$_2$ → CO$_2$R

87. p-NO$_2$C$_6$H$_4$CONHNH$_2$ + HOMe ⎯⎯→ p-NO$_2$C$_6$H$_4$COOMe

A suspension of startg. hydrazide in dry methanol added dropwise to 2 eqs. PhI(OAc)$_2$ in the same solvent at 10°, and stirring continued for ca. 6 h → product. Y 70%. The method is simple, non-toxic, and applicable to aryl and S- or N-hetaryl hydrazides. F.e. **and carboxylic acids** in 10:1 acetonitrile/water s. O. Prakash et al., J. Chem. Res. (S) *1996*, 100-1.

Trimethylsilyl chloride/sodium nitrite/Aliquat 336 ←
Cleavage of oxo compd. N-derivs. C=N< → CO
of oximes s. *17*, 479s*46*; of semicarbazones, also with Me$_3$SiCl/NaNO$_3$, R.H. Khan et al., J. Chem. Res. (S) *1995*, 506-7.

Stannous chloride/magnesium SnCl$_2$/Mg
Ketones from 1,1-nitroethylene derivs. C=C(NO$_2$) → CHCO
with SnI$_2$/HCl cf. *49*, 158; with SnCl$_2$/Mg s. N.B. Das et al., J. Chem. Res. (S) *1996*, 28-9.

Ozone O$_3$
Ozonides from alkoximes ←

88. ⬠=NOMe + (structure: O=C(CF$_3$)-O-C$_6$H$_4$-NO$_2$) ⎯⎯→ (spiro trioxolane structure with CF$_3$ and -O-C$_6$H$_4$-NO$_2$)

A soln. of startg. alkoxime and 2 eqs. *p*-nitrophenyl trifluoroacetate treated with O$_3$ at 0° until reaction complete, and flushed with N$_2$ to remove excess of the oxidant → 3-(*p*-nitrophenoxy)-3-trifluoromethyl-1,2,4-trioxaspiro[4.4]nonane. Y 62%. The procedure obviates the need to prepare the parent olefin. F.e. with electronegatively-substituted carbonyl compds. s. K. Griesbaum et al., Liebigs Ann. *1995*, 1571-4.

Hydrogen peroxide/o,o'-bis(carbamyl)diselenides H$_2$O$_2$/RSeSeR
Oxo compds. from azines ←
with MeCO$_3$H cf. *16*, 254; with H$_2$O$_2$/*o,o'*-bis(carbamyl)diselenides s. J. Mlochowski et al., Synth. Commun. *26*, 291-300 (1996).

Tetra-n-*butylammonium persulfate* $(Bu_4N)_2S_2O_8$
Ketones from tosylhydrazones C=NNHTs → CO
under neutral conditions

89.   Me₂C=CHC=NNHTs   ⟶   Me₂C=CHC=O
              | |
              Me   Me

Startg. tosylhydrazone added rapidly to a stirred soln. of $(n\text{-}Bu_4N)_2S_2O_8$ in 1,2-dichloroethane, refluxed with stirring for 1 h, and poured into water → product. Y 93%. Olefin, acetal and nitrile functionalities remained unaffected. The method is fast, simple and inexpensive, and applicable to cyclic and aromatic tosylhydrazones; work-up is also simple and yields are excellent. F.e.s. F. Chen et al., Synth. Commun. *25*, 3163-72 (1995).

Nicotinium chlorochromate ←
Manganese dioxide MnO_2
Cleavage of oximes C=N< → CO
with pyridinium chlorochromate cf. *34*, 172; with less acidic nicotinium chlorochromate, also cleavage of hydrazones, semicarbazones and azines, s. I.M. Baltork, S. Pouranshirvani, Synth. Commun. *26*, 1-7 (1996); with MnO_2 cf. T. Shinada, K. Yoshihara, Tetrahedron Letters *36*, 6701-4 (1995).

Rhodium(II) actetate $Rh_2(OAc)_4$
Ethers from diazo compds. C=N₂ → CHOR
s. *33*, 191s48; 7-alkoxy-Δ³-cephems s. S. Sályi et al., Synth. Commun. *26*, 445-52 (1996); etherification of a prim. neopentyl system s. S. Bhandaru, P.L. Fuchs, Tetrahedron Letters *36*, 8347-50 (1995); 2-alkoxy- and 2-aryloxy-3,3,3-trifluoropropionates s. G. Shi et al., Tetrahedron *51*, 5011-8 (1995); **polymer-based etherification** s. F. Zaragoza, S.V. Petersen, ibid. 5999-6002 (1996); α-alkoxycarboxylic acid esters **with asym. induction** s. C.J. Moody et al., Tetrahedron Letters *35*, 5949-52 (1994); using a chiral Rh(II)-carboxylate but with zero face selectivity s. ibid. *37*, 107-10 (1996).

Halogen ↑   OC ⇅ Hal

Irradiation ⚡
Carboxylic acid esters from two prim. hypohalite molecules RCO_2CH_2R

90.   2 MeCH₂OCl   ⟶   MeCOOCH₂Me

A soln. of startg. hypochlorite in carbon tetrachloride UV-irradiated under N_2 in a quartz reactor for 4 h with a PRK-375 Hg-lamp → product. Y 80%. F.e.s. R.R. Bikbulatov et al., Zh. Org. Khim. *31*, 952 (1995) (Russ.).

Microwaves s. under NaOH ←
Sodium hydride NaH
O-Benzylation OH → OBn
s. *19*, 260; of aminoalcohols s. X.E. Hu, J.M. Cassady, Synth. Commun. *25*, 907-13 (1995).

Sodium hydroxide/microwaves ←
Phenolethers from phenols OH → OR
in methanol cf. *41*, 199; without solvent s. J.X. Wang et al., Synth. Commun. *26*, 301-5 (1996); diaryl ethers s. A.V. Eltsov et al., Zh. Obshch. Khim. *64*, 1581-2 (1994) (Russ.); fast radiolabelling under microwaves on microscale s. S.A. Stone-Elander et al., J. Labelled Compd. Radiopharm. *36*, 949-59 (1994).

Lithium bis(trimethylsilyl)amide $LiN(SiMe_3)_2$
Enol carbonates from ketones and chloroformates CHCO → C=C(OCOOR)
with $LiNR_2$ cf. *34*, 178; with $LiN(SiMe_3)_2$ s. F. Hénin et al., Tetrahedron Letters *36*, 4795-6 (1995); effect of TMEDA on C- *versus* O-acylation s. L.M. Harwood et al., ibid. *35*, 8027-30 (1994).

Cesium fluoride/lithium perchlorate $CsF/LiClO_4$
α-Glycosides from β-glycosyl fluorides F → OR
with $SnCl_2/AgClO_4$ cf. *37*, 182; with $CsF/LiClO_4$, disaccharides, also from glycosyl

trichloroacetimidates (cf. *44*, 211s*48*) s. G. Böhm, H. Waldmann, Tetrahedron Letters *36*, 3843-6 (1995).

N-Cyclohexyl-N,N',N'',N''-tetramethylguanidine ←
Silver carbonate Ag_2CO_3
Carbonic acid esters from halides and alcohols OH → OCOOR
with K_2CO_3 cf. *49*, 168; with Ag_2CO_3 s. K. Teranishi et al., Biosci. Biotechnol. Biochem. *58*, 1537-9 (1994); with N-cyclohexyl-N,N',N'',N''-tetramethylguanidine cf. W. McGhee, D. Riley, J. Org. Chem. *60*, 6205-7 (1995); alkoxymethyl alkyl carbonates s. Synthesis *1995*, 176-80.

Silver acetate AgOAc
Stereospecific ring expansion of 2-α-bromo-O-heterocyclics s. *51*, 77 ○

Cupric sulfate or Silver oxide $CuSO_4$ or Ag_2O
Alcohols from halides under neutral conditions Hal → OH

91.   BuCl ⟶ BuOH

Butyl chloride added to a soln. of 1 eq. $CuSO_4·5H_2O$ in 20 eqs. water and 10 eqs. DMSO, and heated with stirring to 100-17° for 3 h → butanol. Y 98%. Reaction can take place in neutral or acidic media, and aryl halides (except p-$NO_2C_6H_4Br$) remained unaffected. The order of halide reactivity was I>Br>Cl. F.e. and Cu(II) salts s. L.G. Menchikov et al., Mendeleev Commun. *1995*, 223-4; α-hydroxy- and α-alkoxy-carboxamides with Ag_2O s. S. Cavicchioni, Synth. Commun. *24*, 2223-7 (1994).

Silver oxide/pyridine Ag_2O/C_5H_5N
Silver trifluoromethanesulfonate/2,6-di-tert-butylpyridine ←
Ethers from halides Hal → OR
with Ag(I) s. *2*, 260; with AgOTf/2,6-di-*tert*-butylpyridine cf. R.M. Burk et al., Tetrahedron Letters *35*, 8111-2 (1994); preferential O-alkylation with Ag_2O/py, 2'-O-alkylnucleosides, s. R.P. Hodge, N.D. Sinha, ibid. *36*, 2933-6 (1995).

Zinc Zn
Alcohols from iodides s. *51*, 93 I → OH

Zinc chloride/ethyldiisopropylamine $ZnCl_2$/i-Pr_2NEt
Differentially-protected diols from cyclic formals ←
via regiospecific acylative ring opening

92.

A soln. of 1.2 eqs. freshly distilled acetyl chloride in anhydrous ether added dropwise to a stirred mixture of startg. formal and a little anhydrous $ZnCl_2$ (1 crystal or 2 drops of a 1 M soln. in ether) in the same solvent under N_2, stirring continued at room temp. for 1 h, transferred dropwise via cannula under N_2 to a mixture of 4 eqs. anhydrous methanol and 1.2 eqs. i-Pr_2NEt in anhydrous ether cooled in an ice-bath, the cooling bath removed, and the mixture stirred at room temp. for a further hour → product. Y 95%. The method is simple, general, efficient and highly regioselective (the least hindered oxygen atom of the formal being acylated). F.e. incl. glycol analogs, and regiospecific conversion to monoprotected diols s. W.F. Bailey et al., J. Org. Chem. *60*, 2532-6 (1995).

Di-n-butyl(tert-butyl)tin hydride Bu_2(t-Bu)SnH
Hydroperoxides from halides Hal → OOH
Radical conversion under aerobic conditions

93.

A soln. of 3-iodo-1-phenylheptane in toluene bubbled with dry air at 100 ml/min, cooled to -23°, 1.1 eqs. Bu_2(*t*-Bu)SnH added, the mixture sonicated for 6 h with a 20 kHz 200 W ultrasound generator (equipped with a 15 mm diameter titanium tip operating at 25% power and an internal temp. of -9°), and worked-up with DBU/I_2 → 3-hydroperoxy-1-phenylheptane. Y 54%. The *bulky* hydride was chosen to obviate subsequent reduction to the alcohol. The method is mild, neutral, general for prim., sec., and tert. halides, and the radical nature of the reaction makes it far superior to methods based on alkaline H_2O_2. F.e.s. E. Nakamura et al., Synlett *1995*, 525-6; **from dialkylzinc compds.** with added $ZnBr_2$ in perfluorohexane cf. I. Klement, P. Knochel, ibid. 1113-4; **alcohols from iodides** via alkylzinc iodides cf. F. Chemla, J. Normant, Tetrahedron Letters *36*, 3157-60 (1995).

Lithium pechlorate s. under CsF $\qquad LiClO_4$

Via intermediates $\qquad v.i.$

Prim. alcohols from halides via formic acid esters $\qquad CH_2Hal \rightarrow CH_2OH$
s. *41*, 208; from unactivated halides with labile groups s. J. Alexander et al., Synth. Commun. *25*, 3875-81 (1995).

Sulfur ↑ $\qquad$ OC ⇅ S

Epoxides/fluoroboric acid $\qquad \leftarrow$
Replacement of sulfur by oxygen $\qquad CS \rightarrow CO$
with added AcOH cf. *25*, 167; 1,3-dithiolan-2-ones with added HBF_4 s. M. Barbero et al., J. Chem. Soc. Perkin Trans. I *1996*, 289-94.

Lipase $\qquad \leftarrow$
Deracemization of α-(arylthio)thiolic acid esters $\qquad COSR \rightarrow COOH$
by enzymatic hydrolysis

94.

Racemic α-(arylthio)thiolic acid esters having a highly acidic α-proton can be hydrolyzed *completely* to one enantiomer in an aq. organic media under racemizing conditions, wherein the slower reacting (S)-substrate isomerizes (in the organic phase) to the faster reacting (R)-substrate as reaction proceeds. **E:** A soln. of startg. thioesters and 0.5 eq. *trioctylamine* in toluene added to a solution of Amano lipase PS-30 (from *Pseudomonas cepacia*) in 0.01 *M* PIPES buffer at pH 7, and worked up after 65 h (with periodic addition of 0.2 *M* aq. NaOH to maintain the pH level during reaction) → (R)-product. Conversion >99%; e.e. 96.3%. Details s. D.S. Tan et al., J. Am. Chem. Soc. *117*, 9093-4 (1995).

Trifluoroacetic anhydride/sodium hydrogen carbonate $\qquad (CF_3CO)_2O/NaHCO_3$
***p*-Quinones from *p*-hydroxysulfoxides via Pummerer-type rearrangement** $\qquad \leftarrow$

95.

2,3,5,6-Tetramethyl-*p*-(phenylsulfinyl)phenol treated with 10 eqs. trifluoroacetic anhydride in methylene chloride at 0° for 5 min, and the formed 1:1 mixture of *p*-quinol mono(trifluoroacetate) and quinone treated with aq. $NaHCO_3$ in methanol at room temp. for 1 h → product. Y 84%. With naphthalene derivs. bearing one or two electron-withdrawing groups, the corresponding quinone and quinol were formed in equal amount, the latter being oxidized to the former with MnO_2. F.e.s. S. Akai et al., Chem. Commun. *1995*, 1013-4.

N-Bromosuccinimide $\qquad NBS$
Ketones from 1,3-dithiane 1-oxides s. *51*, 399 $\qquad \subset$

N-Iodosuccinimide/trifluoromethanesulfonic acid NIS/CF_3SO_3H
Glycosides from thioglycosides   SR → OR'
s. *39*, 189s*46*; disaccharides by preferential glycosidation of *hindered* thioglycosides s. G.-J. Boons et al., Tetrahedron Letters *36*, 6325-8 (1995); D-fructofuranosides s. C. Krog-Jensen, S. Oscarson, J. Org. Chem. *61*, 1234-8 (1996); oligosaccharides s. S.V. Ley et al., Tetrahedron Letters *34*, 8523-6 (1993).

Nitrogen dioxide NO_2
Gas-solid reactions with nitrogen dioxide   ←

96.

Gas-solid reactions with NO_2 are superior to the corresponding solution-phase reactions in that products are obtained cleanly and in high, preparatively useful yield, solvents and waste are avoided, and purification procedures are dispensable. **E**: An evacuated sample of startg. thiohydantoin treated with ca. 3 eqs. NO_2 at an initial pressure of 0.2 atm. for 6 h, the excess of NO_2 and evolved NO being condensed in a cold trap at 77 K → product. Y 100%. Clean-up simply requires sublimation of sulfur (at 100°/0.0005 Torr). F.e. and reactions, incl. electron transfer, oxygen atom transfer, oxygenation, and N- and C-nitration s. G. Kaupp, J. Schmeyers, J. Org. Chem. *60*, 5494-503 (1995).

Benzenesulfenyl triflate/2,6-di-tert-butylpyridine   ←
Glycosides from S-glycosyl xanthates   SC(S)OR → OR'
with $Cu(OTf)_2$ cf. *46*, 197; α-sialylation with PhSeOTf/2,6-di-*tert*-butylpyridine s. V. Martichonok, G.M. Whitesides, J. Org. Chem. *61*, 1702-6 (1996).

Trifluoromethanesulfonic acid s.a. under N-Iodosuccinimide CF_3SO_3H

Trifluoromethanesulfonic acid/triethyl phosphite $CF_3SO_3H/(EtO)_3P$
Trifluoromethanesulfonic anhydride/2,6-di-tert-butyl-4-methylpyridine   ←
Glycosides from glycosyl sulfoxides   S(O)R → OR'
with $Me_3SiOTf/(EtO)_3P$ cf. *45*, 106s*49*; disaccharides with $TfOH/(EtO)_3P$ s. I. Alonso et al., Tetrahedron Letters *37*, 1477-80 (1996); β-mannopyranosides with Tf_2O/2,6-di-*tert*-butyl-4-methylpyridine s. D. Crich, S. Sun, J. Org. Chem. *61*, 4506-7 (1996).

Selenium dioxide SeO_2
Tris(1,10-phenanthroline)iron(III) hexafluorophosphate/sodium hydrogen carbonate   ←
Oxo compds. from cyclic mercaptals $C(SR)_2$ → CO
with $(PhSeO)_2O$ cf. *24*, 228s*33*; from 1,3-dithiolanes with SeO_2 s. S.A. Haroutounian, Synthesis *1995*, 39-40; from 1,3-dithianes with tris(1,10-phenanthroline)iron(III) hexafluorophosphate cf. M. Schmittel, M. Levis, Synlett *1996*, 315-6.

Remaining Elements ↑   OC ↓↑ Rem

Irradiation s. under Triphenylpyrylium fluoroborate   *hν*

Zinc bromide $ZnBr_2$
Hydroperoxides from dialkylzinc compds. s. *51*, 93 R_2Zn → ROOH

Lanthanum(III) perchlorate/potassium carbonate $La(ClO_4)_3/K_2CO_3$
Glycosides from glycosyl fluorides and alkoxysilanes   F → OR
with BF_3 cf. *40*, 122; β-glycosides with $La(ClO_4)_3$ [$Ce(ClO_4)_3$ or $Pr(ClO_4)_3$] and K_2CO_3, and α-glycosides with $CaCO_3$, s. M. Shibasaki et al., Tetrahedron Letters *36*, 4443-6 (1995).

Phenyl iodosoacetate
Ethylenecarboxylic acids from 1-siloxybicyclo[n.1.0]alkanes　　　　$PhI(OAc)_2$

97.

1.1 eqs. PhI(OAc)$_2$ added to a soln. of 1-(trimethylsiloxy)bicyclo[4.1.0]heptane in glacial acetic acid, and stirred under inert atmosphere at room temp. for 8 h → product. Y 92%. *endo*-Cyclopropanes gave (Z)-ethylene-derivs., while *exo*-analogs gave the (E)-derivs., no side-products being obtained. The reaction was less efficient with Pb(OAc)$_4$. F.e.s. T. Momose et al., Tetrahedron Letters *36*, 6907-10 (1995).

N-Iodosuccinimide/trifluoromethanesulfonic acid　　　　NIS/CF_3SO_3H
2,4,6-Triphenylpyrylium fluoroborate/irradiation　　　　←
Glycosides from selenoglycosides　　　　SeR → OR'

s. *47*, 178s*50*; trisaccharides by **iterative glycosidation** with retention of thioglycoside groups s. S.V. Ley et al., Synlett *1995*, 781-4; with 2,4,6-triphenylpyrylium fluoroborate under irradiation cf. T. Furuta et al., Chem. Commun. *1996*, 157-8.

*Hydrogen peroxide/potassium or tetra-*n*-butylammonium fluoride*　　　　H_2O_2/F^-
Tamao oxidation of silanes　　　　Si≤ → OH

s. *38*, 829; *39*, 200s*47*; *42*, 448; cleavage of phenyldimethylsilyl groups, chiral 5-vinyl-1,3-cyclohexanediols s. D.F. Taber et al., Tetrahedron Letters *36*, 351-4 (1995); selective cleavage of dimethyl[(1-phenylsulfonyl)cyclopropyl]silyl groups s. R. Angelaud et al., ibid. 3861-4; cleavage of aminosilyl groups, 2-ethylenealcohols, s. K. Tamao et al., Tetrahedron *52*, 5765-72 (1996); phenols, also with (Me$_3$SiO)$_2$, s. J. Dunoguès et al., Bull. Soc. Chim. France *132*, 513-6 (1995); **selective cleavage of disilanyl groups** cf. M. Siginome et al., Synlett *1995*, 941-2.

Trimethylsilyl triflate　　　　$Me_3SiOSO_2CF_3$
Acetals from oxo compds. and alkoxysilanes　　　　C(OR)$_2$

1,3-dioxolanes s. *41*, 221; diallyl acetals s. J.B. Brogan et al., Synth. Commun. *25*, 587-93 (1995).

Trifluoromethanesulfonic acid s. under N-Iodosuccinimide　　　　CF_3SO_3H

Potassium peroxymonosulfate/sodium hydrogen carbonate　　　　←
Pyridinium chlorochromate　　　　$C_5H_5NH\cdot CrO_3Cl$
Alcohols from boronic acids　　　　B(OH)$_2$ → OH

98.

Phenols. 3-Nitrophenylboronic acid vigorously stirred with NaOH and deionized water at room temp. for 5 min, NaHCO$_3$ and acetone added, cooled to 2°, treated with Oxone (buffered with EDTA) *below 8°*, vigorously stirred for *5 min*, and quenched with NaHSO$_3$ in deionized water → 3-nitrophenol. Y 98%. The procedure is generally applicable, rapid, and a useful alternative to methods based on alkaline H$_2$O$_2$. F.e. **and from boronic acid esters** s. K.S. Webb, D. Levy, Tetrahedron Letters *36*, 5117-8 (1995); **α,β-ethylenealdehydes** from β,γ-ethyleneboronic acids with pyridinium chlorochromate s. H.C. Brown, ibid. *34*, 7845-8 (1993).

Hypofluorous acid-acetonitrile　　　　*HOF-MeCN*
α-Hydroxyketones from protected enols　　　　C(OH)CO

99.

Tetralone trimethylsilyl enol ether treated at room temp. with a 2-fold excess of HOF·MeCN (prepared by bubbling F$_2$ in N$_2$ through aq. acetonitrile) for 5-10 min → product. Y >90%. F.e. and from enolethers s. S. Rozen, Y. Bareket, Chem. Commun. *1996*, 627-8.

Tetra-n-*butylammonium fluoride*
Enolesters from enoxysilanes

Bu_4NF
$C(OSi\leqslant)=C \rightarrow C(OCOR)=C$

100.

and carboxylic acid fluorides. A soln. of (trimethylsiloxy)ethene and 1 eq. palmitoyl fluoride in THF treated at 0° with *0.05 eq*. Bu$_4$NF·3H$_2$O, and allowed to react for 2 h → vinyl palmitate. Y 70%. The method avoids the use of stoichiometric amounts of expensive (Me$_2$N)$_3$S$^+$ SiF$_2$Me$_3^-$ [J. Am. Chem. Soc. *105*, 1598 (1983); cf. Synth.Meth. *39*, 877s*40*], and is applicable to aldehyde and ketone derivs. No contaminating by-products were observed. (Z)-Isomers were formed exclusively from simple (Z)-enoxysilanes, whereas (1,3-dien)- or (1,3,5-trien)-oxysilanes gave (E)-isomers predominantly. F.e.s. D. Limat, M. Schlosser, Tetrahedron *51*, 5799-806 (1995); **enol carbonates** from enoxysilanes and 1-carbalkoxyimidazoles with Bu$_4$NF/BF$_3$ s. S. Ohta et al., Heterocycles *41*, 1683-9 (1995).

Carbon ↑   OC ⇅ C

Lithium or sodium
Transesterification of phosphonic acid esters
of phenyl esters with Na$^+$ cf. *24*, 236; of 2,2,2-trifluoroethyl esters with Li or Na s. S. Berté-Verrando et al., J. Chem. Soc. Perkin Trans. I *1995*, 3125-7.

Li or Na
P—OR → P—OR'

Electrolysis
α-Formoxyketones from enol carbonates

COC(OCHO)

101.

A soln. of startg. enol carbonate in DMF added to 2 eqs. LiClO$_4$ in the same solvent contained in the anode compartment of a divided cell fitted with a carbon anode and a stainless steel cathode, and electrolyzed at 15° with stirring at a potential of 1.5 V until 1 F/mol consumed → 2-oxo-2-phenylethyl formate. Y 67%. Only those enol carbonates whose oxidation potential was higher than that of the solvent-supporting electrolyte system failed to react. F.e.s. F. Barba et al., J. Org. Chem. *60*, 5658-60 (1995).

Sodium percarbonate
Carboxylic acids from α-halogenoketones
Oxidative C-cleavage under ultrasonication

←
COC(Br) → COOH

102.   p-NO$_2$C$_6$H$_4$C(O)CH$_2$Br   ⟶   p-NO$_2$C$_6$H$_4$COOH

Arylcarboxylic acids. Water and Na-percarbonate (equivalent to 3 eqs. H$_2$O$_2$) added to a soln. of α-bromo-*p*-nitroacetophenone in acetone, and the water-cooled mixture ultrasonicated for 1 h using a Versonic 400 W high-intensity processor → *p*-nitrobenzoic acid. Y 85% (compared to only 60% after 5 h by the thermal variant). F.e. incl. a δ-dicarboxylic acid, s. G.W. Kabalka et al., Synth. Commun. *25*, 3695-9 (1995).

2,6-Dichloropyridine N-oxide s. under Ru(TMP)CO   ←

Envirocat EPZG   ←
Protection of alcohols as methoxymethyl ethers   OH → OCH$_2$OMe
with TsOH cf. *39*, 214s*46*; with Envirocat EPZG s. B.P. Bandgar et al., J. Chem. Res. (S) *1996*, 90-1.

Diethyl azodicarboxylate s. under Lipase $RO_2CN=NCO_2R$

Lipase   ←
Enzymatic O-acylation   OH → OAc
regioselective O-acylation with 2,2,2-trihalogenoethyl acetates or enol esters s. *43*, 200s*46*-9; survey of procedure, also with oxime esters or carbonates (cf. *46*, 171s*48*; *50*, 97), s. N.B. Bashir et al., J. Chem. Soc. Perkin Trans. I *1995*, 2203-22; ginsenoside derivs. s. B. Danieli et al., J. Org.

Chem. *60*, 3637-42 (1995); vitamin D derivs. s. S. Fernández et al., ibid. 6057-61; regiospecific acylation of glycerol 1-monoethers s. G. Aranda et al., Synth. Commun. *22*, 135-44 (1992); with cyclohexyl palmitate and 2,2′-biphenyl dipalmitate cf. G. Lin et al., ibid. *23*, 2135-8 (1993); O-acylation of ω-subst. 1-alkanols with vinyl acetate (cf. 42, 220*s*46,49) s. K. Nakamura et al., Bull. Chem. Soc. Japan *67*, 3053-6 (1994); of phenols with a supported lipase (on Hyflo Super Cel) s. G. Nicolosi et al., Tetrahedron *48*, 2477-82 (1992); with various vinyl esters s. S.-T. Chen et al., Tetrahedron Letters *35*, 3583-4 (1994).

Lipase ←
Polyethylene glycol monomethyl ether-modified pig liver esterase ←
Asym. enzymatic O-acylation with enolesters OH→OAc
s. *44*, 214*s*46,47; Vol. *48* (p.94) and Vol. *49* (p.90); acceleration by ultrasound s. G. Lin, H.-C. Liu, Tetrahedron Letters *36*, 6067-8 (1995); kinetic resolution of 2,2,2-trifluoroalcohols s. K. Nakamura et al., J. Org. Chem. *61*, 2332-6 (1996); of γ-hydroxy- and α,β-ethylene-γ-hydroxysulfones s. J.C. Carretero, E. Dominguez, ibid. *57*, 3867-73 (1992); of 2-alkanols s. A. Sharma et al., Synth. Commun. *26*, 19-25 (1996); of 1-aroyl-2-hydroxymethylpiperidines s. M. Ors et al., Synlett *1996*, 449-51; of 2-acylamino-3-ethylenealcohols s. D.B. Berkowitz et al., Tetrahedron Letters *35*, 8743-6 (1994); deracemization of 1,1-hydroxythioethers s. *51*, 233; in *organic* media with a polyethylene glycol monomethyl ether-modified pig liver esterase, also asym. hydrolysis (cf. *28*, 13*s*50), s. L. Heiss, H.-J. Gais, Tetrahedron Letters *36*, 3833-6 (1995); preparation of highly active and stable *polymer-based subtilisin* by incorporation of the polyethylene glycol acrylate-modified enzyme into a polyacrylate s. Z. Yang, A.J. Russell, J. Am. Chem. Soc. *117*, 4843-50 (1995).

Lipase/triphenylphosphine/diethyl azodicarboxylate ←
Deracemization of sec. alcohols ←
via asym. enzymatic O-acylation-Mitsunobu reaction

103.

One-pot conversion. Startg. racemic sec. alcohol treated with *Pseudomonas cepacia* lipase in diisopropyl ether containing vinyl acetate under standard conditions until 50% conversion had been reached, the enzyme filtered off, the solvent evaporated, the residue treated with 1.2 eqs. Ph₃P and acetic acid, dissolved in ether, treated at 0° under vigorous stirring with 1.2 eqs. DEAD during 10-15 min, and allowed to stand at room temp. for 1 h → (R)-product. Y 97% (e.e. 97%). F.e. incl. deracemization of α-acoxynitriles, also enzymatic O-deacylation, s. L.T. Kanerva et al., Tetrahedron:Asym. *6*, 1779-86 (1995).

Dimethyl sulfoxide s. under Bi(III)-mandelate Me_2SO

Sulfated stannic oxide ←
Heterogeneous transesterification

104.

of β-ketocarboxylic acid esters. A mixture of startg. ketoester, 1 eq. menthol, and sulfated SnO_2 (10 wt%) in toluene heated at 110° for 6 h while distilling off methanol → product. Y 91%. The method is general, simple, and does not require toxic or expensive catalyst in large amounts. Work-up is easy and the catalyst can be recycled. Simple esters remained unaffected under these reaction conditions. F.e., also transesterification of γ-ketoesters, and preparation of catalyst, s. T. Ravindranathan et al., Tetrahedron Letters *37*, 233-6 (1996); transesterification of β-ketoesters with zeolites cf. B.S. Balaji, B. Chanda et al., Chem. Commun. *1996*, 707-8.

Triphenylphosphine s. under Lipase Ph_3P
Bismuth(III) mandelate/dimethyl sulfoxide ←
Carboxylic acids from α-hydroxyketones COC(OH) → COOH
Oxidative C-cleavage

105. n-C$_8$H$_{17}$COCH$_2$OH ⟶ [n-C$_8$H$_{17}$COCHO] ⟶ n-C$_8$H$_{17}$COOH

5 Mol% Bi(III)-mandelate (readily prepared from Bi$_2$O$_3$ and L-mandelic acid) stirred in anhydrous DMSO at 80° for 30 min under 1 atm. oxygen, startg. ketol added, and hydrolyzed with 0.1 N HCl after 4 h → nonanoic acid. Y 63%. The procedure is simple, clean, inexpensive (being *catalytic* in bismuth), and leaves appended hydroxyl groups, *gem*-diols and olefins unaffected. F.e.s. V. Le Boisselier, E. Duñach et al., Tetrahedron *51*, 4991-6 (1995).

Ozone/triethylamine O_3/Et_3N
2-Cyclopentenone from 3-sulfonyl-1-methylenecyclopentane ring s. *51*, 454 ←

Hydrogen peroxide s. under MeReO$_3$ H_2O_2

o-Nitrobenzenesulfonyl chloride/triethylamine $ArSO_2Cl/Et_3N$
Carboxylic acids from α-hydroxymalonic acids C(OH)(COOH)$_2$ → COOH
Bisdecarbonylation

106.

A soln. of startg. hydroxymalonic acid in methylene chloride treated at -78° with *o*-nitrobenzenesulfonyl chloride and Et$_3$N, warmed to room temp., water-satd. ethyl acetate added, and worked up when the second decarbonylation ceased → product. Y unspecified. Since the startg. m. was prepared via nucleophilic substitution with dibenzyl carbobenzoxymalonate, the latter functions as a novel **carboxyl carbanion equivalent** [HO$_2$C$^-$]. The method is considered generally applicable. F.e.s. B.M. Trost, Z. Shi, J. Am. Chem. Soc. *118*, 3037-8 (1996).

Trimethylsilyl triflate $Me_3SiOSO_2CF_3$
Glycosides from acyl glycosides OAc → OR
s. *34*, 209s*44*; **and tetrahydropyran-2-yl ethers** cf. S. Manfredini et al., Tetrahedron Letters *35*, 5709-12 (1994).

Glycosides from glycosyl trichloroacetimidates OC(=NH)CCl$_3$ → OR
s. *44*, 211s*48*; 5-thio-α-pyranosides, incl. 5′-thiodisaccharides, and thio- or seleno-glycoside analogs, s. B.M. Pinto et al., J. Am. Chem. Soc. *117*, 9783-90 (1995); with CsF/LiClO$_4$ s. *37*, 182s*51*; **also acyl glycosides** from carboxylic acids s. J. Mao, M. Cai et al., Synth. Commun. *25*, 1563-5 (1995); solid-phase synthesis of oligosaccharides s. J. Rademann, R.R. Schmidt, Tetrahedron Letters *37*, 3989-90 (1996); oligosaccharide libraries by random glycosidation s. O. Kanie et al., Angew. Chem. Intern. Ed. *34*, 2720-2 (1995).

Sulfuric acid H_2SO_4
α-Hydroxyketones from 2-alkoxy-Δ^3-oxazolines s. *51*, 418 C

Potassium dichromate/sulfuric acid $K_2Cr_2O_7/H_2SO_4$
Ketones from aminomethylene compds. C=CHN< → CO
with Na$_2$Cr$_2$O$_7$/AcOH cf. *10*, 219; with K$_2$Cr$_2$O$_7$/H$_2$SO$_4$ s. B. Singaram et al., Tetrahedron Letters *36*, 2921-4 (1995).

Hypofluorous acid-acetonitrile HOF-$MeCN$
α-Hydroxyketones from enolethers s. *51*, 99 C=C(OR) → C(OH)CO

Sodium periodate $NaIO_4$
Carboxylic acids from α-diketones s. *51*, 399 COCOR → COOH
Methylrhenium oxide/hydrogen peroxide $MeReO_3/H_2O_2$
Carboxylic acid esters from terminal acetylene deriv. C≡CH → CO_2R
Oxidative cleavage

107. Ph—≡ + HOMe ⟶ PhCOOMe

A mixture of phenylacetylene, 10 mol% $MeReO_3$, and 3 eqs. H_2O_2 in methanol stirred for 2 days at room temp. in a closed reaction vessel → methyl benzoate. Y 96% (88% conversion). Higher prim. alcohols gave a mixture of ester and acid, while sec. and tert. alcohols gave a more complex mixture of products. Internal alkynes gave α-diketones predominantly. F.e.s. Z. Zhu, J.H. Espenson, J. Org. Chem. *60*, 7728-32 (1995).

Ruthenium tetroxide/sodium hypochlorite/acetonitrile ←
Carboxylic acids from ethylene derivs. CH=CH → COOH
with $NaIO_4$ as reoxidant cf. *37*, 220; cleavage of electrophilic olefins (monohalogeno- and 1,2-dihalogeno-derivs.) with NaOCl s. K.C. Hansen et al., Synth. Commun. *25*, 2709-22 (1995).

Carbonyl(tetraarylporphyrinato)ruthenium(VI)/2,6-dichloropyridine N-oxide/hydrogen bromide ←
***p*-Quinones from phenolethers** ←
Selective oxidation

108. [Structure: MeO-substituted biphenyl converting to MeO-substituted quinone]

Alkoxy-*p*-quinones. A soln. of startg. polyalkoxybiphenyl, 2.4 eqs. 2,6-dichloropyridine N-oxide, a little Ru(TMP)CO and 47% HBr, and 4Å molecular sieves in benzene allowed to react at room temp. for 45 h under argon → product. Y 77%. *o*- And *p*-quinol ethers were unaffected. Mechanistically, the oxidizing system resembles that of cytochrome P-450. F.e., also polycyclic quinones from arenes s. T. Higuchi et al., J. Am. Chem. Soc. *117*, 8879-80 (1995).

Elimination ⇑

Hydrogen ↑   OC ⇑ H

Electrolysis/thioanisole ϟ/PhSMe
Ketones from sec. alcohols CHOH → CO
with $MeSC_8H_{17}$-*n*/Et_4NBr as mediators cf. *36*, 234; with thioanisole in acetonitrile/2,2,2-trifluoroethanol s. Y. Matsumura et al., Tetrahedron *51*, 6411-8 (1995).

(S)-3,5-Dihydro-3,3,5,5-tetramethyl-4H-dinaphth[2,1-c:1',2'-e]azepine-N-oxyl/sodium hypochlorite/potassium bromide ←
Ketones from sec. alcohols with kinetic resolution s. *51*, 500
N-Hydroxyphthalimide/cobalt(III) acetoacetonate/oxygen ←
Oxo compds. from alcohols s. *51*, 76
Laccase/2,2'-azinobis(3-ethylbenzothiazoline-6-sulfonic acid) ←
Ar. aldehydes from benzylalcohols under mild conditions CH_2OH → CHO

109. p-$EtC_6H_4CH_2OH$ ⟶ p-EtC_6H_4CHO

A soln. of *p*-ethylbenzyl alcohol and 2,2'-azinobis(3-ethylbenzothiazoline-6-sulfonic acid) diammonium salt (as co-factor) in acetate buffer (pH 4.5) warmed to 40°, treated with laccase (from *Coriolus varsicolor*), flushed with O_2 for 1 min, the vessel closed, allowed to react for 24 h at 40°, further quantities of enzyme and co-factor added if startg. m. still present (GCMS), flushed with O_2 for 5 min, temp. maintained at 40° for a further 24 h, and cooled to room temp. → *p*-ethylbenzaldehyde. Y 92%. The method is clean and ecologically friendly, and yields are excellent.

Overoxidation to the acid does not occur. Aliphatic hydroxyl and amino groups, ethylenes, ethers, esters, halogens, and nitro groups remained unaffected; however, phenolic groups, aromatic and benzylic amino groups, and activated aromatic methyl groups required protection. F.e.s. C.L. Chen et al., Synth. Commun. *26*, 315-20 (1996).

Phenyl iodosoacetate/potassium hydroxide $PhI(OAc)_2/KOH$
Flavones from *o'*-hydroxychalcones   O
with SeO_2 cf. *2*, 288; with $PhI(OAc)_2$/KOH s. S. Antus et al., Liebigs Ann. *1995*, 1711-5.

6-tert-*Butyl-3-pentafluorophenyl-3-trifluoromethyl-1-hydroxy-(3H)-1,2-benziodoxole 1-oxide*   ←
Oxo compds. from alcohols   CHOH → CO
with Dess-Martin periodinane or *o*-iodoxybenzoic acid cf. *44*, 229; *50*, 123; selective oxidation and wider application with 6-*tert*-butyl-3-pentafluorophenyl-3-trifluoromethyl-1-hydroxy-(3*H*)-1,2-benziodoxole 1-oxide s. S.H. Stickley, J.C. Martin, Tetrahedron Letters *36*, 9117-20 (1995).

Pentafluorobenzeneseleninic acid/tert-butyl hydroperoxide $C_6F_5Se(O)OH/t$-$BuOOH$
with $(PhSeO)_2O$ cf. *34*, 222; with $C_6F_5Se(O)OH/t$-BuOOH or 2-pyridineseleninic anhydride N-oxide/$PhI(OAc)_2$, and f. oxidations, s. D.H.R. Barton, T.-L. Wang, Tetrahedron Letters *35*, 5149-52 (1994).

Chromium trioxide/silica CrO_3/SiO_2
with CrO_3/Al_2O_3 cf. *12*, 333s*49*; with CrO_3/SiO_2 s. B. Khadilkar et al., Synth. Commun. *26*, 205-10 (1996).

Bis(trimethylsilyl) chromate/silica $(Me_3SiO)_2CrO_2/SiO_2$
with $(Ph_3SiO)_2CrO_2$ cf. *38*, 221s*39*; with $(Me_3SiO)_2CrO_2/SiO_2$ s. J.G. Lee et al., Synth. Commun. *26*, 543-9 (1996).

n-*Butylammonium chlorochromate/18-crown-6 polyether* $[BuNH_3][CrO_3Cl]/crown$
aryl ketones with $[Bu_4N][CrO_3Cl]$ cf. *39*, 480; selectivity with *n*-butylammonium chlorochromate/18-crown-6 polyether s. H.S. Kasmai et al., J. Org. Chem. *60*, 2267-70 (1995).

Quinolinium fluorochromate   ←
Poly(4-vinylpyridinium bromo- or fluoro-chromate)   ←
allylic and benzylic oxidation with pyridinium bromochromate cf. *42*, 235; with poly(4-vinylpyridinium bromochromate) and the fluorochromate analog s. J. Chem. Res. (S) *1992*, 132-3; with quinolinium fluorochromate and f. oxidations cf. M.K. Chaudhury et al., Bull. Chem. Soc. Japan *67*, 1894-8 (1994).

Ferric β-diketonates or Cobalt(II) acetate/tert-butyl hydroperoxide   ←
benzylic and allylic oxidation with $Co(OAc)_2$/NaBr cf. *26*, 463s*44*; with $Co(OAc)_2/t$-BuOOH and other Co-catalysts ($CoCl_2$, $CoCl_2(PPh_3)_2$) s. S. Iyer, J.P. Varghese, Synth. Commun. *25*, 2261-6 (1995); with Fe(III)-β-diketonates cf. D.H.R. Barton et al., Tetrahedron Letters *35*, 4681-4 (1994).

Cobalt(III) acetoacetonate s. under N-Hydroxyphthalimide $Co(acac)_3$
Ruthenium trichloride/peroxyacetic acid $RuCl_3/MeCO_3H$
Ketones from sec. alcohols
with $RuCl_3/t$-BuOOH cf. *15*, 261s*49*; with $RuCl_3/MeCO_3H$, incl. pyridyl ketones, also carboxylic acids from prim. alcohols, s. S.-I. Murahashi et al., Synlett *1995*, 733-4.

Palladous acetate/oxygen $Pd(OAc)_2/O_2$
Regio- and stereo-specific oxidative ring closures of functionalized ethylene derivs.   O
Molecular oxygen as re-oxidant under mild conditions

110.

Molecular oxygen in DMSO is a useful alternative to *p*-benzoquinone or Cu(II) (cf. *37*, 415) in Pd-catalyzed oxidative ring closures of [cyclo]alkenes possessing internal nucleophiles. **E: Cyclic 2-ethyleneethers.** A soln. of startg. alkene in DMSO purged with O_2, treated with 5 mol% $Pd(OAc)_2$, the reaction vessel evacuated, filled with O_2 (twice), stirred at room temp. (23°) for 24 h, and quenched with water → product. Y 90%. There was no formation of regioisomeric homoallyl deriv. Ring closures of the corresponding N-protected ethyleneamines required a slightly elevated temp. (55°) and 10 mol% Pd(OAc)$_2$. F.e.s. M. Rönn et al., Tetrahedron Letters 36, 7749-52 (1995).

Tetrakis(triphenylphosphine)palladium(0)/air   Pd(PPh$_3$)$_4$/air
Palladous chloride/1,2-dichloroethane/Adogen 464/sodium carbonate   ←
Oxo compds. from alcohols   CHOH → CO
with Pd(OAc)$_2$/PhI/Bu$_4$NCl/NaHCO$_3$ cf. *41*, 245; with PdCl$_2$/Adogen 464/Na$_2$CO$_3$ and 1,2-dichloroethane *as reoxidant* s. S. Aït-Mohand et al., Tetrahedron Letters *36*, 2473-6 (1995); α,β-ethyleneoxo compds. with Pd(PPh$_3$)$_4$ under air s. E. Gómez-Bengoa et al., ibid. *35*, 7097-8 (1994).
Via intermediates   v.i.
Oxo compds. from alcohols via α-ketocarboxylic acid esters
via pyruvates cf. *32*, 212; via 2,4-dimethoxybenzoylformic acid esters s. M.C. Pirrung, R.J. Tepper, J. Org. Chem. *60*, 2461-5 (1995).

α-Hydroxyketones from glycols via O,O-isopropylidene derivs. s. *51*, 114   C(OH)CO

Oxygen ↑   OC ⇑ O

Sodium hydride   NaH
3-Amino-2(5H)-furanones from isoxazolidine-3-carboxylic acid esters   ←

111.

A suspension of 2-methyl-3-carbobutoxy-5-butylisoxazolidine and 1.2 eqs. NaH in anhydrous THF stirred at room temp. under N$_2$ for 5 h, and washed with satd. aq. NH$_4$Cl → 3-methylamino-5-butyl-2(5H)-furanone. Y 75%. This is part of a simple 2-step procedure **from nitrones and ethylene derivs.** F.e.s. F. Casuscelli et al., Tetrahedron *51*, 8605-12 (1995).

Molecular sieves   ←
5,6-Dihydro-2-pyrones from α,β-ethylene-δ-hydroxycarboxylic acid esters s. *51*, 329   O
Amberlyst 15   ←
Chromone ring from *o*-hydroxy-β-diketones
with TsOH cf. *9*, 371; 4'-hydroxyflavones and hydroxychromones using Amberlyst 15 with simultaneous cleavage of methoxymethyl ethers s. T. Patonay et al., Bull. Soc. Chim. France *132*, 233-42 (1995).

Halogen ↑   OC ⇑ Hal

Sodium 2,6-di-tert-butylphenoxide   NaOAr
Diaryl ethers   O
from η^6-chloroarene(η^5-cyclopentadienyl)ruthenium(II) hexafluorophosphates
intermolecular reaction cf. *43*, 749s*48*; cyclic peptidyl diaryl ethers by intramolecular coupling s. J.W. Janetka, D.H. Rich, J. Am. Chem. Soc. *117*, 10585-6 (1995).

tert-*Butyllithium*   t-BuLi
Ring closures of *o*-functionalized β,β-dihalogenostyrenes
via intramolecular nucleophilic substitution of styryl carbenes

112.

An entirely new approach to 5-membered benzo-condensed heteroarenes is based on intramolecular substitution of electrophilic carbenoids by O-, N- or S-nucleophiles in the *o*-position. **E: Benzofurans.** 3 eqs. 1.25 M *t*-BuLi in pentane added to a soln. of 1,1-dichloro-2-(*o*-hydroxy--phenyl)propene in the same solvent at -100°, stirred for 5 h, and quenched with methanol → 3-methylbenzofuran. Y 90%. **Indoles and thianaphthenes** were also accessible by this route. F.e., electrophiles, and limitations s. M. Topolski, J. Org. Chem. *60*, 5588-94 (1995).

Sulfur ↑   OC ↑ S

Trifluoroacetic anhydride/sodium hydrogen carbonate $(CF_3CO)_2O/NaHCO_3$
p-Quinones from p-hydroxysulfoxides s. *51*, 95   ←

Remaining Elements ↑   OC ↑ Rem

Tris[trinitratocerium(IV)] paraperiodate   *Ce(IV)*
Oxo compds. from alkoxysilanes   CH(OSi≤) → CO
with CAN/NaBrO$_3$ cf. *37*, 264; with [(NO$_3$)$_3$Ce]$_3$·H$_2$IO$_6$ s. H. Firouzabadi, F. Shiriny, Synth. Commun. *26*, 423-32 (1996); with PdCl$_2$(MeCN)$_2$/PPh$_3$/2-bromomesitylene/K$_2$CO$_3$ cf. N.S. Wilson, B.A. Keay, J. Org. Chem. *61*, 2918-9 (1996).

Lead tetraacetate $Pb(OAc)_4$
3-Ethylene-1-vinyllactones from bicyclic β′-stannyl-γ-vinyl-γ-lactols s. *51*, 397   ○

Hydrogen peroxide/potassium hydroxide H_2O_2/KOH
Epoxides from 2-hydroxyselenides with inversion   C(OH)C(SeR) → \o/
with TlOEt/CHCl$_3$ cf. *40*, 152; with H$_2$O$_2$/KOH via β-hydroxyselenoxides, cyclohexene oxides, s. P. Ceccherelli et al., Tetrahedron Letters *36*, 5079-80 (1995).

Bis(acetonitrile)dichloropalladium(II)/triphenylphosphine/2-bromomesitylene/
  potassium carbonate   ←
Oxo compds. from alkoxysilanes s. *37*, 264s*51*   CH(OSi≤) → CO

Carbon ↑   OC ↑ C

Without additional reagents   w.a.r.
2-α-Hydroxy-O-heterocyclics from cyclic alkoxyglycol sulfates   OC
Regio- and stereo-specific ring closure

113.

2-α-Hydroxytetrahydrofurans. A 0.1 M soln. of startg. chiral sulfate in acetonitrile containing 50-100 eqs. water refluxed under N$_2$ for 12 h → product. Y 97% (after acetylation); regioselectivity >50:1. Interestingly, the intermediate sulfuric acid monoester was hydrolyzed *in situ* (presumably by generated acid), so that the liberated hydroxyl group becomes available for further manipulation. This is notably evident in the conversion of polyol sulfates, where the liberated hydroxyl group facilitates cascade cyclization **to poly(tetrahydrofurans)**. F.e.s. T.J. Beauchamp et al., J. Am. Chem. Soc. *117*, 12873-4 (1995).

Irradiation   ⁂
Oxo compds. from ethers   CHOR → CO
ketones from 6-phenanthridinylmethyl ethers cf. *48*, 255; also aldehydes from 1-alkoxyanthraquinones s. R.P. Smart et al., J. Org. Chem. *60*, 6852-9 (1995).

Cupric bromide/lithium tert-butoxide $CuBr_2/LiOBu$-t
Ketones from α-hydroxycarboxylic acids s. *23*, 296s*51*   CH(OH)COOH → CO

Dimethyldioxirane
**α-Hydroxyketones from glycols via their O,O-isopropylidene derivs.
Retention of chirality**

114.

ca. 1.2 eqs. ca. 0.1 *M* dimethyldioxirane in acetone added to a stirred soln. of startg. (R,R)-acetonide (e.e. 98%) in methylene chloride at 0°, and worked up after 40 min → (R)-product. Y >96% (>98% conversion; e.e. 98%). Reaction is generally applicable to both *tert,sec*- and *sec,sec*-glycol derivs., methyl(trifluoromethyl)dioxirane being recommended for less reactive substrates. The method is simple and mild. F.e. and regioselectivity s. R. Curci et al., Tetrahedron Letters *37*, 115-8 (1996).

Perfluoro-cis-2-butyl-3-propyloxaziridine
Ketones from ethers

CHOR → CO

115.

2-Adamantanones. 2-Methoxyadamantane treated with 2 eqs. perfluoro-*cis*-2-butyl-3-propyloxaziridine in Freon-11 at room temp. for 2 h → 2-adamantanone. Y 91%. **Iminofluorides** derived from the corresponding polyfluorooxaziridines were isolated as the by-products. Longer reaction times were required for oxidation of benzyloxy-, butoxy, and ethoxy-derivs. Carboxylic acids and esters remained unaffected. F.e. incl. steroidal ketones, reagents, and preferential conversions s. G. Resnati et al., J. Org. Chem. *60*, 2314-5 (1995).

Phosphate buffer
Oxazole-4-carboxylic from β,γ-acetylene-α-acylaminomalonic acid esters

116.

A soln. of startg. malonate in 0.1 *M* phosphate buffer (pH 7.5) warmed at 35-45° for 5 days → product. Y 98%. Ring formation arises by a rare intramolecular 5-*endo*-cyclization of intermediate α-acylamino-α-allenecarboxylic acid esters, which may also be trapped by external nucleophiles. F.e. and bases s. N. Katunuma et al., Tetrahedron Letters *37*, 861-4 (1996).

Sodium periodate/benzyltriethylammonium chloride/dibenzo-18-crown-6 polyether
Oxo compds. from α-hydroxycarboxylic acids   C(OH)COOH → CO
aldehydes with Fe(ClO$_4$)$_3$ cf. *23*, 296s*47*; also ketones from mandelic acids with NaIO$_4$/BnEt$_3$NCl/crown s. A.R. Kore et al., Org. Prep. Proc. Intern. *27*, 373-4 (1995); ketones with CuBr$_2$/LiOBu-*t* cf. T. Takeda et al., Synthesis *1996*, 600-2.

Formation of N-N Bond

Uptake ⇓

Addition to Nitrogen NN ⇓ N

O-Picrylhydroxylamine   ArONH$_2$
N-Aminocyclimmonium salts from N-heterocyclics ←
with mesitylenesulfonyloxylamine cf. *28*, 243; with O-picrylhydroxylamine in organic media s. O.V. Vinogradova et al., Khim. Geterotsikl. Soedin. *1994*, 1364-8 (Russ.).

Exchange ⇅

Hydrogen ↑ NN ⇅ H

Ethylmagnesium bromide/n-butyl nitrate   EtMgBr/BuONO$_2$
N-Nitramines from amines   NH→NNO$_2$
from prim. amines with *n*-BuLi/EtONO$_2$ cf. *21*, 312; from sec. amines with EtMgBr/BuONO$_2$ s. Z. Daszkiewicz et al., Org. Prep. Proc. Intern. *26*, 337-41 (1994).

Trifluoroacetyl nitrate   CF$_3$COONO$_2$
N-Nitration of nucleosides s. *51*, 164

p-*Toluenesulfonic acid*   TsOH
N-Amination   NH → NNH$_2$
with H$_2$NOSO$_3$H/NaOH cf. *25*, 205; with N-(diethoxyphosphoryl)-O-(*p*-nitrophenyl-sulfonyl)hydroxylamine s. A. Koziara et al., Synth. Commun. *25*, 3805-12 (1995).

Oxygen ↑ NN ⇅ O

Magnesium/methanol or Zinc/cadmium chloride   Mg/MeOH or Zn/CdCl$_2$
Sym. azo compds. from nitro compds.   2 NO$_2$ → N=N or N=N(O)
with EtMgBr cf. *38*, 254; with Mg/MeOH s. J.M. Khurana, A. Ray, Bull. Chem. Soc. Japan *69*, 407-10 (1996); sym. azoxy compds. (cf. *23*, 307) with Zn/CdCl$_2$ cf. B. Baruah et al., Chem. Lett. *1996*, 351-2.

Elimination ⇑

Hydrogen ↑ NN ⇑ H

2,4,4,6-Tetrabromo-2,5-cyclohexadienone ←
Triazolium ring closure ○
with NBS cf. *10*, 255; with 2,4,4,6-tetrabromo-2,5-cyclohexadienone (cf. *42*, 273) s. S. Bátori, A. Messmer, J. Heterocyc. Chem. *31*, 1041-6 (1994).

1-Chlorobenzotriazole ←
Tetrazolium salts from formazans
with NBS cf. *10*, 255; 2,3,5-triaryl-2*H*-tetrazolium salts with 1-chlorobenzotriazole s. A.R. Katritzky et al., Heterocycles *39*, 73-80 (1994).

Formation of N-S Bond

Exchange ⇅

Remaining Elements ↑ NS ⇅ Rem

Without additional reagents *w.a.r.*
N-Sulfonyl- from N-silyl-imines s. *51*, 154   C=NSi≼ → C=NSO$_2$R

Formation of N-Rem Bond

Exchange ⇅

Hydrogen ↑ NRem ⇅ H

Diethyl azodicarboxylate *RN=NR*
Phosphine imines from phosphines and amines ≽P → ≽P=NR
s. *34*, 271; protection of nucleoside prim. amino groups as triphenylphosphine imines s. J.-W. Chern et al., Tetrahedron Letters *36*, 7881-4 (1995).

Carbon tetrachloride *CCl$_4$*
Phosphoromonoamidates from dialkyl phosphites P—H → P—N≼
s. *2*, 314; with inversion of configuration, also dinucleoside phosphoramidates, s. I. Tömösközi et al., Tetrahedron *51*, 6797-804 (1995); s.a. R.P. Iyer et al., Tetrahedron Letters *37*, 1543-6 (1996).

Nitrogen ↑ NRem ⇅ N

Sodium hydride *NaH*
Eniminophosphoranes from 1-α-azidobenzotriazoles C=C—N=P≼
via 1-α-(phosphoranylideneamino)benzotriazoles

117.

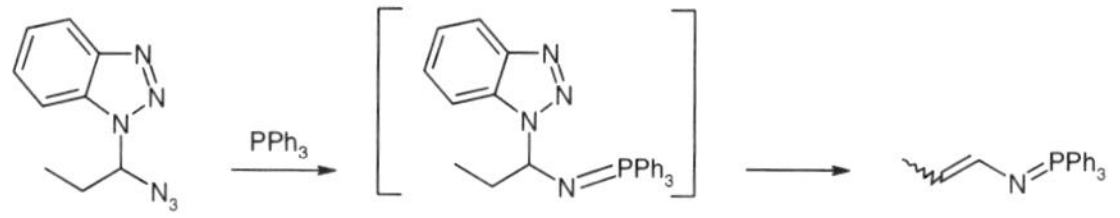

1 eq. triphenylphosphine in THF added dropwise to a stirred soln. of startg. azide in the same solvent, the mixture stirred at room temp. for 1.5 h, 3 eqs. NaH (60% dispersion in mineral oil) added, and refluxed under N$_2$ with vigorous stirring for 2 h → product. Y 94%. The method avoids the use of potentially explosive vinyl azides. F.e. and conversion to pyridines s. A.R. Katritzky et al., J. Org. Chem. *59*, 2740-2 (1994).

Methyl iodide *MeI*
Disilazanes from aminosilanes NHSi≼ → N(Si≼)$_2$

118.   PhNHSiMe$_3$ +   Me$_3$SiNEt$_2$ ⟶   PhN(SiMe$_3$)$_2$

1.2 eqs. Methyl iodide added at room temp. over 1 h to a soln. of N-(trimethylsilyl)aniline and 1.2 eqs. N-(trimethylsilyl)diethylamine in hexane, and stirred at reflux for 20 h → N,N-bis(tri - methylsilyl)aniline. Y 86%. Allyl halides and benzyl bromide also served as promoter. This is an improvement on earlier methods which require the removal of liberated amine to promote reaction. F.e. and from prim. amines s. Y. Hamada et al., J. Organometal. Chem. *510*, 1-6 (1996).

Formation of N-C Bond

Uptake ⇓

Addition to Oxygen and Nitrogen   NC ⇓ ON

Chloro(1,5-cyclooctadiene)rhodium(I) dimer   *[Rh(cod)Cl]$_2$*
Ring expansion of isoxazolidines via regiospecific carbonylation

119.

Tetrahydro-1,3-oxazin-2-ones. A mixture of startg. isoxazolidine and a little [Rh(COD)Cl]$_2$ in dry benzene purged several times with CO in a glass-lined autoclave, pressurized to 65 atm. with CO, and the mixture stirred at 150-170° for 24 h → 5-(acetoxymethyl)-4,6-diphenyl-3-methyltetrahydro-1,3-oxazin-2-one. Y 80%. The reaction appears limited to 3-arylisoxazolidines, 3-alkyl-analogs undergoing ring-expansion re-cyclization. F.e., also tetrahydro-1,3-oxazines via reductive carbonylation, and with iridium catalysts (e.g. IrCl$_3$) s. K. Khumtaveeporn, H. Alper, J. Org. Chem. **60**, 8142-7 (1995).

Addition to Oxygen and Carbon   NC ⇓ OC

Without additional reagents   *w.a.r.*
α-Amino-β-hydroxycarboxylic from glycidic acids   ▽o/ → C(OH)C(N<)
with inversion s. *18*, 369; chiral amide derivs., also addition of azides, s. M. Valpuesta et al., Tetrahedron Letters **36**, 4681-4 (1995).

Hindered peptides from Δ^2-5-oxazolones   C
s. *16*, 384; chiral dipeptides via resolution of diastereomers, also conversion to chiral α,α-disubst. α-aminocarboxylic acids, s. K. Müller et al., Helv. Chim. Acta **78**, 563-80 (1995).

Potassium carbonate   *K$_2$CO$_3$*
N-Hydroxymethylation   NH → NCH$_2$OH
s. *3*, 657; of lactams under ultrasonication s. B. Jouglet et al., Synth. Commun. **25**, 3869-74 (1995).

Ytterbium triisopropoxide/trimethylsilyl azide   *Yb(OPr-i)$_3$/Me$_3$SiN$_3$*
Zirconium(IV) or hafnium(IV) triflate/tetramethylguanidinium azide   ←
2-Azidoalcohols from epoxides   ▽o/ → C(OH)C(N$_3$)
with Me$_3$SiN$_3$/BF$_3$ or ZnCl$_2$ cf. **40**, 176s*41,42*; with Me$_3$SiN$_3$/Yb(OPr-i)$_3$ or Yb(OPr-i)$_3$·3LiOTf, regio- and stereo-selectivity, s. Y. Yamamoto et al., Chem. Commun. *1995*, 1021-2; with tetramethylguanidinium azide and a metal triflate (Zr(OTf)$_4$, Hf(OTf)$_4$, Yb(OTf)$_3$) cf. P. Crotti et al., Tetrahedron Letters **37**, 1675-8 (1996).

Tetraphenylstibonium hydroxide   *Ph$_4$SbOH*
2-Siloxyazides from epoxides   ▽o/ → C(OSi∈)C(N$_3$)
with Me$_3$SiN$_3$/Al(OPr-i)$_3$ cf. **39**, 274s*46*; with Me$_3$SiN$_3$/Ph$_4$SbOH, Ph$_4$SbBr or Ph$_4$SbOTf, regioselectivity, s. M. Fujiwara et al., Tetrahedron Letters **36**, 4849-52 (1995).

Trimethylsilyl triflate $\qquad Me_3SiOSO_2CF_3$
**3-Functionalized 2-aminoalcohols from 2,3-epoxyamines
via 2-α-siloxyaziridinium triflates
Regio- and stereo-specific conversion via Payne-type rearrangement**

120.

1.3 eqs. Trimethylsilyl triflate added to a soln. of 1.1 eqs. *trans*-diallyl(2,3-epoxyhexyl)amine in dichloromethane at -78° under N_2, after 10 min 2.15 eqs. piperidine added, and stirred for a further 12 h → product. Y 67%. Reaction takes place with a range of nucleophiles (even *poorly* nucleophilic ones) with full control of stereochemistry. F.e.s. C.M. Rayner et al., J. Chem. Soc. Perkin Trans. I *1994*, 1363-5; with chiral α-amino-esters as nucleophile s. Synlett *1995*, 1037-9.

Chiral chromium(III) salen complex $\qquad Cr(III)$
2-Siloxyazides from epoxides
Asym. ring opening

121.

Well-defined, readily prepared, *tunable* chiral salen complexes are remarkably efficient in a new role: asym. nucleophilic ring opening of epoxides. E: Startg. epoxide added to 2 mol% chiral Cr(III)-salen complex in ether, stirred for 15 min, 1.05 eqs. trimethylsilyl azide added, stirred at room temp. for 18 h, and the siloxyazide converted to the free alcohol with camphorsulfonic acid in methanol → product. Y 80% (e.e. 98%). Enantioselectivity was very high (95-8%) for epoxides fused to 5-membered rings, but slightly lower (81-8%) for 6 membered ring and acyclic substrates. Ethers, olefins and carbonyl groups were tolerated, indicating that Lewis basic groups do not inhibit the catalyst. **Kinetic resolution of epoxides** can be undertaken by this procedure (e.e. 97-8%), and an efficient *solvent-free* method is available for asym. ring opening of *meso*-epoxides (e.e. 83-94%). It is considered that the catalyst functions by face-selective delivery of the azide group via a $Cr-N_3$ species, rather than by Lewis acid activation of the epoxide by the metal. F.e.s. E.N. Jacobsen et al., J. Am. Chem. Soc. *117*, 5897-8 (1995); kinetic resolution of terminal epoxides and reduction to **chiral 2-aminoalcohols** s. ibid. *118*, 7420-1 (1996); cyclopentane derivs. s. J. Org. Chem. *61*, 389-90 (1996).

Hydrochlorides
Hydroxyalkoximes from cyclic enolethers and alkoxylamines

122.

2,3-Dihydrofuran added dropwise with stirring to 1 eq. startg. O-alkylhydroxylamine hydrochloride in water, stirred for 3 h at room temp., and neutralized with 1 eq. $NaHCO_3$ → product. Y 75%. F.e. and with 2*H*-3,4-dihydropyran, **also hydroxyhydrazones from hydrazines**, s. D. Martin, J. Prakt. Chem. *337*, 599-600 (1995).

Addition to Nitrogen-Nitrogen Bonds NC ⇓ NN

Organolithium compds.   *RLi*
Tetrasubst. hydrazines from azo compds.   N=N → N(R)N(R)
with Li/HMPA cf. *24*, 312; with RLi via 1,2-addition-alkylation s. A.R. Katritzky et al., Synthesis *1995*, 651-3.

Addition to Nitrogen and Carbon NC ⇓ NC

Without additional reagents   *w.a.r.*
α-Acylaminocarboxylic acid amides from 2-iminoaziridines and carboxylic acids   C
Alternative to the Ugi condensation with retention of configuration

123.

Responding to the fact that the Ugi 4-component condensation (*17*, 809) is plagued by steric hindrance and other retarding factors, highly reactive 2-iminoaziridines may be used effectively in place of three (the prim. amine, carbonyl compd. and isocyanide) of the four components. E: A soln. of benzoic acid in ether added dropwise to a stirred soln. of 1 eq. startg. iminoaziridine (R:S 83:17) in the same solvent at 0°, allowed to warm to 20-25° over 1.5 h, and stirring continued for 3.5 h → (R)-2-(N-benzoyl-N-methylamino)-N,3,3-trimethylbutanamide. Y 70% (R:S 81:19). There was *no racemization* and the intermediate O-acyliminoester (presumed to be formed during the Ugi condensation itself) was detectable. F.e. and Mumm rearrangement of the intermediate s. H. Quast, S. Aldenkortt, Chem. Eur. J. *2*, 462-9 (1996).

Sodium azide/ammonium chloride NaN_3/NH_4Cl
Tetrazoles from nitriles   O
with $NaN_3/Et_3N \cdot HCl$ cf. *17*, 503s*43*; with NaN_3/NH_4Cl under ultrasonication s. G.L. Rusinov et al., Khim. Geterotsikl. Soedin. *1994*, 1375-7 (Russ.).

Boron fluoride BF_3
α-*prim*-Aminoamidines from 2-amino-Δ¹-azirines and prim. amines   C

124.

BF$_3$-Etherate in methylene chloride added to a soln. of startg. azirine and 1.5 eqs. aniline in the same solvent at room temp., stirred for 1 day at room temp., treated with 10% aq. K$_2$CO$_3$, and stirring continued for 1 h → 2-amino-2,N¹,N¹-trimethyl-N²-phenylpropanimidamide. Y 69%. F.e.s. M. Hugener, H. Heimgartner, Helv. Chim. Acta *78*, 1823-36 (1995).

Aluminum chloride $AlCl_3$
Dicarboxylic acid amides from dicarboxylic acid imides and amines
under mild conditions

125.

2.5 eqs. N-Methylaniline added to a suspension of 1.3 eqs. AlCl$_3$ in 1,2-dichloroethane at 0°, the mixture allowed to warm to room temp., a soln. of succinimide in the same solvent added dropwise, and stirred at room temp. for 16.5 h → product. Y 79%. Higher temp. (90°) was required for N-alkylimides. F.e. and ring opening of 2-oxazolidones s. E. Bon et al., Tetrahedron Letters *37*, 1217-20 (1996).

Di-n-butyl(chloro)stannane-hexamethylphosphoramide $Bu_2Sn(Cl)H\text{-}(Me_2N)_3PO$
Amines from azomethines via hydrostannylation $C{=}N \rightarrow CHNR$

126.

Tert. amines. Bu_2SnCl_2 added to 1 eq. Bu_2SnH_2 in THF, stirred at room temp. for 10 min, the resulting $Bu_2Sn(Cl)H$ treated with 1 eq. HMPA, followed by 1 eq. N-benzylideneaniline, the mixture stirred at room temp. for 1 h, 1 eq. benzyl bromide added to the *in situ*-formed aminostannane, the mixture stirred at 60° for 3 h, and quenched with methanol → N,N-dibenzylphenylamine. Y 91%. F.e., and **sec. amines** by quenching the intermediate aminostannane with methanol, s. I. Shibata et al., J. Org. Chem. *60*, 2677-82 (1995).

Addition to Carbon-Carbon Bonds   NC ⇓ CC

Without additional reagents *w.a.r.*
Michael addition of amines with asym. induction s. *42*, 290s*51* $C{=}C \rightarrow CHC(N{<})$
Hetero-Diels-Alder reaction with 1,2,4-triazoline-3,5-diones ○
s. *22*, 355s*32*; stereospecific conversion with 2,4-dienols under 1,3-allylic strain s. W. Adam et al., J. Am. Chem. Soc. *117*, 9190-3 (1995); addition of azadienophiles to 5-glycosyloxy-2,4-dienoates **with asym. induction** s. I.H. Aspinall et al., Tetrahedron Letters *35*, 3397-400 (1994); addition of *in situ*-generated 1,3,4-thiadiazoline-2,5-diones s. M. Squillacote, J. De Felippis, J. Org. Chem. *59*, 3564-71 (1994).

1,2,3-Triazoles from acetylene derivs. and azides
s. *8*, 404s*47*; cycloaddition of α-amino-α-azidophosphonic acid esters s. A. Elachqar et al., Synth. Commun. *24*, 1279-86 (1994); cycloaddition to α,β-acetylenealdimines s. Y.L. Piterskaya et al., Zh. Obshch. Khim. *66*, 1180-94 (1996) (Russ.); to α,β-acetyleneacylsilanes s. A. Degl'Innocenti et al., Tetrahedron Letters *36*, 9031-4 (1995); to *in situ*-generated ynamines s. L.M. Beauchamp et al., J. Med. Chem. *39*, 949-56 (1996).

Electrolysis ↯
α-Azidoacetals from enolethers $C(N_3)C(OR)OR'$

127.

Butyl vinyl ether and 1.5 eqs. NaN_3 added to a soln. of tetraethylammonium tosylate in methanol contained in the anode compartment of a divided cell equipped with a cylindrical ceramic diaphragm and carbon rod electrodes, and electrolyzed at room temp. under a constant current (current density 25 mA/cm²) with stirring until 3 F/mol passed → 1-azido-2-butoxy-2-methoxyethane. Y 81%. F.e. and reaction of N-vinyllactams s. I. Nishiguchi et al., Tetrahedron Letters *36*, 7483-6 (1995).

Chiral lithium amides $LiNR_2$
Asym. Michael addition $C{=}C \rightarrow CHC(N{<})$
to enoates s. *42*, 290s*49*; with Li-(α(S)-methylbenzyl)allylamide as a *differentially*-protected **chiral ammonia equivalent** cf. S.G. Davies, D.R. Fenwick, Chem. Commun. *1995*, 1109-10; with a chiral Li-amidocuprate/P(OEt)$_3$ (method s. *48*, 281) cf. N. Sewald et al., Liebigs Ann. *1995*, 925-8; asym. addition of amines to arylmenthol-derived enoates s. F. Dumas et al., J. Org. Chem. *61*; 2293-304 (1996), asym. addition to chiral bicyclic lactams, chiral 3-aminopyrrolidines, s. A.I. Meyers et al., ibid. *60*, 3189-93 (1995).

Sodium azide s. under ↲ NaN₃
Sodium nitrite/ammonium ceric nitrate NaNO₂/(NH₄)₂Ce(NO₃)₆
2-Nitroacylamines from ethylene derivs. and nitriles C(NO₂)C(NHCOR)
via Ritter-type reaction

128. cycloheptene + MeCN $\xrightarrow{NO_2^-}$ 1-NO₂-2-NHCOMe-cycloheptane

5 eqs. NaNO₂ and 2 eqs. (NH₄)₂Ce(NO₃)₆ added to a stirred soln. of cycloheptene in acetonitrile, the mixture stirred at room temp. for 24 h, and quenched with satd. aq. NaHCO₃ → 1-acetylamino-2-nitrocycloheptane. Y 71%. Lowish yields were obtained from acyclic ethylene derivs., while styrenes gave 1,1-nitroethylene derivs. F.e.s. M.V.R. Reddy et al., Tetrahedron Letters *36*, 4861-4 (1995).

Chiral lithium amidocuprates/triethyl phosphite ←
Asym. Michael addition s. *42*, 290s*51* C=C → CHC(NHR)

Ammonium ceric nitrate s. under NaNO₂ (NH₄)₂Ce(NO₃)₆

Chlorobis(cyclopentadienyl)hydridozirconium/mesitylenesulfonyloxylamine ←
Prim. amines from ethylene derivs. C=C → CHC(NH₂)
via regiospecific hydrozirconation

129. n-C₆H₁₃CH=CH₂ $\xrightarrow{Cp_2ZrHCl}$ [n-C₆H₁₃CH₂CH₂ZrCp₂Cl] $\xrightarrow{2,4,6\text{-}Me_3C_6H_2SO_2ONH_2}$ n-C₆H₁₃CH₂CH₂NH₂

1.2 eqs. 1-Octene added to a stirred suspension of Cp₂Zr(H)Cl in THF at room temp. under argon, the yellow soln. cooled in an ice-bath, 1.2 eqs. *freshly prepared* mesitylenesulfonyloxylamine (to be used within 12 h of preparation) in ether added dropwise, stirred for 10 min, and acidified with 1 M HCl → n-octylamine. Y 77%. The method is efficient and applicable to a variety of monosubst. alkenes and alkylidenes. F.e.s. B. Zheng, M. Srebnik, J. Org. Chem. *60*, 1912-3 (1995).

Tetramethylguanidinium azide TMGN₃
Enazides from acetylene derivs. C≡C → CH=C(N₃)

130. ═══COOMe → (N₃)C=CH(COOMe) + (N₃)C=CH-COOMe

β-Azido-α,β-ethylenecarboxylic acid esters. A soln. of 1 eq. tetramethylguanidinium azide in dry chloroform added very slowly to a soln. of startg. acetylenic ester in the same solvent at -10° under N₂, and the mixture stirred at room temp. for 24 h → product. Y 90% (E/Z 7:3). F.e. and from allenes s. F. Palacios et al., Org. Prep. Proc. Intern. *27*, 171-8 (1995).

Bis(dibenzylideneacetone)palladium(0)/triphenylphosphine/triethylamine hydroiodide ←
(E)-2-Ethylene-*tert*-amines from allenes and sec. amines C=CHC(N<)
Regio- and stereo-specific hydroamination

131. pyrrolidine-NH + CH₂=C=CHPh → N-CH₂-CH=CH-Ph (pyrrolidine)

Phenylallene and 1.1 eqs. pyrrolidine added to a mixture of 5 mol% Pd(dba)₂, 0.1 eq. PPh₃, and 0.16 eq. Et₃N·HI in THF (previously mixed for 15 min) under N₂, heated at 60° for 32 h, and hydrolyzed with water → N-(3-phenyl-2-propenyl)pyrrolidine. Y 89% (98% E-isomer). 2-Aminomethyl-1,3-dienes were formed as by-products (8-39%), while allene itself gave 2-methyl-3-pyrrolidinomethyl-1,3-butadiene as the only product. Reaction is thought to involve either a palladium π-allyl intermediate, formed by initial hydropalladation of the allene, or a π-complex formed between the allene and *in situ*-generated HPdI. F.e.s. L. Besson et al., Tetrahedron Letters *36*, 3857-60 (1995); **α-allylation** of active methylene compds. via hydropalladation s. ibid. 3853-6 (1995).

Potassium osmate/1,4-bis(9-O-dihydroquinine)phthalazine/chloramine-T ←
Regiospecific asym. oxyamination of ethylene derivs.   C(NHTs)C(OH)

132. 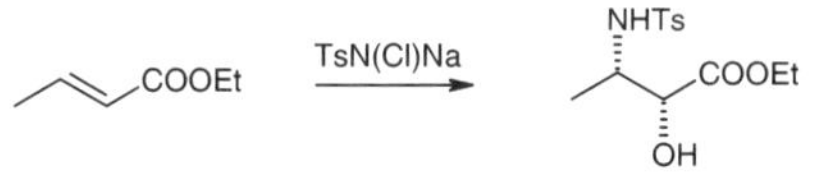
$$\text{CH}_2\text{=CH-COOEt} \xrightarrow{\text{TsN(Cl)Na}} \text{CH(NHTs)-CH(OH)-COOEt}$$

Chiral *cis*-2-tosylaminoalcohols. *trans*-Ethyl crotonate, 3 eqs. Chloramine-T trihydrate, and 4 mol% $K_2OsO_2(OH)_4$ added to a stirred soln. of 5 mol% $(DHQ)_2$-PHAL in 1:1 acetonitrile/water, and allowed to react at room temp. for ca. 90 min → ethyl (2R,3S)-N-(*p*-toluenesulfonyl)-3-amino-2-hydroxybutanoate. Y 52% (e.e. 74%). Similar results were obtained with 1.5 eqs. Chloramine-T in the presence of 1.5 eqs. Et_4NOAc. The products can often be isolated by filtration of the crude reaction mixture, and enantiomeric purity near 100% can often be achieved by recrystallization. Even strongly electron-deficient olefins, such as dimethyl fumarate, react rapidly with high enantioselectivity. F.e. and with $(DHQD)_2$-PHAL (*48*, 142), also in aq. *t*-BuOH of s. K.B. Sharpless et al., Angew. Chem. Intern. Ed. *35*, 451-4 (1996).

Via intermediates   *v.i.*
Lactams from cyclic ketones via 1,1-siloxyazides s. *51*, 191   ○

Rearrangement   ⋂

Oxygen/Carbon Type   NC ⋂ OC

Without additional reagents   *w.a.r.*
Mumm rearrangement of acyl α-aminoimidates s. *51*, 123   ←

Carbon/Carbon Type   NC ⋂ CC

Zeolite   ←
Benzoxazoles   ○
from *o*-hydroxyoxime acetates cf. *12*, 229; from *o*-hydroxyoximes with zeolites s. B.M. Bhawal et al., Synth. Commun. *25*, 3315-21 (1995).

Exchange   ⇅

Hydrogen ↑   NC ⇅ H

*Sodium carbonate/tetra-*n*-butylammonium bromide* Na_2CO_3/Bu_4NBr
Formazans from hydrazones   ←
with py cf. *8*, 421; with Na_2CO_3/Bu_4NBr under phase transfer catalysis, also directly **from aldehydes**, s. A.R. Katritzky et al., Synthesis *1995*, 577-81.

Cesium carbonate s. under TsN_3 Cs_2CO_3

Sodium nitrite/ammonium ceric nitrate/acetic acid
1,1-Nitroethylene from ethylene derivs. $C=CH \rightarrow C=C(NO_2)$
with I_2 as oxidant cf. *42*, 311; with CAN/AcOH, regioselectivity, s. J.R. Hwu et al., Chem. Commun. *1994*, 1425-6.

Sodium nitrite/propionic acid/propionic anhydride $NaNO_2/RCOOH/(RCO)_2O$
***o*-Nitrosation of phenols**   H → NO
with $NaNO_2$ in aq. HOAc cf. *7*, 407; nitroresorcinols in $EtCOOH/(EtCO)_2O$ s. R.J. Maleski et al., Synth. Commun. *25*, 2327-35 (1995).

Potassium nitrite/iodine/18-crown-6 polyether/pyridine/triethylamine   ←
Cyclic 1,1-nitroethylene from ethylene derivs. s. *51*, 135 $C=CH \rightarrow C=C(NO_2)$

Sodium nitrate/hydrogen chloride/acetic anhydride
o-Nitration of phenols in a 2-phase medium H → NO₂
with NaNO₃/HCl/La(NO₃)₃ cf. *38*, 307; with NaNO₃/HCl/Ac₂O s. P. Keller, Bull. Soc. Chim. France *131*, 27-9 (1994).

Cupric nitrate-clay/acetic anhydride *Cu(NO₃)₂-clay/Ac₂O*
Heterogeneous ar. nitration
s. *30*, 239s*44*; of both activated and deactivated arenes, also polynitration, s. B. Gigante et al., J. Org. Chem. *60*, 3445-7 (1995); with 70-90% HNO₃ or *i*-PrONO₂ and H₂SO₄-silica gel (cf. *34*, 320) s. J.M. Riego et al., Tetrahedron Letters *37*, 513-6 (1996).

Zeolites s. under AcONO₂ ←

Ammonium ceric nitrate s. under NaNO₂ *(NH₄)₆Ce(NO₃)₂*

Cyanamide or dicyandiamide/triethylamine ←
Isothiocyanates from amines s. *31*, 336s*51* NH₂ → N=C=S

tert-Butoxyformic anhydride/4-dimethylaminopyridine *(t-BuCO)₂O/DMAP*
Isocyanates from amines under mild conditions NH₂ → N=C=O

133.   2,4,6-Me₃C₆H₂NH₂ $\xrightarrow{\text{(t-BuOCO)}_2\text{O}}$ 2,4,6-Me₃C₆H₂N=C=O

Hindered compds. Solns. of 1 eq. 4-dimethylaminopyridine and startg. amine in anhydrous acetonitrile added successively to a soln. of 1.4 eqs. *tert*-butoxyformic anhydride in the same solvent, the mixture stirred for 10 min at room temp., and worked up by treatment with 7 eqs. H₂SO₄ (40% soln.) in acetonitrile, followed by stirring for 2 min, then hexane extraction → mesityl isocyanate. Y 96% (97% with *0.1 eq.* 4-DMAP). The method is mild and rapid, and avoids the use of COCl₂. Work-up involved column chromatography on silica gel at -30° to -45° (to minimize urethan formation). F.e. incl diisocyanates and hindered alkyl isocyanates s. H.-J. Knölker et al., Angew. Chem. Intern. Ed. *34*, 2497-500 (1995).

1-Azido-1,2-benziodoxoles/benzoyl peroxide ←
Replacement of hydrogen by azido groups H → N₃
with PhIO/Me₃SiN₃ cf. *49*, 277; 3-azido-3-deoxyglycals s. A. Kirschning et al., Synlett *1995*, 767-9; **bridgehead azides** with 1-azido-1,2-benziodoxoles/(PhCOO)₂ s. A.P. Krasutsky et al., ibid. 1081-2.

Tosyl azide/cesium carbonate *TsN₃/Cs₂CO₃*
Diazo group transfer CH₂ → CN₂
with TsN₃/KH cf. *20*, 271s*32*; with TsN₃/Cs₂CO₃, non-aq. work-up, s. J.C. Lee, J.Y. Yuk, Synth. Commun. *25*, 1511-5 (1995).

Acetyl nitrate/zeolites ←
Ar. nitration H → NO₂
with BzONO₂/zeolite cf. *1*, 343s*46*; with AcONO₂/zeolite, *p*-nitration, s. K. Smith et al., Chem. Commun. *1996*, 469-70.

Trimethylsilyl nitrate/chromium trioxide *Me₃SiNO₃/CrO₃*
α-Nitroketones from ethylene derivs. C=CH → C(NO₂)CO
Regiospecific conversion

134.   PrCH₂CH=CH₂ $\xrightarrow{[\text{Me}_3\text{SiNO}_3]}$ PrCH₂COCH₂NO₂

1.1 eqs. AgNO₃ added in the dark to a stirred soln. of 1 eq. trimethylsilyl chloride in dry acetonitrile under N₂ at 0°, the mixture stirred for 1 h, the resulting soln. of trimethylsilyl nitrate decanted (to remove precipitated AgNO₃), added to 1.5 eqs. CrO₃ in acetonitrile with stirring, after 15 min a soln. of 1-hexene in the same solvent added very slowly (CAUTION: vigorous exothermic reaction) with occasional cooling in a cold water-bath, and stirring continued for a further 24 h → 1-nitrohexan-2-one. Y 88%. The method is simple and general. F.e. incl. cyclic and macrocyclic analogs s. M.V.R. Reddy et al., Tetrahedron Letters *36*, 7149-52 (1995).

Nitrogen monoxide/alumina or air $\quad NO/Al_2O_3\ or\ O_2$
(E)-1,1-Nitroethylene derivs. from ethylene derivs. $\quad C=CH \rightarrow C=C(NO_2)$
under mild conditions

135.

A soln. of 4-phenyl-1-butene in 1,2-dichloroethane stirred under 1 atm. NO at room temp., the gas released, activated acidic alumina added, and the slurry stirred under gentle reflux for 30 min → 1-nitro-4-phenyl-1-butene. Y 90%. F.e. incl. β-nitrostyrenes and cyclic derivs., and isolation of the intermediate 2-nitrosonitro compds., s. E. Hata et al., Bull. Chem. Soc. Japan *68*, 3629-36 (1995); cyclic 1,1-nitroethylene derivs. with nitryl iodide generated under ultrasonication from KNO_2/I_2/18-crown-6 and base cf. D. Ghosh, D.E. Nichols, Synthesis *1996*, 195-7; regiospecific **ar. nitration** of anisoles with NO/air s. K. Mizuno et al., J. Chem. Res. (S) *1995*, 284-5.

Nitryl tetrakis(trifluoromethanesulfonato)borate $\quad [NO_2]^+[B(OTf)_4]^-$
Ar. nitration under mild conditions $\quad H \rightarrow NO$

136.

of deactivated arenes. The highly *superacidic* triflatoboric acid ($2CF_3SO_3H$-$B(OSO_2CF_3)_3$) activates nitric acid to generate nitryl tetrakis(trifluoromethanesulfonato)borate, which readily nitrates deactivated arenes with high regioselectivity. **E:** 100% HNO_3 mixed with 1 eq. triflatoboric acid under dry N_2, 0.8 eq. methyl phenyl sulfone added to the partially formed precipitate, and stirred vigorously at 20° for 3 h → product. Y 78%. The *meta*-isomer was formed exclusively. Protosolvated nitronium ion, $[NO_2H]^{2+}$, is considered to be the effective nitrating agent. F.e.s. G.A. Olah et al., J. Org. Chem. *60*, 7348-50 (1995).

(Benzotriazolyloxy)tris(dimethylamino)phosphonium hexafluorophosphate $\quad \leftarrow$
Isothiocyanates from amines and carbon disulfide $\quad NH_2 \rightarrow N=C=S$
with EtMgBr cf. *31*, 336; with BOP s. U. Boas, M.H. Jakobsen, Chem. Commun. *1995*, 1995-6; with cyanamide or dicyandiamide and a little Et_3N cf. T. Yamamoto et al., Org. Prep. Proc. Intern. *24*, 346-9 (1992); *26*, 555-7 (1994).

Sulfur trioxide-trimethylamine/1,8-diazabicyclo[5.4.0]undec-7-ene $\quad SO_3·Me_3N/DBU$
Sym. N,N'-diarylureas from ar. amines $\quad ArNH_2 \rightarrow (ArNH)_2CO$
and CO_2 with $(PhO)_2POH$ cf. *30*, 243; with $SO_3×Me_3N/DBU$ s. C.F. Cooper, S.J. Falcone, Synth. Commun. *25*, 2467-74 (1995).

Sulfuric acid-silica gel $\quad H_2SO_4\text{-}SiO_2$
Heterogeneous ar. nitration s. *30*, 239s*51* $\quad H \rightarrow NO_2$

Chromium trioxide (s.a. under Me_3SiNO_3) $\quad CrO_3$
α-Azidoketones from ethylene derivs. $\quad C=CH \rightarrow C(N_3)CO$
with $Pb(OAc)_4$ cf. *27*, 362; with CrO_3 s. M.V.R. Reddy et al., Tetrahedron Letters *36*, 6751-4 (1995).

Via intermediates $\quad v.i.$
Asym. *prim*-amination of carboxylic acids $\quad H \rightarrow NH_2$
via 5,6-dihydro-1,4-oxazin-2-ones s. *51*, 182

Ar. *prim*-amination via azodicarboxylic acid ester adducts
s. *21*, 349s*49*; with CF_3SO_3H or CF_3CO_2H as catalyst for adduct formation s. J. Org. Chem. *60*, 4268-71 (1995).

Oxygen ↑ NC ⇅ O

Without additional reagents
Alkylation with (triphenylphosphoranylidene)acetic acid esters

w.a.r.
NH → NR

137.

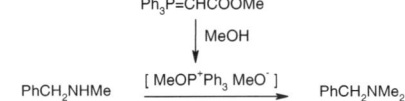

N-Methylation. A soln. of N-methylbenzylamine in 4:1 dichloromethane/*methanol* treated with 5 eqs. methyl (triphenylphosphoranylidene)acetate at room temp. for 5 days → dimethylbenzylamine. Y 70% (80% conversion). Methanol is essential for *in situ*-generation of methoxytriphenylphosphonium methoxide, the effective methylating agent. The corresponding ethyl ester (in ethanol) reacted in the same way but more slowly. F.e. incl. **O-alkylation** of carboxylic acids and phenol, and N-methylation of phthalimide, s. D. Desmaële, Tetrahedron Letters *37*, 1233-6 (1996).

N-Acylation with enolesters NH → NCOR
s. *47*, 273; of α-aminoesters s. S.-T. Chen et al., Tetrahedron Letters *35*, 3583-4 (1994); of acylamines, ureas and urethans s. P.H. Dixneuf et al., Tetrahedron *51*, 10901-12 (1995); N-methacryloylation of acylamines s. Synlett *1995*, 707-8.

N-Trifluoroacetylation NH → NCOCF₃
with CF₃CO₂Et s. *36*, 340; partial conversion and preferential N-trifluoroacetylation of prim. amines, s. K. Prasad et al., Tetrahedron Letters *36*, 7357-60 (1995); N(α),N(ω)-terminal bis-N-trifluoroacetylation of polyamines s. M.C. O'Sullivan, D.M. Dalrymple, ibid. 3451-2.

Carboxylic acid amides from esters COOR → CON<
s. *3*, 722; α,α-dihalogenocarboxylic acid amides in the absence of solvent s. F. Ghelfi et al., Tetrahedron *51*, 12285-92 (1995).

N-Carbalkoxylation with N-carbalkoxyoxydicarboxylic acid imides NH → NCOOR
s. *42*, 339; preferential N-carbo-*tert*-butoxylation s. I. Grapsas et al., J. Org. Chem. *59*, 1918-22 (1994).

Ureas from urethans and amines NHCOOR → NHCON<
s. *13*, 410; polymer-based synthesis from polymer-supported *p*-nitrophenyl carbamates s. S.M. Hutchins, K.T. Chapman, Tetrahedron Letters *35*, 4055-8 (1994).

O-Imidocarbamates as isocyanate substitutes ←
s. *34*, 330; prepn. of O-imidocarbamates, notably ³H-labelled N-methyl-derivs., s. T. Konakahara et al., Synthesis *1993*, 103-6 (1993).

1,2-Diamines from cyclic glycol sulfates via aziridinium ions C
Stereospecific conversion

138.

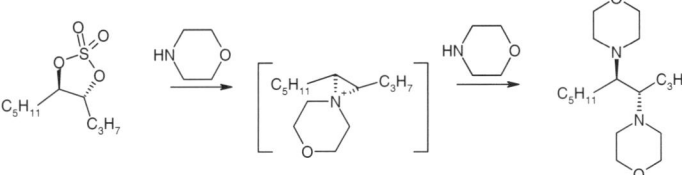

Startg. cyclic sulfate and 5 eqs. morpholine refluxed in the absence of solvent, and worked up with base → *erythro*-product. Y 62%. **Regiospecific formation** of 1,2-diamines can also be effected by using two different amines in sequence. F.e. and preparation of **2-functionalized amines** by appropriate choice of nucleophile in the 2nd step s. K.B. Sharpless et al., Tetrahedron Letters *36*, 9241-4 (1995).

Cyclic amines from ditosylates ○
s. 22, 368; **3-alkoxyazetidines** s. J.M. Chong, K.K. Sokoll, Synth. Commun. 25, 603-11 (1995); chiral morpholines s. K. Bennis, Carbohyd. Res. 264, 33-44 (1994).

Lactams from ethylenecarboxylic acid esters and prim. amines
2-azetidinones cf. 48, 305; **2-pyrrolidones** s. T. Ben Ayed et al., Tetrahedron 51, 9633-42 (1995).

Pyrazoles from β-diketones
s. 2, 368; polymer-based synthesis of pyrazoles and isoxazoles s. A.L. Marzinik, E.R. Felder, Tetrahedron Letters 37, 1003-6 (1996); fatty acid derivs. in water under ultrasonication s. M.S.F. Lie Ken Jie, P. Kalluri, J. Chem. Soc. Perkin Trans. I 1995, 1205-6; 5-perfluoroalkyl-derivs. via Cu(II)-chelates s. V.I. Saloutin et al., Zh. Org. Khim. 31, 266-9 (1995) (Russ.); 5-hydroxy-Δ²-pyrazolines cf. K.N. Zelenin et al., Tetrahedron 51, 11251-6 (1995).

Microwaves s. under p-TsOH and Ni ←
Sodium azide s. under $SiCl_4$ NaN_3
Sodium nitrite s. under C_5H_5N $NaNO_2$
Triethylamine Et_3N

N-Formylation of α-aminocarboxylic acid esters with cyanomethyl formate NH → NCHO
under mild conditions

139.

Startg. aminoacid ester hydrochloride and 1 eq. cyanomethyl formate in methylene chloride cooled to 0° in an ice-bath, a soln. of 1 eq. NEt_3 in the same solvent added, and further methylene chloride added after 12 h at room temp. → product. Y 84%. Cyanomethyl formate is easily prepared and formylation of the amino esters occurs in good yield, the method being applicable to a variety of substrates, incl. t-butyl esters, without racemization. F.e.s. H.-J. Niclas et al., Synthesis 1996, 37-8.

Pyridine/sodium nitrite $C_5H_5N/NaNO_2$
Tetrazoles from iminoesters via N-formylamidrazones ○

140.

5-Subst. tetrazoles. 1.1 eqs. Formylhydrazine added in several portions over 10 min to an ice-cooled suspension of startg. imidate hydrochloride in pyridine, allowed to warm to room temp. over 12 h, cooled in ice, aq. HCl (1:1) added at such a rate that the temp. did not exceed 5°, cooled to -5°, excess of $NaNO_2$ added portionwise so that the temp. remained at or below 0°, and stirring continued for 1 h → 5-phenyltetrazole. Y 95%. Unlike procedures involving nitrosation of amidrazones, there was no intermediate formation of azides, so that scale-up should not be a problem. Furthermore, the method is applicable to *o-subst.* phenyl derivs. which cannot easily be obtained by alternative means. F.e. and from the corresponding thioimidate s. J. Boivin et al., Tetrahedron 51, 11737-42 (1995).

Silver trifluoromethanesulfonate s. under $(ArCO)_2O$ AgOTf
Magnesium/methanol/acetic acid/triethylamine $Mg/MeOH/AcOH/Et_3N$
Sec. amines from prim. amines and oxo compds. NH → NR

141.

4 eqs. Acetic acid added via pipette to a stirred soln. of cyclohexanone, 0.9 eq. phenethylamine, and 3 eqs. Et_3N in methanol with water-cooling, 4.5 eqs. Mg added, the mixture stirred vigorously under reflux for 2 h, a further 4 eqs. acetic acid in methanol added portionwise over 2 h, and

refluxed for 1 h → N-(cyclohexyl)phenethylamine. Y 79%. Significantly, there was no formation of tert. amines or alcohols. The method is inexpensive, relatively rapid, non-toxic, simple, and suitable for large-scale preparations. However, conjugated double and triple bonds, as well as nitro compds., are not compatible since these are also reduced under these conditions. F.e. incl. hindered compds. s. I.V. Micovic et al., J. Chem. Soc. Perkin Trans. I *1996*, 265-9.

Isopropylmagnesium chloride   i-*PrMgCl*
Hydroxamic acid esters from carboxylic acid esters   COOR → CONOR

142.   PhCH$_2$CH$_2$COOEt + HN(Me)OMe·HCl ⟶ PhCH$_2$CH$_2$CON(Me)OMe

A 2 *M* soln. of isopropylmagnesium chloride in THF added over 15 min to a slurry of startg. ester and 1.55 eqs. N,O-dimethylhydroxylamine hydrochloride in THF at -20° under N$_2$ maintaining the temp. below -5°, the mixture aged at -10° for 20 min, then quenched with 20 wt.% aq. NH$_4$Cl → product. Y 97%. The method is simple, general (notably for enolizable and hindered esters), and high-yielding. F.e. and one-pot preparation of ketones from carboxylic acid esters s. J.M. Williams et al., Tetrahedron Letters *36*, 5461-4 (1995).

Magnesium sulfate/zinc chloride   *MgSO$_4$/ZnCl$_2$*
Nitrones from hydroxylamines   CO → C=N(O)R
s. *8*, 429; with ketones under Lewis acid catalysis (MgSO$_4$/ZnCl$_2$) s. P. Merino et al., Synth. Commun. *25*, 2275-84 (1995).

Zinc chloride s. under MgSO$_4$ and NaBH$_4$ *ZnCl$_2$*
Sodium tetrahydridoborate/zinc chloride or titanium tetraisopropoxide   ←
Reductive N-methylation   NH → NMe
with NaBH$_4$/Ti(OPr-*i*)$_4$ cf. *46*, 316s*49*; β-phenethylamines s. S. Bhattacharyya et al., Synlett 1995, 1079-80; tert. ferrocenylamines s. ibid. 1994, 1029-30; tert. dimethylamines from Me$_2$NH·HCl/ Et$_3$N s. J. Org. Chem. *60*, 4928-9 (1995); tert. dimethylamines from prim. amines and tert. methylamines from sec. amines in *aprotic* media with NaBH$_4$/ZnCl$_2$ [Zn[BH$_4$]$_2$], s. Synth. Commun. *25*, 2061-9 (1995).

Trimethylaluminum   *Me$_3$Al*
Hydroxamic acid benzyl esters from carboxylic acid esters   COOR → CONHOBn
with NH$_2$OH/BnCl cf. *17*, 431; with NH$_2$OBn/Me$_3$Al, chiral α-amino-derivs., s. M.C. Pirrung, J.H.-L. Chau, J. Org. Chem. *60*, 8084-5 (1995).

Sodium trihydridocyanoborate   *NaBH$_3$CN*
Reductive N-methylation   NH → NMe
s. *28*, 346; sec. methylamines from prim. amines via N-methylation of N-(*o*-nitrobenzyl)amines s. Y. Gareau et al., J. Org. Chem. *58*, 1582-5 (1993).

Sodium trihydridocyanoborate/acetic acid   *NaBH$_3$CN/AcOH*
Reductive N-cyclopropylation of amines   ←

143.

N,N-Dicyclopropylation. A soln. of startg. amine treated with 10 eqs. acetic acid, 3 Å molecular sieves, and 4-6 eqs. [(1-ethoxycyclopropyl)oxy]trimethylsilane, followed by 3-4.5 eqs. NaBH$_3$CN, and the mixture refluxed overnight → product. Y 91%. Attempted monocyclopropylation of prim. amines [except aniline] gave mixtures. The method is applicable to the prepn. of sterically hindered [and previously unreported] tricyclopropylamine. However, cyclopropylation of very hindered amines, such as dicyclohexylamine, was not possible under these conditions. F.e. incl. cyclopropylation of sec. amines s. B.A. Lefker et al., Tetrahedron Letters *36*, 7399-402 (1995); **reductive N-benzylation** of aminoalcohols s. *19*, 448s*49*.

Sodium hydridotri(acetoxo)borate/acetic acid   *NaBH(OAc)$_3$/AcOH*
Sodium hydridotris(2-ethylhexanoyloxo)borate   *NaBH(OCOR)$_3$*
Reductive N-alkylation   NH → NR
with NaBH(OAc)$_3$/AcOH s. *46*, 317; details s. J. Org. Chem. *61*, 3849-62 (1996); of α-aminoesters s. J.M. Ramanjulu, M.M. Jouillié, Synth. Commun. *26*, 1379-84 (1996); **polymer-based reductive N-alkylation** of the terminal prim. amino group of dipeptides s. D.A. Campbell et al., J. Org. Chem. *61*, 6720-3 (1996); as part of a combinatorial synthesis of 2,5-piperazinediones s. D.W. Gordon, J. Steele, Bioorg. Med. Chem. Lett. *5*, 47-50 (1995); multipin multiple solid phase synthesis of 4-aminoprolines s. A.M. Bray, Tetrahedron Letters *36*, 5081-4 (1995); polymer-based synthesis of peptide N-alkylamides s. W.C. Chan, S.L. Mellor, Chem. Commun. *1995*, 1475-7; influence of ultrasonication s. S.V. Ley et al., Synlett *1995*, 1017-20; *cis*-cyclohexylamines with Na-hydridotris(2-ethylhexanoyloxo)borate cf. J.M. McGill et al., Tetrahedron Letters *37*, 3977-80 (1996).

Bentonite   ←
Azomethines from oxo compds. and prim. amines   CO → C=NR
with molecular sieves cf. *30*, 463; with inexpensive Algerian bentonite s. A. Saoudi et al., Synth. Commun. *25*, 2349-54 (1995).

Molecular sieves   ←
Enaminolactams from enollactones   ←
s. *17*, 416; 1-alkylidenelactams with molecular sieves s. A.D. Abell et al., J. Org. Chem. *60*, 1214-20 (1995).

Diethyl azodicarboxylate s. under Ph$_3$P   *RO$_2$CN=NCO$_2$R*

1-Ethyl-3-(3-dimethylaminopropyl)carbodiimide hydrochloride/4-dimethylaminopyridine ←
N-Acylation with carboxylic acids   COOH → CON—SO$_2$R
of sulfonamides with N,N'-carbonyldiimidazole/DBU cf. *43*, 295; selective N-acylation of nucleosides (with py) via per-O-silylation s. N.D. Sinha et al., Tetrahedron Letters *36*, 9277-80 (1995); N-acylation of sulfonamides with 1-ethyl-3-(3-dimethylaminopropyl)carbodiimide hydrochloride/DMAP s. J.C. Pelletier, D.P. Hesson, Synlett *1995*, 1141-2.

1-Ethyl-3-(3-dimethylaminopropyl)carbodiimide hydrochloride/2-hydroxypyridine N-oxide ←
Peptide synthesis   COOH → CON<
with added 1-hydroxybenzotriazole cf. *12*, 455s*43*; with added 2-hydroxypyridine N-oxide in an aq. 2-phase medium, suppression of racemization, improved work-up, and further additives, e.g. HOBt and 7-aza-analog, s. G.-J. Ho et al., J. Org. Chem. *60*, 3569-70 (1995).

N,N'-Carbonyldiimidazole/pyridine   ←
N-Acylation of nucleosides s. *43*, 295s*51*   NH → NAc

1H-Benzotriazole/2-mercaptoethanol   ←
Phthalimidines from *o*-dialdehydes and prim. amines   O

144.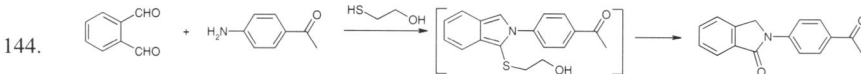

A soln. of 1 eq. startg. amine in acetonitrile, 1 eq. 1*H*-benzotriazole, and a pH 9.6 buffer (0.05 *M* H$_3$BO$_3$-KCl-NaOH) added successively over 1 min periods to a soln. of *o*-phthalaldehyde and 8.6 eqs. 2-mercaptoethanol in the same solvent, and the mixture stirred at room temp. for 13 h → product. Y 84%. 1*H*-Benzotriazole and 2-mercaptoethanol serve as 'dual auxiliaries', the intermediate 1-(2-hydroxyethylthio)isoindole being isolable if the reaction time is short. The method is mild, simple and general for aliphatic as well as aromatic and hindered amines, leaving a variety of functional groups (esters, phenolethers, and ar. halides and nitro groups) unaffected. F.e.s. I. Takahashi et al., Synlett *1996*, 353-5.

Chymotrypsin
Enzymatic peptide synthesis in water NH → NCOR
with N-protected α-aminocarboxylic acid choline esters

145.   Z-Phe-OCH₂CH₂N⁺Me₃ Br⁻ + H-Leu-NH₂ ⟶ Z-Phe-Leu-NH₂

A precooled (0°) soln. of Z-Phe-OCho in water added within 3 min to a stirred soln. of 2 eqs. H-Leu-NH₂ and a little chymotrypsin in water (pH 7.8) at 0°, and stirred for 30 min → product. Y 82%. Slow addition of the acyl donor at 0° was necessary to avoid competing hydrolysis of the ester. F.e. and enzymes s. H.-D. Jakubke et al., Angew. Chem. Intern. Ed. *35*, 106-9 (1996); preparation of benzyl esters and Fmoc derivs. with solubilizing polyethyleneglycol side-chains s. A. Zier et al., Tetrahedron Letters *35*, 1039-42 (1994).

Subtilisin
Enzymatic N-carbalkoxylation with carbonic acid esters NH → NCOOR

146.

Commercially available and inexpensive sym. dialkyl carbonates may be used for the enzymatic protection of amino groups with high chemoselectivity and associated **kinetic resolution or desymmetrization. E:** A soln. of startg. *meso*-diaminotriol and excess of diallyl carbonate in pH 7.8 HEPES buffer treated with subtilisin BPN' (Nagarse, type XXVII) at room temp. for 1 week → product. Y 76% (e.e. >99%). Yields and enantioselectivity were lower with dimethyl or diethyl carbonate, while dibenzyl carbonate was unreactive. The process can be carried out in aq. media for prim. and sec. amines, or in organic media for sec. amines. The protective group can be readily removed, or reduced to the N-methyl-deriv. with LAH. F.e. and with retention of ester groups s. C.-H. Wong et al., J. Am. Chem. Soc. *118*, 712-3 (1996).

Methyl orthoformate HC(OMe)₃
Azomethines from oxo compds. under mild, non-acidic conditions CO → C=N

147.   PhCHO + H₂NCH₂COOMe ⟶ PhCH=NCH₂COOMe

Startg. amine added with stirring to methyl orthoformate, after 5 min 1 eq. benzaldehyde added, and stirred for 8 h → product. Y 92%. Aliphatic aldehydes and ketones gave non-isolable imines, which could be trapped *in situ*. Polymer-supported amines or aldehydes were similarly transformed. F.e.s. G.C. Look et al., Tetrahedron Letters *36*, 2937-40 (1995).

p-Trifluoromethylbenzoic anhydride/titanium tetrachloride/silver trifluoromethanesulfonate
N-Formylation with formic acid NH → NCHO
s. *13*, 442s36; of weakly nucleophilic ar. amines with *p*-trifluoromethylbenzoic anhydride/TiCl₄/AgOTf (cf. *49*, 307) s. T. Mukaiyama et al., Heterocycles *40*, 141-8 (1995).

tert-Butoxyformic anhydride/pyridine (t-BuOCO)₂O/C₅H₅N
N-Unsubst. carboxylic acid amides from carboxylic acids COOH → CONH₂
with 2-ethoxy-N-ethoxycarbonyl-1,2-dihydroquinoline cf. *44*, 312; N-unsubst. α-(carbalkoxyamino)carboxylic acid amides with *tert*-butoxyformic anhydride/py, suppression of racemization, s. V.F. Pozdnev, Tetrahedron Letters *36*, 7115-8 (1995).

Acetic acid AcOH
Imidazoles from α-diketones ○
s. *23*, 423; polymer-based synthesis of a small imidazole library s. S. Sarshar et al., Tetrahedron Letters *37*, 835-8 (1996).

Tetramethylfluoroformamidinium hexafluorophosphate/ethyldiisopropylamine ←
Peptide synthesis NH → NCOR
with tetramethylfluoroformamidinium hexafluorophosphate

148.

A rapid *in situ* version of the acid fluoride route to peptides (cf. 46, 350), either in the soln. or solid phase, has been developed by using tetramethylfluoroformamidinium hexafluorophosphate as a benign substitute for the corrosive cyanuric fluoride. **E: Solid phase synthesis.** A mixture of H-Ile-PEG-PS and 5 eqs. Fmoc-Val-OH in DMF containing 10 eqs. *i*-Pr$_2$NEt and tetramethylfluoroformamidinium hexafluorophosphate subjected to automated coupling using a Biosearch 9050 instrument (programmed for 7 min preactivation) for 10 min → dipeptide. Y 100%. The reagent is easy to prepare, inexpensive, non-hygroscopic, easy to handle, and affords products of high quality. The procedure is also adaptable for **solution phase coupling** with addition of 1 eq. 1-hydroxy-7-azabenzotriazole to prevent extensive epimerization. F.e. and comparison of coupling agents s. L.A. Carpino, A. El-Faham, J. Am. Chem. Soc. *117*, 5401-2 (1995).

N-Bromosuccinimide s. under Ph$_3$P NBS

Silicon tetrachloride/sodium azide SiCl$_4$/NaN$_3$
Tetrazoles from ketones ○
with TiCl$_4$/NaN$_3$ cf. 49, 294; with SiCl$_4$/NaN$_3$, also with ring expansion, s. S.S. Elmorsy et al., Tetrahedron Letters *36*, 7337-40 (1995).

Titanium tetrachloride s. under (ArCO)$_2$O TiCl$_4$

O-Ethylhydroxylamine hydrochloride/sodium hydrogen carbonate/palladium-carbon ←
1,2-Diamines from 1,1-nitroethylene derivs. C═C(NO$_2$) → C(NH$_2$)CH(NH$_2$)
One pot procedure via Michael addition

149.

1.1 eqs. O-Ethylhydroxylamine hydrochloride and 1.1 eqs. NaHCO$_3$ added to a soln. of 1-nitro-2-phenylethene in THF, stirred at room temp. until reaction complete by TLC and/or GC, a little 10% Pd-on-activated carbon and ethanol added, and stirring continued overnight at room temp. under 1 atm. of H$_2$ → 1,2-diamino-1-phenylethane. Y 88%. This simple one-pot method is high-yielding with no polyamine by-product. F.e.s. T. Mukaiyama et al., Chem. Lett. *1996*, 291-2.

Triphenylphosphine/diethyl azodicarboxylate Ph$_3$P/RO$_2$CN═NCO$_2$R
Mitsunobu N-alkylation NH → NR
s. 28, 753s49; of N-benzyltriflamide s. K.E. Bell et al., Tetrahedron Letters *36*, 8681-4 (1995); prepn. of polyamines s. M.L. Edwards et al., Tetrahedron *50*, 5579-90 (1994); of carbocyclic nucleosides s. L.B. Akella, R. Vince, ibid. *52*, 8407-12 (1996); of chiral N-alkylimidazoles s. F. Corelli et al., J. Org. Chem. *60*, 2008-15 (1995); of N-alkylmaleimides in the presence of a non-reacting ('dummy') alcohol s. M.A. Walker, ibid. 5352-5.

Triphenylphosphine/N-bromosuccinimide Ph$_3$P/NBS
N-Alkylation of amines with alcohols

150. t-BuNH$_2$ + HOBn ⟶ t-BuNHBn

N-Bromosuccinimide added over 2-3 min in small portions to a stirred soln. of 1 eq. PPh$_3$ and 1 eq. benzyl alcohol in anhydrous THF at -18°, after ca. 5 min 2.4 eqs. *tert*-butylamine injected via syringe in one portion, stirring continued for a few min, and heated at 80° for 1 h → N-benzyl-*tert*-butylamine. Y 80%. The procedure is suitable for the alkylation of prim. and sec. amines with prim. alcohols, and alkylation of prim. amines with sec. alcohols. Tert. alcohols were unreactive. F.e.s. P. Frøyen, P. Juvvik, Tetrahedron Letters *36*, 9555-8 (1995).

[4-Nitro-6-(trifluoromethyl)benzotriazol-1-yloxy]tris(pyrrolidino)phosphonium ←
hexafluorophosphate/ethyldiisopropylamine
3-(Diethoxyphosphoryloxy)-1,2,3-benzotriazin-4(3H)-one/triethylamine ←
Peptide synthesis COOH → CON<
with PyBop cf. *28*, 144s*50*; coupling *N-methylated* amino-acids with crystalline [4-nitro-6-(trifluoromethyl)benzotriazol-1-yloxy]tris(pyrrolidino)phosphonium hexafluorophosphate s. J.C.H.M. Wijkmans et al., Tetrahedron Letters *36*, 4643-6 (1995); solid and solution phase coupling with 3-(diethoxyphosphoryloxy)-1,2,3-benzotriazin-4(3*H*)-one/NEt₃ cf. C.-X. Fan et al., Synth. Commun. *26*, 1455-60 (1996).

Diphenylphosphinic acid $Ph_2P(O)OH$
Urethans from carbonic acid esters and amines NH → NCOOR
with (EtO)₂CO/NaH cf. *50*, 170; with (PhO)₂CO/Ph₂P(O)OH or (PhO)₂P(O)OH s. M. Aresta et al., Tetrahedron *51*, 8073-88 (1995).

Antimony(III) ethoxide $Sb(OEt)_3$
Carboxylic acid amides from acids or esters s. *51*, 187 COO(H,Me) → CON<

p-Nitrobenzenesulfonyl chloride/triethylamine/4-dimethylaminopyridine ←
Carboxylic acid amides from carboxylic acids COOH → CON<
with ArSO₂Cl/py cf. *11*, 235; with *p*-NO₂C₆H₄SO₂Cl/Et₃N/DMAP, sec. and tert. amides without racemization, s. J.C. Lee et al., Synth. Commun. *25*, 2877-81 (1995).

3,5-Dichloro-2-hydroxybenzenesulfonyl chloride/triethylamine $ArSO_2Cl/Et_3N$
Peptide synthesis
with PhSO₂Cl/py cf. *21*, 446; suppression of racemization and improved work-up with 3,5-dichloro-2-hydroxybenzenesulfonyl chloride/Et₃N or the difluoro-deriv. s. D. Cabaret, M. Wakselman, Tetrahedron Letters *35*, 9561-4 (1994).

p-Toluenesulfonic acid/microwaves ←
Quinoxalines from α-diketones O
with NaOAc cf. *2*, 378; using TsOH without solvent under microwave irradiation s. D. Villemin, B. Martin, Synth. Commun. *25*, 2319-26 (1995).

Hydrogen bromide/acetic acid HBr/AcOH
Solid-phase peptide synthesis COOH → CON<
s. *19*, 33s*46-50*; with novel cross-linked ethoxylate acrylate resin (CLEAR) supports s. M. Kempe, G. Barany, J. Am. Chem. Soc. *118*, 7083-93 (1996); with a tetraethyleneglycol diacrylate-cross linked polystyrene support having a photolabile 2-nitrobenzyl anchor s. M. Renil et al., Tetrahedron Letters *35*, 3809-12 (1994); with bis(2-acrylamidoprop-1-yl)polyethyleneglycol-cross linked dimethyl acrylamide (PEGA) supports s. ibid. *36*, 4647-50 (1995); with *magnetically* manipulable supports s. M.J. Szymonifka, K.T. Chapman, ibid. 1597-600; C-terminal **peptide amides** with a photolabile *o*-nitrobenzhydrylaminopolystyrene support s. A. Ajayaghosh, V.N.R. Pillai, ibid. 777-80; using a xanthenylamide handle cf. Y. Han et al., J. Org. Chem. *61*, 6326-39 (1996); solid-phase peptide syntheses with a universal allyl linker s. X. Zhang, R.A. Jones, Tetrahedron Letters *36*, 3789-90 (1995); prepn. of a high capacity aminomethyl-polystyrene resin s. C.C. Zikos, N.G. Ferderigos, ibid. 3741-4; of a safety-catch resin for direct release of peptides into aq. buffers s. S. Hoffmann, R. Frank, ibid. *35*, 7763-6 (1994); peptide aminoalkylamides using an allyl linker s. K. Kaljuste, A. Undén, ibid. *37*, 3031-4 (1996); with a base-labile linker derived from N-[(9-hydroxymethyl)-2-fluorenyl]succinamic acid, details, s. F. Rabanal et al., Tetrahedron *51*, 1449-58 (1995); minimization of tryptophan alkylation in **Fmoc-based solid-phase peptide synthesis** s. C.G. Fields, G.B. Fields, Tetrahedron Letters *34*, 6661-4 (1993); suppression of diketopiperazine formation s. J. Alsina et al., ibid. *37*, 4195-8 (1996); temporary N(α)-deprotection/ reprotection procedure to facilitate purification s. Y. Nishiyama, Y. Okada, ibid. *35*, 7409-12 (1994); α-hydroxyglycine-extended peptides s. A.R. Brown, R. Ramage, ibid. 789-92; α-aminoglycine-extended peptides s. D.Q.L. René, B. Badet, ibid. *34*, 3861-2 (1993); phosphonopeptides s. D. Maffre-Lafon et al., ibid. *35*, 4097-8 (1994); resin esterification with a Fom-histidine deriv. s. Y.-F. Zhu et al., ibid. 4673-6; enhancement of peptide solvation with oxazolidine derivs. s. T. Wöhr, M. Mutter, ibid. *36*, 3847-8 (1995); stepwise automated solid-phase method with Fmoc-amino acid fluorides (cf. *46*, 350) s. H. Wenschuh et al., J. Org. Chem.

60, 405-10 (1995); solid-phase synthesis of head-to-tail **cyclic peptides** s. Tetrahedron Letters *35*, 9633-6 (1994); rapid method for small cyclic peptides s. L.S. Richter, ibid. 5547-50; cyclic peptide mixtures with an oxime resin s. H. Mihara et al., ibid. *36*, 4837-40 (1995); problems associated with side-chain s. M.-L. Valero et al., ibid. *37*, 4229-32 (1996); a simple multiple release system for **combinatorial synthesis of peptide libraries** s. M. Cardno, M. Bradley, ibid. 135-8; sonication-assisted release of hydrophobic peptides by the multipin method s. A.M. Bray et al., ibid. *35*, 9079-82 (1994); NMR method to identify compds. bound to a single resin bead s. S.K. Sarkar et al., J. Am. Chem. Soc. *118*, 2305-6 (1996); **solution-phase** parallel synthesis of chemical libraries s. D.L. Boger et al., ibid. 2567-73; on a peptidomimetic template s. ibid. 2109-10; small *non-peptide* uncoded libraries on Kemp's triacid scaffolding s. P. Kocis et al., Tetrahedron Letters *36*, 6623-6 (1995); on a cyclopentane scaffold cf. ibid. *35*, 9169-72 (1994).

Nickel/microwaves ←
Sec. ar. amines from prim. ar. amines and alcohols NH → NR

151.

Aniline, 3 eqs. *n*-propanol, and 0.1 eq. Raney nickel W-2 microwave irradiated in a sealed tube for 30 min using a 630 W microwave oven, and cooled to room temp. → product. Y 91%. The method is rapid and efficient, requiring only small amounts of catalyst. Substitution at the ortho position hinders the reaction. F.e.s. Y.L. Jiang et al., Synth. Commun. *26*, 161-4 (1996).

Bis[1,2-bis(diphenylphosphino)butane]nickel(0)/tetra-n-butylammonium ←
 hexafluorophosphate
Nickel-catalyzed allylation with O-allyl derivs. NH → N–C–C=C

152. Et$_2$NH + AcOCH$_2$CH=CH$_2$ ⟶ Et$_2$NCH$_2$CH=CH$_2$

of soft nucleophiles. A soln. of 1 mol% dppb and 2.5 mol% Bu$_4$NPF$_6$ in THF added under N$_2$ to 0.5 mol% Ni(cod)$_2$, stirred for 15 min, allyl acetate, 3 eqs. diethylamine, and heptane (as internal standard) added, and the soln. stirred at 50° until reaction complete by GC (14 min) → product. Conversion 100%; selectivity 100%. Conditions are mild and turnover frequencies high. F.e, promoters, and solvent effects, also allylation with allyl phenyl ether or allyl methyl carbonate, s. H. Bricout et al., Chem. Commun. *1995*, 1863-4.

Palladium-carbon s. under EtONH$_2$·HCl Pd-C

Tris(dibenzylideneacetone)dipalladium/chiral 1,2-bis[o-(diphenylphosphino)benzamido]- ←
 1,2-diphenylethane/triethylamine
Asym. nucleoside fusion synthesis with *meso*-2,5-diacoxy-2,5-dihydrofurans ←

153.

A 1.1:1 mixture of *cis*-2,5-dibenzoyloxy-2,5-dihydrofuran and 6-chloropurine in THF treated with 2 mol% (dba)$_3$Pd$_2$·CHCl$_3$ and 6 mol% (R,R)-1,2-bis[*o*-(diphenylphosphino)benzamido]-1,2-diphenylethane in the presence of Et$_3$N at room temp. → product. Y 85% (e.e. ca. 93%). F.e. and *ent*-nucleosides s. B.M. Trost, Z. Shi, J. Am. Chem. Soc. *118*, 3037-8 (1996).

Via intermediates v.i.
(E)-N-Sulfonylimines from oxo compds. via N-silylimines C=NSO$_2$R

154.

2.5 M *n*-BuLi in hexanes added dropwise to 1.05 eqs. hexamethyldisilazane at 0° under inert atmosphere, 0.1 eq. benzaldehyde added to the resulting soln. of LiN(SiMe$_3$)$_2$, and stirred at 0° for 30 min → N-(trimethylsilyl)benzaldimine, in chloroform refluxed with 1 eq. methanesulfonyl chloride for 1 h under a positive argon pressure → N-(methanesulfonyl)benzaldimine. Conversion 100%. This is an improvement on existing routes which require a large reagent excess, utilization of metals or extremely moisture-sensitive reagents, and which are limited to aldehyde derivs. Work-up is simple as the by-product (Me$_3$SiCl) is volatile. However, the method failed with enolizable substrates. F.e. and solvent effects s. G.I. Georg et al., J. Org. Chem. *60*, 7366-8 (1995).

Cytosines from uracils via tetrazolo[1,5-*c*]pyrimidin-5(6*H*)-ones ←

155.

A mixture of 1-methyluracil, 2 eqs. NaN$_3$, and 5 eqs. POCl$_3$ in acetonitrile refluxed for 10 h → 6-methyltetrazolo[1,5-*c*]pyrimidin-5(6*H*)-one (Y 95%), in methanol shaken with 10% Pd/C under 2 atm. H$_2$ for 10 h → 1-methylcytosine (Y 92%). Attempted reduction with NaBH$_4$ or triphenylphosphine proved fruitless. Reaction of nucleoside derivs. required protection of the hydroxyl groups (as O-acetyl). F.e.s. K. Ciszewski et al., Synthesis *1995* 777-9.

Nitrogen ↑   NC ⇅ N

Without additional reagents   *w.a.r.*
N-Acylation with polymer-based N-acylsulfonic acid amides   NH → NCOR
Activation by N-cyanomethylation

156.

Ar = 3,4,5-(MeO)$_3$C$_6$H$_2$

Acylamines. A soln. of startg. polymer-based N-acylsulfonamide in DMSO treated with iodoacetonitrile in the presence of *i*-Pr$_2$NEt → intermediate N-cyanomethyl deriv., treated with 0.007 *M* benzylamine in DMSO at room temp. (half-life <5 min) → product. Y high. Cleavage of the N-methyl analog was considerably slower (half-life ca. 790 min). Reaction is generally applicable to basic, non-basic and hindered amines, as well as α-amino-esters (with minimal racemization). However, reaction with less nucleophilic amines (e.g. aniline) requires heating. F.e.s. B.J. Backes et al., J. Am. Chem. Soc. *118*, 3055-6 (1996).

Formylation with 1-formylbenzotriazole   NH → NCHO

157.

N-Formylation. A soln. of 1.07 eqs. startg. amine added dropwise to a soln. of 1 eq. N-formylbenzotriazole in THF at 20°, and worked up after 0.15 h → product. Y 78%. Reaction is generally applicable to aliphatic and ar. prim. and sec. amines (incl. deactivated ar. amines). Stable, non-hygroscopic, N-formylbenzotriazole is more simply prepared, and may be handled with greater ease than N-formylimidazole. F.e., **also O-formylation** of alcohols (or phenols), s. A.R. Katritzky et al., Synthesis *1995*, 503-5.

N-Carbalkoxylation with cyclic N-carbalkoxythioureas   NH → NCOOR
using NaOH s. *41*, 367; with cyclic N,N'-dicarbalkoxythioureas s. N. Matsumura et al., J. Chem. Soc. Perkin Trans. I *1995*, 2953-4.

Triethylamine $\quad Et_3N$
Ureas by interchange $\quad NH \rightarrow NCON{<}$
with triethylenediamine cf. 22, 440; with Et$_3$N, limitations, s. K. Ramadas, N. Srinivasan, Org. Prep. Proc. Intern. 25, 600-1 (1993).

Ethyldiisopropylamine $\quad i\text{-}Pr_2NEt$
Guanidines from amines $\quad NH \rightarrow N\text{-}C(NH_2)\text{=}NH$

158.

Equimolar amounts of 40 wt% aq. dimethylamine, benzotriazole-1-carboxamidinium tosylate, and ethyldiisopropylamine diluted with DMF, and stirred at room temp. for 5 h → N,N'-dimethylamine-1-carboxamidinium tosylate. Y 69%. The method is mild and generally applicable to aliphatic and ar. prim. and sec. amines. The tosylate is stable and non-hygroscopic, reaction is faster and higher-yielding than that with pyrazole-1-carboxamidine hydrochloride, and work-up is simple. F.e. and conditions s. A.R. Katritzky et al., Synth. Commun. 25, 1173-86 (1995).

Phthalhydrazides from phthalimides $\quad$ D
s. 13, 472; polymer-based combinatorial synthesis of a small 1,4-phthalazinedione library s. J. Nielsen, P.H. Rasmussen, Tetrahedron Letters 37, 3351-4 (1996).

Silver benzoate/triethylamine $\quad AgOCOPh/Et_3N$
Arndt-Eistert synthesis of carboxylic acid amides $\quad COCHN_2 \rightarrow CH_2CON{<}$
with AgNO$_3$ cf. 2, 216; homopeptides with AgOBz/Et$_3$N s. J. Podlech, D. Seebach, Angew. Chem. Intern. Ed. 34, 471-2 (1995).

Sodium tetrahydridoborate s. under $(CF_3SO_2)_2O$ $\quad NaBH_4$

Boron fluoride $\quad BF_3$
Cyclic iminoesters from azidoalcohols and aldehydes
with H$_2$SO$_4$ cf. 11, 489; with BF$_3$, also from siloxyazides (with Me$_3$SiOTf), s. J.G. Badiang, J. Aubé, J. Org. Chem. 61, 2484-7 (1996).

Boron fluoride/sodium hydrogen carbonate $\quad BF_3/NaHCO_3$
N-ω-Hydroxylactams from cyclic ketones and azidoalcohols via bicyclic iminoester salts

159.

Lewis acid-catalyzed ring expansion of cyclic ketones via azide insertion (Schmidt reaction) is facilitated by using an azidoalcohol, the initially formed hemiketal dehydrating to a reactive oxonium ion which provides the driving force for intramolecular attack by the azido function prior to rearrangement. **E:** A soln. of cyclohexanone and 1.2 eqs. 3-azido-1-propanol in methylene chloride cooled to 0°, 2 eqs. BF$_3$-etherate added dropwise over 5 min (evolution of gas), allowed to warm to room temp. over 30 min, stirring continued for 3 h, concentrated, treated with satd. aq. NaHCO$_3$, and stirred for 30 min → 1-(3'-hydroxypropyl)hexahydroazepin-2-one. Y 90%. F.e., regioselectivity, and isolation of the intermediate iminoester salts, **also with asym. induction** using chiral azidoalcohols, s. V. Gracias et al., J. Am. Chem. Soc. 117, 8047-8 (1995).

Titanocene dichloride Cp_2TiCl_2
Preferential N-acylation of prim. amines with 3-acyl-2-oxazolidones NH → NCOR

160.   Ph-C(=O)-N(oxazolidone) + H$_2$N-CH(CH$_3$)-CH$_2$-NH$_2$ ⟶ Ph-C(=O)-NH-CH$_2$-CH(CH$_3$)-NH$_2$

1.1 eqs. Startg. diamine added at room temp. to a well-stirred soln. of startg. N-acyl-2-oxazolidone in THF containing 20 mol% Cp$_2$TiCl$_2$, and stirred for 8 h → N-(2-aminopropyl)benzamide. Y 83% (85% purity). N-Acylation of α-branched prim. amines is significantly reduced, although the latter could be acylated in high yield with 3-(*p*-trifluoromethylbenzoyl)-2-oxazolidone (and Cp$_2$ZrCl$_2$). Prim. ar. amines, alcohols and acetals were unaffected. F.e. and **with kinetic resolution** s. T. Yokomatsu et al., J. Org. Chem. *59*, 3506-8 (1994).

Tri-n-butylphosphine Bu_3P
Carbo-*tert*-butoxyamines from azides under mild conditions N$_3$ → NHCOOBu-*t*

161.   BnOCOCH$_2$N$_3$ $\xrightarrow{Bu_3P}$ [BnOCOCH$_2$N=PBu$_3$] $\xrightarrow{Boc_2O}$ BnOCOCH$_2$NHBoc

1.1 eqs. Bu$_3$P added dropwise to a stirred soln. of startg. azide in anhydrous ether at room temp. under argon, cooled to -50° when evolution of gas had ceased (1 h), 1.1 eqs. di-*tert*-butyl dicarbonate in the same solvent added dropwise via cannula, stirred for 1 h, quenched with satd. aq. NaHCO$_3$, the cooling bath removed, and the mixture allowed to warm to room temp. → product. Y 81%. More hindered azides required a higher temp. F.e.s. C.A.M. Afonso, Tetrahedron Letters *36*, 8857-8 (1995).

Trifluoromethanesulfonic acid/sodium tetrahydridoborate $(CF_3SO_2)_2O/NaBH_4$
Amines by intermolecular Schmidt reaction of azides with alcohols ←

162.   cyclopentanol-OH + n-BuN$_3$ ⟶ [cyclopentyl-N(Bu)-N$_2^+$] ⟶ [tetrahydropyridinium-Bu] ⟶ N-Bu piperidine

Aliphatic azides undergo intermolecular Schmidt reaction with certain carbocations to form amines via immonium salts (the regioselectivity being difficult to predict). **E: Piperidines from cyclopentanols.** Startg. cyclopentanol treated with *n*-butyl azide and 2.5 eqs. TfOH in methylene chloride, followed by reduction with NaBH$_4$ in methanol → product. Y 95%. F.e. and reaction with benzylic or propargylic alcohols, also details (in Supporting Information), s. W.H. Pearson, W. Fang, J. Org. Chem. *60*, 4960-1 (1995).

Vanadium phosphomolybdate/oxygen ←
Benzyltriethylammonium tetrathiomolybdate $(BnNEt_3)_2MoS_4$
Reduction of azides 2 CHN$_3$ → C=N–CH

163.   2 *p*-MeOC$_6$H$_4$CH$_2$N$_3$ ⟶ *p*-MeOC$_6$H$_4$CH$_2$N=CHC$_6$H$_4$OMe-*p*

Azomethines from two azide molecules. A soln. of 2 eqs. *p*-methoxybenzyl azide in acetonitrile added in one portion to 1.1 eqs. (PhCH$_2$NEt$_3$)$_2$MoS$_4$ in aq. acetonitrile, and the mixture stirred for 20 h → *p*-methoxybenzaldehyde N-(*p*-methoxybenzyl)imine. Y 80%. The method is mild, simple, efficient and high-yielding. F.e. incl. cyclic imines from diazides, and prim. ar. amines from ar. azides, also sulfonic and carboxylic acid amides from their respective azides (nitro groups and aldehydes being unaffected), s. A.R. Ramesha et al., J. Org. Chem. *60*, 7682-3 (1995); ar. N-benzylaldimines with vanadium-based phosphomolybdate under O$_2$ cf. Y. Ishii et al., Chem. Lett. *1993*, 1699-702.

Rhodium(II) acetate $Rh(OAc)_2$
Oxazoles from α-diazocarbonyl compds.
and nitriles s. *23*, 441s*48*; 2-*tert*-aminooxazoles from N,N-disubst. cyanamides s. K. Fukushima, T. Ibata, Heterocycles *40*, 149-54 (1995); oxazoles with Rh(II)-trifluoroacetamide cf. K.J. Doyle, C.J. Moody, Synthesis *1994*, 1021-2.

Via intermediates
15N-Labelled nucleosides by N-nitration-recyclization

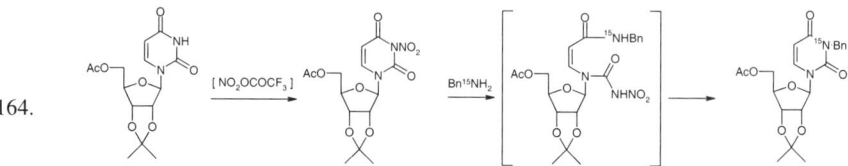

164.

4 eqs. Trifluoroacetic anhydride added to a suspension of 2 eqs. finely powdered NH$_4$NO$_3$ in anhydrous methylene chloride at 0°, vigorously stirred at room temp. until homogeneous (ca. 1 h), re-cooled to 0°, 1 eq. startg. nucleoside added, and stirring continued until reaction complete by TLC (20-60 min) → 5'-O-acetyl-2',3'-O-isopropylidene-3-nitrouridine (Y 94%), in acetonitrile treated at room temp. with ca. 1 eq. ^{15}N-labelled benzylamine and 1 eq. anhydrous K$_2$CO$_3$, allowed to react for 20 h, heated at 80° for 3 h, and poured into phosphate buffer (pH 7) → 3-[^{15}N]-5'-O-acetyl-3-benzyl-2',3'-O-isopropylideneuridine (Y good). F.e. incl. ^{15}N-labelled purine nucleosides, also **^{15}N-labelled N-aminonucleosides** with doubly ^{15}N-labelled hydrazine, and characterization of the intermediate in the 2nd step, s. X. Ariza et al., J. Am. Chem. Soc. *117*, 3665-73 (1995).

Halogen ↑   NC ⇅ Hal

Without additional reagents   *w.a.r.*
N-Substitution of piperazines   NH → NR
s. 6, 465; polymer-based combinatorial synthesis of N-aryl- and N-benzyl-piperazines s. S.M. Dankwardt et al., Tetrahedron Letters *36*, 4923-6 (1995).

**N'-Sulfonylamidines from azomethines
and N,N-dichlorosulfonic acid amides**   C(N≤)═NSO$_2$R

165.   NHPr-i
PhCH=NPr-i + Cl$_2$NSO$_2$Ph ⟶ PhC=NSO$_2$Ph [+ Cl$_2$]

Equivalent amounts of N,N-dichlorobenzenesulfonamide and startg. azomethine in carbon tetrachloride heated for 2 h until evolution of Cl$_2$ ceased → N-isopropyl-N'-phenylsulfonyl-benzamidine. Y 71%. F.e.s. O.V. Pigin et al., Zh. Org. Khim. *31*, 631-2 (1995) (Russ.).

Irradiation s. under Hg   ∭
Electrolysis/tetraethyl ethylenetetracarboxylate   ⚡/(EtO$_2$C)$_2$C═C(CO$_2$Et)$_2$
2,4-Oxazolidiones from α-halogenocarboxylic acid amides s. *51*, 169   ◯

Sodium tert-butoxide s. under Pd catalysts   NaOBu-t

Sodium carbonate   Na$_2$CO$_3$
N^2-(Carbo-9-fluorenylmethoxy)- from N^1-(carbo-*tert*-butoxy)-diamines ←
via N^1-(carbo-*tert*-butoxy)-N^2-(carbo-9-fluorenylmethoxy)diamines

166. BocNHCH$_2$CH$_2$CH$_2$CH$_2$NH$_2$ —Fmoc-Cl→ BocNHCH$_2$CH$_2$CH$_2$CH$_2$NHFmoc —CF$_3$COOH→ CF$_3$COOH · H$_2$NCH$_2$CH$_2$CH$_2$CH$_2$NHFmoc

A soln. of N-(*tert*-butoxycarbonyl)butane-1,4-diamine in THF (1 part) added to a soln. of ca. 3 eqs. Na$_2$CO$_3$ in water (1 part), ca. 1.1 eqs. 9-fluorenylmethyl chloroformate added, and stirred at room temp. for 12 h → intermediate N^1-Boc-N^2-Fmoc-diamine (Y 79%), added to trifluoroacetic acid with stirring, and stirred at room temp. for 1 h → N-(9-fluorenylmethoxycarbonyl)butane-1,4-diamine trifluoroacetate (Y 94%). The products are stable, easily-isolated, crystalline solids which should find widespread application as linkers in bioconjugation chemistry. F.e.s. M. Adamczyk, J. Grote, Org. Prep. Proc. Intern. *27*, 239-42 (1995).

Potassium thiocyanate   *KSCN*
Glycosyl isothiocyanates from glycosyl bromides   Br → N═C═S
with added Bu$_4$NHSO$_4$ cf. *40*, 266; 1,2-*trans*-glycosyl isothiocyanates in the absence of solvent s. T.K. Lindhorst, C. Kieburg, Synthesis *1995*, 1228-30.

Lithium chloride s. under Et₃N   *LiCl*
Potassium salt/tetra-n-butylammonium bromide K^+/Bu_4NBr
Phthalimides from halides   ←
s. *15*, 626; under solid-liq. phase transfer catalysis with Bu₄NBr or KI (from chlorides) s. C. Duboisclard-Gottardi, Y. Fort, Synth. Commun. *25*, 3173-80 (1995).

Triethylamine/lithium chloride $Et_3N/LiCl$
N-Acylation under mild conditions   NH → NCOR

167.

of 2-oxazolidones. 1.2 eqs. Acryloyl chloride added at -20° to a soln. of 1.3 eqs. acrylic acid and 2.5 eqs. Et₃N in THF at -20°, the mixture stirred at ca. -20° for 1 h, *1.1 eqs. LiCl* and 1 eq. startg. 2-oxazolidone added sequentially, allowed to warm to room temp., stirred for 4 h, and quenched with 0.2 *N* HCl → 3-propenoyl-2-oxazolidone. Y 84%. The method is notably applicable to acryloyl derivs. which are difficult to prepare by existing methods. F.e. incl. **N-acylation of sultams** s. G.-J. Ho, D.J. Mathre, J. Org. Chem. *60*, 2271-3 (1995); N-acylation with long-chain acyl chlorides s. G. Liu et al., Synth. Commun. *25*, 3247-53 (1995).

Ethyldiisopropylamine $i\text{-}Pr_2NEt$
N-Cyanomethylation s. *51*, 156   NH → NCH₂CN

Cupric acetoacetonate or Cupric triflate/chiral diamine   *Cu(II)*
N-Sulfonylaziridines from ethylene derivs.
with CuOTf cf. *46*, 355; regio- and stereo-specific conversion of 1,3-dienes with Cu(acac)₂ s. J.G. Knight, M.P. Muldowney, Synlett *1995*, 949-51; with Cu(OTf)₂, asym. induction with axially dissymmetric chiral diamines and diimines s. M. Shi et al., J. Chem. Res. (S) *1996*, 352-3.

Cuprous iodide s. under Pd(OAc)₂   *CuI*

Mercury/irradiation   Hg/*hν*
Partial reductive photoamination of perfluoroalkanes   ←

168.

Purified perfluoroisohexane irradiated in the presence of a drop of Hg at 40° in a quartz tube for 19 h with 254 nm light from a *low-pressure* Hg lamp (Rayonet RMR-600, 32 W) under a flow of NH₃ gas (2 ml/min) → product. Y 95%. The involatility of the latter protects it from overreduction, so that functionalization is controllable. Reaction is thought to proceed by initial elimination of fluoride ion following electron transfer (rather than abstraction of HF by H-atoms). F.e.s. J. Burdeniuc et al., J. Am. Chem. Soc. *117*, 10119-20 (1995).

Superoxide ion O_2^-
Activation of carbon dioxide by electrochemically-generated superoxide   ←
2,4-Oxazolidiones from α-halogenocarboxylic acid amides

169.

A 0.1 *M* soln. of Et₄NClO₄ in DMF electrolyzed at a Hg-cathode while bubbling O₂ and CO₂ into the mixture until 1.5 F/mol consumed at -1.0 V, N₂ introduced for 5 min, startg. amide added, and stirred overnight at room temp. → product. Y 82%. F.e. and without prior generation of superoxide, also from α-tosyloxyamides, s. M.A. Casadei et al., J. Org. Chem. *61*, 380-3 (1996); with tetraethyl ethylenetetracarboxylate cf. Tetrahedron *51*, 5891-900 (1995).

Selenium/sodium o-nitrobenzenesulfonamide ←
2-Ethylenesulfonylamines from ethylene derivs.   C=C-C(NSO$_2$R)
with Chloramine-T/Se cf. *31*, 426; with *soluble* N,N-dichloro-*o*-nitrobenzenesulfonamide/Na-*o*-nitrobenzenesulfonamide/Se s. K.B. Sharpless et al., Angew. Chem. Intern. Ed. *35*, 454-6 (1996).

Palladous acetate/cuprous iodide/sodium carbonate/hexadecyltrimethylammonium bromide ←
Triarylamines from diarylamines and ar. iodides   NH → NAr
in an aq. organic emulsion

170.   Ph$_2$NH + PhI ⟶ Ph$_2$NPh

5 eqs. Na$_2$CO$_3$ and hexadecyltrimethylammonium bromide in 9:1 water/*n*-butanol stirred at 100° until a microemulsion had formed, 2.5 eqs. diphenylamine, 3 eqs. iodobenzene, 0.05 eq. CuI, and 0.025 eq. Pd(OAc)$_2$ added with continued stirring under N$_2$, heated for 24 h, and neutralized with dil. HCl → triphenylamine. Y 90%. Significantly, no high-boiling solvent is required. F.e.s. D.V. Davydov, I.P. Beletskaya, Izv. Akad. Nauk Ser. Khim. *1995*, 1181-2 (Russ.).

Tris(dibenzylideneacetone)dipalladium(0)/tri-o-tolylphosphine/sodium tert-*butoxide* ←
Dichlorobis(tri-o-tolylphosphine)palladium(II)/sodium tert-*butoxide/potassium carbonate* ←
Ar. amines from halides

171.

A mixture of 2-bromonaphthalene, 1.2 eqs. N-methylhomoveratrylamine, 1.4 eqs. NaOBu-*t*, and 2 mol% PdCl$_2$[P(Tol-*o*)$_3$]$_2$ in toluene heated under argon at 100° for 3 h → product. Y 78%. The method is simple, avoids the use of tin or boron reagents, and is generally applicable to ar. bromides bearing electron-withdrawing or -donating substituents or acid-sensitive groups. F.e. and 5-7-membered N-heterocyclics by intramolecular N-arylation with bromides or [preferably] iodides using Pd(PPh$_3$)$_4$/NaOBu-*t* s. A.S. Guram et al., Angew. Chem. Intern. Ed. *34*, 1348-50 (1995); **from ar. iodides** with Pd$_2$(dba)$_3$/P(Tol-*o*)$_3$/NaOBu-*t* cf. J. Org. Chem. *61*, 1133-5 (1996); tert. ar. amines with LiN(SiMe$_3$)$_2$ as base s. J. Louie, J.F. Hartwig, Tetrahedron Letters *36*, 3609-12 (1995).

Via intermediates   *v.i.*
Polymer-based N-alkylation of amines using a handle-less support   NH → NR
Regeneration of the support by Hofmann elimination

172.

The resin-bound acrylate (prepared from hydroxymethyl polystyrene resin and acryloyl chloride) swollen in a mixture of 1,2,3,4-tetrahydroisoquinoline and DMF in a polystyrene tube, the tube agitated (on a rotator) for 18 h at 20°, the Michael adduct-resin washed with DMF, dichloromethane and methanol, dried *in vacuo*, suspended in a soln. of 0.5 eq. allyl bromide in DMF, agitated for 18 h at 20°, washed again with DMF, dichloromethane and methanol, dried *in vacuo*, suspended in DMF containing *i*-Pr$_2$NEt, and agitated for 18 h at 20° → N-allyl-1,2,3,4-tetrahydroisoquinoline. Y 97% (overall). Significantly, the resin is regenerated in the final stage so that no additional functional group linkage is required. The method is mild, simple, general (for prim. and sec. amines), and potentially applicable to Michael addition of other nucleophiles. F.e.s. J.R. Morphy et al., Tetrahedron Letters *37*, 3209-12 (1996).

2-Ethylenetosylamines from β,γ-ethylenehalides   C=C—C-Hal → C(NHTs)—C=C
via 2-ethylenetellurides s. *51, 174*

Sulfur ↑ NC ↑↓ S

Without additional reagents w.a.r.
Thioureas from bisthiuram disulfides and amines NH → NC(S)N<
s. *21*, 504; also cyclic thioureas **and from bisthiouram sulfides** s. K. Ramadas, N. Srinivasan, Synth. Commun. *25*, 3381-7 (1995).

Cupric sulfate-silica gel/triethylamine $CuSO_4$-SiO_2/Et_3N
Guanidines from thioureas >N)$_2$CS → >N)$_2$C=N—
with $CuSO_4$ s. *4*, 443; with $CuSO_4$-SiO_2/Et_3N s. K. Ramadas, N. Srinivasan, Tetrahedron Letters *36*, 2841-4 (1995).

Boron fluoride BF_3
3-Tosyl-3,6-dihydro-2H-1,3-oxazines from β-hydroxyaldehydes and aldehydes

173.

n-Butyraldehyde added to a soln. of 1 eq. N-sulfinyl-p-toluenesulfonamide in dry dichloromethane, stirred under argon for 1 h at room temp., cooled to -20°, 1.44 eqs. BF_3-etherate and 2.05 eqs. 2-ethyl-3-hydroxyhexanal added successively, stirred at -20° for 2 h, and treated with satd. $NaHCO_3$ soln. → 3,6-dihydro-2,6-dipropyl-5-ethyl-3-tosyl-2H-1,3-oxazine. Y 83%. Only the 2,6-*trans*-isomer was formed. F.e., also conversion to 3-(tosylamino)alcohols by hydrogenation followed by acid hydrolysis, and dihydroxylation, s. S.M. Weinreb et al., Synlett *1995*, 527-8.

Remaining Elements ↑ NC ↑↓ Rem

Without additional reagents w.a.r.
Sec. amines from azides and dialkylhalogenoboranes N_3 → NHR
s. *27*, 485; sec. methylamines from Me_2BBr with retention of configuration s. R.L. Dorow, D.E. Gingrich, J. Org. Chem. *60*, 4986-7 (1995).

2-Ethylenetosylamines from β,γ-ethylenehalides ←
via *in situ*-generated 2-ethylenetellurides

174.

Ethanol added via syringe to a mixture of 0.5 eq. diphenyl ditelluride and 1 eq. $NaBH_4$ at 0° under N_2, stirred at 25° for 0.5 h, a soln. of cinnamyl bromide in ethanol added via syringe, stirred for 1-2 h, 1.5 eqs. phenyl(tosylimino)iodinane added portionwise, and stirred at 25° for 20 h → product. Y 85% (and 10% 1-phenylprop-2-en-1-ol). The allyl alcohol was the major product (63%) after 1 h. Reaction proceeds by the novel imination of the intermediate telluride, followed by **[2.3]-sigmatropic rearrangement**. F.e. **and with asym. induction** (using a chiral ditelluride) s. S. Uemura et al., Tetrahedron Letters *36*, 6725-8 (1995); **from 2-ethyleneselenides with asym. induction** s. Chem. Commun. *1995*, 1243-4.

n-Butyllithium *BuLi*
Electrophilic carbalkoxylation RM → RNHCOOR'
with *tert*-butyl N-tosyloxycarbamate cf. *46*, 377; with crystalline allyl N-[(arylsulfonyl)oxy]-carbamates s. J. Org. Chem. *60*, 7010-2 (1995).

N-Silylimines from oxo compds. s. *51*, 154 CO → C=NSi<

Zinc iodide ZnI_2
Aziridines from epoxides and phosphine imines

175. Ph–⟨O⟩ + i-PrN=PPh₃ ⟶ Ph–⟨NPr-i⟩

1.2 eqs. Startg. phosphine imine and 10 mol% ZnI_2 added to a soln. of startg. epoxide in 1,2-dichloroethane, and stirred for 1 h at 80° → product. Y 65%. The reaction is generally applicable to terminal and cyclic epoxides, internal epoxides being less reactive. Chirality is lost on prolonged reaction of optically active epoxides as a result of Zn(II)-induced *cis→trans*-rearrangement. F.e. and Zn-salts s. D. Kühnau et al., J. Chem. Soc. Perkin Trans. I *1996*, 1167-70.

Stannic chloride $SnCl_4$
Vorbrüggen nucleoside synthesis ←

s. *26*, 446s*46-9*; 2'-deoxyribonucleosides with O^2-*m*-trifluoromethylbenzoyl as directing group s. M. Park, C.J. Rizzo, J. Org. Chem. *61*, 6092-3 (1996); pyrimidine 2'-deoxyribonucleosides with various catalysts (e.g. $SnCl_4·2AgOTf$, $SiCl_4·2AgOTf$, or $SbCl_5·2AgClO_4$) s. T. Mukaiyama et al., Chem. Lett. *1996*, 99-100; also purine 2-deoxyribonucleosides from a protected *methyl 2-deoxyriboside* cf. S. Janardhanam et al., Tetrahedron Letters *35*, 3657-60 (1994); β-D- and β-L-dideoxycytidines s. A. Tse, T.S. Mansour, ibid. *36*, 7807-10 (1995); nucleoside 5'-deoxy-5'-difluoromethylphosphonates s. J. Matulic-Adamic, N. Usman, ibid. 3227-30; 1,3-oxathiolane nucleosides s. W. Wang et al., ibid. 4739-42; 6,7-dichloroimidazo[4,5-*b*]quinolin-2-one nucleosides s. Z. Zhu, L.B. Townsend, ibid. *37*, 3263-6 (1996); thienouracil analogs s. F. Jourdan et al., J. Heterocyc. Chem. *32*, 953-7 (1995); S^2-alkyl-2-thiouridines s. E.B. Pedersen et al., Synthesis *1996*, 237-41; s.a. Heterocycles *41*, 2507-18 (1995).

Phosphorus oxide chloride/dimethylformamide $POCl_3/Me_2NCHO$
Peptide synthesis with silyl esters under neutral conditions NSi≡ → NCOR

176. F₃C–C(O)–NH–CH(Pr-i)–C(O)–OSiMe₃ + Me₃SiNH–CH(Pr-i)–C(O)–OMe ⟶ F₃C–C(O)–NH–CH(Pr-i)–C(O)–NH–CH(Pr-i)–C(O)–OMe

A soln. of trimethylsilyl N-trifluoroacetyl-L-valinate and 1 eq. Vilsmeier reagent (prepared from 1:1 $POCl_3$/DMF) in anhydrous dichloromethane stirred under N_2 for 40 min at -20°, a soln. of 1 eq. methyl N-trimethylsilyl-L-valinate in the same solvent added, and stirred for 16 h → methyl N-trifluoroacetyl-L-valyl-L-valinate. Y 90% (D,O 1.3%). There was little racemization under these conditions. F.e.s. A. Benouargha et al., Bull. Soc. Chim. France *132*, 824-8 (1995).

Trimethylsilyl triflate $Me_3SiOSO_2CF_3$
N-Glycosyl-N-heterocyclics from siloxy-N-heterocyclics and unprotected aldoses ←
Improved Vorbrüggen nucleoside synthesis

177.

The Vorbrüggen nucleoside synthesis (cf. *26*, 446) can now be performed more simply and conveniently with *unprotected* aldoses **via per-O-silylation E:** A mixture of *3 eqs.* dry D-ribose and 6-azauracil silylated with hexamethyldisilazane and a little Me_3SiCl for 2 h in boiling abs. MeCN, evaporated *in vacuo*, the residue kept for 2 h at 80°/1 mm, dissolved in boiling abs. MeCN, treated with 1.1 eqs. Me_3SiOTf in MeCN within 45 min at 80°, after 7 h aq. $NaHCO_3$ added at 24°, stirred for 4 h, evaporated, extracted with boiling methanol (to remove silyl groups), and chromatographed on SiO_2 → 6-azauridine. Y 68.3%. Reaction failed with thymine, and yields were lower with $SnCl_4$. F.e.s. H. Vorbrüggen et al., Tetrahedron Letters *36*, 7845-8 (1995); 1,2-*cis*-nucleosides **from 3-methoxy-2-pyridyl glycosides** or glycosyl S-2-pyridyl thiolcarbonates (with AgOTf) cf. S. Hanessian et al., ibid. 5865-8; details s. Tetrahedron *52*, 10827-34 (1996).

Nitridomanganese(V) salen complex
α-(Trifluoroacetylamino)ketones from enoxysilanes ← COC(NHCOR)
Nitrogen atom transfer from nitridomanganese(V) complexes

178.

3 eqs. Pyridine and a soln. of startg. enoxysilane in methylene chloride added sequentially to 2 eqs. nitridomanganese(V) salen complex in the same solvent at -78°, treated dropwise with 2.4 eqs. freshly distilled trifluoroacetic anhydride, allowed to warm slowly to 23° over 3-4 h, silica gel and Celite added with *n*-pentane, and the slurry vigorously stirred at 23° for 30 min → product. Y 78%. Conditions are mild and large quantities of the complex (which is stable to both air and water) can be prepared in high yield. F.e.s. E.M. Carreira et al., J. Am. Chem. Soc. *118*, 915-6 (1996).

Carbon ↑   NC ↕ C

Irradiation s. under TrSNO ⫽

Microwaves ←
Dicarboxylic acid imides from their anhydrides and isothiocyanates ←
under microwave irradiation

179.

A mixture of phenyl isothiocyanate and 1 eq. *cis*-1,2-cyclohexanedicarboxylic anhydride in N,N-dimethylacetamide contained in a covered beaker irradiated in a 2450 MHz microwave oven at 210 W for *2 min*, left to cool to room temp. during 5 min, and irradiated again for a further 2 min → N-phenylhexahydrophthalimide. Y 85%. The procedure is rapid, convenient and high-yielding, and work-up is simple. F.e. incl. N-subst. phthalimides, also with isocyanates, s. M.S. Khajavi et al., J. Chem. Res. (S) *1996*, 96-7; benzimidazoles from *o*-diamines and β-ketoesters on montmorillonite s. K. Bougrin, M. Soufiaoui, Tetrahedron Letters *36*, 3683-6 (1995).

Montmorillonite/microwaves ←
Benzimidazoles from *o*-diamines and β-ketocarboxylic acid esters s. *51*, 179   ○

p-Nitrobenzenesulfonyl azide/1,8-diazabicyclo[5.4.0]undec-7-ene   ArSO$_2$N$_3$/DBU
α-Diazocarbonyl compds. from carbonyl compds.   CHCOR → C=N$_2$
α-diazoketones via α-ethoxalylation cf. *30*, 247; s.a. V.L. Gein et al., Zh. Obshch. Khim. *65*, 1378-80 (1995) (Russ.); α-diazoesters via α-benzoylation using *p*-NO$_2$C$_6$H$_4$SO$_2$N$_3$/DBU s. D.F. Taber et al., J. Org. Chem. *60*, 1093-4 (1995).

Trityl thionitrite/irradiation   TrSNO/⫽
Sym. azo N,N'-dioxides [nitroso compd. dimers]   RN(O)=N(O)R
from 1-acoxy-2-pyridinethiones

180.

Startg. octadecanoic acid deriv. and *1 eq.* trityl thionitrite in degassed benzene irradiated under

N_2 at room temp. with visible light from a 250 W tungsten lamp → *trans*-product. Y 54.7%. Reaction takes place by a radical-chain mechanism with **initiation by sulfenyl radicals**, generated by homolysis of the thionitrite. Conditions are mild and neutral, and reagents inexpensive; however, the method is not applicable to derivs. of tert. carboxylic acids. F.e. and with retention of esters and keto functions s. W.B. Motherwell et al., Chem. Commun. *1995*, 2385-6.

Elimination ⇑

Hydrogen ↑ NC ⇑ H

Monoamine oxidase ←
Azomethines from sec. amines under mild conditions CHNH → C=N
Enzymatic oxidation in organic medium containing a low concentration of water

181.   PhCH$_2$NHMe ⟶ PhCH=NMe

A mixture of N-methylbenzylamine in *n*-octane containing *1 v/v% water* and monoamine oxidase B sonicated for 10 sec in an ultrasonic cleaning bath, and incubated at 25° for 1.5 h → product. Y 99%. As little as 0.1% water was required for an effective conversion. F.e. and solvents s. J.C.G. Woo et al., J. Org. Chem. *60*, 6235-6 (1995).

Oxalyl chloride/dimethyl sulfoxide/triethylamine ←
Azomethines from sec. amines
s. *43*, 918; Δ^1-azirine-3-carboxylic acid esters s. B. Zwanenburg et al., Tetrahedron Letters *36*, 4665-8 (1995); pyrrolo[2,1-*c*][1,4]benzodiazepines s. A. Kamel, N.V. Rao, Chem. Commun. *1996*, 385-6.

Nitrogen monoxide   NO
Pyridines from 1,4-dihydropyridines ←
with HNO$_2$ cf. *18*, 534; pyridine-3,5-dicarboxylic acid esters with NO or NO/O$_2$ s. A. Ohsawa et al., Tetrahedron Letters *36*, 2269-72 (1995).

Selenium dioxide   SeO$_2$
5,6-Dihydro-1,4-oxazin-2-ones from Δ^2-oxazolines   ○
Oxidative ring expansion

182. <chemical structures>

A soln. of (4R)-2-(2,2-dimethylpropyl)-4-phenyl-4,5-dihydrooxazole (readily prepared from 3,3-dimethylbutyric acid and (R)-2-phenylglycinol) in dry dioxane added to a suspension of ca. 2 eqs. SeO$_2$ in the same solvent, and heated at reflux for 2 h → (5R)-3-*tert*-butyl-5-phenyl-5,6-dihydro-2*H*-1,4-oxazin-2-one. Y 84%. This has been applied as the key stage in a multistep **asym. α-amination of carboxylic acids**. F.e.s. C.M. Shafer, T.F. Molinski, J. Org. Chem. *61*, 2044-50 (1996).

Palladous acetate/oxygen   Pd(OAc)$_2$/O$_2$
Regio- and stereo-specific oxidative ring closure of N-protected ethylamines s. *51*, 110 ○

Oxygen ↑ NC ⇑ O

Potassium hydroxide   KOH
2,4-Quinazolinediones from *o*-ureidocarboxylic acid esters   ○
s. *18*, 544; polymer-based synthesis in ethanolic KOH s. B.O. Buckman, R. Mohan, Tetrahedron Letters *37*, 4439-42 (1996).

Potassium tert-*butoxide* KOBu-t
Cyclic amines from phosphinyloxylamines
Intramolecular nucleophilic substitution at nitrogen

183.

Prolinates. The first instance of intramolecular carbanionic substitution *at nitrogen* is reported.
E: Startg. phosphinyloxylamine treated with KOBu-*t* until reaction complete → product. Y 95%.
N-Benzyl derivs. reacted similarly, but poor yields were obtained with N-allyl and N-α-methyl-benzyl derivs. due to competing elimination. F.e. and with LDA s. T. Sheradsky, L. Yusupova, Tetrahedron Letters *36*, 7701-4 (1995).

*Lithium N-(*p-*methoxyphenyl)acetimidate* ←
Multi-step polymer-based synthesis of 1,4-benzodiazepine-2,5-diones

184.

1,4-Benzodiazepine-2,5-diones are obtained by an efficient multi-step, polymer-based procedure from anthranilic acids, α-amino esters and alkylating agents with minimal (<1%) racemization.
E: Startg. anthranilic acid deriv. (prepared in three polymer-based coupling steps) in 1:1 THF/DMF treated with Li-N-(*p*-methoxyphenyl)acetimidate for 30 h at room temp., quenched with ethyl halide, and the resin removed by treatment with 90:5:5 TFA/Me$_2$S/water → product. Y 75%.
The choice of base was critical in order to avoid overalkylation of amide, carbamate or ester functions. F.e. and with Li-phenylacetimidate s. C.G. Boojamra et al., J. Org. Chem. *60*, 5742-3 (1995); **by polymer-based intramolecular aza-Wittig synthesis** (using Bu$_3$P) cf. D.A. Goff, R.N. Zuckermann, ibid. 5744-5.

Triethylamine Et$_3$N
Hydantoins from α-(carbalkoxyamino)carboxylic acid amides
with Bu$_4$NF cf. *31*, 452 (*9*, 568); polymer-based ring closure (with Et$_3$N) s. B.A. Dressman et al., Tetrahedron Letters *37*, 937-40 (1996).

Zinc/acetic anhydride/acetic acid Zn/Ac$_2$O/AcOH
Cyclic hydroxamic acids from nitrocarboxylic acids
with Zn/NH$_4$Cl cf. *20*, 387s47; with Zn/Ac$_2$O/AcOH s. A. Thomas, S. Rajappa, Tetrahedron *51*, 10571-80 (1995); 5-allyl-1-hydroxy-2-pyrrolidone-5-carboxylic acid derivs. s. P. Chittari et al., Tetrahedron Letters *35*, 3793-6 (1994).

Envirocat EPZG ←
Nitriles from aldoximes CH=NOH → CN
with TiCl$_4$/py cf. *27*, 513; with Envirocat EPZG in the absence of solvent s. B.P. Bandgar et al., Synth. Commun. *25*, 2993-8 (1995).

B-Chlorocatecholborane/triethylamine ←
Isocyanates from urethans under mild conditions NHCOOR → N=C=O

185. p-TolNHCOOMe ⟶ p-TolN=C=O

A soln. of startg. urethan in dry toluene treated with 1.2 eqs. Et$_3$N, refluxed for 5 min under N$_2$, 1.2 eqs. B-chlorocatecholborane added, and refluxed for a further 5 min → product. Y 99%. The reagent effectively traps the liberated methanol, thereby preventing the reverse reaction. The method is generally applicable to alkyl, ar., alicyclic and di-isocyanates. F.e.s. V.L.K. Valli, H. Alper, J. Org. Chem. *60*, 257-8 (1995).

Thiourea dioxide/triethylamine $(H_2N)_2CSO_2/Et_3N$
Pyrroles from γ-nitroketones under mild conditions

186.

Pyrrole-2-carboxylic acid esters. A mixture of startg. nitroketone, 4 eqs. formamidinesulfinic acid, and 1 eq. Et$_3$N in isopropanol heated to reflux under inert atmosphere (7-10 h) → product. Y 85%. An electron-withdrawing group (e.g. CO$_2$Et) adjacent to nitro appears essential. F.e. and isolation of the intermediate oximes s. B. Quiclet-Sire et al., Tetrahedron Letters *36*, 9469-70 (1995).

Tris(dimethylamino)phosphine/carbon tetrachloride $(Me_2N)_3P/CCl_4$
2-Azetidinones from β-hydroxycarboxylic acid amides
with N,N'-sulfonyldiimidazole cf. *41*, 427; with (Me$_2$N)$_3$P/CCl$_4$ (cf. *35*, 312), 3-hydroxymethyl derivs., s. C. Selve et al., Chem. Commun. *1995*, 1279-80.

Antimony(III) ethoxide $Sb(OEt)_3$
Regiospecific macrolactamization of tetraaminocarboxylic acid esters

187.

A soln. of startg. ester in dry, freshly distilled benzene treated with 1.2 eqs. Sb(OEt)$_3$ in the same solvent at reflux for 14 h → product. Y 76%. The regioselectivity is explained by a metal template effect. F.e., also carboxylic acid amides from carboxylic acids or methyl esters, s. H. Yamamoto et al., J. Am. Chem. Soc. *118*, 1569-70 (1996).

Tris(dibenzylideneacetone)dipalladium/chiral bis(o-acylaminophosphines)
Palladium-catalyzed asym. allylation
Enhancement and reversal of enantioselectivity by ligand design

188.

Novel chiral ligands derived from 2-(diphenylphosphino)aniline and dicarboxylic acid chlorides effect high enantioselectivity in allylic alkylation of systems which had previously afforded poor results. This is illustrated by **desymmetrization of 2-ene-1,4-diol carbamates.** E: A soln. of startg. enediol bis(N-tosylcarbamate) in THF added to 2.5 mol% Pd$_2$(dba)$_3$·CHCl$_3$ and 5-10 mol% chiral ligand at 0°, and allowed to react for 2-3.5 h → product. Y 99% (e.e. 88%). The enantiomer was obtained (Y 94%; e.e. 88%) by using the previously reported "invertomer" ligand derived from o-(diphenylphosphino)benzoic acid (cf. *43*, 385s47). F.e. incl. asym. C-allylation of malonates with 4-acetoxy-2-pentene s. B.M. Trost et al., Angew. Chem. Intern. Ed. *34*, 2386-8 (1995).

Nitrogen ↑ NC ⇑ N

Tri-n-butyltin hydride/azodiisobutyronitrile $Bu_3SnH/AIBN$
Generation of iminyl radicals from N-(thionocarbalkoxy)hydrazones   ←
Regiospecific ring closure under mild conditions

189.

Δ^1-**Pyrrolines.** Tri-n-butyltin hydride and a little AIBN added over 4 h to a refluxing soln. of startg. hydrazone in cyclohexane → product. Y 92%. The substrates are *crystalline* and easy to prepare. Urethan- and urea-centred radicals were generated similarly, but ring closure onto an appended alkene group was low yielding. F.e., **also lactams via** trapping of **acylaminyl radicals** (cf. *46*, 484), s. A.C. Callier-Dublanchet et al., Tetrahedron Letters *36*, 8791-4 (1995); generation of **iminyl radicals** from N-(benzotriazol-1-yl)imines and their fragmentation to **nitriles with isocyclic ring opening** s. L. El Kaim, C. Meyer, J. Org. Chem. *61*, 1556-7 (1996).

Tri-n-butylphosphine Bu_3P
1,4-Benzodiazepine-2,5-diones via polymer-based intramolecular aza-Wittig synthesis ○
s. *51*, 184

Hydrogen peroxide/2-phenyl-1,2-benziso selenazol-3(2H)-one   ←
Nitriles from hydrazones   CH=N–N< → CN
with H_2O_2 cf. *22*, 408; with H_2O_2/2-phenyl-1,2-benzisoselenazol-3(2*H*)-one or $H_2O_2/o,o$-bis-(carbamyl)diselenides s. J. Mlochowski et al., Synth. Commun. *26*, 291-300 (1996).

Rhodium(II) acetate/triethylamine $Rh(OAc)_2/Et_3N$
Ring closures with diazo compds.   ○
s. *35*, 327; chiral N-protected 3-azetidinones with a little Et_3N s. J. Podlech, D. Seebach, Helv. Chim. Acta *78*, 1238-46 (1995).

Halogen ↑ NC ⇑ Hal

tert-*Butyllithium*   *t-BuLi*
Indoles from *o*-amino-β,β-dihalogenostyrenes s. *51*, 112   ○

Tetrakis(triphenylphosphine)palladium(0)/potassium carbonate/sodium tert-*butoxide*   ←
Palladium-catalyzed intramolecular N-arylation

190.

The intramolecular version of Synth.Meth. *51*, 171 has been applied to the formation of benzocondensed 5-, 6- and 7-membered N-heterocyclics from *o*-bromoarylalkylamines, and 5- and 6-membered rings from the corresponding N-acyl- or N-sulfonyl derivs. **E:** Startg. iodide added to a suspension of 1.6 eqs. NaOBu-*t*, 1.6 eqs. K_2CO_3 and 1 mol% Pd(PPh$_3$)$_4$ in toluene, and heated at 65° for 2 h with stirring → 1-benzylindoline. Y 96%. Reaction was slower with ar. bromides. For amide cyclization, the system Pd$_2$(dba)$_3$/(2-furyl)$_3$P/Cs$_2$CO$_3$ was optimum. F.e., incl. oxindoles and 3,4-dihydrocarbostyrils, s. J.P. Wolfe et al., Tetrahedron *52*, 7525-46 (1996).

Remaining Elements ↑   NC ⇑ Rem

Irradiation
Lactams from cyclic enoxysilanes via 1,1-siloxyazides
Photo-induced Schmidt reaction under neutral conditions

191.

1 eq. Trimethylsilyl azide and 0.2 eq. pyridinium tosylate added sequentially with stirring to a soln. of startg. enoxysilane in anhydrous dichloromethane at room temp. under N_2, stirred for ca. 48 h, and poured into satd. $NaHCO_3$ → intermediate siloxyazide (Y 77%), in cyclohexane irradiated under N_2 for ca. 1 h in a quartz tube at 0° with a UV lamp (exothermic) → product (Y 89%). In the first step, the addition of azide in the presence of a mild acid catalyst avoids the use of hydrazoic acid. F.e. and regioselectivity s. P.A. Evans, D.P. Modi, J. Org. Chem. *60*, 6662-3 (1995).

Carbon ↑   NC ⇑ C

Potassium hydroxide/dimethyl sulfoxide/hydrogen chloride $KOH/Me_2SO/HCl$
Lactam from dicarboxylic acid imide ring

192.

A soln. of startg. bisimide in aq. KOH (85%), DMSO and methanol heated at reflux (100°) for 3 h, stirred into distilled water, further diluted with water, treated with concd. HCl, and left at room temp. for 16 h → product. Y 62%. A second, unaffected imide function, *peri*-fused through an aromatic system, appears essential for reaction. Contraction proceeds via oxidative decarboxylation, aerial O_2 (or DMSO) being essential for the conversion. F.e.s. H. Langhals, P. von Unold, Angew. Chem. Intern. Ed. *34*, 2234-6 (1995).

Cupric bromide/lithium tert-*butoxide* $CuBr_2/LiOBu$-t
Nitriles from α-aminocarboxylic acids $CH(NH_2)COOH → CN$
with NaOCl cf. *13*, 548; with 1:1 $CuBr_2/LiOBu$-*t* s. T. Takeda et al., Synthesis *1996*, 600-2.

Trifluoroacetic acid/sodium iodide CF_3COOH/NaI
Schmidt reaction with azidoketals

193.

Trifluoroacetic acid added to a soln. of 5-azido-1,1-diethoxy-1-phenylpentane in methylene chloride at 5° (vigorous evolution of gas), stirred for 16 h at room temp., solvent replaced by 2 eqs. NaI in dry acetone, and stirred at 70° for 4 h → 1-benzoylpyrrolidine. Y 79%. There are obvious advantages in using protected ketones rather than ketones themselves, and yields are approximately the same. The corresponding intermolecular reaction, however, generally failed. Trimethylsilyl triflate was also effective. F.e. incl. **bicyclic lactams,** also from enolethers, s. C.J. Mossman, J. Aubé, Tetrahedron *52*, 3403-8 (1996).

Manganese dioxide $\qquad$ MnO_2
Aromatization of 1,4-dihydropyridines $\qquad$ ←

194.

with C-dealkylation. A mixture of startg. dihydropyridine and 5 eqs. MnO_2 in dichloromethane stirred at room temp. for 30 min → product. Y 87%. C-Dealkylation at the 4-position takes place if a *sec*-alkyl or benzyl group is present, reaction generally being much faster under ultrasonication (5 min for completion). However, such C-cleavage does not take place with DDQ as oxidant (or with MnO_2 if the 4-substituent is a linear alkyl or (het)aryl group). F.e.s. J.J. Vanden Eynde et al., Tetrahedron *51*, 6511-6 (1995).

Formation of Hal-Rem Bond

Exchange ⇅

Oxygen ↑ $\qquad$ HalRem ⇅ O

Oxalyl chloride/dimethylformamide/quinuclidine $\qquad$ ←
Phosphonochloridates from phosphonic acid esters $\qquad$ $PO(OR)_2 \rightarrow PO(OR)Cl$
with $POCl_3$ cf. *43*, 398s*47*; from benzyl esters with oxalyl chloride/quinuclidine/DMF, also phosphoromonochloridates, s. M. Saady et al., Tetrahedron Letters *36*, 4785-6 (1995); **from phosphonic acid monoesters** (without base) and with $SOCl_2$, s. W.P. Malachowski, J.K. Coward, J. Org. Chem. *59*, 7616-24 (1994).

Phosphorus pentachloride/phosphorus oxide chloride $\qquad$ $PCl_5/POCl_3$
Phosphonic acid dichlorides from esters $\qquad$ $PO(OR)_2 \rightarrow POCl_2$
with PCl_5 cf *31*, 471; improved method with added $POCl_3$ s. C. Patois et al., Bull. Soc. Chim. France *130*, 485-7 (1993).

Formation of Hal-C Bond

Uptake ⇓

Addition to Oxygen and Carbon $\qquad$ HalC ⇓ OC

Silica-supported guanidinium chloride $\qquad$ ←
2-(Chloro)chloroformic acid esters from epoxides $\qquad$ $\triangledown_O \rightarrow C(Cl)C(OCOCl)$
with $COCl_2$ cf. *10*, 412; with $Cl_3COCOCl$ and silica-supported guanidinium chloride, **also 1,2-acoxychlorides**, s. P. Gros et al., J. Org. Chem. *59*, 4925-30 (1994).

Addition to Sulfur and Carbon $\qquad$ HalC ⇓ SC

Tetra-n-butylammonium bromide $\qquad$ Bu_4NBr
2-(Acylthio)chlorides from thiiranes $\qquad$ $\triangledown_S \rightarrow C(Hal)C(SCOR)$
s. *5*, 421; with added onium salt (e.g. Bu_4NBr), also 2-(acylthio)thioethers from thiolic acid esters, s. A. Kameyama et al., Tetrahedron Letters *35*, 4571-4 (1994).

Addition to Carbon-Carbon Bonds   HalC ⇓ CC

Without additional reagents w.a.r.
1,3-Dihalides from cyclopropanes C
s. *29*, 492; polyfluoro-1,3-dihalides s. Z.-Y. Yang et al., J. Am. Chem. Soc. *117*, 5397-8 (1995).

Irradiation s. under CuCl$_2$ ⇊
Cupric halides s.a. under R$_2$BH CuHal$_2$
Cupric chloride/pyridine/irradiation CuCl$_2$/C$_5$H$_5$N/⇊
α-Chloroketones from ethylene derivs. C=CH → C(Cl)CO
with FeCl$_3$ cf. *28*, 481s*36*; more efficiently with CuCl$_2$/py s. Tetrahedron *50*, 7375-84 (1994).

Zinc iodide/lithium hydride/titanocene dichloride ZnI$_2$/LiH/Cp$_2$TiCl$_2$
Regio- and stereo-specific hydrozincation of acetylene derivs. ←

195.

(E)-α,β-Ethyleneiodides. 1 eq. ZnI$_2$ and 2.2 eqs. LiH in THF allowed to react at room temp. for 3 h, cooled to 0°, 0.7-0.8 eq. startg. alkyne and 0.1 eq. Cp$_2$TiCl$_2$ added sequentially, stirred at room temp. for 3 h, and quenched with I$_2$ → (E)-product. Y 78% (88:12 mixture of regioisomers). This is the first instance of hydrozincation of alkynes, reaction taking place via hydrotitanation-transmetalation in a catalytic cycle. F.e., regioselectivity, and electrophiles s. F. Sato et al., J. Org. Chem. *60*, 290-1 (1995).

Dialkylboranes/cupric halides R$_2$BH/CuHal$_2$
Regiospecific hydrohalogenation of acetylene derivs. C≡C → CH=C(Hal)
(E)-α,β-ethylenehalides s. *48*, 426; mixed (E)-dihalogenomethylene compds. from α,β-acetylenehalides s. J. Chem. Soc. Perkin Trans. I *1995*, 2955-6.

N-Chlorosaccharin/hydrogen fluoride/pyridine ←
N-Bromosuccinimide/triethylamine tris(hydrogen fluoride) NBS/Et$_3$N·3HF
1,2-Fluorohalides from ethylene derivs. C=C → C(F)C(Hal)
1,2-bromofluorides with NBS/Et$_3$N·3HF s. *42*, 459; ω-fluoro-(ω-1)-bromoalkanoic acids s. D. Michel, M. Schlosser, Synthesis *1996*, 1007-11; 2-acoxy-2-fluorobromides s. Liebigs Ann. *1995*, 849-53; *anti*-1,2-chlorofluorides with N-chlorosaccharin/HF/py, regioselectivity, s. D. Dolenc, B. Sket, Synlett *1995*, 327-8.

N-Iodosuccinimide NIS
Allyl enolethers from enoxysilanes and 2-ethylenealcohols ←
via O-silyl O-allyl α-iodoacetals
Stereospecific conversion

196.

(Z)-1-(*tert*-Butyldimethylsiloxy)-1-decene in dichloromethane added to a stirred heterogeneous mixture of NIS and startg. allylic alcohol in the same solvent at -78° → *erythro*-1-(hex-2-enyloxy)-1-(*tert*-butyldimethylsiloxy)-2-iododecane (Y 94%; erythro/threo 99:1), treated with 2.4 eqs. *n*-BuLi in DME at -78° → (E)-product (76%; E/Z 96:4, and 18% silyl enol ether). The solvent in the second step is crucial: in DME stereospecific *anti*-elimination takes place, giving (E)- and (Z)-enolethers from *erythro*- and *threo*-acetals, respectively, whereas in hexane *syn*-elimination takes place to afford (E)-enoxysilanes from *erythro*-acetals or (E)-enolethers from *threo*-isomers. F.e.s. K. Utimoto et al., J. Org. Chem. *61*, 2262-3 (1996).

1-Fluoro-4-hydroxy-1,4-diazoniabicyclo[2.2.2]octane bis(fluoroborate) ←
2-Functionalized fluorides from ethylene derivs. C=C → C(X)C(F)
with 1-chloromethyl-4-fluoro-1,4-diazoniabicyclo[2.2.2]octane bis(fluoroborate) cf. *49*, 407; with the 1-fluoro-4-hydroxy-deriv. s. S. Stavber et al., Tetrahedron Letters *36*, 6769-72 (1995).

Chlorobis(cyclopentadienyl)hydridozirconium $Cp_2Zr(H)Cl$
Hydrohalogenation of terminal acetylene derivs. C≡CH → CH=CH(Hal)
α,β-ethylenebromides s. *43*, 625s*46*; (E)-2-halogeneneselenides from 1-acetylene-1-selenides s. L.-S. Zhu et al., J. Chem. Res. (S) *1996*, 112-3.

Titanocene dichloride s. under ZnI_2 Cp_2TiCl_2
Triethylamine tris(hydrogen fluoride) s. under *N-Bromosuccinimide* $Et_3N·3HF$

Potassium dichloroiodate(I) $KICl_2$
1,2-Chloroiodides from ethylene derivs. C=C → C(Cl)C(I)
with $BnMe_3^+ICl_2^-$ cf. *45*, 284s*46*; with $KICl_2$, regioselectivity, s. N.S. Zefirov et al., Synthesis *1995*, 1359-61.

Exchange ⇅

Hydrogen ↑ HalC ⇅ H

Microwaves s. under Al_2O_3 ←

Sodium/ethanol *NaOEt*
***o*-Iodophenols from 2-cyclohexenones** ←
Regiospecific conversion

197.

Startg. cyclohexenone added to a soln. of 6 eqs. NaOEt in ethanol at -78°, treated in small portions with 2 eqs. I_2 after 15 min, stirred for 3 h at the same temp. then overnight at room temp., and acidified with 5% HCl → product. Y 86%. The procedure tolerates a variety of substituents (alkyl, aryl, heterocyclyl, CF_3, esters) in the 3- or 5-position, but appears to be limited to 2-cyclohexenones with an electron-withdrawing group in the 4-position. F.e.s. S.G. Hegde et al., Tetrahedron Letters *36*, 8395-8 (1995).

Sodium peroxide/acetic acid $Na_2O_2/AcOH$
Oxidative ar. halogenation H → Hal

198.

0.04 eq. Na_2O_2 in 6 eqs. glacial acetic acid added dropwise with stirring to 1 eq. mesitylene, 3 eqs. concd. HCl, and 2.5 eqs. glacial acetic acid at 20-5°, and worked up after 4 h → chloromesitylene. Y 78%. The method is more widely applicable than that with H_2O_2 (cf. *3*, 440). F.e. incl. ar. bromination (with HBr), and with K_2O_2 s. N.I. Rudakova et al., Zh. Obshch. Khim. *65*, 315-7 (1995) (Russ.).

Sodium hydrogen carbonate $NaHCO_3$
Iodo-O-heterocyclics from ethylenealcohols ○
s. *33*, 477s*50*; 2-α-iodooxetane-3-carboxylic acid esters s. P. Galatsis, D.J. Parks, Tetrahedron Letters *35*, 6611-4 (1994); 2-(iodomethyl)oxepans with bis(collidine)iodine(I) hexafluorophosphate (cf. *33*, 477s*44*) cf. Y. Brunel, G. Rousseau, Synlett *1995*, 323-4.

Sodium bromide/trimethylsilyl chloride   *NaBr/Me$_3$SiCl*
α-Bromacetals from acetals   H → Br
with NBS cf. *7*, 566 and Br$_2$·dioxane cf. *34*, 485; with NaBr/Me$_3$SiCl s. F. Bellesia et al., Gazz. Chim. Ital. *123*, 629-31 (1993); 2-(α-bromalkyl)-1,3-dioxolanes with Br$_2$ s. S.S. Vershinin et al., Zh. Obshch. Khim. *66*, 1177-9 (1996) (Russ.).

2,6-Lutidine   ←
α-Iodination of carboxylic acid amides   H → I
s. *47*, 428; of N-allylamides s. J. Org. Chem. *60*, 7161-5 (1995).

Cupric halides/sodium hydride   *CuHal$_2$/NaH*
α-Halogenocarbonyl from carbonyl compds.   H → Hal
α-bromoketones cf. *31*, 163; α-bromo- and α-chloro-esters with added NaH s. X.-X. Shi, L.-X. Dai, J. Org. Chem. *58*, 4596-8 (1993).

Cupric bromide/alumina   *CuBr$_2$/Al$_2$O$_3$*
Bromolactonization   ◯

199.

5 eqs. CuBr$_2$-on-alumina added to a soln. of startg. ethylenecarboxylic acid in chloroform, and heated at 65° for 96 h → product. Y 98%. F.e. and with stereospecificity s. G.A. Rood et al., Tetrahedron Letters *37*, 157-8 (1996); δ-bromo-γ-lactones with HBr/H$_2$O$_2$ cf. V.S. Arutyunyan et al., Zh. Org. Khim. *31*, 100-2 (1995) (Russ.).

Mercuric oxide or trifluoromethanesulfonate   *HgO or Hg(OSO$_2$CF$_3$)$_2$*
Ar. monoiodination   H → I
with HgO/HBF$_4$-SiO$_2$ cf. *40*, 345; with Hg(OSO$_2$CF$_3$)$_2$ and other Hg-salts s. M. Yus et al., Tetrahedron *50*, 5139-46 (1994); iodination of phenolethers and ar. diiodination with HgO, cf. K. Orito et al., Synthesis *1995*, 1273-7.

Ethylmagnesium bromide   *EtMgBr*
α,β-Acetyleneiodides from terminal acetylene derivs.   C≡CH → C≡CI
with morpholine cf. *18*, 579; with EtMgBr s. M.L.N. Rao, M. Periasamy, Synth. Commun. *25*, 2295-9 (1995).

Alumina/microwaves   ←
Replacement of hydrogen by bromine   H → Br
ar. bromination without solvent cf. *47*, 430; regiospecific bromination of *p*-quinones and coumarins under microwave irradiation s. V. Bansal, R.N. Khanna, Synth. Commun. *26*, 887-92 (1996).

Zeolites   ←
***p*-Bromination**
s. *41*, 470s*45*; with NaY zeolite s. K. Smith, D. Bahzad, Chem. Commun. *1996*, 467-8; of activated arenes with zeolite HZSM-5/NBS cf. K.V. Srinivasan et al., Tetrahedron Letters *35*, 7055-6 (1994).

Dimethyl sulfide s. under N-Halogenosuccinimides   *Me$_2$S*

Phenyl iodosotrifluoroacetate   *PhI(OCOCF$_3$)$_2$*
Ar. iodination   H → I
with PhI(OH)OTs/NIS cf. *48*, 439; 2-iodination of thiophenes with PhI(OCOCF$_3$)$_2$/I$_2$ s. M. D'Auria, G. Mauriello, Tetrahedron Letters *36*, 4883-4 (1995).

N-Halogenosuccinimides   *NBS or NIS*
Halogeno-N-heterocyclics from ethyleneamines   ◯
3-iodo-1,2,3,4-tetrahydroquinolines s. *48*, 434; 3-halogenomethyl-2,3,4,5-tetrahydro-2,5-methano-1*H*-benzazepines with NBS or NIS s. F.I. Carroll et al., Chem. Commun. *1993*, 758-60; 3-iodo-1-tosylpyrrolidines from 3-ethylene-N-tosylamines, stereoselectivity, s. A.D. Jones, D.W. Knight, ibid. *1996*, 915-6; staurosporine ring by stereospecific intramolecular N-glycosidation with I$_2$/KOBu-*t* s. S.J. Danishefsky et al., J. Am. Chem. Soc. 118, 2825-42 (1996); C-piperidinosides s. O.R. Martin et al., Tetrahedron Letters *37*, 1991-4 (1996); 5-α-iodo-Δ^2-pyrroline-3-carbonyl compds. s. H.M.C. Ferraz et al., J. Org. Chem. *60*, 7357-9 (1995); 2-bromocarbapenams s. H. Horikawa et al., Tetrahedron Letters *35*, 6317-20 (1994).

N-Halogenosuccinimides/dimethyl sulfide   *NBS or NCS/Me$_2$S*
1,1-Halogenohydrazones from hydrazones   CH=NN< → C(Hal)=NN<
with Br$_2$/AcOH cf. *29*, 502; chloro- and bromo-derivs. with NCS/Me$_2$S or NBS/Me$_2$S s. H.V. Patel et al., Tetrahedron *52*, 661-8 (1996).

N-Bromosuccinimide or N-Iodosuccinimide   *NBS or NIS*
Regiospecific ar. bromination under ultrasonication   H → Br

200.

A mixture of nerolin and 1.1 eqs. NBS in carbon tetrachloride ultrasonicated at room temp. for 3 h in a 20 KHz Branson Sonifier 250 with an output control at 5 → 1-bromonerolin. Y 99%. The method is mild, simple and high-yielding, acid or base catalysts and high temp. are not required, reaction times are relatively short, and only one regioisomer is formed. The reaction, however, failed with benzene, toluene, *p*-xylene, *p*-nitrotoluene and naphthalene. F.e. and with acetic acid as solvent s. K.V. Srinivasan et al., Synth. Commun. *25*, 2401-5 (1995); regiospecific bromination of phenolethers *in acetonitrile* without ultrasonication s. M.C. Carreño et al., J. Org. Chem. *60*, 5328-31 (1995); nuclear *vs.* benzylic bromination in CCl$_4$ cf. G.J. Gruter et al., ibid. *59*, 4473-81 (1994); **ar. iodination** with NIS in acetonitrile s. Tetrahedron Letters *37*, 4081-4 (1996).

3-Fluoro-1,2,3-benzoxathiazin-4-one 2,2-dioxide/sodium hydride   ←
C-α-Fluorination of enolates   H → F
with N-fluorosulfonimides s. *39*, 458s*46*; with crystalline 3-fluoro-1,2,3-benzoxathiazin-4-one 2,2-dioxide, also fluorides from Grignard compds., s. I. Cabrera, W.K. Appel, Tetrahedron *51*, 10205-8 (1995); 6-fluorination of Δ^4-3-ketosteroids s. A.J. Poss, G.A. Shia, Tetrahedron Letters *36*, 4721-4 (1995).

Trimethylsilyl azide/pyridine   *Me$_3$SiN$_3$/C$_5$H$_5$N*
α-Halogenation of cyclic α,β-ethyleneketones   H → I
with CCl$_4$/py cf. *38*, 669s*47*; α-iodination with I$_2$/Me$_3$SiN$_3$/py s. C.-K. Sha, S.-J. Huang, Tetrahedron Letters *36*, 6927-8 (1995); with I$_2$/pyridinium dichromate cf. E. McNelis et al., ibid. *35*, 6787-90 (1994).

Trimethylsilyl chloride s. under NaBr   *Me$_3$SiCl*
Nitrosyl hydrogen sulfate/sulfuric acid   *NO$^+$HSO$_4^-$/H$_2$SO$_4$*
p-Bromination of anilines in a strongly acidic medium   H → Br

201.

Aniline, N,N-dialkylanilines, and derivs. with *ortho-* or *meta-*substituents (alkyl or chlorine) undergo exclusive *p*-bromination *in 92-100% H$_2$SO$_4$* in spite of the fact that anilinium salts are normally *meta*-directing. **E:** A soln. of aniline in 92-100% H$_2$SO$_4$ treated with equivalent amounts of Br$_2$ and nitrosyl hydrogen sulfate at 20-35°, and poured into *alkali soln.* → *p*-bromoaniline. Y ca. 95%. There was no reaction in the absence of nitrosyl salts. The key step in the transformation is the conversion of a nitrosoammonium ion into an ion-radical pair, followed by exchange of nitrosyl radical by bromine prior to substitution. F.e. and with the nitrosyl salt in catalytic amount, also *p*-bromodiazonium salts, s. M.V. Gorelik et al., Mendeleev Commun. *1995*, 65-6; f. details s. Zh. Org. Khim. *30*, 553-7 (1995) (Russ.); with HBr/Br$_2$ cf. K.S. Chamberlin, Synth. Commun. *25*, 27-31 (1995).

N-Fluoroditriflimide/sodium carbonate   *(CF$_3$SO$_2$)$_2$NF/Na$_2$CO$_3$*
α,α-Difluoroketones from ketimines under mild conditions   COCF$_2$

202.

A soln. of *2.4 eqs.* N-fluoroditriflimide in anhydrous methylene chloride stirred with anhydrous

Na₂CO₃ for 15 min, a soln. of N-[1-(4-bromophenyl)ethylidene]-*n*-butylamine in the same solvent added dropwise with stirring at room temp., stirred for 5 h, and treated with 1 N aq. HCl → 2,2-difluoro-*p*-bromoacetophenone. Y 78%. Significantly, a strong base is not required. Enamines are thought to be involved in the reaction after initial monofluorination. F.e. and **2-α-fluoropyridines** s. W. Ying et al., Tetrahedron *52*, 15-22 (1996).

Chromium trioxide/acetic anhydride/sulfuric acid ←
Diaryliodonium salts from arenes and ar. iodides s. *51*, 204 Ar_2I^+
Sodium tungstate/hydrogen peroxide/potassium bromide ←
Biomimetic ar. bromination H → Br
with NH₄-molybdate cf. *50*, 272; also with H₂MoO₄ in aq. media cf. M. Bhattercharjee, J. Mukherjee, J. Chem. Res. (S) *1995*, 238-9; with Na₂WO₄ in glacial AcOH cf. J.R. Hanson et al., ibid. 457; *p*-bromination of anisoles s. P. Bezodis et al., ibid. *1996*, 334-5.

Poly(4-vinylpyridinium bromochromate) ←
Ar. bromination
with C₅H₅NH·CrO₃Br cf. *42*, 469; with poly(4-vinylpyridinium bromochromate) s. J. Chem. Res. (S) *1992*, 132-3.

Fluorine/sulfuric acid F_2/H_2SO_4
Ar. iodination s. *51*, 203
Iodine monochloride-pyridine $ICl-C_5H_5N$
Ar. iodination under mild conditions H → I

203.

H₂N—⟨⟩ —ICl-py→ H₂N—⟨⟩—I

Crystalline ICl-py complex is a safe alternative to hazardous ICl for ar. iodination (cf. *1*, 419). E: ICl-py added to a stirred soln. of 1 eq. aniline in 5% aq. HCl, and stirred vigorously for 1 h → *p*-iodoaniline. Y 70%. Activated aromatics (anilines, phenols) can be selectively mono- or polyiodinated, while ar. hydrocarbons and ar. halides are readily monoiodinated in the presence of FeCl₃. Iodination of condensed arenes can be performed in one pot via the corresponding organomercurials (with Hg(OCOCF₃)₂ in methylene chloride). F.e. and method incl. iodination of hetarenes s. H.A. Muathen, J. Chem. Res. (M) *1994*, 2201-31; with I₂/F₂/H₂SO₄, notably for *m*-iodination of deactivated arenes and polyiodination cf. R.D. Chambers et al., Chem. Commun. *1995*, 19; **ar. radioiodination** with labelled IF on sub-micromolar scale s. H.H. Coenen et al., Tetrahedron Letters *35*, 9701-2 (1994).

tert-*Butyl hypochlorite* t-BuOCl
1,1-Chlorazo compds. from hydrazones C=NNH → C(Cl)N=N
with Cl₂ cf. *27*, 558; with *t*-BuOCl s. J.C. Jochims et al., Synthesis *1996*, 274-80.

Pyridinium hydrobromide perbromide ←
Hexamethylenetetramine hydrotribromide ←
3-Halogenochromans from *o*-allylphenols ○
with NIS cf. *36*, 241; 3-bromo-derivs. with hexamethylenetetramine hydrotribromide s. K.C. Majumdar et al., Synth. Commun. *26*, 893-8 (1996); with pyridinium hydrobromide perbromide s. Can. J. Chem. *73*, 1727-32 (1995).

Benzyltrimethylammonium dichloroiodate/sodium hydrogen carbonate $BnMe_3N^+ICl_2^-/NaHCO_3$
α-Iodination of cyclic ketones H → I
with polymer-based ICl₂⁻ cf. *45*, 284; with benzyltrimethylammonium dichloroiodate, β-amino-α,β-ethylene-α-iodoketones, s. K. Matsuo et al., Chem. Pharm. Bull. *42*, 1149-50 (1994).

Bis(collidine)iodine(I) hexafluorophosphate ←
2-(Iodomethyl)oxepans from 6-ethylenealcohols s. *33*, 477s*51* ○
Hydrogen bromide HBr
***p*-Bromination of anilines** s. *51*, 201 H → Br
Hydrogen bromide/hydrogen peroxide HBr/H_2O_2
δ-Bromo-γ-lactones from γ,δ-ethylenecarboxylic acids s. *51*, 199 ○

Oxygen ↑   HalC ↕ O

Without additional reagents   w.a.r.
Diaryliodonium salts from arenes under mild conditions Ar_2I^+

204.

Finely ground iodosylbenzene added to a stirred soln. of 1 eq. SO_3 in dichloromethane at -50°, the mixture warmed to room temp., stirred for 30 min, re-cooled to -50°, 1 eq. nitrobenzene added, and stirring continued at -50° to 20° for a further 30 min → product. Y 54%. In contrast to prior art, the method also works well with arenes containing strongly deactivating functional groups (e.g. fluoro and nitro). F.e. and methods s. N.S. Zefirov et al., Synthesis *1995*, 775-6; **from ar. iodides** with $CrO_3/Ac_2O/H_2SO_4$ cf. P. Kazmierczak, L. Skulski, ibid. 1027-32; sym. bis(N-hetaryl)iodonium salts s. P.J. Stang et al., ibid. 937-8.

Microwaves s. under NaI   ←
Sodium iodide/montmorillonite/microwaves   ←
Iodides from alcohols   OH → I
with $NaI/MeSiCl_3$ cf. *39*, 444; benzyl iodides under solid-supported microwave irradiation in the absence of solvent with NaI/montmorillonite s. J. Singh et al., J. Chem. Res. (S) *1996*, 188-9.

Montmorillonite s. under NaI   ←
p-Methyliodobenzene difluoride $ArIF_2$
Fluorides from xanthates under mild conditions   OCSSR → F

205.

Startg. xanthate added dropwise to a stirred soln. of 1 eq. *p*-methyliodobenzene difluoride in methylene chloride at 0° under inert atmosphere → product. Y 64%. Reaction is generally applicable to prim., sec. and benzylic xanthates. The reagent is readily prepared, crystalline, and soluble in organic solvents. F.e., stereoselectivity, and comparison with DAST s. M.J. Koen et al., Chem. Commun. *1995*, 1241-2.

N-Bromosuccinimide s. under Ph_3P   NBS
Trichloromethyl chloroformate/silica-supported guanidinium chlorides   ←
Carboxylic acid chlorides from carboxylic acids   COOH → COCl
with $(Cl_3CO)_2CO/py$ cf. *43*, 173; with trichloromethyl chloroformate/silica-supported guanidinium chlorides (or $SOCl_2$) s. P. Gros et al., Bull. Soc. Chim. France *130*, 554-61 (1993).

Trimethylsilyl azide/titanium tetrachloride $Me_3SiN_3/TiCl_4$
α-Azidohydroximinochlorides from 1,1-nitroethylene derivs.   C(N$_3$)C(Cl)=NOH

206.

1.04 eqs. 1 M $TiCl_4$ in dry dichloromethane added dropwise to a stirred soln. of startg. nitroethylene and 1.5 eqs. trimethylsilyl azide in the same solvent at room temp. over 30 min under argon, and stirring continued for a further 1 h → product. Y 78%. The method is simple, general, and fast. F.e. and 1,3-dipolar cycloaddition, **also α-cyanohydroximinochlorides** (from nitrostyrenes) with trimethylsilyl cyanide, s. G. Kumaran, G.H. Kulkarni, Synth. Commun. *25*, 3735-40 (1995).

Tetra-n-butylammonium (triphenylsilyl)difluorosilicate   ←
Fluorides from sulfonates s. *51*, 210 $OSO_2R → F$
Titanium tetrachloride s. under Me_3SiN_3 $TiCl_4$

Silica-supported guanidinium chlorides s. under $ClCO_2CCl_3$ ←

Triphenylphosphine/N-bromosuccinimide Ph_3P/NBS
Carboxylic acid bromides from carboxylic acids   COOH → COBr

207.

A vigorously stirred mixture of 2-naphthoic acid and 1 eq. Ph_3P in dioxane treated in small portions with 1.02 eqs. NBS, and heated (90-100°) for 5-10 min → 2-naphthoyl bromide. Y 95%. The method is extremely simple, based on readily available, inexpensive materials, and is generally applicable to aromatic, aliphatic and unsatd. carboxylic acids. Reaction with NIS [or ICN]/Ph_3P proceeds more slowly than that with NBS (or NCS), and gave moderate yields. F.e. and with BrCN in place of NBS, **also bromides and iodides from alcohols**, s. P. Frøyen, Phosphorus, Sulfur and Silicon *102*, 253-9 (1995).

Diethylaminosulfur trifluoride Et_2NSF_3
Replacement of hydroxyl by fluorine   OH → F
s. *31*, 501s*33*; 1,1-dibromo-1,2-difluorides s. M. Kuroboshi et al., Tetrahedron Letters *36*, 6271-4 (1995); 1,1-bis(trifluoromethyl)ethylene derivs. s. H. Lu, D.J. Burton, ibid. 3973-6; $Fe(CO)_3$-complexed tert. 2,4-dien-1-yl fluorides s. D.M. Gree et al., J. Org. Chem. *61*, 1918-9 (1996); β,γ-acetylene-α,α-difluorophosphonic acid esters (cf. *30*, 365) s. F. Benayoud, G.B. Hammond, Chem. Commun. *1996*, 1447-8.

Hydrogen fluoride/pyridine $HF-C_5H_5N$
Glycosyl fluorides from 1H-tetrazol-5-yl glycosides under mild conditions ←

208.

10 eqs. HF·py added to a 0.1 *M* soln. of startg. α-D-mannopyranoside (100% α) in methylene chloride at 0°, and worked up after 10 min → product. Y 76% (100% α). Acid-sensitive protective groups (e.g. O,O-isopropylidene) and the interglycosidic bond of disaccharides were unaffected. However, reaction with O,O-benzylidene derivs. proceeded with concomitant cleavage of the protective group. Reaction with β-glycosides was non-stereoselective. F.e.s. M. Palme, A. Vasella, Helv. Chim. Acta *78*, 959-69 (1995).

Nitrogen ↑   HalC ⇅ N

Without additional reagents   w.a.r.
δ-Iodo-γ-lactones from γ,δ-ethylenecarboxylic acid amides ○
Asym. iodolactonization
s. *39*, 467s*46*; *50*, 500; from polymer-based amides cf. M.J. Kurth et al., Tetrahedron Letters *35*, 8915-8 (1994).

Irradiation s. under CCl_4 ⦀

Cupric bromide/lithium tert-*butoxide* $CuBr_2/LiOBu$-t
1,1-Dibromides from hydrazones   C=NN< → CBr_2
with Br_2 cf. *41*, 434; with $CuBr_2/LiOBu$-*t* from crude hydrazones s. T. Takeda et al., Synlett *1996*, 273-4.

Carbon tetrachloride/irradiation CCl_4/⚡
Regio- and stereo-specific ring closures of unsatd. alkoxyl radicals under neutral conditions

209.

2-α-Chlorotetrahydrofurans. A soln. of N-(5-hexenyl-2-oxy)pyridine-2(1*H*)-thione in anhydrous carbon tetrachloride irradiated with a 150 W Philips Spotline R80 for 15 min at 30° → 2-chloromethyl-5-methyltetrahydrofuran. Y 85% (*trans:cis* 63:37). The *trans:cis* ratio (ranging from 25:75 to 86:14) depended on the substitution pattern and on the bulk of the α-alkyl group. The corresponding unsatd. phenylsulfenates also served as radical precursor. F.e. and **2,5-disubst. tetrahydrofurans** in the presence of Bu$_3$SnH, s. J. Hartung, F. Gallou, J. Org. Chem. *60*, 6706-16 (1995).

Halogen ↑   HalC ⇅ Hal

Tetra-n-*butylammonium (triphenylsilyl)difluorosilicate* $Bu_4N^+[(Ph_3Si)SiF_2]^-$
Tetra-*n*-butylammonium (triphenylsilyl)difluorosilicate as source of fluoride ion ←

210.

Tetra-*n*-butylammonium (triphenylsilyl)difluorosilicate (TBAT) is significantly more effective than traditional fluoride ion sources for nucleophilic fluorination of prim. and sec. substrates in that it is readily soluble in organic solvents, non-hygroscopic, and (unlike Bu$_4$NF) can be prepared in *anhydrous* form; furthermore, reaction produces a *neutral* medium so that base-catalyzed elimination and formation of hydroxylic by-products is not a serious problem. E: A soln. of 1-bromooctane and 6 eqs. TBAT in acetonitrile refluxed for 24 h → 1-fluorooctane. Y 70% (and 30% octene). With Bu$_4$NF, the yield of product was 48% (with 12% octene and 40% octanol). Displacement of chlorine or iodine was less facile, and elimination was a major problem with sec. halides; however, TsO, MsO and TfO groups (at either prim. or sec. sites) were readily substituted. F.e.s. A.S. Pilcher et al., J. Am. Chem. Soc. *117*, 5166-7 (1995).

Sulfur ↑   HalC ⇅ S

Without additional reagents   w.a.r.
Regio- and stereo-specific intramolecular iodoetherification-lactonization ○

211.

A soln. of startg. *endo*-adduct in aq. THF treated with I$_2$ at 25° for 6 h → product. Y 80-5%. This is the first instance of I$_2$-induced sequential heterocyclization, and particularly notable in that the carbonyl oxygen serves *as nucleophile*. F.e. and from esters, **also lactol analogs** from the corresponding γ-diketones, s. H.-J. Wu et al., Chem. Commun. *1996*, 375-6.

1,3-Dibromo-5,5-dimethylhydantoin s. under Bu$_4$NH$_2$F$_3$ ←

Nitrosyl fluoroborate/pyridine-hydrogen fluoride NOBF$_4$/HF-C$_5$H$_5$N
Fluorides from ar. thioethers SAr → F

212.

A soln. of startg. ar. thioether in methylene chloride added dropwise to 1.2 eqs. nitrosyl fluoroborate and 60% pyridinium poly(hydrogen fluoride) in the same solvent at 0° under dry N$_2$ [in a plastic bottle], and stirred at room temp. for 1 h → 1-fluoro-1-phenylethane. Y 89%. F.e., also **1,1-difluorides from 1,3-dithiolanes**, s. C. York et al., Tetrahedron 52, 9-14 (1996).

Tetra-n-butylammonium dihydrogentrifluoride/1,3-dibromo-5,5-dimethylhydantoin ←
1,1-Difluoro-2-ketothioethers COCF$_2$(SR)
from α-hydroxytrithioorthocarboxylic acid esters

213.

Startg. trithioorthocarboxylate added to a 5:4 mixture of *n*-Bu$_4$NH$_2$F$_3$ and 1,3-dibromo-5,5-dimethylhydantoin in dichloromethane at room temp., stirred for 10 min, then poured into satd. aq. Na$_2$SO$_3$ → product. Y 95%. The method is simple and general. F.e., also **1,1-difluoro-1,2-di(thioethers)** (with DAST) from α-hydroxytrithioorthocarboxylic acid esters, s. M. Kuroboshi et al., Tetrahedron Letters 36, 6121-2 (1995).

Hydrogen fluoride/pyridine s. under NOBF$_4$ HF-C$_5$H$_5$N

Remaining Elements ↑   HalC ⇅ Rem

Sodium hydrogen carbonate NaHCO$_3$
2-α-Iodocyclopentanones from 1-siloxy-1-vinylcyclobutanes
Stereospecific ring expansion

214.

A soln. of startg. vinylcyclobutane in ether treated with I$_2$ and NaHCO$_3$ at 0° until reaction complete → product. Y 96% (single stereoisomer). Yields were slightly lower from the corresponding 1-vinylcyclobutanols. F.e., stereoselectivity, and with NIS s. H. Nemoto et al., Tetrahedron Letters 36, 8799-802 (1995); details s. J. Org. Chem. 61, 1347-53 (1996).

Carbon ↑   HalC ⇅ C

Potassium iodide KI
Halogenative ring closures of unsatd. urethans
5-α-halogeno-2-oxazolidones cf. 38, 488s48; 5-(3-iodovinyl)-2-oxazolidones with KI/I$_2$ s. C. Iwata et al., Synlett 1995, 737-8.

Manganese(II) acetate/N-bromosuccinimide Mn(OAc)$_2$/NBS
2,2-Dibromalcohols from α,β-ethylenecarboxylic acids C(OH)CBr$_2$
Degradation with loss of 1 C-atom

215.

A mixture of startg. acid, *2 eqs*. NBS, and 0.1 eq. Mn(OAc)$_2$ in 1:1 acetonitrile/water allowed to react at room temp. for 16 h → product. Y 98%. The proportion of water in the solvent was critical. By using *1 eq.* NBS under the same conditions, **α,β-ethylenebromides** were obtained. The method is mild and does not require a hazardous reagent or low temperature. F.e.s. S. Chowdhury, S. Roy, Tetrahedron Letters *37*, 2623-4 (1996).

Formation of S-S Bond

Exchange ⇅

Hydrogen ↑   SS ⇅ H

Copper/Amberlite IRA-400 tetrahydridoborate $Cu/[BH_4^-]$
Sym. disulfides from mercaptans $2 \text{ RSH} \to (RS)_2$
Disproportionation under mild conditions

216.   2 n-BuSH   ⎯⎯⎯⟶   n-BuSSBu-n

Copper deposited on borohydride exchange resin is a convenient catalyst for the conversion of mercaptans to disulfides *under non-oxidizing conditions*. **E:** A soln. of 10 mol% $CuSO_4$ in methanol added to dry Amberlite IRA-400 tetrahydridoborate, stirred under reflux for 3 h, cooled to room temp. under N_2, 2 eqs. butanethiol in methanol added, and stirred at room temp. for 3 h → dibutyl disulfide. Y 98%. Elimination of H_2 as the reaction proceeds presumably facilitates the otherwise thermodynamically unfavourable process. The method is simple and generally applicable to prim., sec. (3 h) and [more slowly] tert. and ar. mercaptans (6-9 h). Alkene, amino, ester and furyl groups were tolerated. F.e.s. J. Choi, N.M. Yoon, J. Org. Chem. *60*, 3266-7 (1995); **from thiolic acid esters** with Ni_2B/Amberlite IRA-400 tetrahydridoborate s. Synlett *1995*, 1073-4.

Bromine Br_2
Sym. disulfides from mercaptans
in a 2-phase medium cf. *35*, 366; *without* solvent s. R.D. Rieke et al., Synth. Commun. *26*, 191-6 (1996).

Via intermediates   *v.i.*
Disulfides from two different mercaptan molecules   RS-SR'
via sulfenylation with 2-disulfidopyridinium salts cf. *47*, 470; with benzothiazol-2-yl disulfides s. E. Brzezinska, A.L. Ternay, Jr., J. Org. Chem. *59*, 8239-44 (1994).

Oxygen ↑   SS ⇅ O

Titanium tetrachloride/samarium $TiCl_4/Sm$
Sym. disulfides from sulfinic acid esters $2 \text{ RS(O)OR'} \to (RS)_2$
with N_2H_4 cf. *23*, 572; diaryl disulfides with $TiCl_4$/Sm s. J. Wang, Y. Zhang, Synth. Commun. *26*, 135-8 (1996).

Nitrogen ↑   SS ⇅ N

Lithium hydridotriethylborate $LiBHEt_3$
Sym. disulfides from N-(organothio)phthalimides $2 \text{ RSN<} \to (RS)_2$

217.

LiBEt$_3$H added to a soln. of N-[2,4-bis(*t*-butyldimethylsiloxy)phenylthio]phthalimide in dry THF at -78°, kept at this temp. for 15 min, and quenched with satd. NH$_4$Cl soln. → bis[2,4-bis(*t*-butyl dimethylsiloxy)phenyl] disulfide. Y 74%. F.e. and reductants s. G. Capozzi et al., Gazz. Chim. Ital. *126*, 227-32 (1996).

Halogen ↑   SS ⇅ Hal

Lithium   *Li*
Sym. S,S-disulfones RSO_2SO_2R
from sulfinic acids cf. *22*, 591; **from sulfonic acid chlorides** with Li under ultrasonication s. J.-L. Luche et al., Tetrahedron Letters *36*, 3849-50 (1995).

Titanium tetrachloride/samarium $TiCl_4/Sm$
Sym. disulfides from sulfonic acid chlorides $2\ RSO_2Cl \rightarrow (RS)_2$
with AlI_3 cf. *26*, 564s*41*; sym. diaryl disulfides with $TiCl_4$/Sm s. J. Wang, Y. Zhang, Synth. Commun. *26*, 135-8 (1996).

Carbon ↑   SS ⇅ C

Benzyltriethylammonium tetrathiomolybdate $[BnEt_3N]_2MoS_4$
Sym. disulfides from thiocyanates $2\ RSCN \rightarrow (RS)_2$
with $NaBH_4$ cf. *21*, 630; with $[BnEt_3N]_2MoS_4$, selectivity, s. K.R. Prabhu et al., J. Org. Chem. *60*, 7142-3 (1995).

Nickel boride/Amberlite IRA-400 tetrahydridoborate   ←
Sym. disulfides from thiolic acid esters s. *51*, 216 $2\ COSR \rightarrow (RS)_2$

Formation of S-Rem Bond

Exchange   ⇅

Carbon ↑   SRem ⇅ C

Tellurium tetrachloride/lutidine   ←
Monothiolphosphoric acid esters $(RO)_2PO(SR')$
from phosphites and mercaptans
with $BrCCl_3$ cf. *18*, 636; with $TeCl_4$/lutidine or $CaCO_3$ s. Y. Watanabe et al., Synthesis *1995*, 1243-4.

Formation of S-C Bond

Uptake   ⇓

Addition to Oxygen and Carbon   SC ⇓ OC

Lithium bromide   *LiBr*
1,3-Oxathiolane-2-thiones from epoxides
Regiospecific conversion

218.

PhO–[epoxide] $\xrightarrow{CS_2}$ PhO–[1,3-oxathiolane-2-thione]

ca. 1.2 eqs. CS_2 added to a soln. of startg. epoxide and 5 mol% LiBr in *THF*, and stirred at room temp. for 4.5 h → product. Y 97%. Formation of the regioisomer or corresponding 1,3-dithiolan-2-one was minimal under these conditions. The choice of solvent was critical and the method appears limited to mono-subst. epoxides. Benzoate, methacrylate and benzyl groups were unaffected. F.e.s. N. Kihara et al., J. Org. Chem. *60*, 473-5 (1995).

Addition to Sulfur-Sulfur Bonds   SC ⇓ SS

Organolithium compds.   RLi
Di(thioethers) from cyclic disulfides   C̈
with $NaBH_4$/py cf. *32*, 523; unsym. derivs. with RLi/R'Hal, also (acylthio)thioethers with acyl halides, s. K. Smith, M. Tzimas, J. Chem. Soc. Perkin Trans. I *1995*, 2381-2.

Addition to Carbon-Carbon Bonds   SC ⇓ CC

Without additional reagents   *w.a.r.*
Asym. Michael addition to 3-(α,β-ethyleneacyl)-2-oxazolidones   C=C → CHC(S-)
of ar. mercaptans cf. *47*, 487; of thiolic acids s. T.-C. Tseng, M.-J. Wu, Tetrahedron:Asym. *6*, 1633-40 (1995); asym. addition of thiophenol to chiral N-(α,β-ethyleneacyl)-2-pyrrolidones s. K. Tomioka et al., J. Org. Chem. *60*, 6188-90 (1995).

Monothioacetals from enolethers   C=C(OR) → C—C(OR)SR
with $SOCl_2$ cf. *8*, 651; without reagent, or with BF_3 or HCl for less reactive enolethers, also mercaptals, s. L.A. Wessjohann et al., Synth. Commun. *25*, 3155-62 (1995).

Amberlite IRA-400 tetrahydridoborate $[BH_4^-]$
Michael addition of *in situ*-generated mercaptans s. *51*, 234   C=C → CHC(SR)

Thiophene endoperoxides/cobalt(II) tetraphenylporphyrin   ←
Thiiranes from ethylene derivs.   ▽ˢ
Transition metal-catalyzed sulfur atom transfer from monothioozonides

219.

A soln. of startg. endoperoxide (readily prepared by photooxygenation of the corresponding thiophene) and 7 eqs. norbornene in deuteriochloroform allowed to react at -40° for 20 min in the presence of cobalt(II) tetraphenylporphyrin → product. Y 68%. Such metal-assisted sulfur atom transfer is unprecedented and may be applied, notably, for the preparation of fused thiiranes which, like norbornene episulfide, may not readily be prepared by other means. F.e. and thermal conversion (less efficiently) s. W. Adam, S. Weinkötz, Chem. Commun. *1996*, 177-8.

Rearrangement   ꩜

Oxygen/Sulfur Type   SC꩜OS

Aluminum chloride $AlCl_3$
***o*-Hydroxysulfoxides from sulfinic acid aryl esters**   ←
Thia-Fries rearrangement

220.

2 eqs. $AlCl_3$ added in one portion to a soln. of 4-methoxyphenyl phenylsulfinate in dichloromethane under N_2 (exothermic to reflux), stirred for 1 h at 25°, diluted with further solvent, and water added dropwise → 4-methoxy-2-(phenylsulfinyl)phenol. Y 87%. This is the first instance of such

a rearrangement. F.e. and *o/p*-regioselectivity s. M.E. Jung, T.I. Lazarova, Tetrahedron Letters *37*, 7-8 (1996).

Oxygen/Carbon Type SC∩OC

Potassium fluoride/alumina KF/Al_2O_3
2-Ethylenedithiolcarbonic acid esters from 2-ethylenealcohols   ←
via [3.3]-sigmatropic rearrangement

221.

Cinnamyl alcohol in acetonitrile and CS_2 adsorbed on KF-alumina, iodomethane added at 20° after 4 h, and extracted after a further 4 h at room temp. → S-methyl S-[1-(1-phenylprop-2-enyl)] dithiocarbonate. Y 75%. Reaction was slow with tertiary allylic alcohols, although initial addition of alkoxide to CS_2 was improved by addition of DMSO. F.e.s. D. Villemin, M. Hachemi, Synth. Commun. *25*, 2311-8 (1995); **S-(2-ethylene)monothiolcarbonic acid esters** cf. V. Samano, M.J. Robins, Can. J. Chem. *71*, 186-91 (1993).

Tetra-n-*butylammonium bromide* Bu_4NBr
1,3-Oxathiolan-2-ones from 1,3-dioxolane-2-thiones   ←
with Bu_3SnH cf. *42*, 506; with Bu_4NBr s. S.Y. Ko, J. Org. Chem. *60*, 6250-1 (1995).

Exchange ⇅

Hydrogen ↑   SC ⇅ H

Phenyl iodosotrifluoroacetate $PhI(OCOCF_3)_2$
Ar. sulfenylation with mercaptans   SH → SR
with NCS cf. *26*, 592; of phenolethers with $PhI(OCOCF_3)_2$ s. Y. Kita et al., Synlett *1995*, 211-2.

Phenyl iodosotrifluoroacetate/trimethylsilyl isothiocyanate $PhI(OCOCF_3)_2/Me_3SiNCS$
Ar. thiocyanation   H → SCN
with NaSCN/NBS cf. *50*, 307; of phenolethers with $PhI(OCOCF_3)_2/Me_3SiNCS$ s. K.R. Prabhu et al., J. Org. Chem. *60*, 7142-3 (1995).

Oxygen ↑   SC ⇅ O

Without additional reagents   *w.a.r.*
S-Alkylation with isoureas   SH → SR
of mercaptans s. *29*, 547s*35*; of thiolic acids with inversion, also N-alkylation, s. M.A. Poelert et al., Rec. Trav. Chim. Pays-Bas *113*, 365-8 (1994).

Sodium sulfide Na_2S
Dicarboxylic acid thioanhydrides from dicarboxylic acids   ←
via dicarboxylic acid anhydrides

222.

Optimization. A soln. of startg. diacid and 1 eq. N-methylmorpholine in dry THF at 0° under argon treated dropwise over 5 min with 1 eq. methyl chloroformate, stirred at 20° for 15 min, filtered, the filtrate (containing the formed anhydride) treated with *0.5 eq.* Na_2S in water, and stirred for 10-30 min until reaction complete by TLC and GC → product. Y 99% (based on Na_2S). Reaction took place via mixed acyclic anhydride formation, and the remaining diacid salt was readily recovered. F.e. incl. 4-, 6- and 7-membered ring anhydrides, also under phase transfer catalysis (with cetyltrimethylammonium bromide in CH_2Cl_2/H_2O), s. M.J. Kates, J.H. Schauble, J. Heterocyc. Chem. *32*, 971-8 (1995).

Rubidium fluoride/dicyclohexyl-18-crown-6 polyether/O-alkyl O-silyl keteneacetals ←
S-Acylation of mercaptans with N-acoxydicarboxylic acid imides SH → SCOR

223.

Lipophilic S-acylcoenzyme A derivs., previously obtainable in aq. organic media in lowish yield, can now be obtained more efficiently *in organic media* via prior silylation of the coenzyme to confer greater solubility. **E:** A soln. of coenyme A trihydrate (90% pure) in anhydrous acetonitrile treated with 25 eqs. dimethylketene methyl trimethylsilyl acetal at 20° for 14 h, solvent removed (0.1 mmHg; 1 h), dissolved *in anhydrous THF*, desilylated with 25 eqs. RbF and 10 mol% dicyclohexyl-18-crown-6, the mixture added to a soln. of startg. N-acoxyimide in the same solvent, and worked up after 4 h at 20° → product. Y 85%. F.e. and desilylation methods, also with N-acoxy--succinimides, s. K. Lucet-Levannier et al., J. Am. Chem. Soc. *117*, 7546-7 (1995).

Molecular sieves ←
4-Thiazolidones from oxo compds. O
and $(NH_4)_2CO_3$ s. *9*, 672; **polymer-based 3-component synthesis** from supported amines s. C.P. Holmes et al., J. Org. Chem. *60*, 7328-33 (1995).

Montmorillonite (Fe(III)-exchanged) ←
Envirocat EPZG ←
Mercaptals from oxo compds. under heterogeneous catalysis CO → $C(SR)_2$
with zeolites cf. *47*, 493s*48*,50; thioacetalation of aldehydes with Fe(III)-exchanged montmorillonite s. B.M. Chioudary, Y. Sudha, Synth. Commun. 26, 2993-7 (1996); 1,3-dithiolanes with Envirocat EPZG s. B.P. Bandgar et al., ibid. 1579-83.

Ammonium ceric nitrate $(NH_4)_2Ce(NO_3)_6$
1,3-Dithiolanes from oxo compds. O
with $LaCl_3$ cf. *46*, 490; with CAN for conversion of aldehydes and *alicyclic* ketones s. P.K. Mandal, S.C. Roy, Tetrahedron *51*, 7823-8 (1995).

Polymer-based 1-ethyl-3-(dimethylaminopropyl)carbodiimide ←
Thiolic acid esters from carboxylic acids and mercaptans COOH → COSR
with DCCI cf. *17*, 663; with polymer-based 1-ethyl-3-(dimethylaminopropyl)carbodiimide s. M. Adamczyk, J.R. Fishpaugh, Tetrahedron Letters *37*, 4305-8 (1996).

Trifluoroacetic anhydride/stannic chloride $(CF_3CO)_2O/SnCl_4$
Mercaptals from sulfoxides and mercaptans RS(O)CH → RSCSR'
s. *35*, 383; 1-phosphonylmercaptals with added $SnCl_4$ s. T.H. Kim, D.Y. Oh, Synth. Commun. *24*, 2313-8 (1994); chiral *anti*-α-siloxymercaptals with N,O-bis(TMS)acetamide/Me_3SiOTf s. Y. Kita et al., Tetrahedron Letters *36*, 109-12 (1995).

Hexamethyldisilthiane/cobaltous chloride $(Me_3Si)_2S/CoCl_2$
Replacement of carbonyl oxygen by sulfur CO → CS
thioketones s. *37*, 522s*48*; *o*-azidothioaldehydes with added $CoCl_2$ and subsequent heterodiene synthesis s. A. Degl'Innocenti et al., Chem. Lett. *1995*, 147-8.

Stannic chloride s. under $(CF_3CO)_2O$ $SnCl_4$

O,O-Diethyl dithiophosphoric acid (EtO)₂PSSH
Carboxylic acid thioamides from carboxylic acids and amines COOH → C(S)N<

224. PhCOOH + H₂NPh —(EtO)₂P(S)SH→ PhC(S)NHPh

A mixture of benzoic acid, 1 eq. aniline, and *3 eqs*. O,O-diethyl dithiophosphoric acid in toluene refluxed for 6 h → product. Y 94%. The relative amount of the reagent is crucial for the formation of thioamides, the corresponding carboxylic amide being obtained with 1-2 eqs. F.e. and with sec. amines s. N. Borthakur, A. Goswami, Tetrahedron Letters *36*, 6745-6 (1995).

2,4-Bis(p-methoxyphenyl)-1,3,2,4-dithiadiphosphetane 2,4-disulfide ←
5-Functionalized thiazoles from α-acylaminocarboxylic acid derivs. ○

5-alkoxy-derivs. with P₂S₅ cf. *6*, 638; 5-amino-derivs. with Lawesson's reagent s. O. Uchikawa et al., J. Heterocyc. Chem. *31*, 877-87 (1994).

2,4-Bis(p-methoxyphenyl)-1,3,2,4-dithiadiphosphetane 2,4-disulfide/p-toluenesulfonic acid ←
Thiophenes from 2-ethylenesilanes and carboxylic acid chlorides via β,γ-epoxyketones ←

225.

Startg. acid chloride and 1 eq. allylsilane in methylene chloride added slowly at -78° to a soln. of TiCl₄ in the same solvent, and quenched with 3 *N* HCl after 1 h → crude ketone, in dichloromethane treated with 2 eqs. 50% *m*-chloroperoxybenzoic acid at 0°, allowed to react for 17 h, solvent removed, benzene and 1.2 eqs. Lawesson's reagent added, heated to boiling, treated with a little *p*-TsOH, and refluxed for 1 h → product. Y 64%. The order of addition is critical. F.e. and isolation of intermediates s. K.-T. Kang, J.S. U, Synth. Commun. *25*, 2647-53 (1995).

Bismuth(III) sulfate Bi₂(SO₄)₃
Mercaptals from oxo compds. under mild conditions CO → C(SR)₂

226.

1,3-Dithiolanes. *p*-Methoxybenzaldehyde and 2 eqs. ethane-1,2-dithiol added successively to a stirred soln. of *0.1 mol%* Bi₂(SO₄)₃ in acetonitrile at room temp. under air, and stirred for a further 2.5 h → product. Y 99%. The method is simple, high-yielding, economical, and non-toxic, and the Bi(III)-salt is insensitive to moisture. The reagent is also highly chemoselective, aldehydes and cyclic ketones being protected in preference to acyclic ketones. F.e. and with BiCl₃ or BiBr₃, also cyclic mercaptals and monothioacetals from acetals (with BiCl₃), s. N. Komatsu et al., Synlett *1995*, 984-6; mercaptals from aldehydes **or acetals** with LiClO₄ cf. V.G. Saraswathy, S. Sankararaman, J. Org. Chem. *59*, 4665-70 (1994).

Trimethylsilyl triflate Me₃SiOSO₂CF₃
1,3-Oxathiolanes from oxo compds. and reverse reaction ○

227.

1,3-Oxathiolanes and oxo compds. substituted by an aryl group can be interconverted under essentially identical [catalytic, non-aq.] conditions in a 'merry-go-round' fashion. E: Benzophenone and 2-mercaptoethanol stirred at room temp. for 10 min in the presence of a little trimethylsilyl triflate → 2,2-diphenyl-1,3-oxathiolane (Y 72%), treated with a little trimethylsilyl triflate at room temp. for 10 min (cf. *50*, 110) → benzophenone (Y 82%). Selective cleavage of the oxathiolane in the presence of the corresponding 1,3-dithiolane is also possible. F.e.s. T. Ravindranathan et al., Tetrahedron Letters *36*, 2285-8 (1995).

Lithium perchlorate LiClO₄
Mercaptals from aldehydes or acetals s. *51*, 226 C(SR)₂

Cobaltous chloride s. under *(Me₃Si)₂S* CoCl₂

Tris(dibenzylideneacetone)dipalladium/(R,R)-1,2-bis[o-(diphenylphosphino)
benzoylamino]cyclohexane ←
Asym. synthesis of β,γ-ethylenesulfones from acoxy-2-ethylenes $C(SO_2R)C=C$
s. *37*, 527s*41*; also desymmetrization of 2-ene-1,4-diol esters with Pd_2dba_3/(R,R)-1,2-bis[*o*-(diphenylphosphino)benzoylamino]cyclohexane s. B.M. Trost et al., J. Am. Chem. Soc. *117*, 9662-70 (1995); also from β,γ-ethylenechlorides with $Pd(PPh_3)_4$/chiral 2-*o*-phosphinoaryl-Δ^2-oxazolines cf. H. Eichelmann, H.-J. Gais, Tetrahedron:Asym. *6*, 643-6 (1995).

Nitrogen ↑   SC ⇅ N

Sodium mercaptides   *NaSR*
Thioiminoesters from 1-imidoylbenzotriazoles s. *51*, 85 $C(=NR)SR'$
Boron fluoride BF_3
Thiazoles from α-diazoketones   ○
with HBr cf. *20*, 456; thiazole-4-carboxylic acid esters from thioamides with BF_3 or $AlCl_3$ s. H.-S. Kim et al., J. Heterocyc. Chem. *32*, 937-9 (1995).

Halogen ↑   SC ⇅ Hal

Potassium hydride/n-butyllithium   *KH/n-BuLi*
Ethynyl thioethers from mercaptans s. *51*, 392 $SH \rightarrow SC\equiv CH$
Sodium hydroxide/polyethylene glycol 400   *NaOH/PEG*
Sym. disulfides from halides under phase transfer catalysis $2 \text{ RHal} \rightarrow (RS)_2$
with Li_2S/Aliquat 336 cf. *38*, 524; with $NaOH/S_8$/PEG-400 s. J.-X. Wang et al., Synth. Commun. *25*, 3573-81 (1995).
Potassium hydroxide   *KOH*
Potassium hydroxide/benzyltriethylammonium chloride/imidazol-1-ylacetonitrile ←
Sym. trithiocarbonic acid esters from halides $2 \text{ RHal} \rightarrow (RS)_2CS$
with K_2S cf. *16*, 671; with excess of KOH s. M. Leung et al., J. Chem. Res. (S) *1995*, 478-9; also cyclic trithiocarbonates with imidazol-1-ylacetonitrile/KOH/$BnEt_3NCl$ cf. V.J. Ram et al., ibid. *1996*, 64-5.
Sodium disulfide/polyethylene glycol-400 Na_2S_2/*PEG*
Sym. acyl disulfides from carboxylic acid chlorides $2 \text{ RCOCl} \rightarrow RCOSSCOR$
with $Na_2S_2/R_4P^+Br^-$ cf. *5*, 452s*37*; with PEG-400 s. J.-X. Wang et al., Synth. Commun. *25*, 889-98 (1995).
Sodium thiosulfate/potassium salt $Na_2S_2O_3/K^+$
Disulfides from halides   Hal → SSR
with $NaOH/S_8$ in isopropanol cf. *34*, 541; from K-thiolates with $Na_2S_2O_3$ via thiosulfuric acid monoesters s. P. Hiver et al., Tetrahedron Letters *35*, 9569-72 (1994).
Cuprous iodide/sodium salt CuI/Na^+
Diaryl sulfones from arenesulfinic acids and unactivated ar. iodides   ArI → $ArSO_2Ar'$
under neutral, non-oxidative conditions

228.

A mixture of phenyl *p*-iodophenyl sulfide, 1.6 eqs. commercial Na-*p*-toluenesulfinate tetrahydrate, and 1.5 eqs. CuI in DMF stirred at 110° for 6 h under argon → *p*-tolyl *p*-(phenylthio)phenyl sulfone. Y 62%. The method is simple and general, a variety of functional groups are compatible (aldehydes, acetylamines, phenolethers), and the substrates are readily available. F.e. incl. bis(diaryl sulfones) from arenediiodides, also with Cu(II)-bis(*p*-toluenesulfinate), s. H. Suzuki, H. Abe, Tetrahedron Letters *36*, 6239-42 (1995).

Zinc sulfide ZnS
Thiolic acid esters from halides Hal → SCOR
with K⁺ cf. *5*, 444; with ZnS (or ZnCl₂/K₂CO₃), notably for conversion of *hindered* tert., allylic, and benzylic chlorides or bromides (prim. and sec. halides being unreactive) s. K.N. Gurudutt et al., Tetrahedron *51*, 3045-50 (1995).

Tris(dibenzylideneacetone)dipalladium/1,1'-bis(diphenylphosphino)ferrocene/triethylamine ←
Ar. thioethers from halides ArHal → ArSR
with Pd(PPh₃)₄/NaOBu-*t* cf. *36*, 561; chiral α-acylamino-β-(arylthio)carboxylic acid esters with Pd₂dba₃/dppf/Et₃N s. P.G. Ciattini et al., Tetrahedron Letters *36*, 4133-6 (1995).

Tetrakis(triphenylphosphine)palladium(0)/chiral 2-o-phosphinoaryl-Δ²-oxazoline ←
Asym. synthesis of β,γ-ethylene-sulfones from -chlorides s. *37*, 527s*51* C(SO₂R)C=C
Via intermediates *v.i.*
Mercaptans from halides via thiolic acid esters Hal → SH
One-pot conversion

229. BnCl →[(P)-N⁺Me₃ AcS⁻]→ [BnSAc] → BnSH

A soln. of 0.5 eq. Pd(OAc)₂ in methanol and *3 eqs.* borohydride-exchange resin (prepared from NaBH₄ and the chloride-form of Amberlite IRA-400) added under N₂ to a stirred soln. of benzyl chloride and 2 eqs. thioacetate-exchange resin (prepared by stirring K-thioacetate with the same resin in the same solvent at room temp. for 1 h), refluxed with stirring for 1 h, and treated with 2 *N* HCl → benzyl mercaptan. Y 95%. A variety of halides are thus converted to the corresponding thiolic acid esters, and thence to mercaptans under *neutral conditions* by novel Pd-catalyzed methanolysis (cf. *50*, 234). Secondary halides required longer reaction times. F.e. and **from** the intermediate **thiolic acid esters** s. J. Choi, N.M. Yoon, Synth. Commun. *25*, 2655-63 (1995).

Sulfur ↑ SC ⇅ S

Potassium salt K⁺
2-Functionalized acylamines from 2-acylaminoalcohols C
via 3-acyl-1,2,3-oxathiazolidine 2,2-dioxides
Regio- and stereo-specific ring opening

230. [structure: Ph-fused dioxane with OBn, NHAc, OH] ⋯→ [cyclic sulfamidate with NAc, OBn] →[KSAc]→ [AcS, NHAc, OBn product]

3-Thioglucosamine derivs. A soln. of startg. cyclic sulfamidate (readily obtainable from the parent 2-acylaminoalcohol and N,N'-sulfonyldiimidazole) in DMF treated with KSAc at room temp. for 30 min, worked up, and treated with H₂SO₄ in aq. THF at room temp. for 30 min → product. Y 82%. There was no displacement of the corresponding triflates or tosylates. F.e. and nucleophiles (RSH, NaN₃) s. B. Aguilera, A. Fernández-Mayoralas, Chem. Commun. *1996*, 127-8.

Anion exchanger-supported tetrahydridoborate [BH₄⁻]
Ar. thioethers from disulfides and halides RSSR → RSR'
with N₂H₄ cf. *6*, 658s*47*; with anion exchanger-supported tetrahydridoborate under essentially non-basic conditions s. N.M. Yoon et al., J. Org. Chem. *59*, 3490-3 (1994).

Diisobutylaluminum mercaptides/diethylaluminum chloride i-Bu₂AlSR/Et₂AlCl
Phenyl iodosoacetate/sodium azide PhI(OAc)₂/NaN₃
Thiolic acid esters from aldehydes and disulfides s. *51*, 256 CHO → C(O)SR

Remaining Elements ↑   SC ⇅ Rem

Sodium hydrogen carbonate $NaHCO_3$
Sulfonic acid derivs. from stannanes $\geqslant Sn \rightarrow S(O)X$
ar. sulfonamides with $ClSO_2NCO$ cf. *47*, 512; sulfonic acid salts with Me_3SiOSO_2Cl s. Chem. Ber. *128*, 575-80 (1995).

Carbon ↑   SC ⇅ C

Without additional reagents   w.a.r.
Benzothiazoles from *o*-(phosphoranylideneamino)thioethers   ○
and carboxylic acid chlorides

231.

1 eq. Benzoyl chloride added to a soln. of startg. iminophosphorane in dry, hot benzene under N_2, and refluxed for 24 h → 2-phenylbenzothiazole. Y 65%. F.e.s. M. Takahashi, M. Ohba, Heterocycles *41*, 455-60 (1995).

Potassium hydroxide   KOH
Thiazolidine-2-thiones from oxazolidine-2-thiones   ←

232.

A mixture of (S)-4-benzyloxazolidine-2-thione and 5 eqs. CS_2 in 1 *N* aq. KOH stirred at 100° for 16 h → (S)-4-benzylthiazolidine-2-thione. Y 80%. The conversion is restricted to *5-unsubst.* derivs. F.e.s. D. Delaunay et al., J. Org. Chem. *60*, 6604-7 (1995).

Triethylamine Et_3N
2-Hydroxythioethers from 1,3-dioxolan-2-ones   C
s. *32*, 536; α-functionalized β-hydroxycarboxylic acid esters with inversion of configuration s. S.-K. Kang et al., J. Chem. Soc. Perkin Trans. I *1994*, 3513-4; **from 1,3-dioxolane-2-thiones**, regioselectivity, s. S.Y. Ko, J. Org. Chem. *60*, 6250-1 (1995).

Dilithium di-tert-butylcyanocuprate $t\text{-}Bu_2Cu(CN)Li_2$
Thioethers from thiocyanates   SCN → SR
and Grignard compds. cf. *23*, 628; aryl *tert*-butyl sulfides with $t\text{-}Bu_2Cu(CN)Li_2$ s. F.D. Toste, I.W.J. Still, Tetrahedron Letters *36*, 4361-4 (1995).

Resin-supported tetrahydridoborate s. under $Pd(OAc)_2$ $[BH_4^-]$
Samarium diiodide/hexamethylphosphoramide $SmI_2/(Me_2N)_3PO$
α,β-Ethylenethiolic acid esters from 5-alkylidene-1,3-dioxane-4,6-diones   C
s. *51*, 261

Enzyme (lipase)/silica   ←
Chiral 1,1-acoxythioethers from aldehydes and mercaptans   ←
via asym. O-acylation of 1,1-hydroxythioethers
Dynamic kinetic resolution

233.

A mixture of startg. aldehyde and 1 eq. octanethiol in 3:1 *tert*-butyl methyl ether/vinyl acetate stirred vigorously at 30° for 6 days with *Pseudomonas fluorescens* lipase in the presence of SiO_2 → (S)-product. Y 85% (e.e. >95%). Chemical yields in excess of 50% are achieved by a dissociation-recombinative mechanism, the SiO_2 effectively epimerizing the intermediate, unreactive (R)-1,1-hydroxythioether. F.e.s. S. Brand et al., Tetrahedron Letters *36*, 8493-6 (1995).

Palladous acetate/Amberlite IRA-400 tetrahydridoborate
**Michael addition with *in situ*-generated mercaptans
under neutral conditions**

C=C → C(SR)CH   ←

234.

[HSC$_6$H$_{13}$] + AcSC$_6$H$_{13}$ → /\COOMe → SC$_6$H$_{13}$-CH-CH$_2$-COOMe

A soln. of 0.05 eq. Pd(OAc)$_2$ in methanol added to 1 eq. borohydride exchange resin (BER), stirred under reflux for 3 h, methyl crotonate and 1.05 eqs. hexyl thioacetate added, and refluxed for a further 2 h → methyl 3-(hexylthio)butyrate. Y 93%. This one-pot procedure combines Pd-catalyzed methanolysis of the thioacetate (cf. *50*, 229) with BER-catalyzed Michael addition, and is particularly useful for reactions of unstable mercaptans. F.e.s. D.W. Lee et al., Synth. Commun. *26*, 2189-96 (1996).

Elimination ⇑

Oxygen ↑   SC ⇑ O

Methanesulfonyl chloride/triethylamine
Ene-1,2-di(thioethers) from α-hydroxymercaptals s. *51*, 487
Methyl(carboxysulfamoyl)triethylammonium hydroxide inner salt
Diethylaminosulfur trifluoride
Δ^2-Thiazolines from 2-(thioacylamino)alcohols

MeSO$_2$Cl/Et$_3$N
C(SR)=C(SR)
←
Et$_2$NSF$_3$
○

with CF$_3$COOH cf. *25*, 458; with Et$_2$NSF$_3$, inversion of configuration, s. P. Lafargue et al., Synlett *1995*, 171-2; peptide derivs. with Burgess reagent *without* epimerization cf. P. Wipf, P.C. Fritch, Tetrahedron Letters *35*, 5397-400 (1994).

Halogen ↑   SC ⇑ Hal

tert-*Butyllithium*
Benzo[*b*]thiophenes from *o*-mercapto-β,β-dihalogenostyrenes s. *51*, 112

t-*BuLi*
○

Formation of Rem-Rem Bond

Rearrangement ⋒

Oxygen/Remaining Elements Type   RemRem ⋒ ORem

Tetrakis(triphenylphosphine)palladium(0)
**Skeletal rearrangement of (alkoxy)oligosilanes
via regiospecific silylene group migration**

Pd(PPh$_3$)$_4$
←

235.

MeO–Si(OMeOMe CD$_3$)–Si(OMe)–Si(CD$_3$)–OMe
      Me   Me   CD$_3$

→ [MeO–Si(OMe CD$_3$)–Si(OMe)–PdL$_2$ Me–Si(OMe)–Si(CD$_3$)–OMe]
         Me

→ MeO–Si(OMe)–Si(OMe)=PdL–Si(CD$_3$)(CD$_3$)
         Me

→ MeO–Si(OMe CD$_3$ OMe)–Si–Si–OMe
         Me CD$_3$ Me

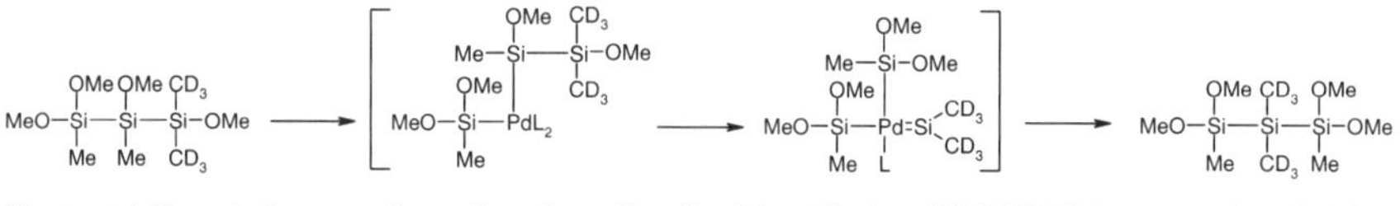

Startg. trisilane in benzene heated under reflux for 2 h with 4 mol% Pd(PPh$_3$)$_4$ → product. Y 90%. Reaction proceeds via oxidative addition of Pd(0) to the Si-Si bond possessing the more alkoxy groups, followed by migratory insertion of silylene. Tetrasilane derivs. reacted in the same way. F.e.s. K. Tamao et al., J. Am. Chem. Soc. *117*, 8043-4 (1995).

Exchange ⇅

Hydrogen ↑   RemRem ⇅ H

Tris(trinitratocerium(IV)) paraperiodate $Ce(NO_3)_3H_2IO_6$
Sym. diselenides from selenols   2 RSeH → (RSe)$_2$
with FeCl$_3$/O$_2$ cf. *22*, 652; with tris(trinitratocerium(IV)) paraperiodate, also sym. disulfides (cf. *40*, 133), s. N. Iranpoor et al., Org. Prep. Proc. Intern. *27*, 216-9 (1995).

Carbon ↑   RemRem ⇅ C

Lithium hydridotriethylborate or Diisobutylaluminum hydride   ←
Sym. diselenides from selenocyanates   2 RSeCN → (RSe)$_2$
with NaBH$_4$ cf. *22*, 655s*41*; with LiEt$_3$BH or DIBAL (for diketodiselenides) s. P. Salama, C. Bernard, Tetrahedron Letters *36*, 5711-4 (1995).

Formation of Rem-C Bond

Uptake ⇓

Addition to Oxygen and Carbon   RemC ⇓ OC

Lithium diisopropylamide   i-Pr_2NLi
Lithium aluminum (R)-1,1'-bi-2-naphthoxide   ←
α-Oxyphosphonic acid derivs. from aldehydes with asym. induction   ←
s. *49*, 510s*50*; chiral *a*-hydroxyphosphonic acid diamides s. V.J. Blazis et al., Tetrahedron:Asym. *5*, 499-502 (1994); f. details s. J. Org. Chem. *60*, 931-40 (1995); nucleoside 3'-phosphonates s. W.L. McEldoon, D.F. Wiemer, Tetrahedron *51*, 7131-48 (1995); chiral α-hydroxyphosphonates with Li-Al-(R)-BINOL cf. T. Arai et al., J. Org. Chem. *61*, 2926-7 (1996).

Sodium phosphate dodecahydrate $Na_3PO_4·12H_2O$
α-Hydroxyphosphonic acid esters from aldehydes   CHO → CH(OH)PO(OR)$_2$
with KF-Al$_2$O$_3$ cf. *41*, 556; with Na$_3$PO$_4$·12H$_2$O, natural phosphate or KF-doped phosphate s. S. Sebti et al., Tetrahedron Letters *37*, 3999-4000 (1996).

Addition to Nitrogen and Carbon   RemC ⇓ NC

Without additional reagents   w.a.r.
α-Aminophosphonous acids from N-protected aldimines   CH(NH$_2$)PH(O)OH
from N-tritylaldimines cf. *49*, 556; from polymer-based azomethines s. E.A. Boyd et al., Tetrahedron Letters *37*, 1647-50 (1996); addition of HP(OSiMe$_3$)$_2$ to cyclic azomethines s. M. Hatam, J. Martens, Synth. Commun. *25*, 2553-9 (1995).

Lanthanum triisopropoxide/(R)-1,1'-bi-2-naphthol/potassium bis(trimethylsilyl)amide   ←
α-Aminophosphonic acid esters from aldimines   CH(NH)PO(OR)$_2$
Asym. conversion

236.   cyclopentyl-CH=NCHPh$_2$ + HPO(OMe)$_2$ ⟶ cyclopentyl-CH(NHCHPh$_2$)PO(OMe)$_2$

Lanthanoid-*potassium*-BINOL (LPB) complexes are excellent chiral auxiliaries for asym. hydrophosphonylation of imines. **E:** A soln. of 20% LPB complex (prepared from 1:1:0.33:0.33

(R)-BINOL, KN(SiMe₃)₂, La(OPr-*i*)₃ and water in 7:1 toluene/THF by stirring the mixture for 1 h at room temp. under argon) added slowly to startg. imine and 5 eqs. dimethyl phosphite *at room temp.* under argon, stirred for 90 h, and quenched with water → product. Y 87% (e.e. 85%). The corresponding lithium complexes were less effective. F.e. and with the praseodymium or gadolinium complexes s. M. Shibasaki et al., J. Org. Chem. *60*, 6656-7 (1995).

Addition to Carbon-Carbon Bonds   RemC ⇓ CC

(Triethylphosphine)gold(I) chloride/1,2-bis(dicyclohexylphosphino)ethane   AuCl(PEt₃)/dcpe
1,2-Di(boronic acid esters) from ethylene derivs.

237.

1-Aryl-1,2-di(boronic acid esters). A mixture of 4-vinylanisole, 1.5 eqs. bis(catecholborane), and 8 mol% gold(I) complex (generated from 1:2 AuCl(PEt₃)/1,2-bis(dicyclohexylphosphino)-ethane) in [D₈]THF heated at 80° for 48 h → product. 100% selectivity. This is the first example of catalytic diborylation of an alkene. The highest yield under Rh-catalysis was 44%. F.e. and activated alkenes s. R.T. Baker et al., Angew. Chem. Intern. Ed. *34*, 1336-8 (1995).

Diisobutylaluminum hydride   i-Bu₂AlH
(E)-Enetellurides from acetylene derivs. s. *50*, 343s*51*   C≡C → CH=C(TeR)

Yttrium pentamethylcyclopentadiene complexes   ←
Lanthanide-catalyzed hydrosilylation
of alkenes cf. *48*, 543; regiospecific hydrosilylation **of internal acetylene derivs.** with Cp*₂YMe·THF s. Organometallics *14*, 4570-5 (1995).

Chlorobis(cyclopentadienyl)hydridozirconium   Cp₂Zr(H)Cl
Functionalized ethylene from acetylene derivs.   C≡C → CH=C(X)
via regio- and stereo-specific hydrozirconation
(E)-enetellurides cf. *50*, 343; **(E)-eneselenides** (cf. *36*, 584) s. X. Huang, L.-S. Zhu, J. Chem. Soc. Perkin Trans. I *1996*, 767-8; (E)-eneboronates (cf. *47*, 567) with pinacolborane, *catalytic* procedure, cf. S. Pereira, M. Srebnik, Organometallics *14*, 3127-8 (1995); **(E)-enetellurides** s. M.J. Dabdoub et al., Tetrahedron Letters *36*, 7623-6 (1995); via hydroalumination (*i*-Bu₂AlH) cf. Tetrahedron *51*, 12971-82 (1995).

Zirconium tetrachloride/triethylamine   ZrCl₄/Et₃N
Enestannanes from acetylene derivs.   C≡C → C(Sn≤)=CH
(E)-isomers with Et₃B cf. *42*, 554; (Z)-isomers with ZrCl₄/Et₃N s. Y. Yamamoto et al., Chem. Commun. *1995*, 2405-6; details and divinylstannanes with Bu₂SnH₂ s. J. Org. Chem. *61*, 4568-71 (1996).

Dichlorotris(2,2'-bipyridyl)ruthenium(II)   Ru(bpy)₃Cl₂
2-Sulfonylselenides from ethylene derivs.   C=C → C(SeR)C(SO₂R')
with BF₃ cf. *36*, 582; with Ru(bpy)₃Cl₂ under visible light irradiation, α-arylseleno-β-sulfonylesters and α-sulfonylmonoselenoacetals, s. D.H.R. Barton et al., Tetrahedron Letters *35*, 2869-72 (1994); using *p*-TolS(O)ONa/PhSeSePh cf. C.-P. Chuang, S.-F. Wang, Synth. Commun. *25*, 3549-63 (1995).

Chlorotris(triphenylphosphine)rhodium(I) $RhCl(PPh_3)_3$
Regiospecific hydroboration
←

238.

Improved procedure. Pinacolborane is more stable than catecholborane and a more effective *stoichiometric* reagent in Rh-catalyzed hydroboration (cf. *44*, 128s*47-9*). Unlike catechol adducts, pinacolboronates are air-, moisture- and chromatographically-stable compds., so that the separation of isomers is more facile. **E:** A soln. of 1-octene, *1.1 eqs.* pinacolborane, and 1 mol% Rh(PPh$_3$)$_3$Cl in methylene chloride stirred at 25° for 10 min → product. Y 99%. Reaction with catecholborane normally requires an excess of the reagent. F.e. and less efficiently with HZrCp$_2$Cl, also Kharasch reaction in carbon tetrachloride, s. S. Pereira, M. Srebnik, J. Am. Chem. Soc. *118*, 909-10 (1996); hydroboration of internal alkenes and alkynes s. Tetrahedron Letters *37*, 3283-6 (1996).

Bis(cyclooctadiene)rhodium(I) fluoroborate/triphenylphosphine   ←
Rhodium trichloride/Tröger's base   ←
Enesilanes from acetylene derivs.   C≡C → CH=C(Si≦)
with Pt(II)-halides cf. *15*, 494s*44*; (E)-enesilanes from terminal alkynes with Rh(COD)$_2$BF$_4$/2PPh$_3$, and one-pot conversion of propargyl alcohols to **β-silylketones**, s. R. Takeuchi et al., J. Org. Chem. *60*, 3045-51 (1995); with RhCl$_3$/Tröger's base cf. Y. Goldberg, H. Alper, Tetrahedron Letters *36*, 369-72 (1995).

Palladous acetate/1,1,3,3-tetramethylbutyl isocyanide $Pd(OAc)_2$/RNC
2-Silylenestannanes from acetylene derivs.   C≡C → C(Si≦)=C(Sn≦)
regio- and stereo-selectivity s. *41*, 564; (Z)-1-alkoxy- and (Z)-1-arylthio-2-silylenestannanes s. P. Kocienski et al., Synthesis *1994*, 1301-9; (Z)-2-alkoxy-2-silylenestannanes with Pd(OAc)$_2$/ 1,1,3,3-tetramethylbutyl isocyanide s. M. Murakami et al., Organometallics *12*, 4223-7 (1993).

Tris(dibenzylideneacetone)dipalladium/triphenylphosphine Pd_2dba_3/Ph_3P
Regio- and stereo-specific hydrostannylation   C≡C → CHC(Sn≦)
of oxabicycloalkenes s. *51*, 28

cis-*Dimethylbis(methyldiphenylphosphine)palladium* $PdMe_2(PPh_2Me)_2$
α-Methylenephosphonic acid esters from terminal acetylene derivs.   C(=CH$_2$)PO(OR)$_2$
via oxidative addition of palladium(0) to phosphorus-hydrogen bonds

239.

Equimolar amounts of dimethyl phosphite and startg. acetylene deriv. in THF heated at 67° for 15-20 h with 3 mol% *cis*-PdMe$_2$(PPh$_2$Me)$_2$ → product. Y 89% (96:4 mixture of regioisomers). The reaction is generally applicable to both aliphatic and aromatic alk-1-ynes, and leaves olefinic functionality unaffected. The initial adduct, (EtO)$_2$P(O)Pt(H)(PEt$_3$)$_2$, was isolated when Pt(PEt$_3$)$_4$ was used as catalyst. F.e.s. L.-B. Han, M. Tanaka, J. Am. Chem. Soc. *118*, 1571-2 (1996).

Bis(allylpalladium chloride)/(S)-2-(diphenylphosphino)-1,1'-binaphthyl   ←
Asym. hydrosilylation   C≡C → CHC(Si≦)
s. *47*, 542; of styrenes with (S)-2-(diphenylphosphino)-1,1'-binaphthyl and conversion to chiral benzylalcohols s. Chem. Commun. *1995*, 1533-4; of 1,3-dienes, chiral (Z)-1-aryl-2-ethylenesilanes, s. Y. Hatanaka et al., Tetrahedron Letters *35*, 7981-2 (1994).

Via intermediates   *v.i.*
2-Hydroxyselenides from ethylene derivs.   C=C → C(OH)C(SeR)
via 2-acoxyselenides cf. *29*, 601; **via 2-(arylseleno)nitric acid esters** with *in situ*-generated PhSeONO$_2$, regio- and stereo-selectivity, s. L.-B. Han, M. Tanaka, Chem. Commun. *1996*, 475-6.

Exchange ⇅

Oxygen ↑   RemC ⇅ O

Without additional reagents   *w.a.r.*
**α-Aminophosphonic acids from aldehydes and amines   ←
via α-aminophosphonous acids**
Asym. induction

240.

(1S)-α-Methylbenzylamine added dropwise to anhydrous *hypophosphorous acid* in dry ethanol with cooling at such a rate that the temp. remained ≤ 25°, the crude amine ·H_3PO_3 salt separated, and refluxed with a slight excess of isobutyraldehyde in ethanol for 4 h → intermediate (R)-α-aminophosphonous acid (Y 45%; single diastereomer), treated with bromide-water at 70° for 6 h → (R)-product (Y 88%). The 2nd step conveniently effects simultaneous oxidation and deprotection. F.e.s. R. Hamilton et al., Tetrahedron Letters *36*, 4451-4 (1995).

n-*Butyllithium/boron fluoride*   *BuLi/BF₃*
α,β-Acetyleneborinic acid esters from terminal acetylene derivs. $B(OR)_2 \to B(OR)C\equiv C$
s. *42*, 562; borination using 10-methoxy-9-oxa-10-borabicyclo[3.3.2]decane with added BF_3 s. J.A. Soderquist et al., Tetrahedron Letters *36*, 6847-50 (1995).

n-*Butyllithium/zirconocene dichloride*   *BuLi/Cp₂ZrCl₂*
Stannanes from alkoxystannanes via organozirconium(IV) complexes   H → Sn≤

241.

Acetylenestannanes. A mixture of startg. alkynylzirconocene chloride (prepared *in situ* from trimethylsilylacetylene by lithiation followed by treatment with Cp_2ZrCl_2) and ethoxy(tri-*n*-butyl)stannane in THF heated at 40° for 12 h → product. Y 78%. Other tin reagents, e.g. Bu_3SnCl, Bu_3SnSPh and $(Bu_3Sn)_2O$, gave poor yields. F.e. incl. **enestannanes, 2-ethylenestannanes** and cyclic stannanes, s. S. Kim, K.H. Kim, Tetrahedron Letters *36*, 3725-8 (1995).

n-*Butyllithium/methanesulfonyl chloride*   *BuLi/MeSO₂Cl*
2-Ethylenestannanes from 2-ethylenealcohols via mesyloxy-2-ethylenes   OH → Sn≤

242.

1 eq. 1.78 M *n*-BuLi in cyclohexane added to a soln. of startg. alcohol in THF at -78°, 1 eq. methanesulfonyl chloride added after 20 min, the mixture allowed to react for 35 min, a soln. of 1 eq. *n*-Bu₃SnLi (prepared from LDA and *n*-Bu₃SnH) in THF added dropwise via cannula, after 2 h at -78° the mixture allowed to warm to room temp., and quenched with water after 12 h 5 2-(tri-*n*-butylstannylmethyl)hept-1-ene. Y 85%. Reaction, which is notably suitable for preparing α-unsubst. derivs., proceeds **with retention of olefin configuration**. Silyl ethers and phenylthio groups remained unaffected. F.e.s. S. Weigand, R. Brückner, Synthesis *1996*, 475-82.

Sodium hydrogen selenide/sodium selenide   *NaHSe/Na$_2$Se$_2$*
Sym. diselenides from aldehydes   2 RCHO → (RCH$_2$Se)$_2$

243.
$$2 \text{ p-ClC}_6\text{H}_4\text{CHO} \xrightarrow[\text{[Na}_2\text{Se}_2/\text{NaHSe]}]{\text{Se/NaBH}_4} \text{p-ClC}_6\text{H}_4\text{CH}_2\text{SeSeCH}_2\text{C}_6\text{H}_4\text{Cl-p}$$

Ethanol added to an ice-cooled mixture of 4 eqs. Se-powder and 3.3 eqs. NaBH$_4$ under N$_2$, the mixture heated at 60-70° for 1 h, after vigorous evolution of gas had subsided a soln. of startg. aldehyde in DMF (EtOH:DMF 1:2) injected at 60-70°, and the mixture stirred at 120-130° for 7 h → product. Y 92%. This simple, one-pot method is relatively mild and safe (avoiding the use of poisonous H$_2$Se gas), affords better yields than those previously reported, and is applicable to both aliphatic and aromatic aldehydes. F.e.s. X. Huang et al., Org. Prep. Proc. Intern. 27, 492-4 (1995).

Boron fluoride s. under BuLi   *BF$_3$*
Zirconocene dichloride s. under BuLi   *Cp$_2$ZrCl$_2$*
Methanesulfonyl chloride s. under BuLi   *MeSO$_2$Cl*

Nitrogen ↑   RemC ⇅ N

Sodium hydride   *NaH*
Nucleophilic ring opening of 1-nitroxido compds.   C
s. 42, 176s46,48; cyclic β-ketophosphonic acid esters from dialkyl phosphites s. D.Y. Kim, M.S. Kong, J. Chem. Soc. Perkin Trans. I 1994, 3359-60.

Halogen ↑   RemC ⇅ Hal

Without additional reagents   *w.a.r.*
Arylseleno-O-heterocyclics from ethylenealcohols   O
s. 32, 591s46; 3-arylselenotetrahydrofurans with (2,4,6-triisopropylphenyl)selenenyl bromide s. B.H. Lipshutz, T. Gross, J. Org. Chem. 60, 3572-3 (1995); from 2-functionalized 3-ethylenealcohols s. Y. Landais, D. Planchenault, Synlett 1995, 1191-3; 3-arylseleno-4-silyltetrahydrofurans, stereoselectivity, s. Y. Landais et al., Tetrahedron Letters 36, 2987-90 (1995).

Electrolysis   ↯
Acylsilanes from 1-acylimidazoles under mild conditions   C(O)Im → C(O)Si≤

244.
<chemical reaction scheme: N-octanoylimidazole + Me$_3$SiCl → octanoyltrimethylsilane>

A stirred soln. of N-octanoylimidazole and 5 eqs. trimethylsilyl chloride in THF containing 0.5 M Bu$_4$NClO$_4$ electrolyzed in the cathode chamber of an H-type cell fitted with Pt-electrodes under a constant current of 20 mA at room temp. until 2.5 F/mol passed → octanoyltrimethylsilane. Y 73%. Alkyl chloride and carbomethoxy groups were unaffected, but conjugated double bonds were reduced. F.e.s. J. Yoshida et al., Tetrahedron Letters 36, 8839-42 (1995).

Sodium hydroxide/poylethylene glycol 400   *NaOH/PEG*
Sym. diselenides from halides   2 RHal → (RSe)$_2$
sym. diacyl diselenides cf. 32, 578s46; sym. dibenzyl diselenides s. J.-X. Wang et al., J. Chem. Soc. Perkin Trans. I 1994, 2341-3.

Potassium hydroxide/hydrazine   *KOH/N$_2$H$_4$*
Sym. diselenides from halides   2 RHal → R$_2$Se
with Se/CO/H$_2$O/DBU cf. 47, 563; with Se/N$_2$H$_4$/KOH, **also sym. tellurides** (cf. 44, 531s48), s. N.A. Korchevin et al., Zh. Obshch. Khim. 65, 99-101 (1995) (Russ.).

Tellurides from two different halide molecules RTeR′

245. MeI $\xrightarrow{[K_2Te_2]}$ [MeTeTeMe] $\longrightarrow$ [MeTeK] $\xrightarrow{EtBr}$ MeTeEt

Finely powdered Te added portionwise to 1 eq. KOH in hydrazine hydrate at 80°, heated for 2 h at 80-90°, cooled, 1 eq. methyl iodide added at such a rate that the temp. remained below 40°, treated with a further equivalent of Te and 3 eqs. KOH, stirred vigorously at 80-90° for 2-3 h, cooled to 25°, and 1 eq. ethyl bromide added at such a rate that there was no significant increase in temp. → ethyl methyl telluride. Y 82%. F.e.s. E.N. Deryagina et al., Zh. Obshch. Khim. 65, 1145-7 (1995) (Russ.).

Triethylamine Et_3N
β-Ketothiophosphorus(V) compds. SH → P(S)<
from α-mercaptoketones and phosphorus(III) acid chlorides

246. MeCOCH$_2$SH + ClP(OEt)$_2$ $\longrightarrow$ MeCOCH$_2$P(S)(OEt)$_2$

Diethyl chlorophosphite added dropwise to a stirred soln. of startg. mercaptoketone and 1 eq. Et$_3$N in dry benzene (or ether) at 0-5°, and left for 12 h → product. Y 78%. F.e., also S-phosphorylation (with enol phosphites), S-silylation, and S-stannylation s. M.A. Pudovik et al., Izv. Akad. Nauk Ser. Khim. 1995, 353-7 (Russ.).

Magnesium Mg
Dithiophosphonic acid O-monoesters >P—Cl → >P—R
from alcohols and Grignard compds. via 1,3,2-dithiaphospholane-2-thiones

247. PhMgX + [chlorodithiaphospholane] $\longrightarrow$ [Ph-P dithiaphospholane] $\xrightarrow{S_8}$ [Ph-P(S) dithiaphospholane] $\xrightarrow{HO\sim N^+Me_3\ TsO^-}$ [product]

A soln. of phenylmagnesium halide in THF added dropwise to a stirred soln. of 1.1 eqs. 2-chloro-1,3,2-dithiaphospholane in the same solvent at -78° over 10 min, warmed to room temp. after 1 h, worked-up, the crude dithiaphospholane in benzene treated with elemental S$_8$, and stirred at room temp. for 10 h or refluxed for 3 h → intermediate dithiaphospholane 2-thione (Y 82%), stirred with startg. alcohol and DBU in acetonitrile (or THF or CH$_2$Cl$_2$) at room temp. for 1 h → product (Y 67%). Organo-cerium and -lithium reagents failed to react. F.e.s. S.F. Martin et al., J. Org. Chem. 59, 7957-8 (1994).

Reductive silylation of carbonyl compds. CO → C(OSi≤)Si≤
of esters s. 34, 591; also 1-siloxy-1-silanes from oxo compds. s. I. Nishiguchi et al., Chem. Lett. 1995, 829-30.

Magnesium/1,2-dibromoethane ←
Arylstannanes from ar. bromides under ultrasonication Br → Sn≤

248. m-Cl-C$_6$H$_4$-Br + (Bu$_3$Sn)$_2$O $\longrightarrow$ m-Cl-C$_6$H$_4$-SnBu$_3$

A soln. of m-chlorobromobenzene, 2.3 eqs. Mg-powder, 1.1 eqs. 1,2-dibromoethane, and 1 eq. bis(tri-n-butyltin) oxide in anhydrous THF sonicated for 1 h in a commercial ultrasonic cleaning bath (Crest 575-D, 39 kHz) at ca. 45° → m-chlorophenyl(tri-n-butyl)stannane. Y 94%. The method is quick, simple, high-yielding and free from by-products. Reaction failed with ar. chlorides and did not proceed, or yields were low, in the absence of 1,2-dibromoethane. F.e. incl. hetaryl stannanes s. A.S.-Y. Lee, W.-C. Dai, Tetrahedron Letters 37, 495-8 (1996).

Diorganozinc compds. R_2Zn
Tert. phosphines from chlorophosphines >P—Cl → >P—R

249. (C$_8$H$_{17}$)$_2$Zn + ClPPh$_2$ $\xrightarrow{BH_3.Me_2S}$ C$_8$H$_{17}$PPh$_2$.BH$_3$

Chlorodiphenylphosphine allowed to react with 0.5 eq. dioctylzinc in THF at 0° for 1 h, and treated with BH$_3$·Me$_2$S at the same temp. → diphenyloctylphosphine·BH$_3$. Y 96%. The method is convenient and in one-pot, allowing the synthesis of a range of **functionalized and chiral tert. phosphines** in high yield by using the appropriately functionalized diorganozincs or organozinc halides. Both alkyl groups of the dialkylzinc are transferred, and halogen, cyano, ester and enone groups were unaffected. F.e. and with dichlorophosphines and PCl_3 s. F. Langer, P. Knochel, Tetrahedron Letters *36*, 4591-4 (1995).

Sodium tetrahydridoborate $\qquad\qquad NaBH_4$
4-(Arylseleno)isoxazolidines from O-allyloximes
Stereospecific selenylative ring closure

250.

1.1 eqs. Crystalline phenylselenenyl bromide added portionwise to a soln. of startg. (E)-O-allyloxime in methylene chloride at room temp., after 1 h the soln. diluted with methanol, treated with 1.5 eqs. $NaBH_4$, and stirred at room temp. for 1 h → *trans*-product. Y 93%. F.e. incl. tetrasubst. derivs. s. M. Tiecco et al., Tetrahedron *51*, 1277-84 (1995); **6-α-(arylseleno)tetrahydro-1,2-oxazines** from γ,δ-ethylene- or 4-ethylene-nitrones s. ibid. *52*, 6811-22 (1996); N-unsubst. 4-(arylseleno)isoxazolidines with $PhSeSePh/(NH_4)_2S_2O_8/CF_3SO_3H$ s. Chem. Commun. *1995*, 235-6; **selenylative ring closures of allyl hydroxamates** cf. ibid. 237-8; **of allyloxylamines** s. Tetrahedron Letters *36*, 163-6 (1995).

Samarium diiodide $\qquad\qquad SmI_2$
2-Ketoselenides from α-bromoketones and selenylhalides $\qquad$ Br → SeR

251. $\qquad PhC(O)CH_2Br \ + \ BrSePh \ \longrightarrow \ PhC(O)CH_2SePh$

A soln. of startg. α-bromoketone in acetonitrile added to a soln. of 2.1 eqs. SmI_2 in the same solvent, 1.1 eqs. benzeneselenyl bromide in acetonitrile added, allowed to react for 1 h, and treated with 0.1 M HCl → 2-phenylselenoacetophenone. Y 66%. The reaction is believed to occur via **samarium(III) enolates**. The method is mild and neutral. F.e. incl. alkylseleno derivs., also with SmI_3, s. T. Ying et al., Synth. Commun. *26*, 1517-23 (1996).

Samarium diiodide/hexamethylphosphoramide $\qquad\qquad SmI_2/HMPA$
Sym. diacyl selenides from carboxylic acid chlorides $\qquad$ 2 RCOCl → $(RCO)_2Se$
with $Se/CO/H_2O/DBU$ cf. *47*, 563; with $Se/SmI_2/HMPA$ s. Y. Zhang et al., Org. Prep. Proc. Intern. *25*, 681-3 (1993).

Selenoamides $\qquad\qquad RC(Se)NH_2$
Sym. diselenides from halides under mild, neutral conditions $\qquad$ 2 RHal → $(RSe)_2$

252. $\qquad\qquad 2 \ PhCH_2Cl \ \xrightarrow{PhC(Se)NH_2} \ PhCH_2SeSeCH_2Ph$

Selenobenzamide added to 1 eq. benzyl chloride in anhydrous ethanol under N_2, stirred at room temp. for 30 min, then heated at 70° for 4 h → dibenzyl diselenide. Y 91%. The method is inexpensive and the reagents are stable. This is the first example of selenium transfer from a primary selenoamide. F.e. and with selenourea s. W.-Q. Fan et al., J. Organometal. Chem. *485*, 19-24 (1995).

Triorganostannyllithium compds. $\qquad\qquad R_3SnLi$
Imidoylstannanes from iminochlorides $\qquad$ N=C(Cl) → N=C(Sn≤)

253.

1 eq. Startg. triorganostannyllithium soln. added slowly via cannula to startg. imidoyl chloride in THF at -78°, and stirred at this temp. for 1 h then at room temp. for a further 1 h → α-(2,6-xylylimino)benzylidene(trimethyl)stannane. Y 86%. The method is applicable to both N-aryl-C-

alkyl- and N-aryl-C-aryl-imidoyl chlorides, and the R_3Sn residue may be Me_3Sn or Ph_3Sn. F.e. and reactions s. B. Jousseaume et al., J. Chem. Soc. Perkin Trans. I *1994*, 2283-8.

Hydrazine hydrate s. under KOH	N_2H_4
Ammonium hypophosphite/hexamethyldisilazane	$H_2POONH_4/(Me_3Si)_2NH$
Cyclic phosphinic acids from dihalides	◯

254.

A mixture of 1,5-dibromopentane, 5 or 10 eqs. hexamethyldisilazane, and 2 or 5 eqs. ammonium hypophosphite in mesitylene refluxed overnight under argon, cooled, concentrated, and hydrolyzed with brine at room temp. → 1-hydroxyphosphorinane 1-oxide. Y 75%. The reaction is generally applicable to the formation of 4- to 8-membered rings. F.e.s. J.-L. Montchamp et al., J. Org. Chem. *60*, 6076-81 (1995); dialkylphosphinic acids from halides by intermolecular process s. E.A. Boyd et al., Tetrahedron Letters *35*, 4223-6 (1994).

Palladous acetate/diphenyl(m-*sulfophenyl*)*phosphine sodium salt/sodium hydroxide/*
tetra-n-butylammonium iodide ←
Arylphosphorus(V) acid esters from halides $Hal \to P(O){\lt}$
arylphosphinates s. *38*, 584; H-arylphosphonates with $(Ph_3P)_2PdCl_2$ s. A.W. Schwabacher, A.D. Stefanescu, Tetrahedron Letters *37*, 425-8 (1996); arylphosphonic acid esters with $Pd(OAc)_2/$ diphenyl(*m*-sulfophenyl)phosphine sodium salt under solid-liq. phase transfer catalysis with $NaOH/Bu_4NI$ or K_2CO_3/Bu_4NCl (without solvent) s. Z.S. Novikova et al., Zh. Org. Khim. *31*, 142 (1995) (Russ.).

Remaining Elements ↑   RemC ↕ Rem

Irradiation s. under 1,5-Dimethoxynaphthalene	⫽
Sodium hydride	*NaH*
(E)-α,β-Ethylenephosphonic acid esters	$C{=}C\text{-}PO_3R_2$
from β-oxophosphonic acid esters and dialkyl phosphites	

255.

1.2 eqs. Diethyl phosphite added dropwise at room temp. under N_2 to a suspension of 1.5 eqs. NaH (80% oil dispersion) in THF, diethyl 1-formylbenzylphosphonate in the same solvent added after 10 min, stirred for 30 min at room temp., and quenched with aq. NH_4Cl → (E)-diethyl 2-phenylethenephosphonate. Y 94%. The startg. β-oxophosphonates thus serve as **vinyl cation equivalents**. Conditions are mild for aldehyde derivs., but reactions of β-ketophosphonates required harsher conditions for only moderate yields. F.e.s. Y.J. Koh, D.Y. Oh, Synth. Commun. *25*, 2587-90 (1995).

Silver trifluoromethanesulfonate s. under Br_2	$AgOSO_2CF_3$
Diisobutylaluminum hydride/diethylaluminum chloride	*i-Bu$_2$AlH/Et$_2$AlCl*
Tellurolic acid esters from aldehydes via Tishchenko reaction	$CHO \to C(O)TeR$

256.

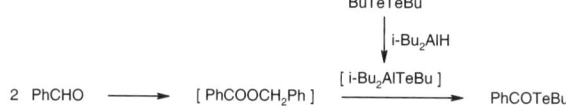

1 N *i*-Bu_2AlH in hexane added under argon to dibutyl ditelluride, stirred at 25° for 1 h, diluted

with THF, cooled to -23°, 4 eqs. benzaldehyde and 1 N Et$_2$AlCl in hexane injected into the solution, gradually warmed to 25°, stirred for a further 1 h, and poured into satd. NH$_4$Cl → Te-butyl tellurobenzoate. Y 71%. The procedure is generally good for conversion of aryl and aliphatic aldehydes (except pivalaldehyde), but the corresponding prim. alcohols were obtained with diisobutylaluminum *aryl*tellurides. Et$_2$AlCl improves the yield significantly, but its mode of action is uncertain. **Selenolic and thiolic acid esters** were obtained similarly from the appropriate diisobutylaluminum selenides or mercaptides. F.e.s. N. Kambe et al., J. Org. Chem. *59*, 5824-7 (1994); selenolic and thiolic acid esters with PhI(OAc)$_2$/NaN$_3$ cf. M. Tingoli et al., Synlett *1995*, 1129-30.

1,5-Dimethoxynaphthalene/ascorbic acid/irradiation
**Silyl radical-mediated ring closur-es of ethylenehalides
via photo-sensitized electron transfer to silyl selenides**

257.

Under conditions for photo-sensitized electron transfer, *tert*-butyldiphenylsilyl phenylselenide undergoes fission to a silyl radical which initiates ring closure of ethylenehalides as a useful alternative to the standard tin-hydride method (cf. *29*, 970s*46-50*). A key distinction, however, is that under such sensitization, *in situ*-generated diphenyl diselenide terminates the process **with introduction of the arylseleno group. E: γ'-Arylseleno-γ-lactolides.** A 1:1.12:0.4:1.12 soln. of startg. ethylenebromide, *tert*-butyldiphenylsilyl phenylselenide, 1,5-dimethoxynaphthalene, and ascorbic acid in acetonitrile irradiated (without removal of dissolved oxygen) through Pyrex with a 450 W Hanovia medium-pressure Hg vapour lamp at room temp. for 7 h (70% bromide conversion by HPLC) → product. Y 81% (*cis/trans* 16:84). The regio- and stereo-chemistry are in accord with Beckwith rules. F.e. incl. 3-methylenetetrahydrofuran, 2,3-dihydrobenzofuran and cyclopentane rings s. G. Pandey, K.S.S.P. Rao, Angew. Chem. Intern. Ed. *34*, 2669-71 (1995).

Ammonium persulfate/trifluoromethanesulfonic acid   *(NH$_4$)$_2$S$_2$O$_8$/CF$_3$SO$_3$H*
Selenylative ring closures of allyloxylamines and allyl hydroxamates s. *51*, 250
Bromine/silver(I) salt   *Br$_2$/Ag(I)*
Asym. intramolecular oxyselenation

258.

Asym. selenolactonization. A soln. of bis[2,6-bis(1(R)-ethoxyethyl)phenyl] diselenide in methylene chloride treated at -78° with 1.08 eqs. 1 *M* Br$_2$ in carbon tetrachloride, after 5 min, 2.2 eqs. 2 *M* AgOTf *in methanol* added, stirred for 5 min, 2.6 eqs. *trans*-6,6-dimethyl-4-heptenoic acid added, and after stirring for 15 min neutralized with *s*-collidine → 5(R)-[(S)-1-[2,6-bis(1(R)-ethoxyethyl)phenylseleno]-2,2-dimethylpropyl]-2(3*H*)-dihydrofuranone. Y 84% (10:1 facial selectivity). F.e. and regioselectivity, also chiral arylseleno-O-heterocyclics from ethylenealcohols s. R. Déziel, E. Malenfant, J. Org. Chem. *60*, 4660-62 (1995); cf. T. Wirth, Angew. Chem. Intern. Ed. *34*, 1726-8 (1995); f. chiral diselenides with AgPF$_6$ cf. S. Tomoda et al., Chem. Commun. *1995*, 1641-2; with AgBF$_4$, **also asym. intramolecular aminoselenation**, s. S. Uemura et al., ibid. 2321-2.

Dichloro[1,1'-bis(diphenylphosphino)ferrocene]palladium(II)/potassium acetate
Arylboronic acid esters from ar. halides   Hal → B(OR)$_2$

259.

DMSO and *p*-iodo-N,N-dimethylaniline added to a N$_2$-purged mixture of 1.1 eqs. diboronic acid bis(pinacolate), 3 mol% PdCl$_2$(dppf), and 3 eqs. KOAc, and stirred at 80° for 6 h → product. Y 90%. The method is simple, convenient, high-yielding, and general (for ar. halides with either electron-donating or -withdrawing substituents, hetaryl and sterically hindered halides). A variety of functional groups on the aromatic ring (e.g. keto, alkoxy, nitro or cyano groups) remained unaffected. F.e. incl. cross-coupling with ar. bromides s. T. Ishiyama et al., J. Org. Chem. *60*, 7508-10 (1995).

Carbon ↑   RemC ↕ C

Without additional reagents   *w.a.r.*
Phosphonic acid benzyl esters from halides   Hal → PO(OBn)$_2$
Double Arbusov-Michaelis rearrangement

260.

A mixture of 4 eqs. tribenzyl phosphite and benzyl bis(chloromethyl)phosphinate stirred for 10 h at 140° (*6-10 Torr*) → tetrabenzyl [(benzyloxy)phosphoryl)bis(methylene)]bis(phosphonate). Y 71%. The reduced pressure ensures the effective removal of benzyl chloride so that reaction can proceed to completion (the conventional thermal route normally being applicable only to lower trialkyl phosphites where the corresponding alkyl halide is volatile and easily removed on distillation). F.e. and reactions s. M. Saady et al., Helv. Chim. Acta *78*, 670-8 (1995).

Samarium diiodide/hexamethylphosphoramide   SmI$_2$/(Me$_2$N)$_3$PO
α,β-Ethyleneselenolic acid esters from 5-alkylidene-1,3-dioxane-4,6-diones   C

261.

Selenolcinnamic acid esters. HMPA and startg. diaryl diselenide added to 2.2 eqs. SmI$_2$ in THF, stirred at room temp. for ca. 30 min, isopropylidene 5-benzylidenemalonate in the same solvent injected into the mixture, heated at 50° for 3 h under N$_2$, 0.1 *M* HCl added dropwise, and stirred for a short time → 3-phenylpropenoic selenophenyl ester. Y 80%. Yields were high if there were no substituents at the 3- or 4-position of the benzene ring. F.e. and **α,β-ethylenethiolic acid esters** with disulfides, s. W. Bao, Y. Zhang, Synth. Commun. *25*, 143-8 (1995).

N-Phenylthiourea/hydrogen chloride   PhNHCSNH$_2$/HCl
α-Aminophosphonic acids from aldehydes   CHO → C(NHR)P(O)(OH)$_2$
and urethans with PCl$_3$ cf. *36*, 608; α-aminodi(phosphonic acids) with PhNHC(S)NH$_2$/(PhO)$_3$P s. Z.H. Kudzin et al., J. Organomet. Chem. *479*, 199-205 (1994).

Trimethylsilyl chloride   Me$_3$SiCl
α-Acylaminophosphine oxides from aldehydes and acylamines   NH → NCH(R)P(O)<
s. *49*, 560; also acylaminomethylphosphonic acid esters s. Tetrahedron Letters *36*, 2483-6 (1995);

from azomethines and acyl chlorides, **also α-acylaminophosphonic acid esters**, cf. Synthesis *1994*, 953-6; via N-α-acylaminobenzotriazoles (with ZnBr₂) cf. A.R. Katritzky et al., Synth. Commun. *25*, 1187-96 (1995).

Phosphorus trichloride PCl_3
α-Alkoxyphosphonic acid esters from acetals $C(OR)_2 \rightarrow C(OR)PO(OR')_2$
with BF₃ s. *39*, 576; α-alkoxy-β,γ-ethylenephosphonic acid esters with PCl₃ s. J. Chem. Soc. Perkin Trans. I *1995*, 2123-7.

Elimination ⇑

Oxygen ↑ RemC ⇑ O

Palladous acetoacetonate/1,1,3,3-tetramethylbutyl isocyanide/n-butyllithium ←
(E)-2-Ethylenesilanes from disilanyloxy-2-ethylenes $C(Si\leqslant)C=C$

262.

with 1,3-chirality transfer. A soln. of startg. (R)-disilanyloxy-2-ethylene (e.e. 99.7%) in toluene refluxed for 1 h with 2 mol% Pd(acac)₂ and 8 mol% 1,1,3,3-tetramethylbutyl isocyanide, and treated with *n*-BuLi in THF at 0° → (S)-(E)-product. Y 87% (e.e. 97.3%). (Z)-Substrates gave the enantiomeric products. F.e.s. M. Suginome et al., J. Am. Chem. Soc. *118*, 3061-2 (1996).

Formation of C-C Bond

Uptake ⇓

Addition to Hydrogen and Carbon CC ⇓ HC

Palladous acetate/potassium persulfate/trifluoroacetic acid ←
Arylcarboxylic acids from arenes $ArH \rightarrow ArCOOH$
oxidative carbonylation with Na₂S₂O₈ cf. *23*, 757s*38*; with K₂S₂O₈ in TFA s. Y. Fujiwara et al., Chem. Lett. *1995*, 345-6.

Addition to Oxygen and Carbon CC ⇓ OC

Without additional reagents w.a.r.
α-Siloxynitriles from aldehydes $CHO \rightarrow CH(OSi\leqslant)CN$
Uncatalyzed conversion

263. PhCHO + Me₃SiCN ⟶ PhCH(OSiMe₃)CN

with retention of ketones. A 1:1 mixture of benzaldehyde and acetophenone in dry *acetonitrile* treated with 1 eq. trimethylsilyl cyanide, and refluxed under N₂ for 2 h → α-trimethylsiloxyphenylacetonitrile. Y 100%. Acetophenone was recovered unchanged. Reaction also occurs at a comparable rate in nitromethane, but there was no reaction in dichloromethane or benzene. F.e.s. K. Manju, S. Trehan, J. Chem. Soc. Perkin Trans. I *1995*, 2383-4.

Sodium hydride NaH
Functionalized lactams from bicyclic iminoester salts C
Regiospecific ring opening

264.

Schmidt reaction with azidoalcohols (*51*, 159) has been elaborated by regiospecific nucleophilic ring opening of the intermediate iminoester salts. **E:** 1 eq. NaH (60% mineral oil dispersion) in THF added at 0° to bis(phenylsulfonyl)methane in the same solvent, stirred for 15 min, a soln. of startg. iminoester salt in DMF added, allowed to warm to room temp. over 1 h, stirring continued for 17 h, and poured into water → product. Y 54%. Most heteroatom-based nucleophiles and C-nucleophiles reacted in this fashion, but sodiomalononitrile and hydride ion reacted at the C-fusion. Reaction could also be conducted in one pot from *in situ*-generated iminoester salts, but yields were lower. F.e.s. V. Gracias et al., J. Org. Chem. *61*, 10-11 (1996).

Potassium tert-*butoxide* KOBu-t
2-Acetylenealcohols from oxo compds. CO → C(OH)C≡C

with NaOH/Bu$_4$NBr cf. *19*, 734s*38*; from ketones with a *little* (20 mol%) KOBu-*t* s. J.H. Babler et al., J. Org. Chem. *61*, 416-7 (1996); α,β-acetylene-γ-hydroxycarboxylic acid amides from propiolic acid amides with LDA (cf. *16*, 722) via N,C-dianions s. G.M. Coppola, R.E. Damon, J. Heterocyc. Chem. *32*, 1133-9 (1995); trifluoropropynylcarbinols s. T. Yamazaki et al., J. Org. Chem. *60*, 6046-56 (1995).

Organolithium compds. or Magnesium/cerium(III) chloride RLi or Mg/CeCl$_3$
Synthesis of alcohols from oxo compds. CO → C(OH)R

s. *37*, 623s*48*; addition to 2-oxazolidone-4(5)-carboxaldehydes **with asym. induction** s. A.G.H. Wee, F. Tang, Tetrahedron Letters *37*, 6677-80 (1996); addition to 17-keto-steroids s. X. Li et al., ibid. *35*, 1157-60 (1994); addition to chiral cyclic ketones using *catalytic* or substoichiometric amounts of CeCl$_3$ s. ibid. 6713-6; stereospecific addition to achiral cyclic ketones s. N. Greeves et al., ibid. 285-8; addition to ketonucleosides s. P.M.J. Jung et al., ibid. *36*, 1031-4 (1995); effect of drying procedure on activity of anhydrous CeCl$_3$ s. V. Dimitrov et al., ibid. *37*, 6787-90 (1996).

Organolithium compds./cerium(III) chloride RLi/CeCl$_3$
Stereospecific synthesis of trisubst. ethylene derivs.
from β-ketophosphine oxides via β-hydroxyphosphine oxides

265.

Startg. β-ketophosphine oxide added dropwise to a suspension of 3.5 eqs. BuLi·CeCl$_3$ in THF at -78°, stirred for 2 h, and quenched with 5% aq. acetic acid → intermediate β-hydroxy-deriv. (Y 93%; R*,R*/R*,S* 93/7), in anhydrous DMF treated with excess of KH, and heated at 50° for 30 min → product (Y 94%; E/Z 93/7). The procedure is mild, high-yielding, and notably applicable to methyl and ethyl derivs. for which the Horner approach is limited. F.e.s. G. Bartoli et al., Angew. Chem. Intern. Ed. *34*, 2046-8 (1995); s.a. Tetrahedron Letters *35*, 8453-6 (1994).

Organolithium compds./cerium(III) chloride/chiral 1,3-dioxolane-4,5-dimethanols ←
**Asym. synthesis of sec. alcohols from aldehydes
with *in situ*-generated organocerium(III) alkoxides**   CHO → CH(OH)R

266.

CeCl$_3$·7H$_2$O heated *in vacuo* at 135-140°/0.5 mmHg for 2 h, argon introduced, cooled in an ice-bath, dry ether added via syringe, sonicated for 1 h using a Camlab transonic T460/H bath, stirred overnight at room temp., cooled at -78°, n-BuLi soln. added dropwise via syringe, a soln. of 1 eq. (4R,5R)-2,2-dimethyl-α,α,α',α'-tetraphenyl-1,3-dioxolane-4,5-dimethanol in ether added dropwise after 1 h, stirred for 1 h at -78°, cooled to -100°, 0.4 eq. cyclohexanecarboxaldehyde in dry ether added over 1 h via syringe, stirred at -100° for 2 h, and quenched with satd. NH$_4$Cl → (R)-product. Y 65% (e.e. 70%). The structure of the TADDOL has a significant effect on enantioselectivity: C$_2$-symmetric TADDOLs afforded higher selectivity than those with no symmetry (C$_1$), as did TADDOLs bearing aliphatic substituents at C$_2$. F.e.s. N. Greeves et al., Tetrahedron Letters *37*, 2675-8 (1996).

n-Butyllithium (s.a. under i-Pr$_2$NLi)   BuLi
β-Hydroxyketones from α-iodoketones   CHO → CH(OH)C—CO
via organotin(IV) enolates cf. *48*, 616; **via lithium enolates** with *n*-BuLi, also 1,3-diol monoesters from α-*tert*-iodoketones, and f. organometallics s. Y. Aoki et al., Chem. Lett. *1995*, 463-4.

n-Butyllithium/zinc chloride   BuLi/ZnCl$_2$
Terminal *threo*-3-acetylenealcohols from iodoallenes   C(OH)C—C≡C

267.

1.2 eqs. 1.58 *M* n-BuLi in hexane added to a soln. of 1-iodo-1,2-octadiene in toluene at -78°, stirred for 5 min, treated with a suspension of 1.2 eqs. ZnCl$_2$ in THF, stirred for 10 min, 1.1 eqs. startg. aldehyde in toluene added, and stirred for a further 10 min → product. Y 65% (>99% *threo*-isomer). F.e. and with Et$_2$Zn s. K. Utimoto et al., Tetrahedron *51*, 11681-92 (1995).

n-Butyllithium/(-)-B-chlorodiisopinocampheylborane   BuLi/(-)-ClB(Ipc)$_2$
Asym. synthesis of 2-allenealcohols from acetylene derivs. and aldehydes   C(OH)C=C=C

268.

1.05 eqs. 2-Hexyne added to a stirred soln. of 1 eq. 2.5 *M* n-BuLi (in hexane) in THF at -10°, the mixture stirred for 10 min, allowed to warm to room temp., stirred for 30 min, re-cooled to -78°, treated dropwise with a soln. of 1.1 eqs. (-)-B-chlorodiisopinocampheylborane in ether (pre-cooled to 0°) via cannula, stirred for 10 min, allowed to warm to room temp., held at this temp. for 30 min, solvent replaced by ether, re-cooled to -100°, a cooled (-78°) soln. of isobutyraldehyde in ether added slowly via cannula, the mixture allowed to warm to room temp. after 2 h, and treated with 30% H$_2$O$_2$ and 3 *M* NaOH for 4 h with vigorous stirring → (R)-product. Y 80% (e.e. 96%). The isomeric 2-acetylenealcohols were not detected even by NMR. Since both (+)- and (-)-

chlorodiisopinocampheylborane are commercially available, either enantiomer of the product may be readily prepared. F.e.s. S.V. Kulkarni, H.C. Brown., Tetrahedron Letters 37, 4125-8 (1996); **chiral 2-silyl-2-allenealcohols** from 1-(trimethylsilyl)propyne with *t*-BuLi/(-)-MeOB(Ipc)$_2$/BF$_3$ s. J. Org. Chem. 60, 8130-1 (1995).

tert-*Butyllithium*	t-*BuLi*
2-Aminoalcohols from azomethines	CO → C(OH)C—N═CR$_2$

via 2-(benzylideneamino)alcohols cf. 25, 488; *syn*-products via 2-(diphenyl methyleneamino)alcohols (with *t*-BuLi) s. C.Z. Ding, Tetrahedron Letters 37, 945-8 (1996).

Lithium diisopropylamide	i-*Pr$_2$NLi*
Stereospecific aldol condensation via lithium enolates	C(OH)C—CO

s. 29, 894s32; with bridged bicyclic 4-cycloheptenones s. I. Stohrer, H.M.R. Hoffmann, Helv. Chim. Acta 76, 2194-209 (1993); with *endo*-2-methylbicyclo[2.2.1]hept-5-enyl ethyl ketone s. M. Ahmar et al., Tetrahedron Letters 33, 2501-4 (1992); **asym. induction** with chiral α-oxyketones, reversal of diastereoselectivity by O-protective groups, s. A. Choudhury, E.R. Thornton, ibid. 34, 2221-4 (1993); double stereodifferentiation s. D.A. Evans et al., ibid. 37, 1957-60 (1996).

Lithium diisopropylamide or n-Butyllithium	i-*Pr$_2$NLi or BuLi*
β-Hydroxynitriles from oxo compds.	CO → C(OH)C—CN

with LiNEt$_2$ cf. 29, 959; *anti*-α-aryl-β-hydroxynitriles with LDA s. P.R. Carlier et al., J. Org. Chem. 60, 7511-7 (1995); improved procedure with *n*-BuLi s. R.B. Silverman et al., ibid. 2261-2.

β-Hydroxysulfoxides from oxo compds. CO → C(OH)C(SOR)

with NaH cf. 17, 733; 1-sulfinylcyclobutylcarbinols s. L. Fitjer et al., Tetrahedron Letters 36, 4985-8 (1995).

Lithium diisopropylamide/chiral tetradentate lithium amide	←
Asym. aldol condensation via chiral lithium enolates	CO → C(OH)C—CO

chiral β-hydroxyketones with a chiral Li-diamide cf. 43, 555; chiral *anti*-β-hydroxyesters with LDA and a chiral tetradentate lithium amide s. M. Uragami et al., Tetrahedron:Asym. 6, 701-4 (1995).

Lithium diisopropylamide/cerium(III) chloride	i-*Pr$_2$NLi/CeCl$_3$*
Aldol condensation via cerium(III) enolates	

β-hydroxyketones s. 39, 592; γ,δ-ethylene-β-hydroxycarboxylic acid amides s. X. Shang, H.-J. Liu, Synth. Commun. 25, 2155-9 (1995).

Lithium diisopropylamide/chlorotitanium triisopropoxide	i-*Pr$_2$NLi/ClTi(OPr-i)$_3$*
Asym. aldol condensation	CO → C(OH)C—CON<

with 3-acyltetrahydro-1,3-oxazin-2-ones cf. 48, 588; **with 3-acyl-2,3-dihydro-4H-1,3-benzoxazin-4-ones** s. T. Miyake et al., Tetrahedron Letters 37, 3129-32 (1996).

Lithium dicyclohexylamide/B-methoxydiisopinocampheylborane/boron fluoride	←
syn-3-Ethylene-2,1-chlorhydrins from aldehydes	CH(OH)C(Cl)C═C
Asym. synthesis with addition of three C-atoms	

269.

A soln. of Li-dicyclohexylamide in THF added via cannula to a stirred soln. of 0.76 eq. (-)-B-methoxydiisopinocampheylborane and 1 eq. allyl chloride in freshly distilled ether at -95°, 2 eqs. BF$_3$-etherate added via syringe after 30 min, followed by 0.76 eq. cyclohexanecarboxaldehyde, stirring continued for 4 h, and allowed to warm to room temp. → *syn*-product. Y 72% (e.e. 93%). The corresponding **chiral *cis*-3-ethyleneoxido compds.** were obtained by treatment with NaOH/H$_2$O$_2$ (1 eq.) in THF, and could also be prepared from the aldehydes without isolation of the chlorhydrins. This nicely complements the Sharpless epoxidation, which is only efficient for producing chiral *trans*-derivs. The high regioselectivity is thought to be a consequence of the bulk of the borane group and the presence of an excess of BF_3. F.e.s. S. Jayaraman et al., Tetrahedron Letters 36, 4765-8 (1995).

Triethylenediamine *dabco*
Baylis-Hillman reaction ←
s. 39, 593s46, 48-50; with vinyl ketones, details s. D. Basavaiah et al., J. Chem. Res (S) 1995, 267; s.a. F.R. van Heerden et al., Synth. Commun. 24, 2863-72 (1994); addition of aryl acrylates s. P. Perlmutter et al., Tetrahedron Letters 37, 1715-8 (1996); with non-enolizable α-dicarbonyl compds. s. G.M. Strunz et al., Can. J. Chem. 73, 1666-74 (1995).

Cuprous chloride/ethyldiisopropylamine/trichlorosilane $CuCl/i\text{-}Pr_2NEt/Cl_3SiH$
2-Allenealcohols and 3-acetylenealcohols from β,γ-acetylenechlorides ←
via 2-acetylenesilanes and allenesilanes, respectively

270.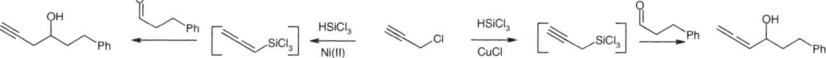

Synth.Meth. 50, 365 has been extended to synthesis of 2-allene- or 3-acetylene-alcohols from the *same* propargyl chloride *in one pot* by appropriate choice of catalyst for the inital reaction with Cl_3SiH. **E:** 1.1 eqs. Ethyldiisopropylamine in 10:1 ether/propionitrile added at room temp. to 3.3 mol% CuCl suspended in the same solvent, treated with equimolar amounts of propargyl chloride and Cl_3SiH in the same solvent, stirred for 10 h, diluted with DMF, cooled to 0°, 0.83 eq. startg. aldehyde in DMF added, stirred for 12 h at 0°, and quenched with 1 N HCl → 6-phenylhexa-1,2-dien-4-ol. Y 77% (containing <3.3% 6-phenylhex-1-yn-4-ol). With bis(ethyl acetoacetato)nickel(II) and 1,2,2,6,6-pentamethylpiperidine (in place of CuCl and $i\text{-}Pr_2NEt$) in THF/DMF, the regioisomeric 6-phenylhex-1-yn-4-ol was obtained (Y 69% and <3.3% allenealcohol). F.e. and with isolation of the intermediate silanes, also with subst. propargyl chlorides, s. S. Kobayashi, K.Nishio, J. Am. Chem. Soc. 117, 6392-3 (1995); γ-hydroxy-α-methylenecarboxylic acid esters s. Synlett 1996, 153-4.

Cupric chloride s. under Mg $CuCl_2$

Magnesium Mg
Synthesis of alcohols from oxo compds. CO → C(OH)R
with asym. induction s. 9, 741s49; Grignard addition to chiral N-protected oxazolidine-4-carboxaldehydes s. L. Williams et al., Tetrahedron 52, 11673-94 (1996); to 2-acyloxazolidines s. ibid. 51, 4043 (1995); addition of 2-bromomagnesiomethyl-1,3-dioxolane to ketosugars s. M. Schmeichel, H. Redlich, Synthesis 1996, 1002-6; addition to ketoazasugars s. I.K. Khanna et al., Tetrahedron Letters 37, 1355-8 (1996); new perspectives on formation of Grignard reagents s. ibid. 35, 5857-60 (1994); method of determining the concentration of Grignard reagents s. Synth. Commun. 24, 2503-6 (1994).

Asym. synthesis of prim. amines from oxazolidines C
chiral 3-ethylene-*prim*-amines s. 46, 601; chiral 2-*prim*-aminoalcohols via addition of isopropoxydimethylsilylmethylmagnesium chloride s. K. Higashiyama et al., Heterocycles 41, 2007-17 (1995).

Magnesium/cupric chloride $Mg/CuCl_2$
Barbier-type synthesis of 3-ethylenealcohols CO → C(OH)C-C=C
with $Mg/AlCl_3$ cf. 38, 623; with $Mg/CuCl_2 \cdot 2H_2O$ s. N.B. Das et al., Tetrahedron Letters 36, 7119-22 (1995).

Zinc (s.a. under $BiCl_3$) Zn
Grignard-type synthesis with zinc CO → C(OH)R
s. 18, 736; α,β-ethylene-γ-hydroxyboronic acid esters s. E. Jehanno, M. Vaultier, Tetrahedron Letters 36, 4439-42 (1995).

Barbier-type synthesis of 3-ethylenealcohols CO → C(OH)C—C=C
with allyl shift s. 34, 614s42; with *commercial unactivated* zinc s. B.C. Ranu et al., Tetrahedron Letters 36, 4885-8 (1995); with *activated* Zn (from $ZnCl_2$ and Na in liq. NH_3 cf. M. Makosza, K. Grela, ibid. 9225-6; with Zn powder in liq. NH_3 cf. Synth. Commun. 26, 2935-40 (1996).

Zinc/chiral 2-aminoalcohols ←
α-Difluoro-β-hydroxycarboxylic acid esters from aldehydes ←
s. *49*, 585; asym. synthesis with added (1R,2S)-N-methylephedrine (cf. *49*, 584) s. M. Braun et al., Liebigs Ann. *1995*, 1447-50.

Zinc/ammonium chloride or di-n-butyltin dichloride Zn/NH_4Cl or Bu_2SnCl_2
Barbier-type synthesis of 3-ethylenealcohols CO → C(OH)C—C≡C
in aq. media s. *40*, 567s*46*; from β- or γ-ketoesters, stereoselectivity with added NH_4Cl, s. M. Ahonen, R. Sjöholm, Chem. Lett. *1995*, 341-2; in a 2-phase medium, and under catalysis with Bu_2SnCl_2, cf. D. Marton et al., J. Org. Chem. *61*, 2731-7 (1996); **3-acetylenealcohols** from propargyl bromide s. I. Yavari, F. Riazi-Kermani, Synth. Commun. *25*, 2923-8 (1995).

Zinc/trimethylsilyl chloride Zn/Me_3SiCl
Synthesis of unsatd. alcohols from oxo compds. CO → C(OH)R
3-ethylenealcohols s. *43*, 562; 2-arylalcohols from benzyl bromides and electro-deposited Zn s. C. Gosmini et al., Tetrahedron Letters *35*, 5637-40 (1994).

Isopropylmagnesium bromide s. under $Ti(OPr-i)_4$ i-PrMgBr
Dialkylzinc/chiral aminoalcohols or supported derivs. ←
Dialkylzinc/chiral Schiff bases or 1,1'-bi-2-naphthol-3,3'-dicarboxamides ←
Asym. synthesis of sec. alcohols from aldehydes
s. *42*, 616s*46-50*; with (2S,3S)-1,4-diamino-2,3-butanediol derivs. as ligand s. B.T. Cho, N. Kim, Synth. Commun. *26*, 2273-80 (1996); with carbohydrate-based 2-aminoalcohols s. ibid. 855-65; with α-D-xylose-based 3-aminoalcohols s. Tetrahedron Letters *35*, 4115-8 (1994); with chiral 3- or α-hydroxy-1,2,3,4-tetrahydro-β-carbolines s. W.-M. Dai et al., ibid. *37*, 5971-4 (1996); with chiral 3-hydroxymethyl-2-azabicyclo[3.3.0]octanes s. K. Stingl, J. Martens, Liebigs Ann. Chem. *1994*, 491-6; with chiral ferrocenyl aminoalcohols s. G. Nicolosi et al., Tetrahedron:Asym. *5*, 1639-42 (1994); s.a. M. Watanabe, Synlett *1995*, 1050-1; with a *soluble* polymer-based 2-aminoalcohol s. C. Dreisbach et al., J. Chem. Soc. Perkin Trans. I *1995*, 875-8; with a chiral Schiff base complex s. P.G. Cozzi et al., Tetrahedron Letters *37*, 4613-6 (1996); with a chiral 1,1'-bi-2-naphthol-3,3'-dicarboxamide s. H. Kitajima et al., Chem. Lett. *1996*, 343-4; asym. addition of diisopropylzinc s. K. Soai et al., J. Org. Chem. *59*, 7908-9 (1994); of mixed diorganozincs s. E. Laloë, M. Srebnik, Tetrahedron Letters *35*, 5587-90 (1994); **asym. autocatalysis** s. L. Shengjian et al., J. Chem. Soc. Perkin Trans. I *1993*, 885-6; **continuous asym. synthesis** in a membrane reactor s. U. Kragl, C. Dreisbach, Angew. Chem. Intern. Ed. *35*, 642-4 (1996); desymmetrization of ($η^4$-2,4-diene-1,6-dial)tricarbonyliron(0) complexes s. Y. Takemoto et al., Tetrahedron Letters *37*, 3345-6 (1996).

Dialkylzinc/chiral pyrimidin-5-ylcarbinols ←
Catalytic asym. automultiplication ←

271.

Synthesis of 5-α-hydroxypyrimidines. Asym. automultiplication, wherein an optically active product is generated without the assistance of an ancilliary chiral reagent, can take place with high enantioselectivity. **E:** 20 Mol% (S)-2-methyl-1-(5-pyrimidyl)-1-propanol (e.e. 92.6%) in toluene and 1.2 eqs. 1 M $Zn(Pr-i)_2$ in the same solvent stirred at 0° for 30 min, 1 eq. pyrimidine-5-carboxaldehyde in toluene added, stirred for 66 h, and quenched with 1 N HCl and satd. aq. $NaHCO_3$ at the same temp. → (S)-2-methyl-1-(5-pyrimidyl)-1-propanol. Y 63% (exclusive of initially added catalyst); e.e. 90.4%. Such automultiplication provides one of the most direct and economical routes to chiral molecules. In certain instances, it is even possible to secure the product *without* any loss of e.e. F.e.s. K. Soai et al., J. Am. Chem. Soc. *118*, 471-2 (1996).

*Dialkylzinc/titanium tetraisopropoxide/chiral 1,3-dioxolane-4,5-dimethanols or 1,2-
di(sulfonylamino)cyclohexanes* ←
Asym. synthesis of sec. alcohols from aldehydes   CO → C(OH)R
s. *44*, 565s*47-9*; prepn. of chiral silylethynylcarbinols s. N. Oguni et al., Synlett *1995*, 1043-4; of
C_2-symmetric 1,4-diols s. S. Vettel, P. Knochel, Tetrahedron Letters *35*, 5849-52 (1994).

Diethylzinc/tris(dibenzylideneacetone)dipalladium/triphenylphosphine ←
Regio- and stereo-specific synthesis of 3-ethylenealcohols   C(OH)C—C=C
from aldehydes and allyl phosphates

272.

$$\text{Ph}\overset{O}{\underset{H}{\diagdown}} + \text{Ph}\diagup\diagdown\diagup\text{OPO(OEt)}_2 \longrightarrow \text{Ph}\overset{OH}{\underset{Ph}{\diagdown}}\diagup$$

Startg. allyl phosphate and 2 eqs. 1 *M* Et_2Zn added to a stirred soln. of $Pd_2(dba)_3\cdot CHCl_3$ and 20 mol% PPh_3 in dry THF under N_2, followed by the addition of 1 eq. benzaldehyde, and stirred at room temp. for 30 min → 1,2-diphenyl-3-butenol. Y 68% (*anti:syn* 93:7). The reaction proceeds via a nucleophilic allylzinc species which reacts regiospecifically to yield a single isomer. F.e.s. S.-K. Kang et al., Synth. Commun. *26*, 1493-8 (1996); 3-acetylenealcohols s. *51*, 289.

Zinc chloride s. under BuLi $ZnCl_2$
Zinc bromide s. under RCp_2ZrCl $ZnBr_2$
Indium   In
Indium/lanthanum(III) triflate $In/La(OTf)_3$
Barbier-type synthesis of 3-ethylenealcohols   CO → C(OH)C—C=C
in aq. media s. *40*, 567s*47,49,50*; regio- and diastereo-selectivity s. M.B. Isaac, T.-H. Chan, Tetrahedron Letters *36*, 8957-60 (1995); stereospecific addition to α- and β-hydroxyaldehydes s. L.A. Paquette, T.M. Mitzel, ibid. 6863-6; to α- and β-oxyaldehydes, solvent effect, s. J. Am. Chem. Soc. *118*, 1931-7 (1996); addition to cyclic ketones s. ibid. 1917-30; 4-hydroxy-2-methylenesilanes s. R. Remuson et al., Synlett *1996*, 37-8; *anti*-β-hydroxy-α-vinylcarboxylic acid esters with $In/La(OTf)_3$ s. S.-C.H. Diana et al., ibid. 263-4.

B-Methoxydiisopinocampheylborane s. under $LiNR_2$ $MeOB(Ipc)_2$

Sodium dihydridodiethylaluminate $NaAlH_2Et_2$
2-Acetylenealcohols from oxo compds.   CO → C(OH)C≡C

273.   2 PhC≡CH $\xrightarrow{NaAlH_2Et_2}$ $[(PhC≡C)_2AlEt_2]^- Na^+$ $\xrightarrow{PhCHO}$ PhC≡CCH(OH)Ph

0.5 *M* Benzaldehyde in toluene added to a soln. of 1.1 eqs. 0.5 *M* sodium diethylbis(phenylethynyl)aluminate (freshly prepared under N_2 by stirring vigorously $NaAlH_2Et_2$ with ca. 2 eqs. phenylacetylene in toluene for 3 h at room temp. until elimination of H_2 complete) at 0°, and hydrolyzed after 1 h with satd. NH_4Cl → 1-(phenylethynyl)-1-phenylmethanol. Y 95%. The conditions are mild, and the alkynylating reagent is more selective than the corresponding Li- or Mg-acetylides, leaving halide, epoxide, ester, amide and nitrile groups unaffected. It also reacts with α,β-ethyleneoxo compds. by exclusive 1,2-addition. F.e. and with aliphatic oxo compds. s. J.H. Ahn et al., J. Org. Chem. *60*, 6173-5 (1995); selective addition to aldehydes with GaI_3/Bu_3N cf. Y. Han, Y.-Z. Huang, Tetrahedron Letters *36*, 7277-80 (1995); ethynylcarbinols with *in situ*-generated Na-trimethylethynylaluminate cf. M.J. Joung et al., J. Org. Chem. *61*, 4472-5 (1996).

Dialkylborinyl triflates/tert. amines R_2BOTf/R_3N
Asym. aldol condensation with 3-acyl-2-oxazolidones   C(OH)C—C(O)N<
s. *38*, 632s*46-9*; addition to α,α-difluoro- and α,α,α-trifluoro-carbonyl compds., reversal of stereoselectivity, s. K. Iseki et al., Tetrahedron *52*, 71-84 (1996); *anti*-selective addition to acetaldehyde and new work-up protocol s. B.C. Raimundo, C.H. Heathcock, Synlett *1995*, 1213-4; with camphor-derived 2-oxazolidones, *anti*-selectivity with added $TiCl_4$ or $SnCl_4$, s. Y.-C. Wang et al., J. Org. Chem. *61*, 2038-43 (1996); with 2-oxazolidones based on diphenylalaninol s. M.P. Sibi et al., Tetrahedron Letters *36*, 8965-8 (1995); **with camphor-based N-acyl-2-oxazolidinethiones** cf. T.-H. Yan et al., J. Org. Chem. *59*, 8187-91 (1994).

Diethylborinyl triflate/ethyldiisopropylamine Et_2BOTf/i-Pr_2NEt
Asym. aldol condensation with bicyclic N-acyllactams   CO → C(OH)C—CO
Effect of dipole alignment in the transition state on enantioselectivity

274.

Chiral enolates in which steric and dipole alignment factors are complementary provide higher levels of diastereofacial selectivity than those where such effects are opposed. E: A soln. of startg. chiral N-acyllactam in methylene chloride treated with 1 eq. Et_2BOTf at 0° followed by Hünig's base, cooled to -40°, 3 eqs. isobutanal added, and worked up after 48 h → *syn*-product. Y 90% (2R:2S>98:2). With the corresponding camphor-derived bicyclic lactam (cf. *50*, 368), enantioselectivity was lower as dipole and non-bonded interactions in the transition state are opposed. F.e.s. R.K. Boeckman, Jr., B.T. Connell, J. Am. Chem. Soc. *117*, 12368-9 (1995).

Asym. aldol condensation with N-acylsultams
s. *47*, 598s*48,49*; asymmetrization of dialdehydes (with Et_2BOTf) s. W. Oppolzer et al., Tetrahedron Letters *36*, 4413-6 (1995).

Dicyclohexylborinyl triflate/triethylamine R_2BOTf/Et_3N
Asym. aldol condensation with carboxylic acid esters   C(OH)C—COOR
via boron ester enolates

275.

Chiral *syn*-β-hydroxycarboxylic acid esters. A soln. of startg. chiral ester in methylene chloride treated with 1.3 eqs. dicyclohexylborinyl triflate and 1.5 eqs. Et_3N *at 0°* for 1 h, cooled at -78°, 1.2 eqs. isobutyraldehyde added, and allowed to react at this temp. for 1 h, then at 0° for 1 h → product. Y 58% (*syn:anti* 84:16; d.e. for *syn*-isomer 98:2). The result explodes the myth that aldol condensation with esters cannot take place via boron enolates. Enolization at or below 0° (under *kinetic* control) favours the (E)-isomer, which leads to predominant formation of the *anti*-adduct. Reaction with amides was less efficient. F.e. with achiral esters, and influence of the dialkylboron triflate-base combination on stereoselectivity, s. A. Abiko et al., J. Org. Chem. *61*, 2590-1 (1996).

Boronic acid chlorides or Dialkylborinyl triflates/tert. amine   ←
Stereospecific aldol condensation via enol borinates   C(OH)C—CO
s. *32*, 614s*49*; condensation of chiral methyl ketones with chiral β-oxyaldehydes, diastereoselectivity, s. W.R. Roush et al., Tetrahedron Letters *36*, 3443-6 (1995); with chiral β-hydroxyketones, and reversal of diastereoselectivity by $TiCl_4$/i-Pr_2NEt, s. G.P. Luke, J. Morris, J. Org. Chem. *60*, 3013-9 (1995); with chiral α-methylene-β-alkoxyaldehydes s. I. Paterson et al., Tetrahedron Letters *34*, 4393-6 (1993); with lactate-derived ketones s. ibid. *35*, 9083-6 (1994).

Trialkylsilyl tetrakis(triflyloxy)borates   ←
Stereospecific aldol-type condensation   COC—C(OSi≤)

276.

Trialkylsilyl tetrakis(triflyloxy)borates are supersilylating agents which catalyse the Mukaiyama addition to unfunctionalised α-subst. aldehydes with unprecedented levels of Cram-type selectivity. E: A mixture of startg. enoxysilane, aldehyde, and 5 mol% i-$Pr_3SiB(OTf)_4$ (prepared from *i*-

Pr$_3$SiOTf and B(OTf)$_3$ in methylene chloride allowed to react at -80° for 1 h, and quenched at low temp. with satd. aq. NaHCO$_3$ → product. Y 71% [d.r. (Cram:anti-Cram) 97:1] Selectivity correlates with the bulk of the silyl group. F.e.s. A.P. Davis, S.J. Plunkett, Chem. Commun. *1995*, 2173-4.

Hydrotalcite
α-**Siloxynitriles from oxo compds. under heterogeneous conditions**　　CO → C(OSi≤)CN
with Fe(III)-exchanged montmorillonite cf. *49*, 589; with hydrotalcite, also β-siloxynitriles from oxido compds., s. B.M. Choudary et al., Synth. Commun. *25*, 2829-36 (1995); with Bu$_2$SnCl$_2$ under homogeneous conditions (cf. *34*, 616s*46*) s. J.K. Whitesell, R. Apodaca, Tetrahedron Letters *37*, 2525-8 (1996); with Bu$_3$SnCN cf. J. Org. Chem. *59*, 7178-9 (1994).

(-)-B-Chlorodiisopinocampheylborane s. under BuLi　　　　　　　　　　　*(-)-ClB(Ipc)$_2$*
Gallium iodide/tri-n-butylamine　　　　　　　　　　　　　　　　　　　　*GaI$_3$/Bu$_3$N*
2-Acetylenealcohols from aldehydes s. *51*, 273　　　　　　　　　CHO → CH(OH)C≡C
Scandium(III) triflate/trimethylsilyl cyanide　　　　　　　　　　　　　*Sc(OTf)$_3$/Me$_3$SiCN*
Asym. synthesis of α-alkoxynitriles from oxo compds. via cyclic acetals　　CO → C(OR)CN

277.

One-pot procedure. 1.3 eqs. (2R,3R)-Butanediol and benzaldehyde added successively at room temp. to a suspension of 0.2 eq. Sc(OTf)$_3$ in methylene chloride, stirred vigorously at this temp. for 2 days, 6 eqs. trimethylsilyl cyanide added during 30 min at 0°, stirred for 3 h, and treated with 0.1 *M* methanolic HCl → product. Y 66% (d.e. 95:5). The catalyst effects both acetalation and silylcyanation. F.e.s. S. Fukuzawa et al., Synlett *1995*, 1077-8.

Ytterbium　　　　　　　　　　　　　　　　　　　　　　　　　　　　　　　　*Yb*
Sym. 1,2-disiloxy compds. from ketones s. *45*, 377s*51*　　　　C(OSi≤)C(OSi≤)
Lanthanum(III) triflate s. under In　　　　　　　　　　　　　　　　　　　*La(OTf)$_3$*
Chiral yttrium(III) β-diketonate/trimethylsilyl cyanide
Asym. hydrocyanation of aldehydes　　　　　　　　　　　　　　　　CHO → CH(OH)CN
with Cl$_2$Ti(OPr-*i*)$_2$/chiral diol cf. *43*, 576; with Y$_5$(O)(OPr-*i*)$_{13}$/(R,R)-1,3-bis(2-methylferrocenyl)propane-1,3-dione s. A. Abiko, G. Wang, J. Org. Chem. *61*, 2264-5 (1996).

Samarium diiodide/hexamethylphosphoramide　　　　　　　　　　　*SmI$_2$/(Me$_2$N)$_3$PO*
2-Acetylenealcohols from α,β-acetyleneiodides and oxo compds.　　　C(OH)C≡C
Alkynylsamarium(III) iodides as intermediates　　　　　　　　　　　　　　←

278.

'Samarium-Grignard' addition (*48*, 818) is also applicable to α,β-acetyleneiodides **E**: Iodododecyne and 1.5 eqs. dibutyl ketone in *benzene* added in one portion at room temp. under N$_2$ to 4 eqs. SmI$_2$ in benzene containing 10% HMPA, and quenched after 20 min with satd. NaHCO$_3$ → 5-butylpentadec-6-yn-5-ol. Y 76%. This Barbier-type procedure is generally applicable to aliphatic ketones and aldehydes. However, for addition to aryl and vinyl ketones, a Grignard-type procedure is available in HMPA or THF. Alkynylsamarium(III) iodides are considered to be intermediates, these being formed via H-atom abstraction from THF when the latter is used as solvent. F.e. and with retention of ester groups s. S. Tani et al., Tetrahedron Letters *36*, 3707-10 (1995).

Glycols from two different oxo compd. molecules　　　　　　　　　　C(OH)C(OH)
s. *43*, 571s*46*; α,β-dihydroxyketones s. N. Miyoshi et al., Chem. Lett. *1993*, 959-62; hydroxyl-mediated stereospecific grayanotoxin B-ring closure s. T. Kan et al., Synlett *1991*, 391-2; *threo*-selective pinacol coupling of Cr(CO)$_3$-complexed benzaldehydes, and with asym. induction, s. M. Uemura et al., J. Org. Chem. *61*, 6088-9 (1996).

2-Ene-1,5-diols from 3-ethyleneepoxides and ketones ←
Regio- and geo-specific reductive ring opening

279.

5 eqs. HMPA added to 2 eqs. SmI$_2$ (prepared from Sm and I$_2$ or 1,2-diiodoethane) in THF, stirred for 30 min, cooled to -50°, a soln. of startg. epoxide and 1 eq. cyclohexanone in THF added, and the mixture allowed to react until reaction complete → (E)-product. Y 87% (100% E). Regioisomeric 2-vinyl-1,3-diols were formed as by-products in the absence of HMPA, and E/Z selectivity was much lower. F.e.s. J.M. Aurrecoechea, E. Iztueta, Tetrahedron Letters 36, 7129-32 (1995).

Cerium(III) chloride s. under RLi and i-Pr$_2$NLi	*CeCl$_3$*
Chiral 1,3-dioxolane-4,5-dimethanols s. under RLi and R$_2$Zn	*TADDOLS*
Enzymes	←

Cyanohydrins from aldehydes with asym. induction   CHO → CH(OH)CN
s. *22*, 693s*47-9*; using (S)-hydroxynitrile lyase from *Hevea brasiliensis* s. H. Griengl et al., Tetrahedron *52*, 7833-40 (1996); with a recombinant lyase cf. F. Effenberger et al., Angew. Chem. Intern. Ed. *35*, 437-9 (1996); chiral β,γ-ethylene-α-hydroxynitriles s. J. Brussee et al., Rec. Trav. Chim. Pays-Bas *115*, 20-4 (1996); chiral ω-bromo-derivs. and **kinetic resolution of ketone cyanohydrins** s. V. Gotor et al., Chem. Commun. *1995*, 989-90.

Aldolase   ←
Enzymatic aldol condensation   CHO → CH(OH)C—CO
s. *48*, 607; with 2-keto-3-deoxy-6-phosphogluconate aldolases, details, s. E.J. Toone et al., J. Am. Chem. Soc. *118*, 2117-25 (1996); deoxysugars via aldolase-catalyzed synthesis of 1-deoxy-1-thioketoses s. R. Duncan, D.G. Drueckhammer, J. Org. Chem. *61*, 438-9 (1996); L-β-hydroxy-α-amino acids with L-threonine aldolase s. V.P. Vassilev et al., Tetrahedron Letters *36*, 4081-4 (1995); stabilization of crystalline fructose diphosphate aldolase by cross-linking s. S.B. Sobolov et al., ibid. *35*, 7751-4 (1994); multienzymatic oxidation-aldol condensation s. O. Eyrisch et al., ibid. 9013-6; *tandem* aldol condensation s. Angew. Chem. Intern. Ed. *34*, 1639-41 (1995).

Amberlyst A 21   ←
2-Nitroalcohols from aliphatic nitro compds.   CHO → CH(OH)C(NO$_2$)
and aldehydes s. *34*, 610s*51*

Trichlorosilane s. under CuCl	*Cl$_3$SiH*
1-Trimethylsilylbenzotriazole/lithium diisopropylamide/hydrogen chloride	←

β-Functionalization of cyclic α,β-ethyleneketones   ←
via 1,4-addition with 1-trimethylsilylbenzotriazole

280.

Cyclic α,β-ethylene-γ'-hydroxyketones from oxo compds. An equimolar mixture of 1-trimethylsilylbenzotriazole and 2-cyclohexenone stirred at room temp. under N$_2$ for 30 min, THF added, cooled to -78°, 1.04 eqs. benzophenone added followed by dropwise addition of 1.04 eqs. LDA, and the mixture treated with 5% HCl after 1 h at -78° → product. Y 70%. The method is simple and efficient. F.e. and electrophiles s. A.R. Katritzky et al., Tetrahedron Letters *36*, 5491-4 (1995).

Trimethylsilyl cyanide s. under Sc(OTf)$_3$ *Me$_3$SiCN*

Chiral titanium(IV)-1,1'-bi-naphthol complex
Asym. ene reaction   C=C—C—C(OH)
s. *44*, 568s*48,49*; with allylsilanes s. Tetrahedron Letters *35*, 3133-6 (1994); with chiral homoallyl ethers s. ibid. 7793-6; with fluoral s. Tetrahedron *52*, 85-98 (1996); with fluoral and vinyl sulfides s. Synlett *1996*, 837-8; with methyl glyoxylate using the (R)-*6,6'-dibromo*-1,1'-bi-2-naphthol complex cf. Tetrahedron Letters *35*, 6693-6 (1994).

β-Siloxythiolic acid esters from O-silylketene S,O-acetals   C(OSi≤)C—COSR
Asym. aldol-type condensation
s. *46*, 605s*50*; enhanced yield and enantioselectivity from 1-(1-alkylthiovinyloxy)silacyclobutanes s. S. Matsukawa, K. Mikami, Tetrahedron:Asym. *6*, 2571-4 (1995).

Titanium tetraisopropoxide/isopropylmagnesium bromide $Ti(OPr\text{-}i)_4/i\text{-}PrMgBr$
Regio- and stereo-specific synthesis of 3-ethylenealcohols from oxo compds.   C(OH)C-C=C
Generation of 2-ethylenetitanium(IV) alkoxides via oxidative addition of titanium(II)

281.

1.9-2 eqs. *i*-PrMgBr added to a mixture of 1-phenyl-2-propenyl ethyl carbonate and 1 eq. Ti(OPr-*i*)$_4$ in ether at -50°, stirred for 1 h at -40° to -50°, 0.7 eq. benzaldehyde added at -40°, and stirring continued for 30 min → product. Y 73% (*anti:syn* >97:3). A variety of other allyl derivs. (allyl halides, acetate, sulfonates, phosphates and allyl phenyl ethers) reacted in the same way, but the *carbonates* (and acetates) are particularly noteworthy as they can be prepared regio- and stereoselectively from the parent alcohols. A variety of functional groups are compatible (e.g. esters, ar. bromine) and aldehydes react faster than ketones. Reaction appears to involve generation of a (η²-propene)titanium(II) species which undergoes olefin exchange with the substrate, followed by oxidative insertion to give intermediate allyltitanium(IV) species. F.e.s. F. Sato et al., J. Am. Chem. Soc. *117*, 3881-2 (1995); **3-acetylenealcohols and 2-allenealcohols** from β,γ-acetylenehalides or 2-acetylenecarbonates s. Tetrahedron Letters *36*, 3207-10 (1995).

Bromomagnesium phenyl(tetraisopropoxy)titanate $BrMg^+[PhTi(OPr\text{-}i)_4]^-$
Regio- and stereo-specific aldol condensation   C(OH)C—CO
Organotetraalkoxytitanium ate complexes as Lewis acid

282.

Aldol condensation with *hindered* aliphatic aldehydes affords *anti*-adducts in high yield by using *in situ*-generated titanium ate complexes as Lewis acid. E: 1 eq. Phenylmagnesium bromide soln. added at 0° under inert conditions to Ti(OPr-*i*)$_4$ in abs. toluene, warmed to room temp., stirred for 1 h, 1.5 eqs. diethyl ketone added slowly, followed by careful addition of 1 eq. isobutyraldehyde, and worked up after 6 h at room temp. → *anti*-5-hydroxy-4,6-dimethyl-3-heptanone. Y 72% (>98% *anti*-isomer). There was no addition of the organometallic reagent to the carbonyl group, and no dehydration to the enone. F.e. and steric effects of the ligands s. R. Mahrwald, Tetrahedron *51*, 9015-22 (1995).

Bis(cyclopentadienyl)titanium(III) chloride Cp_2TiCl
Sym. *threo*-glycols from aldehydes   2 CHO → CH(OH)CH(OH)

283.

Benzaldehyde in THF added dropwise to a green soln. of 1.1 eqs. titanocene chloride dimer in the same solvent at -78°, allowed to warm to room temp. after the colour had changed to red-brown, and hydrolyzed after 1 h with aq. NaOH → hydrobenzoin. Y 95% (98:2 *dl:meso*). Unlike most conventional low-valent complexes, [Cp$_2$TiCl]$_2$ is also effective **in aq. media**. Thus, the same reaction in 4:1 THF/H$_2$O contaning 62 eqs. *NaCl* at 0° to room temp. over 5 h gave the same product in 84% yield (94:6 *dl:meso*). The method is applicable to ar. and α,β-unsatd. aldehydes, as well as methyl glyoxylate and glyoxylic acid (affording **tartaric acid derivs.**). Notably, carboxyl groups and ar. halides were unaffected. F.e.s. M.C. Barden, J. Schwartz, J. Am. Chem. Soc. *118*, 5484-5 (1996).

Organozirconocene chlorides/zinc bromide $RCp_2ZrCl/ZnBr_2$
Synthesis of sec. alcohols from aldehydes   CHO → CH(OH)R

284.

A *0.8* M soln. of benzaldehyde and 1.1 eqs. *n*-octylzirconocene chloride in THF treated with *0.2 eq.* ZnBr$_2$ at 25° for 3 h, and the mixture hydrolyzed → product. Y 88% . Reaction is generally applicable to aliphatic and ar. aldehydes, the intermediate alkoxyzirconocene chlorides being oxidized *in situ* in the presence of excess of aldehyde to give the corresponding **ketones.** The concentration of the substrate is critical. F.e. incl. 2-ethylenealcohols from vinylzirconocene chlorides s. B. Zheng, M. Srebnik, J. Org. Chem. *60*, 3278-9 (1995).

Germanium diiodide/zinc iodide GeI_2/ZnI_2
Barbier-type synthesis of 3-ethylenealcohols s. *41*, 810s*51*   CO → C(OH)C—C=C
Titanium trichloride $TiCl_3$
Pinacols   2 CO → C(OH)C(OH)
s. *30*, 561s*40*; in anhydrous *methylene chloride* s. A. Clerici et al., Tetrahedron Letters *37*, 3035-8 (1996).

Titanium tetrachloride/ethyldiisopropylamine $TiCl_4$ /*i*-Pr_2NEt
Asym. aldol condensation with carboxylic acid esters via titanium(IV) (Z)-enolates   C(OH)C-COOR

285.

Chiral *anti*-β-hydroxycarboxylic acid esters. A soln. of startg. chiral propionate (based on (1R,2S)-1-tosylaminoindan-2-ol) in methylene chloride treated with 1.2 eqs. TiCl$_4$ at 0-23° for 15 min, 4 eqs. *i*-Pr$_2$NEt added at 23°, stirred for 1 h, the mixture added to 2 eqs. isovaleraldehyde *precomplexed* with 2.4 eqs. TiCl$_4$ in methylene chloride at -78°, stirred for 1 h, and worked up with aq. NH$_4$Cl → product. Y 97% (*anti:syn* >99:1). There was no reaction with uncomplexed aldehydes suggesting that *two* titanium atoms are implicated in the transition state. *anti*-Selectivity was very high with a variety of α,β-ethylenealdehydes and non-conjugated aldehydes (not benzaldehyde). F.e.s. A.K. Ghosh, M. Onishi, J. Am. Chem. Soc. *118*, 2527-8 (1996).

Tri-n-butyltin hydride/azodiisobutyronitrile $Bu_3SnH/AIBN$
Cyclic glycols from dioxo compds.
Stereospecific radical ring closure

286.

A soln. of startg. dialdehyde, 1.2 eqs. Bu$_3$SnH, and a little AIBN in benzene heated to reflux, further initiator added at 3 hourly intervals, and worked up after 12 h → product. Y 67% (*cis:trans* >99:1). Ths is the first example of such metal hydride-mediated pinacol coupling. Dialdehydes reacted more rapidly than ketoaldehydes or diketones, thereby providing scope for selectivity. Ph$_3$SnH and (TMS)$_3$SiH were less effective. F.e. incl. cyclization of 1,6-dioxo compds., also under UV-initiation, s. D.S. Hays, G.C. Fu, J. Am. Chem. Soc. *117*, 7283-4 (1995).

Tri-n-butyltin cyanide $\qquad$ Bu_3SnCN
α-Acoxynitriles from aldehydes and acylcyanides $\qquad$ CHO → CH(CN)OCOR

287. $\quad$ C_6H_{13}CHO $\;+\;$ MeOCOCN $\longrightarrow$ C_6H_{13}CH(CN)OCOOMe

Cyanohydrin carbonates. *n*-Heptanal added to a soln. of 5 *mol*% Bu_3SnCN in 1.25 eqs. methyl cyanoformate, and stirred at room temp. for 1 h → product. Y 92%. Reaction is generally applicable to aliphatic, ar. and α,β-ethylenic aldehydes, including hindered derivs. Ketones were unaffected so that selective addition to ketoaldehydes may be effected. F.e. and with pyruvonitrile s. M. Scholl et al., J. Org. Chem. *60*, 6229-31 (1995).

Tri-n-butyltin cyanide or Di-n-butyltin dichloride $\qquad$ Bu_3SnCN or Bu_2SnCl_2
α-Siloxynitriles from oxo compds. s. *34*, 616s*51* $\qquad$ CO → C(OSi≤)CN

Stannous bromide $\qquad$ $SnBr_2$
Barbier-type synthesis of 3-ethylenealcohols $\qquad$ CO → C(OH)C—C=C
with $SnCl_2$/NaI or Cu(I) cf. *41*, 810s*48,49*; with $SnBr_2$ (cf. *37*, 632), effect of temperature on regioselectivity, s. Y. Masuyama et al., Chem. Commun. *1995*, 1405-6; with GeI_2/ZnI_2, regioselectivity, s. Y. Hashimoto et al., Tetrahedron Letters *35*, 4805-8 (1994).

Tin or Lead $\qquad$ Sn or Pb
Barbier-type synthesis of 3-ethylenealcohols $\qquad$ CO → C(OH)C—C=C
with $PbBr_2$/Al cf. *42*, 826; γ-hydroxy-α-methylenecarboxylic acid esters with Pb s. S.-H. Wu et al., Synth. Commun. *26*, 2397-406 (1996); with Sn cf. S.E. Drewes et al., ibid. *25*, 321-9 (1995); with Sn and Me_3SiCl as promotor s. ibid. 3081-7.

Bismuth trichloride/zinc $\qquad$ $BiCl_3/Zn$
Glycols from two different oxo compds. stereospecifically $\qquad$ RC(OH)C(OH)R'
1-arylglycols with VCl_3/Zn cf. *45*, 385; α,β-dihydroxycarbonyl compds. with $BiCl_3$/Zn (in THF or aq. media) s. N. Miyoshi et al., Chem. Lett. *1995*, 999-1000.

Chromium(II) chloride/lithium iodide $\qquad$ $CrCl_2/LiI$
3-Ethylenealcohols from oxo compds. $\qquad$ CO → C(OH)C—C=C
and β,γ-ethylenehalides s. *34*, 614; also from allyl phosphates with added LiI, stereoselectively, s. P. Knochel et al., J. Org. Chem. *60*, 2762-72 (1995); from N-protected α-aminoaldehydes s. P. Ciapetti et al., Tetrahedron *52*, 7379-90 (1996).

Chromium(II) chloride/nickel(II) chloride/manganese/trimethylsilyl chloride $\qquad$ ←
Catalytic Nozaki-Hiyama-Kishi reaction $\qquad$ ←

288. PhI $\xrightarrow{CrCl_2}$ [$PhCrCl_2$] $\xrightarrow{RCHO}$ [$RCH(OCrCl_2)Ph$] $\xrightarrow{Me_3SiCl}$ [$RCH(OSiMe_3)Ph$] $\longrightarrow$ RCH(OH)Ph

R = Cl(CH_2)$_5$

Unsatd. sec. alcohols from aldehydes. Excess of toxic and expensive Cr(II) as used in the Nozaki-Hiyama reaction (*34*, 614; *39*, 798) can be avoided by conducting the reaction with a catalytic amount (7-15 mol%) of $CrCl_2$ (doped with $NiCl_2$ *a lá* Kishi) in the presence of an excess of *inexpensive Mn* and a chlorosilane. **E**: A mixture of startg. aldehyde, 2 eqs. iodobenzene, 2.4 eqs. Me_3SiCl, 15 mol% $CrCl_2$ (doped with a little $NiCl_2$), and 1.7 eqs. Mn powder in 20:3 DMF/DME heated at 50° until reaction complete, and worked up with aq. Bu_4NF → product. Y 66%. Yields were comparable with those obtained with excess of Cr(II), and alkyl chlorides and esters were unaffected. The method is generally applicable to the reaction of ar. and vinyl iodides, vinyl triflates, and allyl bromides (without $NiCl_2$) with aliphatic or ar. aldehydes. Zn dust was ineffective due to the loss of aldehyde by enoxysilane formation. F.e. and **α-methylene-γ-lactones** s. A. Fürstner, N. Shi, J. Am. Chem. Soc. *118*, 2533-4 (1996).

Tetra-n-butylammonium fluoride $\qquad$ Bu_4NF
2-Nitroalcohols from aliphatic nitro compds. and aldehydes $\qquad$ CHO → CH(OH)C(NO_2)
s. *34*, 610s*41*; *40*, 449; asym. induction with diphenylbenzylamino as stereo-directing group, and conversion to chiral *anti,anti*-2,2'-diaminoalcohols, s. S. Hanessian, P.V. Devasthale, Tetrahedron Letters *37*, 987-90 (1996); with Amberlyst A 21 (cf. *34*, 610s*39*) s. R. Ballini et al., Tetrahedron

52, 1677-84 (1996); C-linked disaccharides (with KF) s. W.R. Kobertz et al., J. Org. Chem. *61*, 1894-7 (1996).

Nickel chloride s. under $CrCl_2$ $NiCl_2$

Bis(ethyl acetoacetato)nickel(II)/1,2,2,6,6-pentamethylpiperidine/trichlorosilane   ←
3-Acetylenealcohols from β,γ-acetylenechlorides s. *51*, 270   CHO → CH(OH)C—C≡C

Manganese s. under $CrCl_2$ Mn

Cationic thiolate-bridged diruthenium complex
Sym. 1,2-disiloxy compds. from oxo compds.   C(OSi≤)C(OSi≤)
with Zn/Me₃SiCl cf. *45*, 377; sym. 1,2-diaryl-1,2-disiloxy compds. with a cationic thiolate-bridged diruthenium complex/R₃SiH s. M. Hidai et al., Chem. Lett. *1995*, 671-2; from aliphatic ketones with Yb cf. Y. Taniguchi et al., Tetrahedron Letters *35*, 4111-4 (1994).

Tris(dibenzylideneacetone)dipalladium s. under Et_2Zn $Pd_2(dba)_3$

Tetrakis(triphenylphosphine)palladium(0)/diethylzinc $Pd(PPh_3)_4/Et_2Zn$
3-Acetylenealcohols from acoxy-2-acetylenes and aldehydes   C(OH)C—C≡C

289.

Propargylpalladium complexes, which ordinarily serve as electrophiles towards hard [organometallic] carbon nucleophiles, may serve as *nucleophilic* species towards carbonyl compds. via Pd→Zn-transmetalation. E: Benzaldehyde and 3.6 eqs. 2 *M* Et₂Zn in dry THF added via syringe to a soln. of 1.2 eqs. startg. acetylene deriv. and 5 mol% Pd(PPh₃)₄ in the same solvent under N₂, and stirred at room temp. for 3 h → product. Y 70% (55:45 mixture of diastereomers). There was no cross-coupling of the Pd-complex with diethylzinc. 2-Allenealcohols were isolated as by-products from certain substrates. F.e. incl. silylacetylene derivs., and leaving groups, s. Y. Tamaru et al., Angew. Chem. Intern. Ed. *35*, 878-80 (1996); 3-ethylenealcohols s. *51*, 272.

Via intermediates   *v.i.*
Regio- and stereo-specific synthesis of (E)-acoxy-3-ethylenes   C
from 3-ethyleneepoxides via iron η³-(acoxymethyl)allyl complexes

290.

A 1:1:1 mixture of Bu₄N[Fe(CO)₃NO], acetyl chloride, and pyridine in methylene chloride stirred at room temp. for 30 min under argon, 1 eq. 3,4-epoxy-1-phenyl-1-butene added, and stirred at room temp. for 3 h → (η³-1-acetoxymethyl-3-phenylallyl)Fe(CO)₂NO (Y 58%), in THF allowed to react with dimethyl sodiomalonate at 60° for 15 h, and acidified with 4 *M* HCl → product (Y 78% overall). Reaction takes place preferentially at the C-atom bearing the acetoxymethyl group. F.e. and stereochemistry s. K. Itoh et al., Tetrahedron Letters *36*, 5211-4 (1995).

Addition to Nitrogen and Carbon   CC ⇓ NC

Without additional reagents   *w.a.r.*
β-Aminoketones from ketimines and azomethinium salts   COC—C(N≤)
Regio- and stereo-specific Mannich-type reaction under mild conditions

291.

Startg. iminium salt added in one portion to a soln. of 1 eq. startg. ketimine in anhydrous methylene chloride at -80° under argon, the mixture stirred for 3-4 h with warming to ca. -30°, stored in a freeezer for ca. 15 h at this temp., treated with 2 N aq. acetic acid, diluted with ether, stirred at 25° for 3-4 h, then 6 N HCl added, and stirring continued for 10 min → product. Y 62% (regioselectivity ≥ 99:1; *anti*-selectivity ≥ 99:1). The method is simple and convenient, based on readily available substrates, and applicable to amino*alkyl*ation. In general, reaction takes place at the sterically *less* hindered α position [via enamine formation], and high regioselectivities are achieved even when the two α-C-atoms are only marginally different (cf. *32*, 715). F.e.s. M. Arend, N. Risch, Angew. Chem. Intern. Ed. *34*, 2639-40 (1995).

2-Subst. N-acyl-N-heterocyclics from cyclic azomethines   ←
s. *18*, 744; 1-subst. N-carbalkoxy-1,2,3,4-tetrahydroisoquinolines s. A.P. Venkov, S. Statkova-Abeghe, Synth. Commun. *26*, 2135-44 (1996).

Electrolysis/isopropanol   ⇵/i-PrOH
2-(Alkoxylamino)alcohols from alkoximes s. *46*, 612s*51*   CO → C(OH)—C(NHOR)
Sodium hydroxide/benzyltriethylammonium chloride   NaOH/BnEt₃NCl
β-Aminoketones from azomethines   C=N → C(NH)C—CO
with MeOMgOCOOMe cf. *28*, 620; α,β-diaryl-β-arylaminoketones with NaOH/BnEt₃NCl s. V. Dryanska et al., J. Chem. Res. (S) *1995*, 418-9.

Organolithium compd. or magnesium/cerium(III) chloride   RLi or Mg/CeCl₃
Asym. syntheses via 1,2-addition to aldehyde hydrazones   ←
chiral prim. amines cf. *42*, 621; **chiral 2-acylaminoalcohols** s. Angew. Chem. Intern. Ed. *34*, 1219-22 (1995); chiral N-(3-ethylene)urethans via N-carbalkoxylative Grignard addition in the presence of CeCl₃ s. Synlett *1994*, 795-7.

Organolithium compd./cerium(III) chloride   RLi/CeCl₃
Asym. synthesis of sec. amines from aldimines   CH=NR → CH(R')NHR
s. *47*, 616; chiral 2-acetylene- and 2-ethylene-*prim*-amines s. D. Enders, J. Schankat, Helv. Chim. Acta *78*, 970-92 (1995).

Barium or Magnesium or Zinc   Ba or Mg or Zn
Regiospecific synthesis of 3-ethyleneamines from azomethines   CH(NH)C—C=C

292.

with 2-ethylenebarium chlorides. A soln. of 3 eqs. prenylbarium chloride (generated from prenyl chloride and Ba in THF) treated with N-benzylbenzaldimine *at -78°* for 1 h → product. Y 94% (α:γ <1:99). At 0°, the regioselectivity was >99:1 (Y 74% after 1.5 h). The regioselectivity of addition to aldimines is remarkably *temperature-dependent*, the kinetically favoured γ-adduct isomerizing to the thermodynamically more stable α-adduct at higher temp. F.e. **and with asym. induction** s. H. Yamamoto et al., Chem. Commun. *1996*, 367-8; with Mg or unactivated Zn cf. D.-K. Wang et al., Tetrahedron Letters *37*, 4187-8 (1996).

Magnesium/cuprous iodide   Mg/CuI
Ring opening of N-phosphinylaziridines   C
s. *49*, 610; **prim. amines** with N-(diethoxyphosphoryl)aziridine via phosphoromonoamidates s. K. Osowska-Pacewicka, A. Zwierzak, Synthesis *1996*, 333-5; regiospecific ring opening of N-phosphinyl-2-vinylaziridines s. J.B. Sweeney et al., Synlett *1996*, 847-9.

Zinc/iodide Zn/I_2
β-Aminocarboxylic acid esters   C=NR → C(R′)NHR
from α-bromocarboxylic acid esters and azomethines
Reformatskii-type synthesis with asym. induction

293.

7.3 eqs. Zn-dust and 0.2 eq. I_2 in anhydrous dioxane refluxed for 1 h, cooled to room temp., the flask immersed in an ultrasonic bath, a mixture of 1.2 eqs. startg. (R)-bromoacetate and *p*-benzyloxybenzylidene(*p*-fluoroaniline) added, and the mixture sonicated at room temp. for 48 h → (R)-product. Y 55% (99% e.e. determined after conversion to the corresponding 2-azetidinone). F.e.s. B.B. Shankar et al., Tetrahedron Letters *37*, 4095-8 (1996); from chiral $Cr(CO)_3$-complexed ar. aldimines cf. P. Del Buttero et al., Tetrahedron *52*, 4849-56 (1996).

Isopropylmagnesium chloride s. under Ti(OPr-i)₄   i-PrMgCl
Zinc iodide ZnI_2
Asym. synthesis of sulfenamides and sulfinamides   C=N—S— → C(R)NH—S—
by Grignard addition cf. *50*, 380; chiral α-sulfinylamino- and α-sulfenamido-nitriles with Et_2AlCN and Me_3SiCN, respectively, also hydrolysis to chiral α-*prim*-aminocarboxylic acids, s. (respectively) F.A. Davis et al., J. Org. Chem. *61*, 440-1 (1996); L. Yan et al., Synth. Commun. *26*, 63-6 (1996); addition of ester enolates s. J. Org. Chem. *61*, 2222-5 (1996); reversal of enantioselectivity with Ti(IV) or aluminum ester enolates s. T. Fujisawa et al., Tetrahedron Letters *37*, 3881-4 (1996).

Aluminum or bismuth/potassium hydroxide   Al/KOH or Bi/KOH
Sym. 1,2-diamines from azomethines   RC(NHR′)C(NHR′)R
with In/NH_4Cl cf. *43*, 584s*48*; with Al/KOH (cf. *19*, 757) or Bi/KOH s. B. Baruah et al., Tetrahedron Letters *36*, 6747-50 (1995).

Indium   In
3-Ethyleneamines from azomethines   C=NR → C(NHR)C—C=C
s. *48*, 626; 3-ethylene-*tert*-amines from tert. enamines, also β-*tert*-aminocarboxylic acid esters via Reformatskii-type synthesis, and with $Al/InCl_3$ (cat.) s. P. Mosset et al., Tetrahedron Letters *36*, 6055-8 (1995).

Aluminum chloride $AlCl_3$
Synthesis of 2-acyl-1-aryl-1,2,3,4-tetrahydroisoquinolines   ←
from 3,4-dihydroisoquinolines and arenes s. *18*, 744s*51*

Cerium(III) chloride s. under RLi $CeCl_3$
Samarium diiodide/tert-butanol SmI_2/t-BuOH
2-(Alkoxylamino)alcohols from alkoximes   CO → C(OH)—C(NHOR)
s. *46*, 612s*47*; intramolecular variant for cyclic analogs [aminocyclopentitol derivs.] stereoselectively s. J.L. Chiara et al., J. Org. Chem. *60*, 6010-11 (1995); under electrolysis in *i*-PrOH, also coupling with hydrazones, cf. T. Shono et al., ibid. *59*, 1730-40 (1994).

Chiral cyclic dipeptide
**α-Aminocarboxylic acids from N-(diphenylmethyl)aldimines
via α-(diphenylmethylamino)nitriles
Asym. Strecker synthesis**

CH=NR → CH(NHR)CN

294.

A soln. of startg. imine (preformed from benzaldehyde and benzhydrylamine) and 2 mol% cyclo[(S)-phenylalanyl-(S)-α-amino-γ-guanidinobutyric acid] in methanol treated with 2 eqs. HCN at -25° → intermediate (S)-α-aminonitrile (Y 95%; e.e. >99%), hydrolyzed in 6 N HCl at 60° for 6 h → (S)-phenylglycine (Y 92% from benzaldehyde; e.e. >99%). Enantioselectivity was low with aliphatic or hetaryl aldehydes, and with ar. aldehydes substituted by electron-withdrawing groups. Poor results were obtained by the direct route from benzaldehyde, HCN and NH$_3$ in the presence of the same catalyst. F.e.s. M. Lipton et al., J. Am. Chem. Soc. *118*, 4910-1 (1996).

Titanium tetraisopropoxide/isopropylmagnesium chloride
**3-Ethyleneamines
from aldimines and 2-ethylenecarbonic acid esters
Regiospecific synthesis with 1,3-asym. induction**

Ti(OPr-i)$_4$/i-PrMgCl
CH(NHR)C—C=C

295.

3.2 eqs. *i*-PrMgCl added to a mixture of 1.6 eqs. startg. carbonate and 1.6 eqs. Ti(OPr-*i*)$_4$ in ether at -50°, stirred at -50° to -40° for 1 h, 1 eq. startg. chiral aldimine added, and warmed gradually to -10° over 2 h → product. Y 92% (d.r. 94:6). The method is simple, general, high-yielding, and based on readily available substrates. F.e. and with allyl bromides or aryloxy-2-ethylenes s. Y. Gao, F. Sato, J. Org. Chem. *60*, 8136-7 (1995).

Cobaltous chloride/acetyl chloride
3-Component synthesis of acylamines

CoCl$_2$/MeCOCl
CHO → CH(NHCOR)R'

296.

A mixture of 2,4-pentanedione, 1 eq. *p*-nitrobenzaldehyde, 2 eqs. acetyl chloride, and a little CoCl$_2$ in *acetonitrile* heated under N$_2$ until reaction complete → 3-(1-acetylamino-*p*-nitrobenzyl)-2,4-pentanedione. Y 75%. Remarkably, the same reaction with aliphatic aldehydes required an *O$_2$ atmosphere* rather than N$_2$. F.e., **also 3-α-acylaminofurans from γ-diketones**, s. M.M. Reddy et al., Tetrahedron Letters *36*, 4877-80 (1995); β-acylaminoketones s. Chem. Commun. *1994*, 713-4.

Dichlorobis(triphenylphosphine)nickel(II)
**Synthesis of sec. amines from aldimines
Addition of CH-acidic compds. under transition metal catalysis**

NiCl$_2$(PPh$_3$)$_2$
CH=N → CH(R)NH

with RhH(CO)(PPh$_3$)$_3$ cf. *50*, 383; from N-activated imines with NiCl$_2$(PPh$_3$)$_2$ s. Tetrahedron Letters *36*, 5023-6 (1995); under high pressure cf. Synlett *1993*, 465-6.

Addition to Sulfur and Carbon   CC ⇓ SC

Magnesium
Thiolic acids from Grignard compds.
Synthesis with addition of one C-atom

Mg
RMgX → RCOSH

297.

Carbon oxide sulfide passed through a 0.5 M soln. of phenylmagnesium bromide in 1:1 ether/ THF at 0° for 0.5 h, the mixture stirred at 20° for 0.5 h, re-cooled to 0°, and treated dropwise with 12 M HCl, followed by water → benzoic acid. Y 99%. **Thiolic acid esters** were obtained by quenching the adduct with 5 eqs. alkyl halide, followed by refluxing overnight. The method appears quite general and simple. F.e., **also carboxylic acid amides** with amines in the presence of 1-trifluoromethylsulfonylbenzotriazole, s. A.R. Katritzky et al., Org. Prep. Proc. Intern. 27, 361-6 (1995); **carboxylic acid thioamides** with CS_2 s. Synlett *1995*, 99; with CS_2/Tf_2O cf. Synthesis *1995*, 1497-505.

Addition to Remaining Elements and Carbon   CC ⇓ RemC

Potassium tert-*butoxide/18-crown-6 polyether*
1,3-Diols from siliranes and aldehydes via 1,2-oxasilacyclopentanes

KOBu-t /*18-crown-6*

298.

Stereospecific conversion. A soln. of startg. *trans*-silirane and benzaldehyde in THF treated with *t*-BuOK/18-crown-6 (25 mol%) at 22° until reaction complete → intermediate (yield 54% after chromotographic clean-up), in DMF heated at 75° with *t*-BuOOH, $CsOH·H_2O$ and Bu_4NF → product (Y 64%). The catalyzed aldehyde insertion proceeds with 96-9% *inversion* of configuration, in dramatic contrast with predominant retention in the corresponding thermal reaction. Significantly, the standard Tamao desilylation (H_2O_2, KF, $KHCO_3$) failed in this instance. F.e. and regioselectivity s. K.A. Woerpel et al., J. Am. Chem. Soc. *117*, 10575-6 (1995); ring expansion of 3-methylenesilacyclobutanes s. K. Oshima et al., Tetrahedron Letters *36*, 8067-70 (1995).

Addition to Carbon-Carbon Bonds   CC ⇓ CC

Without additional reagents
Asym. Diels-Alder reaction with ketene equivalents
s. *37*, 669s*46,49*; with 1-(phenylseleno)-2-(*p*-toluenesulfonyl)ethyne s. T.G. Back, D. Wehrli, Synlett *1995*, 1123-4.

Diels-Alder reaction with furans
s. *1*, 529; with 2,5-disubst. furans s. B. Alcaide et al., Heterocycles 36, 1795-802 (1993); with Li-furan-3-olates s. D. Caine, R.F. Collison, Synlett *1995*, 503-4; **asym. induction** with chiral 3-aminofurans s. R.H. Schlessinger et al., ibid. 536-7.

Diels-Alder reaction with vinylheterocyclics
s. *37*, 646s*48*; with 3-vinyl-coumarins and -3-chromenes s. T. Minami et al., J. Org. Chem. *57*, 167-73 (1992); with 5-vinyl-2(3*H*)-furanones s. F. Rouessac et al., Tetrahedron *47*, 5481-90 (1991); with 2-vinylbenzothiazoles s. M. Sakamoto et al., ibid. *52*, 733-42 (1996); with 2-vinylindoles s. B. Danieli et al., ibid. 11291-6; with 3-(β-oxyvinyl)indoles s. U. Pindur, M. Rogge, Heterocycles

41, 2785-93 (1995); with 5-vinylimidazoles s. M.A. Walters, M.D. Lee, Tetrahedron Letters *35*, 8307-10 (1994); with a 1-(3-furyl)enoxysilane s. A. Benitez et al., J. Org. Chem. *61*, 1487-92 (1996).

Diels-Alder reaction with quinones ○
s. *21*, 725s*46,48*; with benzoyl-*p*-quinones s. B. Chaudhary et al., Pharmazie *48*, 943-4 (1993); propellane derivs. from cyclobuteno-*p*-quinones s. M.N. Paddon-Row et al., Tetrahedron Letters *36*, 1129-32 (1995); with 2-sulfinyl-*p*-quinone monoimines s. M.C. Carreño et al., ibid. *37*, 3187-90 (1996); with sulfinylnaphthazirins s. ibid. *35*, 3789-92 (1994); cycloaddition of *o*-quinones to fulvenes s. V. Nair et al., ibid. *36*, 1605-8 (1995); s.a. Chem. Lett. *1995*, 383-4; Tetrahedron *51*, 9155-66 (1995); **asym. induction** with sulfinyl-*p*-quinones s. M.C. Carreño et al., J. Org. Chem. *61*, 503-9, 2980-5 (1996); s.a. S. Paul et al., Tetrahedron Letters *37*, 4055-8 (1996); cycloaddition **with electrogenerated *o*-quinones** *in situ* s. K. Chiba, M. Tada, Chem. Commun. *1994*, 2485-6.

Asym. Diels-Alder reaction with chiral amino-1,3-dienes
with 2-amino-1,3-dienes cf. *49*, 622; with 1-acylamino-1,3-dienes s. P.J. Stevenson et al., Tetrahedron Letters *36*, 9533-6 (1995).

Diels-Alder reaction with enazomethines
with unactivated derivs. cf. *42*, 292; with electron-deficient derivs. s. F. Palacios et al., J. Org. Chem. *60*, 2384-90 (1995).

1,3-Dipolar cycloaddition with nitrile oxides
with asym. induction s. *16*, 888s*49*; chiral 6,8-dioxabicyclo[3.2.1]oct-3-ene s. R.M. Paton et al., J. Chem. Soc. Perkin Trans. I *1993*, 75-9; to chiral 2-alkoxy-2-(perfluoroalkyl)vinyl sulfoxides s. P. Bravo et al., J. Chem. Res. (S) *1996*, 348-9; s.a. A. Arnone et al., ibid. 198-9; to unsatd. carbohydrates s. J.A. Galbis et al., Tetrahedron *51*, 6349-62 (1995).

Irradiation (s.a. under $(Bu_3Sn)_2$, $Co_2(CO)_8$ and $Co(acac)_2$)
Oxetane ring from ethylene derivs. and oxo compds.
3-siloxyoxetanes s. *19*, 764s*49*; details s. Liebigs Ann. *1995*, 855-65; **with asym. induction**, substituent effects, s. Tetrahedron *52*, 10861-78 (1996); s.a. Tetrahedron Letters *35*, 5845-8 (1994); 2-alkoxy-2-siloxyoxetanes s. M. Abe et al., ibid. *37*, 5901-4 (1996); 3-acylaminooxetanes s. Angew. Chem. Intern. Ed. *35*, 884-6 (1996).

Photochemical [6+2]-cycloaddition ○
with tricarbonylchromium-complexed cycloheptatrienes s. *46*, 631; with complexed cyclooctatrienes and -tetraenes s. Tetrahedron Letters *36*, 8569-72 (1995); **asym. induction** with sultam derivs. s. J. Am. Chem. Soc. *117*, 8851-2 (1995).

Electrolysis/nickel bromide $\frac{4}{7}/NiBr_2$
Electrochemical 1,4-addition $C=C \rightarrow CHC(R)$
of ar. halides at a zinc cathode under Ni(II) catalysis cf. *47*, 450; also of alkyl and vinyl bromides at a Ni-cathode with $NiBr_2/Bu_4NBr/Bu_4NI/py$ s. S. Condon-Gueugnot et al., J. Org. Chem. *60*, 7684-6 (1995).

Organolithium compds./(-)-sparteine
Asym. carbolithiation

299.

of cinnamyl derivs. (E)-Cinnamyl alcohol added to a soln. of *n*-BuLi in *cumene* containing 1 eq. (-)-sparteine at -10°, and the mixture hydrolyzed → product. Y 82% (e.e. 80%). *syn*-Isomers were formed with electrophiles other than H⁺ (e.g. PhSSPh, MeI), reaction proceeding **with inversion of configuration** at the benzyllithium site. (Z)-Cinnamyl derivs. gave the enantiomeric products. F.e. incl. carbolithiation of cinnamyl ethers and amines with prim. and sec. organolithiums, and catalytic procedures with 5% (-)-sparteine (in somewhat reduced yield) s. J.-F. Normant et al., J. Am. Chem. Soc. *117*, 8853-4 (1995); generation of benzyllithiums from styrenes s.a. X. Wei, R.J.K. Taylor, Chem. Commun. *1996*, 187-8.

Organolithium compds./aluminum tris(2,6-diphenylphenoxide) $RLi/Al(OAr)_3$
Nucleophilic 1,6-addition to aryloxo compds.
Protection of the carbonyl group by aluminum complexation

300.

Unprecedented nucleophilic 1,6-addition of organolithiums to aromatic aldehydes and ketones can be effected by screening the carbonyl function (from 1,2- and 1,4-addition) by aluminum complexation. **E:** A soln. of benzaldehyde in 1:1 toluene/THF treated with 1.5 eqs. Al-tris(2,6-diphenylphenoxide), 2 eqs. *t*-BuLi in pentane added at -78°, and quenched with concd. HCl after 3 h → product. Y 81% (>99% purity). There was no 1,4-adduct (except on addition of MeLi to the methyl 1-naphthyl ketone complex), but the proportion of aromatized product may increase depending on the choice of solvent and quenching conditions. Reaction appears general for prim., sec. and tert. alkyllithiums. F.e.s. K. Maruoka et al., J. Am. Chem. Soc. *117*, 9091-2 (1995).

n-*Butyllithium or* tert-*Butyllithium/cuprous iodide*   *BuLi or* t-*BuLi/CuI*
Regiospecific asym. Michael addition $C=C \to CHC(R)$
with β,γ-ethylenephosphonic diamides cf. *49*, 630; with 2-allyl-1,3,2-oxazaphosphorinane 2-oxides, asym. γ-addition to cyclic enones using t-BuLi/CuI, s. S.E. Denmark, J.-H. Kim, J. Org. Chem. *60*, 7535-47 (1995); with β,γ-ethylenephosphonic esters as acetaldehyde carbanion equivalent s. K. Tanaka et al., ibid. 8036-43.

n-*Butyllithium/magnesium*   *BuLi/Mg*
Regio- and stereo-specific Diels-Alder reaction
based on a temporary vinyl-magnesium or -aluminum alkoxide tether

301.

in situ-Generated vinyl-magnesium or -aluminum alkoxide tethers, in spite of their *carbanionic* character, greatly facilitate diene synthesis by comparison with the familiar vinylsilyloxy tethers (cf. *46*, 696s*48*). **E:** A soln. of startg. 2,4-dienol in THF treated with 1 eq. *n*-BuLi at -78°, warmed to room temp., 1 eq. 2 *M* vinylmagnesium bromide in THF added, heated *at 80° for 1 h* in a sealed tube, and acidified with dil. HCl prior to workup → product. Y ca. 70%. Cycloaddition of the analogous siloxy-tethered substrate required heating at 160° for 3 h, while the alkoxy-analog required 10 h at 160°. Excellent regioselectivity was recorded with substituted vinylmagnesium derivs. and, remarkably, terminal disubstitution of the diene is tolerated. An aluminum tether was equally viable. F.e.s. G. Stork, T.Y. Chan, J. Am. Chem. Soc. *117*, 6595-6 (1995).

n-*Butyllithium/trimethylsilyl chloride/hexamethylphosphoramide* ←
1,4-Addition with organolithium compds. C=C → CHC(R)
to α,β-ethyleneketones s. *33*, 639; also addition to enals and acrylates **via enoxysilanes**, prevention of polymerization (with added Me$_3$SiCl/HMPA), s. H. Liu, T. Cohen, Tetrahedron Letters *36*, 8925-8 (1995).

Lithium diisopropylamide i-*Pr$_2$NLi*
Michael addition with carboxylic acid ester enolates
glutaric acid esters cf. *36*, 652s*39*; addition to α-methylene-γ-lactones and subsequent conversion to (E)-γ,δ-ethylenecarboxylic acid esters s. W. Adam et al., J. Org. Chem. *60*, 3879-86 (1995); addition to γ-alkoxyenones, reversal of diastereoselectivity with Ti(IV)-enolates, s. A. Bernardi et al., Tetrahedron *52*, 3497-508 (1996).

Asym. Michael addition with hydrazones
s. *39*, 630; *44*, 599; chiral δ-ketophosphonic esters via addition to α,β-ethylenephosphonic esters s. Liebigs Ann. *1995*, 1177-84.

Asym. Michael addition with N-acyl-N-heterocyclics
with 3-acyl-2-oxazolidones cf. *47*, 664; **with 3-acyloxazolidines** via lithium (Z)-enolates s. S. Kanemasa et al., Tetrahedron *51*, 10463-76 (1995).

Sequential Michael addition ○
bicyclo[2.2.2]octan-2-ones s. *32*, 641; **polymer-based conversion** s. S.V. Ley et al., Synlett *1995*, 1017-20.

Triethylenediamine *dabco*
2-Methylene-1,5-diketones from two α,β-ethyleneketone molecules COC(=CH$_2$)C—CHCO
with RhCl(PMe$_3$)$_3$ cf. *45*, 422; with dabco s. D. Basavaiah et al., J. Chem. Res. (M) *1995*, 1656-73.

Lithium diorganocuprates/trimethylsilyl chloride R$_2$CuLi/Me$_3$SiCl
Lithium organo(cyano)cuprates RCu(CN)Li
Lithium 2-thienyl(cyano)cuprate/organolithium compd. ←
Asym. 1,4-addition to α,β-ethylenecarboxylic acid esters C=C → CHC(R)
with CuI/PhLi/BF$_3$ cf. *37*, 657s*40*; asym. 1,4-addition to chiral 2-(2-carbalkoxyvinyl)imidazolidines with R$_2$CuLi s. A. Alexakis et al., J. Am. Chem. Soc. 117, 10767-8 (1995); to chiral 4-(2-carbalkoxyvinyl)oxazolidines with added Me$_3$SiCl s. S. Hanessian et al., Tetrahedron Letters 37, 7477-80 (1996); to chiral O-protected α,β-ethylene-γ-hydroxyesters s. ibid. 7473-6; to α,β-unsatd. 2,3-dideoxyaldonolactones s. P.C. Raveendranath et al., Carbohyd. Res. 253, 207-23 (1994); asym. 1,4-addition of RCu(CN)Li to chiral bicyclic Δ^3-2-pyrrolone-3-carboxylic acid esters s. A.I. Meyers et al., J. Org. Chem. *57*, 3814-9 (1992); ibid. *58*, 36-42 (1993); asym. 1,4-addition to cyclic α,β-ethylene-β'-ketocarboxylic acid esters with lithium 2-thienyl(cyano)cuprate/RLi s. E. Urban et al., Tetrahedron *51*, 11149-64 (1995); chiral 6-subst. 2-oxocyclohexanecarboxylic acid esters s. Tetrahedron Letters *36*, 4773-6 (1995).

Organocopper compds./lithium iodide/trimethylsilyl iodide ←
Asym. 1,4-addition to N-(α,β-ethyleneacyl)-2-pyrrolidones s. *37*, 657s*51*

Cupric trifluoromethanesulfonate/chiral bis(Δ^2-oxazolines) ←
Asym. Diels-Alder reaction ○
s. *46*, 662s*48*; to 3-(α,β-ethyleneacyl)-2-oxazolidones with chiral magnesium-Δ^2-oxazoline complexes cf. T. Fujisawa et al., Tetrahedron Letters *36*, 5031-4 (1995); **asym. hetero-Diels-Alder reaction** with glyoxylic acid esters s. M. Johannsen, K.A. Jørgensen, J. Org. Chem. *60*, 5757-62 (1995); solvent effects s. Tetrahedron *52*, 7321-8 (1996); *endo*-selectivity with

conformationally rigid 2,2'-methylenebis(Δ^2-oxazolines), and with the Mg(II)-salt, s. A.K. Ghosh et al., Tetrahedron Letters *37*, 3815-8 (1996).

Magnesium/cuprous bromide or iodide   Mg/CuBr or CuI
Asym. 1,4-addition   C≡C → CHC(R)
to 3-(α,β-ethyleneacyl)-2-oxazolidones cf. *37*, 657s50; chiral α-*prim*-aminocarboxylic acids via 1,4-addition to α-(2-oxazolidon-3-yl)acrylic acid esters s. P.A. Lander, L.S. Hegedus, J. Am. Chem. Soc. *116*, 8126-32 (1994); asym. 1,4-addition to 3-(α,β-ethyleneacyl)-4-imidazolidones with BrMgCuAr$_2$ s. D. Seebach et al., Helv. Chim. Acta *78*, 1185-206 (1995); to N-(α,β-ethyleneacyl)-2-pyrrolidones with RCu/LiI/Me$_3$SiI s. M. Nilsson et al., Tetrahedron Letters *36*, 3227-30 (1995).

Magnesium/zinc bromide   Mg/ZnBr$_2$
Regio- and stereo-specific synthesis of ethylene derivs.   C(R)═CH(R')
from lithium acetylides via ene-1,1-di(organometallics)

302.

1,2-Disubst. 1,4-dienes. A soln. of startg. Li-acetylide in ether treated with *allylmagnesium bromide* and ZnBr$_2$-etherate *at -10°* for 30 min, cooled to -50°, 2 eqs. ethereal TsCN added, allowed to warm to room temp., and quenched with acid at -20° after 2 h → product. Y 63% (single stereoisomer). The differing reactivity of the 2 metals in the intermediate allows chemoselective functionalization by electrophilic species, and chelation with the *tert*-butoxy group is responsible for the stereoselectivity. F.e. **and 1,1,2-trisubst. 1,4-dienes** s. J.-F. Normant et al., Tetrahedron Letters *36*, 7451-4 (1995).

Magnesium/manganese(II) iodide   Mg/MnI$_2$
Regio- and stereo-specific manganese-catalyzed allylmagnesiation of acetylene derivs. ←

303.

An ethereal soln. of startg. homopropargyl ether treated with 1.5 eqs. allylmagnesium bromide and 3 mol% MnI$_2$ at 25° for 3 h → product. Y 83%. Other Mn-catalysts were less effective, and transition metal catalysts such as PdCl$_2$(MeCN)$_2$, NiCl$_2$(PPh$_3$)$_2$, CrCl$_3$ and RuCl$_3$ were ineffective. The intermediate vinylmagnesium bromide was trappable with C-electrophiles (aldehydes, allyl bromide). Propargyl ethers, however, gave **1,2,5-trienes**, while homopropargyl vinyl ethers gave **functionalized 1,3,6-trienes**. F.e. and synthesis of **1,4,7-trienes by catalyzed diallylation** on exposure to air s. K. Okada et al., J. Am. Chem. Soc. *118*, 6076-7 (1996).

Zinc s.a. under TaCl$_5$ Zn
Zinc/boron fluoride/trimethylsilyl chloride   Zn/BF$_3$/Me$_3$SiCl
1,4-Addition of *sec*- and *tert*-alkylzinc bromides to α,β-ethyleneketones ←
in a heterogeneous medium without copper(I) catalysis

304.

Methyl 3-bromobutyrate added with stirring to Rieke zinc (freshly prepared from ZnCl$_2$ and Li-naphthalenide and carefully washed with THF to remove deactivating Li-salt) in THF, refluxed for 2 h, allowed to cool to room temp. and the zinc to settle (4 h), the resulting 1 *M* soln. of alkylzinc bromide added dropwise during 20 min to 0.55 eq. 2-cyclohexen-1-one, 1.45 eqs. BF$_3$-

etherate, and 1.98 eqs. Me₃SiCl in *pentane* (pentane/THF 9:1) at -30°, the heterogeneous mixture stirred for 3.5 h at -30°, and quenched with satd. NH₄Cl → methyl 3-(3'-oxocyclohexyl)butanoate. Y 51%. Neither copper(I) salt nor cyanide ion is required in this economical and environmentally friendly procedure based on a mixed *polar/non-polar* solvent system. The method is general, will tolerate functional groups, and permits transfer of bulky residues. *prim*-Alkyl- and aryl-zinc bromides, however, were unreactive. F.e.s. M.V. Hanson, R.D. Rieke, J. Am. Chem. Soc. *117*, 10775-6 (1995).

Methylmagnesium iodide/chiral 2-(o-tosylaminophenyl)-4-phenyl-Δ²-oxazoline/iodine ←
Asym. Diels-Alder reaction with 3-(α,β-ethyleneacyl)-2-oxazolidones s. *46*, 662s51 ○

Isopropylmagnesium halides s. under Ti(OPr-i)₄ and Cp₂TiCl₂ ←

Organozinc compds./bis(1,5-cyclooctadiene)nickel(0) R₂Zn/Ni(cod)₂
Regio- and stereo-specific alkylative ring closure of 2,n-enynones via intramolecular carbonickelation

305.

Conjugate addition of an organonickel(II) species to an enone functionality in a catalytic cycle affords an intermediate which can be trapped by an appended alkynyl group via intramolecular carbonickelation, rather than undergo elimination to give the 1,4-adduct. E: **α-(2-Alkylidenecyclopentan-1-yl)ketones**. A soln. of startg. enynone and Bu₂Zn/BuZnCl in THF treated with 5 mol% Ni(cod)₂ at 0° until reaction complete → product. Y 68%. Interestingly, good yields were obtained even with alkylzinc compds. bearing a β-hydrogen. Reaction takes place with both internal and terminal acetylene functions; however, reductive elimination is effected in the presence of PPh₃. F.e. incl. spirocyclic derivs. s. J. Montgomery, A.V. Savchenko, J. Am. Chem. Soc. *118*, 2099-100 (1996); ring closure **of bis(α,β-ethyleneketones)** via intermolecular-intramolecular 1,4-addition s. J. Org. Chem. *61*, 1562-3 (1996).

Magnesium bromide-etherate MgBr₂
α-Silyl-β-lactones from aldehydes
with BF₃ cf. *44*, 139; asym. induction with chiral α- or β-benzyloxyaldehydes using MgBr₂-Et₂O s. R. Zemribo, D. Romo, Tetrahedron Letters *36*, 4159-62 (1995).

Triethylborane Et₃B
Radical ring closure of acetylenealkoximes ○
s. *46*, 656; chiral aminocyclopentitol derivs. from carbohydrate derivs. with Et₃B as initiator s. J. Marco-Contelles et al., Tetrahedron:Asym. *6*, 1547-50 (1995).

Borane-N,N-diethylaniline/cobaltous chloride BH₃·Et₂NPh/CoCl₂
Sym. ketones from ethylene derivs. 2 C=C → CHC—CO—CCH
with B₂H₆/Et₃COLi/Cl₂CHOMe cf. *30*, 495; with BH₃·PhNEt₂/CoCl₂/CO s. M.L.N. Rao, M. Periasamy, Tetrahedron Letters *36*, 9069-70 (1995).

Pinacolborane s. under Rh(PPh₃)₃Cl ←

3,5-Bis(trifluoromethyl)phenylboronic acid/(R)-3-(2-hydroxy-3-biphenylyl)- ←
2,2'-dihydroxy-1,1'-binaphthyl
Brønsted acid-assisted Lewis acid-catalyzed asym. Diels-Alder reaction ○
with borate-derived complexes cf. *49*, 655; acceleration of addition to both α-subst. *and* α-unsubst. enals with boronate-derived complexes, also intramolecular process, s. J. Am. Chem. Soc. *118*, 3049-50 (1996).

Aluminum tris(2,6-diphenylphenoxide) s. under RLi Al(OAr)₃
Lithium aluminum bis[(R,R)-1,1'-bi-2-naphthoxide] ←
Asym. Michael addition and Michael addition-aldol condensation s. *51*, 309

Zeolite (HSZ-360)
1,1-Diarylethylenes from arylacetylenes and arenes   C═C(Ar)Ar'

306.

100 wt% Zeolite HSZ-360 (preliminarily heated at 120° for 10 h under dry N_2) added to *p*-cresol and 1 eq. phenylacetylene in dry *o*-dichlorobenzene, and heated at 110° for 14 h → product. Y 90% (conversion 94%; selectivity 96%). Exclusive *o*-substitution takes place with phenols, in keeping with an electrophilic process. There was no disubstitution, competitive isomerization, or transalkylation, and work-up is simple. F.e.s. G. Sartori et al., Tetrahedron Letters *36*, 9177-80 (1995).

Chiral 1,3,2-oxazaborolidines
Asym. 1,3-dipolar cycloaddition with nitrones s. *51*, 317
Lewis acids (BF_3, $SnCl_4$, $TiCl_4$)
5-Hydroxy-2,3-dihydrobenzofurans from *p*-quinones and ethylene derivs.
with $TiCl_4$/Ti(OPr-*i*)$_4$ cf. *44*, 624s*47*; *trans*-2,3-diaryl-derivs. s. J. Org. Chem. *60*, 3700-6 (1995); with $LiClO_4$ cf. J. Asano et al., J. Chem. Res. (M) *1995*, 862-77; 5-sulfonylamino-2,3-dihydrobenzofurans or **5-hydroxy-1-sulfonylindolines** from *p*-quinone mono-N-sulfonylimines s. Tetrahedron Letters *36*, 2713-6; 7003-6 (1995); s.a. G.K. Trivedi et al., Tetrahedron *52*, 2217-28 (1996).

Fluoroboric acid-dimethyl etherate HBF_4-OMe_2
Activation of α,β-ethyleneketones
by intermediate formation of 4-vinyl-1,3-dioxolanium ions
with asym. induction on subsequent Diels-Alder reaction

307.

A soln. of startg. chiral dienophile and 2 eqs. isoprene in toluene treated with 0.02 eq. HBF_4-dimethyl etherate at -20° for 16 h, and hydrolyzed by addition of TsOH and ethanol → product. Y 72% (200:1 mixture of diastereomers). In addition to its protecting role, the 1-ethoxyethoxy residue serves as an activating group by virtue of intermediate formation of a conjugated cyclic oxocarbenium ion, the chirality of which affords sufficient face selectivity to ensure diastereoselectivity up to 200:1. F.e. and with Me_2AlCl s. T. Sammakia, M.A. Berliner, J. Org. Chem. *60*, 6652-3 (1995); activation of α,β-ethylenecarboxylic acid N-allylamides via 2-vinyl-5-iodomethyl-Δ^2-oxazolinium salts s. T. Taguchi et al., Tetrahedron Letters *36*, 593-6 (1995).

Diethylaluminum chloride (s.a. under Nickel complexes) Et_2AlCl
Asym. Diels-Alder reaction with 3-(α,β-ethyleneacyl)-2-oxazolidones
s. *36*, 667s*50* (Volume *50,* p.301); *endo-* selectivity with rigid, tricyclic derivs., s. I.W. Davies et al., Tetrahedron Letters *36*, 7619-22 (1995).

Dimethylaluminum chloride Me_2AlCl
Asym. [2+2]-cycloaddition
with $ZnCl_2$ cf. *38*, 682; chiral 3-azabicyclo[3.2.0]heptan-2-ones s. A.I. Meyers et al., J. Org. Chem. *60*, 4359-62 (1995).

Aluminum chloride/tetraoctylammonium bromide $AlCl_3/R_4NBr$
Chromans from phenols and 1,3-dienes   ○
from K-salts cf. *34*, 650; with added $(n\text{-}C_8H_{17})_4NBr$ s. M. Matsui, H. Yamamoto, Bull. Chem. Soc. Japan *68*, 2663-8 (1995); with TsOH or BF_3-etherate s. ibid. 2657-61.

Montmorillonite- or silica-supported aluminum chloride
Friedel-Crafts alkylation with ethylene derivs.   ArH → ArR
with metal-doped montmorillonite cf. *43*, 703; monoalkylation with $AlCl_3$-K10 or $AlCl_3$-SiO_2 s. J.H. Clark et al., Chem. Commun. *1995*, 2037-40.

Bis(η^5-pentamethylcyclopentadienyl)samarium $(Me_5Cp)_2Sm$
Synthesis of 1,3-diol esters   CH(OCOR)C—CH(OCOR')
from enolesters and two aldehyde molecules

308.

A mixture of vinyl acetate and 1 eq. cyclohexanecarboxaldehyde added to 10 mol% (η^5-$Me_5Cp)_2Sm(THF)_2$ in toluene at 0° (the colour of the soln. immediately changing from dark purple to light yellow), and stirred at that temp. for 0.5 h then at 25° for 2.5 h → 1-cyclohexanoyloxy-1-cyclohexyl-3-acetoxypropane. Y 77%. The method is applicable to aliphatic or aromatic aldehydes, but trimethylacetaldehyde gave a complex mixture. F.e. and with SmI_2 s. Y. Ishii et al., J. Org. Chem. *60*, 4974-5 (1995).

Lanthanum(III) trisodium tris[(R)-1,1'-bi-2-naphthoxide]
Asym. induction with multifunctional heterobimetallic catalysts ("chemzymes")

309.

Asym. Michael addition. In Michael addition of enolates to enones, the basic lanthanum(III) trisodium tris[(R)-1,1'-bi-2-naphthoxide] ((R)-LSB) serves the dual role of generating the required enolate and, through the central Lewis acidic metal, activating the enone, while the asymmetric environment ensures the necessary face selectivity to effect adduct formation in up to 92% e.e. **E:** Equivalent amounts of 2-cyclohexenone and dibenzyl malonate added successively at 0° to a stirred soln. of 10 mol% (R)-LSB (generated from a 3:3:1 mixture of (R)-BINOL, NaOBu-*t* and La(OPr-*i*)$_3$ in THF at 0°), stirring continued for 24 h, and treated with 1 *N* HCl → (R)-3-[bis(benzyloxycarbonyl)methyl]cyclohexanone. Y 97% (e.e. 88%). Enantioselectivity with the alkali metal-free catalyst was low. The parallel with enzyme activation was underscored. F.e.s. M. Shibasaki et al., J. Am. Chem. Soc. *117*, 6194-8 (1995); enhanced enantioselectivity with lithium aluminum bis[(R,R)-1,1'-bi-2-naphthoxide], also **asym. Michael addition-aldol condensation**, s. Angew. Chem. Intern. Ed. *35*, 104-6 (1996).

Tris(6,6,7,7,8,8,8-heptafluoro-2,2-dimethyl-3,5-octanedionato)europium $Eu(fod)_3$
Asym. hetero-Diels-Alder reaction under high pressure   ○
with aldehydes s. *38*, 688s*44*; chiral 2-alkoxy-3,4-dihydro-2*H*-pyrans from α,β-ethylenealdehydes and enolethers s. D.A.L. Vandenput, H.W. Scheeren, Tetrahedron *51*, 8383-8 (1995).

Scandium(III) triflate $Sc(OTf)_3$
α,β-Ethylene-N-sulfonylthioiminoesters
from N-sulfonylimines and 1-acetylene-1-thioethers

310.

N-Benzylidene-*p*-toluenesulfonamide in acetonitrile and 1.5 eqs. 1-methylthio-1-propyne in the same solvent added at room temp. to 20 mol% Sc(OTf)$_3$ in acetonitrile, stirred for 15 h, and treated with water → product. Y 80%. F.e., Lewis acids, and with *in situ*-generated N-sulfonylimines, also intramolecular conversion, s. H. Ishitani et al., J. Org. Chem. *61*, 1902-3 (1996).

Ytterbium(III) triflate s. under Bu$_3$SnH $Yb(OTf)_3$

Yttrium-zirconium-sulfur Y-Zr-S
Stereospecific Diels-Alder reaction under heterogeneous Lewis acid catalysis

311.

with 1,4-naphthoquinones. A mixture of 3 eqs. cyclopentadiene, 1,4-naphthoquinone and the Y-Zr-S catalyst [50 wt% with respect to dienophile; composition 82.6 mol% Zr, 15.6 mol% Y and 1.8 mol% S; prepared by treating a vigorously stirred, aq. mixture of Y(NO$_3$)$_3$ and zirconyl nitrate (16:84) with aq. ammonia, (pH 8.5), collecting the precipitate, washing with deionized water, drying at 110°, treating with 2 *N* H$_2$SO$_4$, drying again at 120° for 24 h, heating to 500° at a rate of 2°/min, then calcining for 3 h] in methylene chloride stirred at room temp. for 5 h → product. Y 93% (100% *endo*). This is the first yttrium-based *strong Lewis acid* catalyst. In contrast, Yb(OTf)$_3$ failed to accelerate the reaction (cf. *48*, 661, 659), and under thermal conditions lower *endo*-selectivity was observed (95:5). F.e. and hetero-Diels-Alder reaction, s. A.V. Ramaswamy et al., Angew. Chem. Intern. Ed. *34*, 2143-5 (1995).

1,4-Cyclohexadiene/triethylamine
Double ring closure of enediynes via Myers-type 1,4-diradicals

312.

2,3-Dihydrobenz[*e*]indenes. A pressure vial containing a soln. of startg. enediyne in anhydrous benzene degassed with dry N$_2$ for 30 min, Et$_3$N and 1,4-cyclohexadiene added via syringe, the vial sealed with a Teflon screw-cap, and heated *at 37°* for 14 h → product. Y 76%. In contrast, the tandem enediyne cyclization via *p*-diradicals proceeds at a much higher temp. (cf. *48*, 665). F.e.s. J.W. Grissom, D. Klingberg, Tetrahedron Letters *36*, 6607-10 (1995); **indans from 1,2,4,10-tetraen-6-ynes** s. Angew. Chem. Intern. Ed. *34*, 2037-9 (1995).

Amberlyst A27
Michael addition under heterogeneous catalysis without solvent C=C → CHC(R)
with Al$_2$O$_3$ cf. *38*, 668s*47*; addition of aliphatic nitro compds. with Amberlyst A27 s. R. Ballini et al., J. Org. Chem. *61*, 3209-11 (1996).

Phenylsilane/bis(tributyltin) oxide/azodiisobutyronitrile/ethanol
Cyclic alcohols from ethyleneoxo compds.
Tin hydride-catalyzed radical ring closure

313.

with subsequent lactonization. 15 Mol% $(Bu_3Sn)_2O$, 0.5 eq. phenylsilane, 2 eqs. ethanol, and 10 mol% AIBN in benzene added to a soln. of startg. enal in the same solvent contained in a Schlenk tube under N_2, the tube sealed, shaken, heated at 80° for 6.5 h, treated with 1 M Bu_4NF in THF at room temp. for 1 h, worked up with 2 M HCl, and the crude mixture of *trans*-hydroxy-ester and *cis*-lactone stirred in methylene chloride with a little TsOH at room temp. for 20 h to complete lactonization → product. Y 74% (1.1:1 mixture of *cis*- and *trans*-stereoisomers). Reaction, which is generally applicable to cyclopentanol and cyclohexanol ring closure, is achieved with a catalytic amount of *in situ*-formed Bu_3SnH, regenerated by the silane in a catalytic cycle. F.e.s. D.S. Hays, G.C. Fu, J. Org. Chem. *61*, 4-5 (1996).

Diethylzirconocene Cp_2ZrEt_2
Syntheses via stereospecific vinylzirconation of unactivated acetylene derivs.

314.

with vinyl ethers. 5-Decyne added to a soln. of 1.25 eqs. Cp_2ZrEt_2 (prepared at -78° from Cp_2ZrCl_2 and 2 eqs. EtMgBr in THF), warmed to 0°, stirred for 2 h, startg. vinyl ether added at room temp., stirred for 6 h at 50°, and quenched with I_2 → *cis*-4-iodo-3-butyl-1,3-octadiene. Y 85%. Unsym. alkynes gave a mixture of regioisomers, while subst. vinyl ethers were unreactive. F.e. and with H^+ or D^+ as electrophile s. T. Takahashi et al., J. Am. Chem. Soc. *117*, 5871-2 (1995); **1,3-dienes from two different acetylene molecules** via zirconacyclopentadienes with Cp_2ZrCl_2/ethylene s. J. Org. Chem. *60*, 4444-8 (1995).

Bis(cyclopentadienyl)bis(trimethylphosphine)titanium(II) $Cp_2Ti(PMe_3)_2$
Ring closures of ethyleneoxo compds. via 1,2-oxatitanacyclopentanes

315.

Bicyclic γ-lactones. 5-Hexenal added to 1 eq. $Cp_2Ti(PMe_3)_2$ in pentane, stirred for 2 h at room temp., filtered, the resulting reddish soln. covered with a balloon of CO (1 atm) for 12 h, diluted with ether, the mixture exposed to air (to effect demetalation), and stirred for 4 h → product. Y 60.3%. This exemplifies the novel **carbonylation of 1,2-oxatitanacyclopentanes**. There was no reaction with the corresponding unsatd. azomethines or with acetylenealdehydes. F.e. and with isolation of intermediates s. W.E. Crowe, A.T. Vu, J. Am. Chem. Soc. *118*, 1557-8 (1996); O-silylated **cyclopentanols**, and 2-alkylidenecyclopentanols from acetyleneoxo compds., via reductive ring opening of the 1,2-oxatitanacyclopentanes with $(EtO)_3SiH$ s. ibid. *117*, 6787-8 (1995); also 1-indenols and 5-membered heterocyclic alcohols s. N.M. Kablaoui, S.L. Buchwald, ibid. 6785-6; **ring closures of unsatd. azomethines via 1,2-azatitanacyclopent(a,e)nes** with less expensive *(η²-propene)titanium diisopropoxide* generated from 1:2 Ti(OPr-*i*)$_4$/*i*-PrMgCl cf. F. Sato et al., Chem. Commun. *1996*, 533-4; ring closures of 1,3-dien-8-ynes and trapping of the intermediate fused vinyltitanacyclopentenes with aldehydes s. Tetrahedron Letters *37*, 1253-6 (1996).

Titanium tetraisopropoxide/isopropylmagnesium bromide   *Ti(OPr-i)₄/i-PrMgBr*
2-β-Hydroxycyclopropanols from acoxy-3-ethylenes via 1,4-O→C-acyl migration

4 eqs. 1.3 M i-PrMgBr in ether added to a stirred soln. of 3-butenyl acetate and 2 eqs. Ti(OPr-i)₄ in ether at -45°, the mixture stirred at -45° to -40° for 1 h, warmed to 0° over 1.5 h, 3:1.3 THF/water added, and stirred at 20° for 30 min → 1-methyl-2-(2-hydroxyethyl)cyclopropanol. Y 93% (58% *trans*-isomer). The reaction takes place via intramolecular nucleophilic acyl substitution-intramolecular 1,2-addition, the proportion of *trans*-isomer increasing as the temp. is raised. F.e.s. A. Kasatkin, F. Sato, Tetrahedron Letters *36*, 6079-82 (1995).

Titanium tetraisopropoxide/chiral 1,1'-bi-2-naphthols
Asym. Diels-Alder and hetero-Diels-Alder reaction s. *36*, 667s*51*

Ditosylatotitanium(IV) (R,R)-2,2-dimethyl-α,α,α',α'-tetraphenyl-1,3-dioxolane-4,5-dimethoxide
Asym. 1,3-dipolar cycloaddition with nitrones

Chiral isoxazolidines. A soln. of 0.5 eq. ditosylatotitanium(IV) (R,R)-2,2-dimethyl-α,α,α'α'-tetraphenyl-1,3-dioxolane-4,5-dimethoxide (freshly prepared by stirring the free TADDOL with ca. 1 eq. Ti(OPr-i)₂Cl₂ and 3 eqs. AgOTs in toluene at room temp. for 0.5 h under N₂) added via syringe to a stirred soln. of startg. alkene and 1.2 eqs. startg. nitrone in the same solvent containing 4 Å molecular sieves at 0° (preliminarily stirred for 15 min), allowed to warm to room temp. over 24 h, 5% methanol in methylene chloride added (total reaction time 48 h), and stirred for 10 min → (3'S,4'R,5'S)-3-[(5'-methyl-2',3'-diphenylisoxazolidin-4'-yl)carbonyl]-1,3-oxazolidin-2-one. Y 61% (*endo:exo* >95:<5; e.e. of *endo*-isomer 93%). Enantioselectivity was lower with N-alkyl- and N-benzyl-nitrones. Reaction involves *endo*-α-*Re*-attack of the nitrone on the alkene, coordinated to the chiral complex in such a way that the critical *tosylato* ligands are placed *trans* at the central atom. F.e.s. K.V. Gothelf et al., J. Am. Chem. Soc. *118*, 59-64 (1996); chiral 5,5-dialkoxyisoxazolidines with chiral 1,3,2-oxazaborolidines s. J.-P.G. Seerden et al., Tetrahedron Letters *35*, 4419-22 (1994).

Bis(cyclopentadienyl)titanium(III) chloride   *Cp₂TiCl*
Titanium(III)-mediated radical ring closures of acetyleneepoxides

Triquinanes. 2.2 eqs. [Cp₂TiCl]₂ in THF added dropwise over ca. 5 min to a stirred soln. of racemic (1α,2α,2aα,5aα)-2-(3-butynyl)-1a,2,4,4-tetramethyloctahydropentaleno[1,2-*b*]oxirene in the same solvent, stirring continued overnight, and quenched with 10% H₂SO₄ → racemic 3aα,3bβ,6aβ,7α,7aα-decahydro-3a,5,5,7a-tetramethyl-1-methylene-(1*H*)-cyclopenta[*a*]pentalen-7-ol. Y 82%. This is a useful alternative to stannyl-mediated ring closures of enynes. F.e. incl. **propellanes** s. D.L.J. Clive et al., J. Org. Chem. *61*, 2095-108 (1996).

Titanocene dichloride/isopropylmagnesium chloride $Cp_2TiCl_2/i\text{-}PrMgCl$
Regiospecific syntheses with 1,3-dienes via titanium(III) η³-allyl complexes ←

N,N-Disubst. β,γ-ethylenecarboxylic acid amides. A soln. of isoprene in THF treated with 1 eq. Cp_2TiCl_2 and 2 eqs. *i*-PrMgCl at room temp. under argon, cooled to -20°, 1 eq. dimethylcarbamoyl chloride added slowly via syringe, stirred for 1 h at -20°, and quenched with satd. $NaHCO_3$ soln. → product. Y 75%. Reaction is synthetically equivalent to an anti-Markownikov addition of dimethylformamide, and generally applicable to both aliphatic and cyclic (incl. congested) 1,3-dienes. F.e.s. J. Szymoniak et al., Tetrahedron Letters *37*, 33-6 (1996); **β,γ-ethylenecarboxylic acid esters** with methyl chloroformate, **and 3-ethylenealcohols from aldehydes** s. Synlett *1996*, 46-8; **β-hydroxyketones from 2-siloxy-1,3-dienes**, being equivalent to a regio- and stereo-specific reductive aldol condensation, s. Synthesis *1995*, 815-9.

Dichlorobis(1-neomenthylindenyl)zirconium ←
Regiospecific asym. carboalumination of ethylene derivs. $C=C \rightarrow C(R)C(OH)$

Asym. synthesis of prim. alcohols from terminal ethylene derivs. A soln. of 1-octene and 1 eq. Me_3Al in 1,2-dichloroethane treated with 8 mol% dichlorobis(1-neomenthylindenyl)zirconium at 22° for 12 h under 1 atm argon, O_2 bubbled through the mixture for 30 min at 0°, stirred for 6 h under O_2, and treated with 15% aq. NaOH → (2R)-2-methyl-1-octanol. Y 88% (72% e.e.). Silyl-, hydroxyl- and amino-groups were tolerated. Reaction is thought to involve Al-promoted four-centred *syn*-addition of a methylzirconium species to the *Re*-face of the alkene. F.e.s. D.Y. Kondakov, E. Negishi, J. Am. Chem. Soc. *117*, 10771-2 (1995); with higher trialkylalanes s. ibid. *118*, 1577-8 (1996).

Chiral metallocene bis(triflates) ←
Dihalogenotitanium diisopropoxide/chiral α,α,α',α'-tetraaryl-1,3-dioxolane- ←
4,5-dimethanols
Asym. Diels-Alder reaction ○
with Ti(OPr-*i*)$_4$/chiral 1,1'-bi-2-naphthols cf. *36*, 667s49 (Volume *49*, p.263); **estrane ring** from 4-vinyl-1,2-dihydronaphthalenes with $Cl_2Ti(OPr-i)_2$/chiral α,α,α',α'-tetraaryl-1,3-dioxolane-4,5-dimethanols s. G. Quinkert et al., Helv. Chim. Acta *78*, 1345-91 (1995); study of asym. catalysis with Ti(IV)-TADDOLates s. C. Haase et al., J. Org. Chem. *60*, 1777-87 (1995); comparison of BINOL and TADDOL ligands s. D. Seebach et al., ibid. 1788-99; asym. cycloaddition with 3-methylthiofuran s. I. Yamamoto, K. Narasaka, Chem. Lett. *1995*, 1129-30; asym. cycloaddition of electron-rich alkenes to 2-pyrone-3-carboxylic acid esters (with Ti(OPr-*i*)$_4$/(R)-1,1'-bi-2-naphthol) s. G.H. Posner et al., J. Org. Chem. *61*, 671-6 (1996); detrimental effect of molecular sieves s. K. Mikami et al., J. Am. Chem. Soc. *116*, 2812-20 (1994); with chiral 6,6'-dibromo-1,1'-bi-2-naphthol as ligand, **also asym. hetero-Diels-Alder reaction**, s. Synlett *1995*, 967-8; chiral 2-alkoxy-3,4-dihydro-2*H*-pyrans with $Br_2Ti(OPr-i)_2$ and chiral 1,3-dioxolane-4,5-dimethanols cf. E. Wada et al., Chem. Lett. *1994*, 1637-40; asym. Diels-Alder reaction with bidentate dienophiles using chiral metallocene bis(triflates) s. S. Collins et al., Organometallics *14*, 1079-81 (1995); with water-tolerant diaquo[(S)-biphenacene]titanium(IV) bis(triflate) s. W. Odenkirk, B. Bosnich, Chem. Commun. *1995*, 1181-2.

Titanium tetrachloride $TiCl_4$
3-Silylcyclopentenyl ketones from α,β-ethyleneketones
cyclopent-3-enyl derivs. from 2-acetylenesilanes via 1,2-silyl group migration s. *48*, 674;
cyclopent-2-enyl derivs. **from allenyldisilanes** via 1,2-disilanyl group migration s. Y. Ito et al., Synlett *1995*, 941-2.

Tri-n-butyltin hydride/irradiation or azodiisobutyronitrile $Bu_3SnH/hν$ or AIBN
Radical 1,4-addition with asym. induction   C=C → CHC(R)
s. *39*, 646s*50*; asym. addition to chiral 3-acyl-4-methylene-5-oxazolidones s. J.R. Axon, A.L.J. Beckwith, Chem. Commun. *1995*, 549-50; to chiral β-amino-β-aryl-α-methylenecarboxylic acid esters s. E.P. Kündig et al., Tetrahedron Letters *36*, 4047-50 (1995).

Tri-n-butyltin hydride/triethylborane/oxygen/ytterbium(III) triflate   ←
Asym. radical 1,4-addition to 3-(α,β-ethyleneacyl)-2-oxazolidones

321.

Record levels of acyclic diastereoselection are reported for radical 1,4-addition in the presence of a *mild* Lewis acid and by using a simple, readily available, and easily removable chiral auxiliary. **E:** 2 eqs. Yb(OTf)$_3$ in THF added to a soln. of startg. chiral 2-oxazolidone in methylene chloride under N$_2$ at -78°, followed by 10 eqs. 2-iodopropane, 5 eqs. Bu$_3$SnH, and 10 eqs. 1 *M* Et$_3$B in hexane, and O$_2$ introduced (in 1 ml amounts) at the commencement and after 1 and 2 h → product. Y 93% (25:1 mixture of diastereomers). The Lewis acid serves the dual role of chelating agent (to direct face selectivity) and activator (the yield being 60% and diastereomeric ratio 1.3:1 without the Lewis acid). Asym. induction was also efficient with *30 mol%* Yb(OTf)$_3$, which was active in the presence of limited amounts of water (a distinct advantage over ionic 1,4-addition or reaction with conventional Lewis acids). F.e., Lewis acids, and solvent and substituent effects s. M.P. Sibi et al., J. Am. Chem. Soc. *117*, 10779-80 (1995).

Tri-n-butyltin hydride/azodiisobutyronitrile $Bu_3SnH/AIBN$
Reductive ring closures of α,β-ethyleneketones via allylic O-stannylketyls   ←

322.

Reductive aldol-type condensation. A mixture of 2-cyclohexenone, 1.1 eqs. Bu$_3$SnH, and 0.2 eq. AIBN in benzene degassed for 20 min with argon, refluxed for ca. 4 h, *cooled to 10°*, 2 eqs. benzaldehyde added, stirred for 8-12 h, and worked up (with DBU/I$_2$) → product. Y 73% (*erythro:threo* 6:1). This is a mild and regioselective alternative to classical procedures via metal enolates requiring strong bases. The stereoselectivity is highly temperature-dependent. F.e. and dehydration (TsOH) to **(E)-α-alkylideneketones** s. E.J. Enholm, P.E. Whitley, Tetrahedron Letters *36*, 9157-60 (1995); stereospecific *intramolecular* cycloalkanol annelation s. J. Org. Chem. *60*, 1112-3 (1995); **ring closures via intramolecular radical Michael addition** (cf. *50*, 412) s. ibid. 4850-5.

Bis(tri-n-butyltin) oxide s. under PhSiH$_3$ $(Bu_3Sn)_2O$

Hexabutyldistannane/irradiation $(Bu_3Sn)_2$/⇃⇂
Unimolecular chain transfer reactions of organosilicon hydrides ←

323.

Bimolecular radical reactions involving chain propagation, such as the Giese reaction (*39*, 646s*50*) can now be conducted by unimolecular chain transfer with the aid of a silicon hydride-functionalized substrate. The advantage is that the initially generated radical cannot compete with the intermediate adduct radical for H-atom (cf. bimolecular reaction) so that the reaction is cleaner and does not require a high-dilution technique. E: A 0.5 M soln. of startg. enoate and 1 eq. ethyl 2-bromo-2-methylpropionate in *tert*-butanol containing 10 mol% $(Bu_3Sn)_2$ and 2% water irradiated under normal conditions, then heated with 2 eqs. Et_3N at 80° for 12 h → product. Y 69%. The tin-hydride route fails completely with α-bromoesters. F.e.s. D.P. Curran, J. Xu, J. Chem. Soc. Perkin Trans. I *1995*, 3061-3; 3049-59.

Triphenylphosphine Ph_3P
Synthesis of γ-subst. (E)-α,β-ethylenecarboxylic acid esters via inverse addition ←
to α,β-acetylenecarboxylic acid esters cf. *50*, 411; to α-allenecarboxylic acid esters s. C. Zhang, X. Lu, Synlett *1995*, 645-6.

Tantalum pentachloride/zinc $TaCl_5/Zn$
Benzene ring from diynes and acetylene derivs. ○
with $RhCl(Ph_3P)_3$ cf. *26*, 725; **C-aryl glycosides** s. F.E. McDonald et al., J. Am. Chem. Soc. *117*, 6605-6 (1995); with $TaCl_5$/Zn via tantalum-alkyne complexes s. K. Takai et al., Chem Lett. *1995*, 851-2.

Lithium trifluoromethanesulfonimide $LiN(SO_2CF_3)_2$
Stereospecific Diels-Alder reaction s. *46*, 659s*51*

Trimethylsilyl triflate $Me_3SiOSO_2CF_3$
β-Siloxylaminocarboxylic acid esters from nitrones RN(OSi≤)C—C—COOR
and O-silyl O-alkyl keteneacetals s. *38*, 648; vinylogs from vinylketeneacetal derivs. with Me_3SiOTf s. C. Trombini et al., Tetrahedron Letters *36*, 7293-6 (1995).

Lithium perchlorate $LiClO_4$
5-Hydroxy-2,3-dihydrobenzofurans from *p*-quinones s. *44*, 624s*51* ○

Lithium perchlorate/camphorsulfonic acid $LiClO_4/RSO_3H$
Diels-Alder reaction
with $LiClO_4$ cf. *46*, 659s*47,50*; addition to α,β-ethylene-ketals and -orthoesters with added camphorsulfonic acid s. P.A. Grieco et al., Synlett *1995*, 1155-7; more safely with $LiNTf_2$ cf. ibid. 565-7; short review of reactions with $LiClO_4$ in organic solvents. A. Flohr, H. Waldmann, J. Prakt. Chem. *337*, 609-11 (1995).

Manganese(II) iodide s. under Mg MnI_2

Iron or Iron/cuprous bromide Fe or Fe/CuBr
Addition of polyhalides to ethylene derivs. ←
γ-bromo-α-chlorocarboxylic acid esters s. *18*, 776s*51*; γ-bromo-α-chlorocarboxylic acid esters with Fe/CuBr s. L. Forti et al., Tetrahedron Letters *36*, 2509-10 (1995); s.a. Synth. Commun. *26*, 1699-710 (1996); α-fluoro-γ-iodocarboxylic acid esters s. C. Zhi, Q.-Y. Chen, J. Chem. Soc. Perkin Trans. I *1996*, 1741-7.

Cobalt carbonyl $Co_2(CO)_8$
Cyclopent-1-enyl ketones from 1,6-enynes
Interrupted Pauson-Khand reaction

324.

The Pauson-Khand reaction of dicobalt hexacarbonyl-complexed 1,6-enynes can be interrupted *under an atmosphere of air* by oxidative functionalization of the metal-carbon bond, thereby diverting the intermediate to cyclopent-1-enyl ketones (or hetero-analogs) rather than the familiar carbonylative bicyclization (cf. *40*, 475s*41,42*). **E:** A 0.003 *M* soln. of startg. enyne complex in toluene added dropwise during 20 min to toluene (preliminarily exposed to air at room temp. for 10 min) at 90°, and stirred at this temp. for 1 h under an atmosphere of air → product. Y 67%. The 3-atom tether appears critical as 1,7-enynes underwent the conventional bicyclization. F.e. incl. a 3-acyl-1-tosyl-Δ^3-pyrroline s. M.E. Krafft et al., J. Am. Chem. Soc. *118*, 6080-1 (1996).

Cobalt carbonyl/irradiation $Co_2(CO)_8/hv$
Cobalt(II) acetoacetonate/sodium tetrahydridoborate $Co(acac)_2/NaBH_4$
Catalytic Pauson-Khand reaction
with 1 mol% 1,5-cyclooctadiene(indenyl)cobalt(I) cf. *50*, 415; with 0.05 mol% Co(acac)$_2$ and 0.1 mol% NaBH$_4$ s. Tetrahedron Letters *37*, 3145-8 (1996); with 5 mol% Co$_2$(CO)$_8$ under irradiation s. B.L. Pagenkopf, T. Livinghouse, J. Am. Chem. Soc. *118*, 2285-6 (1996).

Cobalt carbonyl/tert. amine oxide $Co_2(CO)_8/R_3NO$
Pauson-Khand reaction
s. *40*, 475s*46-9*; with *gaseous* alkenes s. W.J. Kerr et al., Tetrahedron *52*, 7391-420 (1996); with electron-deficient alkenes s. M. Costa, A. Mor, Tetrahedron Letters *36*, 2867-70 (1995); cyclopropyl derivs. s. A. de Meijere et al., Synthesis *1993*, 998-1012.

Cobalt carbonyl/chiral phosphines/N-methylmorpholine N-oxide ←
Asym. Pauson-Khand reaction
with chiral alkoxyacetylenes cf. *49*, 674; under mild conditions with chiral pentacarbonylcobalt-phosphine complexes s. W.J. Kerr et al., Organometallics *14*, 4986-8 (1995); s.a. Synlett *1995*, 1085-6.

Cobalt(II) acetoacetonate/pyridine/xanthone/irradiation ←
γ-Lactones from 2-ethylenealcohols
Photochemical carbonylation

325.

A soln. of startg. allyl alcohol, 5 mol% Co(acac)$_2$, pyridine, and 10 mol% xanthone in THF irradiated in a Pyrex Hanovia-type reactor equipped with a 450 W medium-pressure Hg-lamp under slow purging with 9:1 CO/H$_2$ (1 atm) for 10-15 h at room temp. → product. Y 80%. F.e. and from internal allyl alcohols with co-formation of δ-lactones s. Y.L. Chow et al., Can. J. Chem. *73*, 740-2 (1995).

Cobalt(II) chloride s. under BH$_3$ $CoCl_2$
Bis(1,5-cyclooctadiene)nickel(0) s. under R$_2$Zn $Ni(cod)_2$

Hexakis(acetonitrile)nickel(II) bis(fluoroborate)/triphenylphosphine/diethylaluminum chloride
Allylarenes from styrenes
Regiospecific hydrovinylation with ethylene under mild conditions

326.   Ph—CH=CH₂ + CH₂=CH₂ ⟶ Ph—CH(CH₃)—CH=CH₂

A mixture of styrene, 0.25 mol% [Ni(MeCN)$_6$][BF$_4$]$_2$, 1 mol% PPh$_3$, and 1.25 mol% 1.8 M Et$_2$AlCl in toluene stirred in dichloromethane at room temp. under argon for 30 min under 10 atm of ethylene in a stainless steel autoclave → 3-phenyl-1-butene. Y 90% (regioselectivity 97%). F.e. and limitations s. A.L. Monteiro et al., Tetrahedron Letters *37*, 1157-60 (1996).

Nickel bromide s. under ↓   NiBr$_2$
Dodecacarbonyltriruthenium   Ru$_3$(CO)$_{12}$
o-Acylation of N-heteroarenes with ethylene derivs.   H → COR
2-acylpyridines s. *48*, 681; 4-acylimidazoles s. S. Murai et al., J. Am. Chem. Soc. *118*, 493-4 (1996).

Ruthenium carbene complex   ←
(Z)-1,5-Dienes from cyclobutenes and terminal ethylene derivs.   C
Ring opening-cross metathesis

327.

1,2-Divinylcyclopentanes. A mixture of startg. cyclobutene and 2 eqs. 1-octene in methylene chloride added *over 2 h* to 1 mol% (Cy$_3$P)$_2$Cl$_2$Ru=CHCH=CPh$_2$ in the same solvent under argon, and worked up when GC indicated completion of reaction → product. Y 63% (*cis:trans* 2.3:1). The proportion of sym. 1,5-dienes and self-metathesis products was minimal, and only significant when all the cyclobutene had been consumed. With sym. olefins, polymerization of the cyclobutene predominated. F.e.s. M.L. Randall et al., J. Am. Chem. Soc. *117*, 9610-1 (1995); 1,3-divinylcyclopentanes and 2,5-divinyltetrahydrofurans from sym. olefins s. M.F. Schneider, S. Blechert, Angew. Chem. Intern. Ed. *35*, 411-2 (1996).

Carbonyldihydridotris(triphenylphosphine)ruthenium(II)   Ru(CO)H$_2$(PPh$_3$)$_3$
Stereospecific C-β-alkylation of electron-deficient ethylenes with ethylene derivs.   ←
via regiospecific oxidative addition of ruthenium(0) to the vinyl carbon-hydrogen bond

328.

A soln. of startg. enoate and triethoxyvinylsilane in toluene treated with 5 mol% (Ph$_3$P)$_3$RuH$_2$(CO) at reflux for 20 h in the presence of 1 eq. dioxane → product. Y 86%. The catalyst is activated by prior dehydrogenation with the alkene component to give a highly coordinatively unsatd. species - (Ph$_3$P)$_3$RuLn - which oxidatively inserts into the β-hydrogen of the vinyl group to initiate the catalytic cycle. A variety of functions were unaffected: isolated alkene groups, aryl rings (cf. *49*, 679), acetoxy groups, silyl ethers, ketals, mercaptals, amides, epoxides and halides. However, the corresponding enones reacted more slowly. F.e. incl. cyclic analogs s. B.M. Trost et al., J. Am. Chem. Soc. *117*, 5371-2 (1995); β-alkylation of cyclic enones s. S. Murai et al., Chem. Lett. *1995*, 679-80.

o-Alkylation with ethylene derivs. via cyclometalation   H → R
of ar. ketones cf. *49*, 679s*50*; of arylcarboxylic acid esters s. Chem. Lett. *1996*, 109-10; of ar. azomethines s. ibid. 111-2; of oxyarylketones s. P.W.R. Harris et al., J. Organometal. Chem. *506*,

339-41 (1996); *o*-vinylation with alkynes s. Chem. Lett. *1995*, 681-2; β-alkylation of 2-vinylpyridines with RhCl(PPh$_3$)$_3$ cf. Y.-G. Lim et al., Chem. Commun. *1996*, 585-6.

Dihydridotetrakis(triphenylphosphine)ruthenium(II)/1,1′-bis(diarylphosphino)ferrocene ←
1,3-Enynes from terminal acetylene derivs. C≡C—C≡C

and 1,3-dienes cf. *37*, 675; and allenes with added 1,1′-bis(diarylphosphino)ferrocene, 1,3-enyn-4′-ols, s. M. Yamaguchi et al., Synlett *1995*, 1181-2.

Chloro(1,5-cyclooctadiene)cyclopentadienylruthenium(II)   RuCl(cod)Cp
3-γ-Keto-5,6-dihydro-2-pyrones
from α,β-acetylene-δ-hydroxycarboxylic acid esters and 2-ethylenealcohols
via anti-Alder-ene reaction

329.

The Ru-catalyzed anti-Alder-ene reaction (*49*, 681) has been extended to α,β-acetylene-δ-hydroxycarboxylic acid esters and allyl alcohols. **E**: A soln. of startg. allyl alcohol and 5-hydroxy-2-alkynoate in 1:1 DMF/water (degassed) containing 5 mol% CpRu(cod)Cl stirred vigorously at 100° for 2 h under N$_2$ → intermediate mixture of hydroxyketoesters (Y 65%), in benzene heated at reflux under N$_2$ with 3Å molecular sieves for 24 h → product (Y 55% after chromatographic purification). The procedure is **equivalent to Michael addition of α,β-ethylenelactones** (at the β-site) to enones. F.e.s. B.M. Trost et al., J. Am. Chem. Soc. *117*, 1888-99 (1995).

Bis(cyclooctadiene)rhodium(I) hexafluorophosphate/1,2-bis(diphenylphosphino)ethane ←
2-Alkylidenecyclopentenones from 1,2,4-trienes
Rhodium-catalyzed carbonylative [4+1]-cycloaddition

330.

A mixture of startg. vinylallene, 2 mol% [Rh(cod)$_2$]PF$_6$ and 2.4 mol% dppe in toluene stirred at 80° for 20 h under 15 atm CO in an autoclave → 2-isopropylidene-4-phenylcyclopent-3-enone. Y 56% (and 26% 4-cyclopentenone isomer). The ratio of 3-cyclopentenone:4-cyclopentenone increased with increasing reaction time, suggesting that the latter forms by double-bond isomerization (and/or protodesilylation where appropriate). The pronounced reactivity of vinylallenes (compared to 1,3-dienes) is ascribable both to the substantial facility for ligand exchange and the significant contribution from a metallocyclo-3-pentene resonance form in the resultant η^4 complex. F.e. under 1 atm CO s. M. Murakami et al., Angew. Chem. Intern. Ed. *34*, 2691-4 (1995).

Rhodium phosphine complexes ←
Hydroformylation   C≡C → CHC—CHO
α-arylaldehydes from styrenes with [Rh(cod)OAc]$_2$ cf. *4*, 667s*49*; with [Rh(NBD)(C$_6$F$_5$)$_2$PCH$_2$CH$_2$P(C$_6$F$_5$)$_2$]BF$_4$ s. A.S.C. Chan et al., Chem. Commun. *1995*, 2031-2; with bis(phosphino)dibenzoheterocyclics as ligand, regioselectivity, s. P.W.N.M. van Leeuwen et al., Organometallics *14*, 3081-9 (1995); hydroformylation of linear and cyclic mono- and di-olefins with Rh(acac)(P(OPh)$_3$)$_2$/P(OPh)$_3$ or Rh(acac)(CO)(PPh$_3$)/PPh$_3$ s. A.M. Trzeciak, J.J. Ziókowski, J. Organometal. Chem. *479*, 213-6 (1994); of 1,3-dienes with Rh·mesitylene/dppe s. S. Bertozzi et al., ibid. *487*, 41-5 (1995); of vinylpyrroles with Rh$_4$(CO)$_{12}$ s. R. Settambolo et al., ibid. *506*, 337-8 (1996); of 1,1-diarylethenes s. J. Org. Chem. *60*, 6612-5 (1995); with [m^3-tris(tricarbonylcobaltio)methylidyne]silanetriol PEG ethers in a two-phase medium cf. U. Ritter et al., Angew. Chem. Intern. Ed. *35*, 524-6 (1996).

Chiral rhodium phosphine or phosphinite complexes
Asym. hydroformylation   C=C → CHC—CHO
s. 20, 55s49; with carbohydrate-based bis(phosphinite) ligands s. T.V. RajanBabu, T.A. Ayers, Tetrahedron Letters 35, 4295-8 (1994); with ligands based on pydiphos and its P-oxide s. M. Marchetti et al., J. Organometal. Chem. 488, C10-21 (1995); with [(R)-2-diphenylphosphino-1,1′-dinaphthalen-2′-yl][(S)-1,1′-dinaphthalene-2,2′-diyl]phosphiterhodium(I), chiral β,γ-ethylenealdehydes from 1,3-dienes, regioselectivity, s. K. Nozaki et al., Chem. Commun. 1996, 155-6; with chiral Pt(II)-bis(binaphthophosphole) complexes (cf. 20, 55s48) s. S. Gladiali et al., J. Organometal. Chem. 491, 91-6 (1995); asym. hydroformylation of styrene s. S. Naili et al., Organometallics 14, 401-6 (1995).

Acetoacetonato(dicarbonyl)rhodium(I)/tris(m-sulfophenyl)phosphine trisodium salt/per(2,6-di-O-methyl)-β-cyclodextrin
Acetoacetonato(dicarbonyl)rhodium(I)/bis(triaryl phosphites)
Regiospecific hydroformylation
of ethylene derivs. cf. 48, 686; of terminal ethylene derivs. in water with tris(m-sulfophenyl)phosphine trisodium salt and per(2,6-di-O-methyl)-β-cyclodextrin as inverse phase transfer catalyst s. E. Monflier et al., Tetrahedron Letters 36, 9481-4 (1995); (E)-α,β-ethylenealdehydes from acetylene derivs. s. Angew. Chem. Intern. Ed. 34, 1760-1 (1995); in aq. methanol for hydroformylation of higher olefins with simplified work-up s. S. Kanagasabapathy et al., J. Prakt. Chem. 337, 446-50 (1995).

Acetoacetonato(dicarbonyl)rhodium(I)/bis(cyclic phosphite)
Regiospecific ring closures of ethyleneamines via hydroformylation   O
s. 46, 669s48,49; 46, 670s47; 6-alkoxy-1-carbalkoxypipecolic acid esters with Rh(acac)(CO)$_2$/BIPHEPHS s. I. Ojima et al., J. Org. Chem. 60, 7078-9 (1995); 3,4-dihydro-2H-1,3-benzoxazines with Rh$_2$(OAc)$_4$ s. Chem. Commun. 1994, 2763.

Zwitterionic rhodium complexes
Regio- and stereo-specific silylformylation of acetylene derivs.   C(Si≤)=C—CHO
s. 46, 672; with [Rh(OCOC$_3$F$_7$)$_2$]$_2$ s. J.S. Panek et al., Tetrahedron Letters 36, 8723-6 (1995); with an (η^2-methylenecyclopropane)rhodium complex s. I.P. Beletskaya et al., Mendeleev Commun. 1995, 220-1; (Z)-α,β-ethylene-β-germylaldehydes by **germylformylation** with zwitterionic rhodium complexes s. F. Monteil, H. Alper, Chem. Commun. 1995, 1601-2.

Carbonylhydridotris(triphenylphosphine)rhodium(I)   RhH(CO)(PPh$_3$)$_3$
(E)-1,3-Enynes from allenes and terminal acetylenes s. 51, 334   CH=C(CH$_3$)C≡C
α,β-Ethylene-β-(organothio)aldehydes from acetylene derivs.   C(SR)=C—CHO
Regiospecific thiolformylation

331.   PhSH + n-C$_6$H$_{13}$—≡   →^{CO}   n-C$_6$H$_{13}$\C=C/CHO with PhS

A soln. of 1.5 eqs. 1-octyne and 1 eq. benzenethiol in acetonitrile containing 3 mol% RhH(CO)(PPh$_3$)$_3$ purged 3 times with CO in a stainless steel autoclave, charged with 30 atm CO, and stirred for 5 h at 120° → (Z)-3-(phenylthio)-2-nonenal. Y 69% (and 13% (E)-isomer). Prolonged heating resulting in (Z)→(E)-isomerization. Reaction with alkanethiols required a longer time and was accompanied by thioester formation. Palladium catalysts were ineffective. F.e. and with RhCl(PPh$_3$)$_3$ s. A. Ogawa et al., J. Am. Chem. Soc. 117, 7564-5 (1995).

Chlorotris(triphenylphosphine)rhodium(I)   RhCl(PPh$_3$)$_3$
β-Alkylation of 2-vinylpyridines with ethylene derivs. s. 49, 679s51   H → R
Chlorotris(triphenylphosphine)rhodium(I)/pinacolborane
Addition of polyhalides to ethylene derivs.
of CCl$_4$ with AIBN cf. 18, 776; with Rh(PPh$_3$)$_3$Cl/pinacolborane, regioselectivity (cf. 18, 776s46), s. S. Pereira, M. Srebnik, J. Am. Chem. Soc. 118, 909-10 (1996); γ-bromo-α-chlorocarboxylic acid esters with Fe (cf. 18, 776s41-3) s. L. Forti et al., Tetrahedron Letters 37, 2077-8 (1996).

Tetrakis(tert-*butyl isocyanide*)*rhodiumtetracarbonylcobalt*   (*t*-BuNC)$_4$RhCo(CO)$_4$
Regio- and stereo-specific intramolecular silylformylation of acetylene derivs. ○
Cyclic α,β-ethylene-β-silylaldehydes

332.

Unlike intermolecular silylformylation (46, 672) which affords 1-silyl-2-formyl-1-alkenes, the corresponding intramolecular process yields the regioisomeric 3-silyl-2-alken-1-als. **E:** A stirred soln. of startg. ω-(dimethylsiloxy)alkyne in dry toluene containing 0.5 mol% (*t*-BuNC)$_4$RhCo(CO)$_4$ heated at 70° in a stainless-steel autoclave under 10 atm CO [at room temp.] for 14 h → product. Y 42%. Reaction is applicable to the formation of 4-7 membered O,Si-heterocyclics. Other Rh and Rh-Co complexes were less effective. F.e.s. I. Ojima et al., J. Am. Chem. Soc. *117*, 6797-8 (1995).

Palladous acetate/cupric acetate/lithium chloride   ←
α-(1-Halogenoalkylidene)-β-(1-halogenalkyl)-γ-lactones
from α,β-acetylenecarboxylic acid allyl esters
with PdCl$_2$(PhCN)$_2$ and Cu(II) in excess cf. *48*, 692; details, also bromine analogs with Pd$_2$dba$_3$/CuBr$_2$/LiBr, s. J. Org. Chem. 60, 1160-9 (1995); with Pd(OAc)$_2$ and a *catalytic* amount of Cu(OAc)$_2$ under O$_2$ s. J. Organometal. Chem. *508*, 83-90 (1996).

Palladous acetate/tris(2,6-dimethoxyphenyl)phosphine/1,8-diazabicyclo[5.4.0]
   undec-7-ene   ←
4-Alk-1-ynyl-2(5*H*)-furanones or furan-3-ylacetic acid esters
from α,β-acetylene-γ-hydroxycarboxylic acid esters and terminal acetylene derivs.
via palladium-catalyzed Michael-type addition

333.

Phenylacetylene and 1 eq. startg. alkynoate added to *2 mol%* tris(2,6-dimethoxyphenyl)phosphine (TDMPP) and *5 mol%* Pd(OAc)$_2$ in benzene at room temp., followed by 0.75-1.5 eqs. DBU → ethyl 5-phenylfuran-3-ylacetate. Y 87%. Increasing the ligand:Pd ratio favours the formation of the corresponding 2(5*H*)-furanone, but a more effective approach to the latter was revealed by addition of a little Bu$_3$SnOAc to enforce the transesterification. **E:** 20 mol% Bu$_3$SnOAc added to *3 mol%* Pd(OAc)$_2$ and *3 mol%* TDMPP in THF, followed by 1 eq. each of startg. alkynoate and alkyne, and worked up after 16 h at room temp. → 4-(phenylethynyl)-2(5*H*)-furanone. Y 58%. F.e.s. B.M. Trost, M.C. McIntosh, J. Am. Chem. Soc. *117*, 7255-6 (1995).

Tris(dibenzylideneacetone)dipalladium/1,4-bis(diphenylphosphino)butane Pd$_2$dba$_3$/dppb
Regiospecific addition of CH-acidic compds. ←

334.

to allenyl ethers. 1.1 eqs. Methylmalononitrile and THF added to a mixture of startg. alkoxyallene, 5 mol% Pd$_2$(dba)$_3$, and 26 mol% dppb under argon, and kept at 75° for 20-4 h until reaction complete by TLC → product. Y 80%. γ-Adducts were obtained from bulky nucleophiles. F.e.s. Y. Yamamoto, M. Al-Masum, Synlett *1995*, 969-70; addition to arylallenes s. Tetrahedron Letters *36*, 2811-4 (1995); (E)-thioenolethers by **addition to allenyl thioethers** s. Chem. Commun. *1996*, 831-2; **synthesis of allenes from 1,3-enynes** (with dppf as ligand) s. ibid. 17-8; **(E)-1,3-enynes from allenes** and terminal acetylene derivs. with HRh(CO)(PPh$_3$)$_3$/PEt$_3$ s. M. Yamaguchi et al., Tetrahedron Letters *35*, 5689-92 (1994); s.a. *51*, 131.

Tetrakis(triphenylphosphine)palladium(0) Pd(PPh$_3$)$_4$
Styrenes from two 1,3-enyne molecules ○
Cyclodimerizaton

335.

2 Mol% Pd(PPh$_3$)$_4$ added under argon at room temp. to a soln. of startg. 1,3-enyne in dry toluene, and heated at 65° for 1 h → product. Y 77%. The mechanism of the reaction can only be speculative at present. F.e. and **paracyclophanes** by intramolecular process s. Y. Yamamoto et al., J. Am. Chem. Soc. *118*, 3970-1 (1996).

Rearrangement ∩

Hydrogen/Oxygen Type   CC ∩ HO

Without additional reagents w.a.r.
Thermal oxy-Cope rearrangement ←
s. *21*, 744; dolabellane skeleton s. G. Mehta et al., Tetrahedron Letters *35*, 2761-2 (1994); chiral 2,3-dioxy-2,3,6,7-tetrahydro-4(5*H*)-oxoninones s. A.V.R.L. Sudha, M. Nagarajan, Chem. Commun. *1996*, 1359-60.

*Molybdenyl acetoacetonate/di-n-butyl sulfoxide/4-*tert-*butylbenzoic acid* ←
Meyer-Schuster rearrangement   C(OH)C≡CH → C=CHCHO
with (PrO)$_3$VO/Ph$_3$SiOH/PhCOOH cf. *31*, 665 and *31*, 665s*49*; from unhindered substrates with MoO$_2$(acac)$_2$/*n*-Bu$_2$SO/*p*-*t*-BuC$_6$H$_4$COOH s. C.Y. Lorber, J.A. Osborn, Tetrahedron Letters *37*, 853-6 (1996).

Hydrogen/Nitrogen Type   CC ∩ HN

Without additional reagents w.a.r.
Tricyclic oxazolidines from hydroxylaminoenynes ○
via intramolecular 1,3-dipolar cycloaddition

336.

A soln. of startg. hydroxylamine in toluene refluxed for 9 h → product. Y 71-6%. The intermediate nitrones were not isolated. F.e. and stereospecificity s. E.C. Davison et al., Tetrahedron Letters 36, 9047-50 (1995).

Hydrogen/Carbon Type   CC ∩ HC

Without additional reagents   w.a.r.
Ene reaction s. *51*, 375   ←
Lithium diisopropylamide   i-Pr$_2$NLi
Pyrroles from 2-acetyleneamines via Δ^2-pyrrolines   ○

337.

3-Cyanopyrroles. A stirred soln. of 1.3 eqs. LDA in THF treated with 1 eq. startg. acetyleneamine at -78° under N$_2$, stirred at 25° for 24 h, and quenched with aq. NH$_4$Cl → intermediate dihydropyrrole (Y 50%), in benzene treated with 1.1 eqs. DDQ at 25°, stirred for 2 h, and quenched with satd. aq. NaHCO$_3$ → product (Y 95%). F.e. incl. 2-cyano- and 2-carbalkoxy-pyrroles s. J.-L. Wang et al., Tetrahedron Letters 36, 2823-6 (1995).

Lithium diisopropylamide/N,N,N',N'-tetramethylethylenediamine   i-Pr$_2$NLi/TMEDA
1,2-Carbalkoxy group migration s. *51*, 344   ←
Potassium bis(trimethylsilyl)amide   KN(SiMe$_3$)$_2$
1-Naphthols from o-acetyleneketones   ○

338.

A soln. of o-octynylacetophenone in toluene added dropwise over 10 min to a slurry of 1.2 eqs. KN(SiMe$_3$)$_2$ in the same solvent at -78° under argon, heated at 80°, and after 1 h the mixture acidified with 1 M H$_2$SO$_4$ → 3-hexyl-1-naphthol. Y 74%. KOBu-t and KH also promoted cycloaromatization but sodium and lithium bases failed, suggesting that the potassium counterion is essential. F.e. incl. 2,3-dialkyl-derivs. s. F. Makra et al., Tetrahedron Letters 36, 6815-8 (1995); with camphorsulfonic acid cf. M.A. Ciufolini, T.J. Weiss, ibid. 35, 1127-30 (1994).

1,8-Diazabicyclo[5.4.0]undec-7-ene   DBU
Migration of carbon-carbon double bonds   C=C-CH → CHC=C
β,γ- from α,β-ethylenesulfones cf. 41, 500; (Z)-3-alkylidene- from 3-vinyl-phthalides s. Y. Ogawa et al., Heterocycles 39, 47-50 (1994).

Cuprous chloride   CuCl
1,2-Dihydroquinolines from N-propargylanilines   ○

339.

A mixture of startg. N-propargylaniline and *a little* (10% by weight) CuCl in toluene refluxed under N$_2$ for 1 h → 2,2,6-trimethyl-1,2-dihydroquinoline. Y 75%. Electron-donating groups in the *para* position favour ring closure, but bulky α-substituents in the side chain have a detrimental effect (especially in the presence of an electron-withdrawing substituent). F.e. and chlorination s. N.M. Williamson et al., Tetrahedron Letters 36, 7721-4 (1995).

Lewis acids   ←
Intramolecular ar. alkylation with ethylene derivs.
s. 25, 527; ring closures of aryl-subst. dienones, hydrophenanthrenes, s. G. Majetich et al., Tetrahedron Letters 36, 4749-52 (1995); furan derivs. s. ibid. 35, 4887-90 (1994); 2-silatetralins

s. O. Hoshino et al., Heterocycles *38*, 883-96 (1994); 1,2,3,4-tetrahydroquinolines from ar. 3-ethyleneamines with H_2SO_4 s. V.V. Kuznetsov et al., Khim. Geterotsikl. Soedin. *1994*, 73-8 (Russ.).

Bis(η^5-pentamethylcyclopentadienyl)bis(trimethylsilyl)methylsamarium(III) Cp^*_2SmR
N-Condensed Δ^2-pyrroline ring from 4-azaenynes   ○
Regiospecific cycloisomerization via intramolecular aminometalation-carbometalation

340.

Tetrahydropyrrolizines. A soln. of startg. enyne and ca. 2 mol% (η^5-Me_5C_5)$_2$SmCH(SiMe$_3$)$_2$ in benzene allowed to react at 21° until reaction complete → product. Y 75% (catalyst turnover 777 h^{-1}). The organolanthanide centre mediates an unusual tandem sequence of insertive C-N and C-C bond formation in a single catalytic cycle. **N-Condensed pyrrolidines** were obtained similarly from the corresponding 4-azadienes, and the 4-alkylidene-Δ^2-pyrroline ring from 4-azadiynes. F.e. incl. **hexahydroindolizines**, and with $Me_3Si(\eta^5$-$Me_5C_5)NdCH(SiMe_3)_2$ s. Y. Li, T.J. Marks, J. Am. Chem. Soc. *118*, 707-8 (1996).

Propionic acid   *EtCOOH*
Isomerization of γ,δ-ethyleneketones by intramolecular ene-retroene reaction   ←

341.

γ-Aryl-γ,δ-ethyleneketones. A soln. of 4-methyl-4-phenylhex-5-en-2-one in toluene containing a little propionic acid heated in a sealed tube at 250-70° for 48 h, cooled, and poured into water 5 4-methyl-5-phenylhex-5-en-2-one. Y 70%. F.e.s. A. Srikrishna et al., J. Chem. Soc. Perkin Trans. I *1995*, 2033-7.

3-Benzyl-5-(2-hydroxyethyl)-4-methylthiazolium chloride/triethylamine   ←
Intramolecular hydroacylation of electron-deficient ethylene derivs.   ○

342.

4-Chromanone-3-acetic acid esters. A mixture of 4-methyl-(2-formylphenoxy)but-2-enoate, 0.06 eq. 3-benzyl-5-(2-hydroxyethyl)-4-methylthiazolium chloride, and 0.43 eq. Et$_3$N in DMF stirred at room temp. for 1 h, further amounts of catalyst and Et$_3$N added, and stirring continued for 30 min → methyl 3,4-dihydro-4-oxo-2*H*-1-benzopyran-3-acetate. Y 86%. This is a rare example of an **intramolecular Stetter reaction** (cf. *28*, 648s*32*). F.e. and 3(2*H*)-benzofuranone-2-acetic acid esters s. E. Ciganek, Synlett *1995*, 1311-4.

Sulfuric acid H_2SO_4
1,2,3,4-Tetrahydroquinolines from ar. 3-ethyleneamines s. *25*, 527s*51*

Chlorotris(triphenylphosphine)rhodium(I)/n-butyllithium $Rh(PPh_3)_3Cl/BuLi$
Vinyl from allyl glycosides   RO–CHC=C → RO–C≡C–CH
with an Ir(I)-phosphine complex cf. *47*, 18; with Rh(PPh$_3$)$_3$Cl/*n*-BuLi, selectivity, s. G.-J. Boons et al., Chem. Commun. *1996*, 141-2.

Bis(dibenzylideneacetone)palladium(0)/triphenylphosphine/acetic acid   ←
(E,E)-1,3-Dienes from acetylene derivs.   CHCHC≡C → C=C–CH=CH
(E,E)-2,4-dienones with Pd(OAc)$_2$/dppb cf. *43*, 650; (E,E)-1-polyfluoroalkyl-1,3-dienes with Pd(dba)$_2$/PPh$_3$/AcOH s. Z. Wang, X. Lu, Tetrahedron *51*, 11765-74 (1995).

Tris(dibenzylideneacetone)dipalladium/chiral ferrocenyldi(phosphines)/acetic acid   ←
Asym. cycloisomerization of enynes   ○
with chiral di(phosphines) s. *49*, 699; chiral N-protected 3-methylene-4-vinylpyrrolidines with (S,S)-(R,R)-*p*-CF$_3$C$_6$H$_4$-TRAP s. Y. Ito et al., Angew. Chem. Intern. Ed. *35*, 662-3 (1996).

Oxygen/Carbon Type   CC ∩ OC

Without additional reagents   w.a.r.
o-Claisen rearrangement   ←
s. 2, 621; *4,* 671; 1,1-bis(*o*-hydroxybenzyl)ethylenes by **double *o*-Claisen rearrangement** s. K. Hiratani et al., Tetrahedron Letters *36*, 5567-70 (1995).

Microwaves s. under AlCl₃   ←
n-*Butyllithium*   *BuLi*
Asym. Wittig rearrangement of allyl ethers   ←
s. *39*, 658s*44,46,48*; chiral γ,δ-ethylene-α-hydroxyphosphonamidates s. S.E. Denmark, P.C. Miller, Tetrahedron Letters *36*, 6631-4 (1995); chiral γ,δ-ethylene-α-hydroxyphosphonic acid esters based on (–)- or (+)-menthol s. M. Gulea-Purcarescu et al., ibid. 6635-8; racemic *anti*-derivs. with LDA s. T. Yokomatsu et al., Synlett *1995*, 1035-6; chiral γ,δ-ethylene-α-hydroxyhydrazones s. D. Enders, D. Backhaus, ibid. 631-3; chiral 3-methylene-4-siloxyalcohols with remote asym. induction s. K. Tomooka et al., Tetrahedron Letters *36*, 2789-92 (1995).

Isopropyllithium/(–)-sparteine   ←
Bicyclic alcohols from *meso*-epoxides   OC
Desymmetrization

343.

2.45 eqs. (–)-Sparteine (freshly distilled) added dropwise over 0.5 h to a stirred soln. of 2.4 eqs. *i*-PrLi in ether at -98°, after stirring for 1 h at this temp. cyclooctene epoxide in ether added dropwise over 0.5 h, stirring continued at -98° for 5 h, warmed slowly to room temp. overnight (15 h), cooled to 0°, and treated dropwise with 2 *N* aq. HCl → product. Y 86% (e.e. 84%). A *sec*-alkyllithium was essential to obtain a good level of enantioselectivity. F.e.s. D.M. Hodgson, G.P. Lee, Chem. Commun. *1996*, 1015-6.

Lithium diisopropylamide/N,N,N′,N′-tetramethylethylenediamine   *i-Pr₂NLi/TMEDA*
5-β-Hydroxy-2-cyclopentenones from α-allyl-γ-lactones   ←

344.

Startg. lactone treated with 3 eqs. LDA in 10:1 THF/TMEDA at -78°, and the mixture stirred at room temp. overnight → product. Y 71%. α-Carbalkoxy-derivs. underwent 1,2-carbalkoxy group migration, as also did allylmalonic acid esters to give **α-vinylsuccinic acid esters**. F.e. and application to the preparation of **spiro[4.5]dec-2-ene-1,6-diones** s. W. Jaivisuthunza et al., Tetrahedron Letters *37*, 3199-202 (1996).

Lithium bis(trimethylsilyl)amide/aluminum isopropoxide/quinine   ←
Chelation-controlled Claisen rearrangement   ←
with LDA/ZnCl₂ cf. *49*, 703; chiral γ,δ-ethylene-α-(trifluoroacetylamino)carboxylic acids by **asym. Claisen rearrangement** with LiN(SiMe₃)₂/Al(OPr-*i*)₃ and added quinine s. Angew. Chem. Intern. Ed. *34*, 2012-4 (1995).

*Dicyclohexyl(methyl)amine/*tert-*butyldimethylsilyl triflate*   *R₃N/t-BuMe₂SiOSO₂CF₃*
Stereospecific Claisen rearrangement of allyl esters   ←
à la Ireland with LiNR₂/Me₃SiCl cf. *28*, 683; enhanced 1,4-asym. induction and (E)-selectivity with (c-C₆H₁₁)₂NMe/*t*-BuMe₂SiOTf s. T. Nakai et al., Tetrahedron Letters *37*, 3005-8 (1996); rearrangement of α-functionalized esters s. K. Hattori, H. Yamamoto, Tetrahedron *50*, 3099-112 (1994).

4-Dimethylaminopyridine   *DMAP*
o-Acetonylphenols via Carroll rearrangement s. *51*, 459   ←
Isopropylmagnesium bromide s. under ClTi(OPr-i)$_3$ *i-PrMgBr*
Aluminum isopropoxide s. under LiN(SiMe$_2$)$_2$ *Al(OPr-i)$_3$*
Triisobutylaluminum (s.a. under Pyridinium tosylate)   *i-Bu$_3$Al*
3-Hydroxytetrahydrofurans from 1,3-dioxolan-4-ones via 4-alkylidene-1,3-dioxolanes   ←
Stereospecific 1,3-sigmatropic rearrangement

345.

A soln. of startg. 1,3-dioxolan-4-one in THF treated with dimethyltitanocene at 65° according to Synth.Meth. *41*, 885 → intermediate 4-alkylidene deriv. (Y 55%), in toluene treated with 2 eqs. *i*-Bu$_3$Al at 0° until reaction complete → product (Y 90%; >99% *syn*). The rearrangement proceeds in a stepwise fashion **via a reactive aluminum enolate** which undergoes intramolecular addition to the generated oxocarbenium moiety in a rare example of a 5-(enol-*endo*)-*endo*-trig cyclization. The *in situ*-formed oxolan-3-one is reduced to the product with the trialkylalane. However, with less-hindered trialkylalanes, the corresponding tert. alcohols were formed by a terminating carbophilic addition. Interestingly, the stereochemistry at the acetal carbon is retained. F.e. and with dibenzylidenetitanocene s. N.A. Petasis, S.-P. Lu, J. Am. Chem. Soc. *117*, 6394-5 (1995); **4-hydroxytetrahydropyrans** from 1,3-dioxan-4-ones via 4-alkylidene-1,3-dioxanes s. Tetrahedron Letters *37*, 141-4 (1996).

Tris(pentafluorophenyl)boron or Boron fluoride   *(C$_6$F$_5$)$_3$B or BF$_3$*
Aldehydes from epoxides
with MgBr$_2$ cf. *23*, 735s*36*; chiral α,α-disubst. β,γ-ethylenealdehydes with inversion of configuration using BF$_3$-etherate and other Lewis acids s. M.E. Jung, D.C. D'Amico, J. Am. Chem. Soc. *117*, 7379-88 (1995); α,α-dialkylaldehydes with (C$_6$F$_5$)$_3$B by selective migration, also ketones (cf. *26*, 149), s. K. Ishihara et al., Synlett *1995*, 721-2.

Aluminum tris(4-bromo-2,6-diphenylphenoxide)   *Ar(OAr)$_3$*
Stereospecific Claisen rearrangement of allyl vinyl ethers   ←
with MeAl(OAr)$_2$ cf. *44*, 654s*45*; enhanced (E)-selectivity with aluminum tris(4-bromo-2,6-diphenylphenoxide) s. H. Yamamoto et al., Synlett *1996*, 720-2; effect of hindered Lewis acids on remote asym. induction s. R.K. Boeckman, Jr. et al., Tetrahedron Letters *36*, 803-6 (1995).

Di-n-butylborinyl triflate/tetra-n-butylammonium iodide   *n-Bu$_2$BOSO$_2$CF$_3$/Bu$_4$NI*
Siloxycyclobutane ring by intramolecular Michael addition-aldol condensation   □
with Et$_3$N cf. *47*, 694; 3,4-fused 2-siloxycyclobutane-1-carboxylic acid esters with *n*-Bu$_2$OSO$_2$CF$_3$/Bu$_4$NI s. Tetrahedron Letters *36*, 8071-4 (1995).

Aluminum chloride/microwaves
Fries rearrangement s. *51*, 407   ←
Aluminum chloride/sodium chloride   *AlCl$_3$/NaCl*
7-Hydroxy-1-indanones from acrylic acid aryl esters via Fries rearrangement   ○

346.

A mixture of startg. ester, AlCl$_3$ and NaCl heated at 100° for 15 min then at 155° for 45 min → 4-chloro-5-methyl-7-hydroxyindan-1-one. Y 79-85%. The chlorine (or bromine) substitution served to block the *para*-position and was readily removable. F.e.s. B. Muckensturm, F. Diyani, J. Chem. Res. (M) *1995*, 2544-56.

Scandium(III) or hafnium(IV) triflate or Zirconium tetrachloride ←
Fries rearrangement s. *51* 407 ←

Pig liver esterase ←
6-Formylbicyclo[3.1.0]hex-2-ene-1,2-dicarboxylic acid 1-monoesters ←
from 5,6-*exo*-epoxybicyclo[2.2.1]hept-2-ene-2,3-dicarboxylic acid esters
Asym. hydrolysis-Meinwald-type rearrangement

347.

Pig liver esterase added to a soln. of dimethyl 4,5-*exo*-epoxybicyclo[2.2.1]hept-2-ene-2,3-dicarboxylate in pH 8 phosphate buffer containing acetone, incubated at 30° for 6 h, and acidified with 2 *N* HCl → 6-formyl-1-(methoxycarbonyl)bicyclo[3.1.0]hex-2-ene-2-carboxylic acid. Y ca. 100% (e.e. 47%). Although the enantioselectivity is not high, the method is none-the-less interesting in that 3 different functional groups can be introduced regiospecifically and simultaneously under *basic* conditions, with the potential of generating chiral quaternary and tertiary carbon centres. F.e.s. S. Niwayama et al., J. Am. Chem. Soc. *116*, 3290-5 (1994).

Chlorotitanium triisopropoxide/isopropylmagnesium bromide   *ClTi(OPr-i)₃/i-PrMgBr*
(E)-α-Alkylidene-ω-hydroxyketones from acoxyacetylenes ←
via O→C-acyl migration

348.

4.6 eqs. *i*-PrMgBr in ether added at -78° to a soln. of startg. ester and 2.3 eqs. ClTi(OPr-*i*)₃ in the same solvent, warmed to -50° over 30 min, stirred for 1 h at -50° to -40°, treated with 3 *N* HCl at -40°, and warmed to room temp. over 30 min → (E)-4-(trimethylsilyl)-3-(phenylcarbonyl)but-3-en-1-ol. Y 76%. F.e.s. F. Sato et al., J. Am. Chem. Soc. *118*, 2208-16 (1996).

Pyridinium tosylate/triisobutylaluminum   *C₅H₅NHOTs/i-Bu₃Al*
4-Ethylene- from 2-ethylene-alcohols and enolethers ←
via regiospecific reductive ketal-Claisen rearrangement

349.

1.5 eqs. 2-Methoxypropene and 1 mol% pyridinium tosylate added to a soln. of geraniol in methylene chloride at 0°, stirred at room temp. until ketal formation complete by TLC, 3 eqs. 1 *M* *i*-Bu₃Al added dropwise, the mixture stirred at 0° for 2 h, and poured into ice-cold 2 *N* HCl → product. Y 88%. The method proceeds under very mild conditions. The regioselectivity complements that of the standard acid-catalyzed thermal rearrangement, favouring the linear product with unsym. ketals. Generally a mixture of stereoisomers was obtained although acyclic sec. allyl alcohols favoured (E)-products. F.e.s. S.D. Rychnovsky, J.L. Lee, J. Org. Chem. *60*, 4318-9 (1995).

Nitrogen/Carbon Type   CC ∩ NC

Without additonal reagents   w.a.r.
Intramolecular 1,3-dipolar cycloaddition with nitrones   ○
s. *20*, 284s*46*,*47*,*49*; tricyclic isoxazolidines s. H.G. Aurich et al., Tetrahedron Letters *37*, 841-4 (1996); M. Kobayakawa, Y. Langlois, ibid. *33*, 2353-6 (1992); piperidine systems s. U. Chiacchio et al., Tetrahedron *52*, 14311-22 (1996); isoxazolidino[4,3-c]chroman-4-ones and -quinolin-4-ones s. G. Zecchi et al., J. Chem. Res. (S) *1995*, 282-3; **asym. induction** with carbohydrate derivs., chiral tetrahydrofurans, s. A. Bhattacharjee et al., Tetrahedron Letters *36*, 7729-30 (1995); chiral pyrans s. ibid. 4677-80; s.a. ibid. *37*, 7635-6 (1996); precursors of chiral carbocyclic nucleosides s. S.B. Mandal et al., Tetrahedron *52*, 11265-72 (1996).

Intramolecular 1,3-dipolar cycloaddition with oximes
s. *44*, 663s*49*; fused 2-pyrrolidones s. U. Chiacchio et al., Tetrahedron *52*, 7875-84 (1996); steric acceleration s. P.G. Sammes et al., J. Chem. Soc. Perkin Trans. I *1995*, 2551-5; fused coumarins s. M. Noguchi et al., ibid. 1857-62.

Lithium diisopropylamide   i-Pr_2NLi
4-Imino-Δ²-pyrroline ring by Ziegler cyclization
with NaOR cf. *35*, 496; N¹-protected 3-iminoindoline-2-carboxylic acid derivs. with asym. induction (using LDA) s. D.L. Boger et al., J. Am. Chem. Soc. *118*, 2301-2 (1996).

Stereospecific intramolecular 1,3-dipolar cycloaddition of nitrones to enolates

350.

A soln. of startg. nitrone in THF added dropwise to 1.33 eqs. LDA in the same solvent at -78°, stirred for 1 h, warmed to room temp. during 4 h, and quenched with satd. aq. NH_4Cl → (1R,4S,5R,8S)-7-benzyl-2-*tert*-butyl-4,8-dimethyl-4-hydroxy-3-oxa-2,7-diazabicyclo[3.3.0]-octane. Y 65%. This is the first instance of using an enolate group as dipolarophile in such a cycloaddition. F.e.s. H.G. Aurich, H. Köster, Tetrahedron *51*, 6285-92 (1995).

Cadmium chloride/sodium hydride $CdCl_2$/NaH
Trimethylaluminum Me_3Al
Cycloisomerization of nitriles
4-aminopyridine ring with NaOEt cf. *36*, 711; with NaH/$CdCl_2$, Bu_2Cd, BuCdOTf, $ZnHal_2$ or CuOAc s. J.B. Campbell, J.W. Firor, J. Org. Chem. *60*, 5243-9 (1995); 4-aminopyrazolo[3,4-*b*]-pyridines with Me_3Al s. Synth. Commun. *26*, 981-90 (1996).

Zwitterionic Claisen rearrangement under mild conditions   ←
with ring expansion s. *47*, 611; γ,δ-ethylenecarboxylic acid amides with asym. induction (using Me_3Al) s. U. Nubbemeyer, J. Org. Chem. *60*, 3773-80 (1995).

Boron fluoride (s.a. under s-BuLi) BF_3
3,4-Dihydroisocarbostyril ring from 2-arylisocyanates   ○
with $POCl_3$/$SnCl_4$ cf. *33*, 694; with BF_3 s. L. Töke et al., Synthesis *1995*, 1373-5.

Titanium tetrapropoxide $Ti(OPr)_4$
Pictet-Spengler ring closure
Asym. induction with α-phthalimidocarboxylic acid chlorides

351.

Chiral 2-acyl-1,2,3,4-tetrahydro-9*H*-pyrid[3,4-*b*]indoles. Startg. tryptamine deriv. and N,N-phthaloylamino acid chloride allowed to react in methylene chloride at room temp. in the presence of Ti(OPr-*n*)$_4$ for several (up to 8) days → product. Y 60% (diastereoselectivity >99:1). Ti(OPr-*i*)$_4$

may also be used. With imines derived from aliphatic aldehydes, reaction time was reduced to 5-60 min. The level of stereoselectivity was consistently lower with Fmoc- or Z-protected amino acid chlorides; furthermore, N,N-phthaloyl derivs. are not prone to racemization, and the formed Pictet-Spengler adducts are highly crystalline compds., from which the major isomer may be obtained pure by a single recrystallization. F.e. and reductive N-deacylation s. H. Waldmann et al., Angew. Chem. Intern. Ed. *34*, 2402-3 (1995).

Di-n-butyltin oxide Bu_2SnO
**N-Unsubst. isoxazolidine ring from ethylenealdehydes
via stereospecific 1,3-dipolar cycloaddition
with *in situ*-generated N-(tetrahydropyran-2-yl)nitrones**

352.

An equimolar mixture of startg. ethylenealdehyde, 5-hydroxypentanal oxime, and 5 mol% Bu_2SnO in toluene heated for several h under reflux, then the adduct hydrolyzed with 2 N HCl in ethanol at room temp. → product. Y 90%. The method is general and highly regio- and stereo-selective. F.e., also *isolated* isoxazolidines by intermolecular cycloaddition, and large-scale conversion, s. A. Abiko, Chem. Lett. *1995*, 357-8; intramolecular cycloaddition of *in situ*-generated N-benzylnitrones cf. H.G. Aurich, J.-L.R. Quintero, Tetrahedron *50*, 3929-42 (1994).

Sulfur/Carbon Type CC ∩ SC

Without additional reagents w.a.r.
**Transannular hetero-Diels-Alder reaction
with *in situ*-generated α,β-ethylenethioketones**

353.

A mixture of startg. macrocyclic enone and 0.7 eq. Lawesson's reagent in dry benzene refluxed for 2 h under argon → product. Y 95%. This is the first example of a transannular hetero-Diels-Alder reaction. F.e. and desulfurization to benzoxonin derivs. s. T. Saito et al., Synlett *1996*, 72-4.

Remaining Elements/Carbon Type CC ∩ RemC

Irradiation s. under $(Bu_3Sn)_2$ ⚡
Boron fluoride BF_3
**1,1-(Homoallyloxy)silanols from Si-allyl-1,1-hydroxysilanes and aldehydes
via stereospecific intramolecular allylation**

354.

Benzaldehyde and 1.4 eqs. (α-hydroxyhexyl)dimethylallylsilane in methylene chloride treated dropwise via syringe with 5 eqs. BF_3-etherate at room temp., stirred for 10 min at this temp., quenched with water, worked up, the crude intermediate fluorosilane dissolved in a mixture of

10% methanolic KOH, water, and THF, and stirred overnight at room temp. → *syn*-4-phenyl-4-[[1-(hydroxydimethylsilyl)hexyl]oxy]-1-butene. Y 93% (*syn* selectivity >120:1). Diastereoselectivity is extraordinarily high, except for electron-rich aromatic aldehydes. The reaction proceeds via synclinal addition of the allyl group to an intermediate oxocarbenium ion, the corresponding intermolecular reaction taking place by attack of the nucleophile from the opposite face. F.e.s. R.J. Linderman, K. Chen, J. Org. Chem. *61*, 2441-53 (1996).

Hexabutyldistannane/irradiation $(Bu_3Sn)_2$/ℋℋ
Regiospecific radical ring closure of unsatd. tellurides with aryltelluro group migration

355.

3-α-Aryltellurocyclopentanols. A soln. of startg. telluride (0.05 mol dm^{-3}) in refluxing benzene containing 40 mol% $(Bu_3Sn)_2$ irradiated with a sunlamp for 2-6 h under N_2 → product. Y 68% (*cis:trans* 1.3:1). F.e. (63%, 73%), also regiospecific formation of 3-α-aryltellurotetrahydrofurans from 2-allyloxytellurides (5-α-oxy-derivs. being formed with high stereospecificity) and 3-(α-aryltelluroalkylidene)tetrahydrofurans from propargyloxy-derivs., s. L. Engman, V. Gupta, Chem. Commun. *1995*, 2515-6.

Carbon/Carbon Type   CC ∩ CC

Without additional reagents   *w.a.r.*
Enoxysilanes by oxy-Cope-type rearrangement   ←
s. *35*, 501a; 7-siloxy-2,6-dienecarboxylic acid imides, stereoselectivity, s. C. Schneider, M. Rehfeuter, Synlett *1996*, 212-4; substituent effect s. Tetrahedron *53*, 133-44 (1997).

(E)-Alkylidenecyclobutenes from 1,2,4-trienes
Regio- and stereo-specific cycloisomerization

356.

A ca. 0.0001 *M* soln. of startg. (E,E)-divinylallene in benzene-d_6 heated under argon in a sealed, thick-walled NMR tube at 100° until reaction complete → product. Y 92% (E/Z >99:1; half-life 24.1 h). Ring closure takes place at the sterically more congested vinylallene residue to afford *trisubst.* alkylidenecyclobutenes, the bulky C_4-substituent determining the high (E)-selectivity. F.e.s. S. López, A.R. de Lera et al., J. Am. Chem. Soc. *118*, 1881-91 (1996); 3-alkylidene-4-silylcyclobutenes s. Y. Ito et al., Angew. Chem. Intern. Ed. *34*, 1476-7 (1995); torquoselectivity s. Tetrahedron Letters *36*, 4669-72 (1995).

Regio- and stereo-specific intramolecular Diels-Alder reaction
s. *34*, 693s*46-9*; 1-*epi*-baccatin III steroid skeleton s. C.A. Alaimo et al., Tetrahedron Letters *35*, 6603-6 (1994); azadirachtin skeleton s. A. Murai et al., Synlett *1995*, 895-7; *cis*-octalins s. T. Kitahara et al., ibid. 909-11; bicyclo[n.3.1]alkenes s. K.J. Shea et al., Tetrahedron Letters *36*, 7177-80 (1995); with *in situ*-generated eneboranes s. D.A. Singleton, Y.-K. Lee, ibid. 3473-6; with fused 1,3-cyclohexadienes s. K. Harano et al., Tetrahedron *50*, 13395-408 (1994); with a 2,3-ethylene-4-ketoglycoside s. Y. Chapleur et al., Chem. Commun. *1995*, 937-8.

... with α,β-ethylenesulfonates
s. *44*, 669s*49*; details s. P. Metz et al., Tetrahedron *51*, 711-32 (1995); asym. induction with chiral 2-alkoxy-1,3-dienes s. G. Galley, M. Pätzel, J. Chem. Soc. Perkin Trans. I *1996*, 2297-302.

... with tethered 1,3-dienes
s. 46, 696s48; 34, 693s47,48; 6-vinylcyclohex-3-ene-1,2-diols from Si,Si'-bis(1,3-dienyl)-disiloxanes s. R.-M. Chen et al., J. Org. Chem. 60, 3272-3 (1995); with silyl acetal-tethered trienes, details, s. D. Craig et al., Tetrahedron 51, 11601-22 (1995); with acetal-tethered 1,3-dienes s. ibid. 52, 8937-46 (1996); **with asym. induction** based on chiral siloxy tethers s. K.J. Shea, D.R. Gauthier, Jr., Tetrahedron Letters 35, 7311-4 (1994); with a phenylboronate tether cf. K. Narasaka et al., Synthesis 1991, 1171-2.

Transannular Diels-Alder reaction
s. 43, 944s48; with a tetrasubst. non-activated dienophile s. J.Y. Roberge et al., Can. J. Chem. 72, 1820-9 (1994); with 14-membered macrocyclic trienes s. A. Ndibwami et al., ibid. 71, 695-713, 714-25 (1993); prepn. of 1,7-dimethyl-A.B.C.[6.6.6]tricyclics s. Y.-C. Xu et al., ibid. 1152-68, 1169-83; with (E,E,E)-cyclotetradeca-2,8,10-trienones s. W.R. Roush et al., Tetrahedron Letters 34, 4427-30 (1993); with (E,E,E)-13-trideca-2,8,10-trienolactones s. S.-H. Jung et al., ibid. 36, 1051-4 (1995); *trans-anti-cis*-tricyclics s. P. Deslongchamps et al., Bull. Soc. Chim. France 131, 142-51, 271-8 (1994).

Irradiation
Regio- and stereo-specific intramolecular [2+2]-cycloaddition
s. 22, 761s46-50; temperature effects s. D. Becker, Y. Cohen-Arazi, J. Am. Chem. Soc. 118, 8278-84 (1996); benzofuran-thymidine adducts s. W.R. Kobertz, J.M. Essigmann, ibid. 7101-7; **enecyclobutane ring** with allenes s. E.M. Carreira et al., ibid. 116, 6622-30 (1994); asym. addition to the benzene ring, effect of chiral ligands, s. P.J. Wagner, K. McMahon, ibid. 10827-8; **asym. intramolecular [2+2]-cycloaddition** in aq. media containing a *chiral 1,3-dioxolane-4,5-dimethanol* and surfactant s. F. Toda et al., Chem. Commun. 1995, 621-2; Cu(I)-catalyzed photochemical ring closure of 1,6-dienes s. K. Langer, J. Mattay, J. Org. Chem. 60, 7256-66 (1995); s.a. S. Ghosh et al., Tetrahedron Letters 37, 2073-6 (1996).

Cyclobutene ring from cyclic 1,3-dienes
s. 17, 801; 2,4-bridged bicyclo[3.2.0]hept-6-enes s. J.H. Rigby et al., Tetrahedron Letters 37, 2553-6 (1996).

3,5-Bis(trifluoromethyl)phenylboronic acid/(R)-3-(2-hydroxy-3-biphenylyl)-2,2'-dihydroxy-1,1'-binaphthyl
Asym. intramolecular Diels-Alder reaction
under Brønsted acid-assisted Lewis acid catalysis s. 49, 655s51

Tris(p-bromophenyl)aminium hexachloroantimonate/sodium carbonate
Cyclopentane ring from ethylenecyclopropanes
with Et$_2$AlCl cf. 42, 654; 3-arylthiobicyclo[3.3.0]octanes by stereospecific cation radical-catalyzed intramolecular [3+2]-cycloaddition s. C. Iwata et al., Tetrahedron Letters 36, 4085-8 (1995).

Chromium alkoxycarbene complexes
1-Vinylcycloalkenes from enynes
with [RuCl$_2$(CO)$_3$]$_2$ cf. 50, 444; with a little chromium alkoxycarbene complex s. M. Mori et al., Organometallics 13, 4129-30 (1994)

Bis(1,5-cyclooctadiene)nickel(0)/tri-o-biphenylyl phosphite
Nickel-catalyzed intramolecular Diels-Alder reaction with silylacetylenes
s. 45, 439; fused N-heterocyclics from substrates with sulfonamide or acylamine tethers s. P.A. Wender, T.E. Smith, J. Org. Chem. 61, 824-5 (1996).

Exchange ⇅

Hydrogen ↑   CC ⇅ H

sec-*Butyllithium/(-)-sparteine/cupric pivalate*   ←
Asym. deprotonation of phosphines via borane complexation   ←

357.

and subsequent oxidative coupling. 1.1 eqs. 1.32 M s-BuLi in cyclohexane added to a soln. of 1.1 eqs. (-)-sparteine in *ether* at -78°, stirred for 10 min, 1.eq. dimethylphenylphosphine-borane in ether added via cannula, treated with 3 eqs. Cu(II)-pivalate after 3 h, warmed to -20°, and worked up by flash chromatography → (S,S)-product. Y 54% (e.e. 98%). The borane residue, which is readily removable on treatment with excess of Et$_2$NH at 55° during 12-18 h, confers crystallinity so that purification can be carried out easily. F.e. and electrophiles, **also asym. deprotonation of phosphine sulfides**, s. A.R. Muci et al., J. Am. Chem. Soc. *117*, 9075-6 (1995).

Sodium nitrite/trifluoromethanesulfonic acid/air   ←
Cuprous chloride/N,N,N',N'-tetramethylethylenediamine/oxygen   ←
1-Oxo-2,2,6,6-tetramethylpiperidinium fluoroborate/potassium hydrogen carbonate   ←
Oxidative dimerization of arenes   2 ArH → Ar—Ar
aridines with NaNO$_2$/HF cf. *31*, 719; sym. 1,1'-binaphthyls with NaNO$_2$/CF$_3$SO$_3$H s. M. Tanaka et al., J. Org. Chem. *61*, 788-92 (1996); sym. 1,1'-bi-2-naphthols with CuCl/TMEDA/air cf. M. Noji et al., Tetrahedron Letters *35*, 7983-4 (1994); oxidative coupling of phenols with 1-oxo-2,2,6,6-tetramethylpiperidinium fluoroborate/KHCO$_3$ s. J.M. Bobbit, Z. Ma, Heterocycles *33*, 641-8 (1992).

Ammonium ceric nitrate $(NH_4)_2Ce(NO_3)_4$
2,3-Dihydrofurans from ethylene derivs.   O
s. *50*, 449; comparison with Mn(III) acetate s. J. Chem. Soc. Perkin Trans. I *1996*, 1487-92; 2-aryl-3-carbalkoxy-derivs. from cinnamic acid esters s. S.C. Roy, P.K. Mandal, Tetrahedron *52*, 2193-8 (1996); with cyclic enolethers s. ibid. 12495-8.

1-Cyano-1,2-benziodoxol-3-one   ←
α-Cyanation of tert. amines   H → CN
with FeCl$_3$/PhCOCN cf. *47*, 715s*49*; N-cyanomethyl-N-methylanilines with 1-cyano-1,2-benziodoxol-3-one s. V.V. Zhdankin et al., Tetrahedron Letters *36*, 7975-8 (1995).

Dichlorobis(triphenylphosphine)nickel(II)/n-butyllithium $Cl_2Ni(PPh_3)_2/BuLi$
Bis(dibenzylideneacetone)palladium(0)/allyl bromide/tetra-n-butylammonium bromide/
   sodium hydroxide   ←
Sym. 1,3-diynes from terminal acetylene derivs.   C≡C-C≡C
Phase transfer catalysis

358.

A degassed mixture of 50% aq. NaOH (1 part), methylene chloride (4 parts), 10 mol% Bu$_4$NBr, startg. 1-alkyne and 8 mol% allyl bromide treated with 5 mol% Pd(dba)$_2$, and stirred at room temp. under N$_2$ for 24 h → 4,6-decadiyne. Y 80% (by GC). The proposed mechanism involves initial reaction of the alkyne with allyl bromide to give an enyne which undergoes palladium π-allyl complex formation, followed by **nucleophilic displacement of the π-allylpalladium residue** by alkyne anion. F.e.s. M. Vlassa et al., Tetrahedron *52*, 1337-42 (1996); with Cl$_2$Ni(PPh$_3$)$_2$/BuLi cf. E.H. Smith, J. Whittall, Organometallics *13*, 5169-72 (1994).

Oxygen ↑   CC ⇅ O

Without additional reagents   w.a.r.
Ugi 4-component condensation   ←
s. *17*, 809; solid-phase parallel synthesis of a 96-member library s. R.W. Armstrong et al., Angew. Chem. Intern. Ed. *35*, 640-2 (1996).
4-Component synthesis of 1,1′-iminodicarboxylic acid derivs.   ←
with asym. induction

359.

Isobutanal and 1 eq. *tert*-butyl isocyanide (each in methanol) added to a soln. of 1 eq. startg. chiral α-amino acid in the same solvent at -30°, after 3 h the mixture allowed to warm to room temp., and worked up when the soln. became clear (3 h to 2 days) → product. Y 98% (88% d.e.). There is clear potential in combinatorial synthesis and adaptation by using nucleophiles other than alcohols. F.e.s. A. Demharter, I. Ugi et al., Angew. Chem. Intern. Ed. *35*, 173-5 (1996); details s. Tetrahedron *52*, 11657-64 (1996).

α-(2-Azetidinon-1-yl)carboxylic acid amides   ⌐N
from β-aminocarboxylic acids, aldehydes and isonitriles

360.

Ugi reaction. 1 eq. Each of startg. aldehyde and isonitrile added with vigorous stirring to a soln. of L-aspartic acid α-benzyl ester in methanol at -10°, warmed to room temp. after 1 h, and stirring continued for 12 h → 4-benzyloxycarbonyl-1-[(N-*tert*-butylcarbamoyl)(2-nitrophenyl)methyl]-azetidin-2-one. Y 85%. F.e. and diastereoselectivity s. K. Kehagia, I.K. Ugi, Tetrahedron *51*, 9523-30 (1995).

Pictet-Spengler ring closure   ○
s. *8*, 823; *cis*-1,3-disubst. tetrahydro-β-carbolines **with asym. induction** using TFA s. P.D. Bailey et al., Tetrahedron Letters 35, 3587-8 (1994); trans-1,3-disubst. derivs. with TsOH cf. P. Zhang, J.M. Cook, ibid. 36, 6999-7002 (1995); **polymer-based synthesis** s. K. Kaljuste, A. Undén, ibid. 9211-4; s.a. R. Mohan et al., ibid. *37*, 3963-6 (1996); L. Yang, L. Guo, ibid. 5041-4; use of gel-phase ^{19}F NMR to evaluate reactions in solid phase organic synthesis s. A. Svensson et al., ibid. 7649-52.

Polymer-supported Biginelli synthesis of 3,4-dihydro-2(1*H*)-pyrimidinones

361.

4 eqs. Each of benzaldehyde and startg. β-ketoester added to a suspension of startg. polymer-supported urea in THF, treated with 4:1 THF/concd. HCl, stirred at 55° for 36 h, the resin filtered, washed with THF, hexanes, methanol, and methylene chloride, and the resin removed by washing with TFA followed by methylene chloride → product. Y 88%. Products are isolated by simple filtration; no crystallization or chromatographic purification is necessary and excess of reagents

are removed by rinsing with THF. Yields are higher than those obtained in solution phase, and the method is adaptable for parallel synthesis of single compd. dihydropyrimidine libraries. F.e.s. P. Wipf, A. Cunningham, Tetrahedron Letters *36*, 7819-22 (1995).

Microwaves s. under SiO$_2$ ←
*Lithium/4,4'-di-*tert-*butylbiphenyl* ←
Allyl- and benzyl-lithiums from allyl and benzyl silyl ethers, respectively ←

362. PhCH(Me)OSiMe$_3$ ⟶ [PhCH(Me)Li] —Et$_2$CO→ PhCH(Me)CEt$_2$OH

(1-Methylbenzyloxy)trimethylsilane and 1.2 eqs. diethyl ketone in THF added during 30 min at 0° to a suspension of 14 eqs. Li-powder and 10 mol% 4,4'-di-*tert*-butylbiphenyl in the same solvent, and the mixture hydrolyzed after 10 min → 3-ethyl-2-phenyl-3-pentanol. Y 55%. Allyllithium compds. were obtained similarly. Yields were generally lower from the corresponding benzyl or allyl alcohols. F.e.s. E. Alonso et al., Tetrahedron *51*, 11457-64 (1995).

Sodium hydride NaH
α-Hydroxymethylenation CH$_2$ → C=CHOH
s. *9*, 816; *10*, 573; *16*, 796; of β-phosphonyl-esters and -nitriles (with NaH) s. N. Collignon et al., Synth. Commun. *25*, 3443-55 (1995); of β,γ-ethylenephosphonic esters s. Synthesis *1995*, 1401-4.

Potassium tert-*butoxide* KOBu-t
2-Azetidinones from aldimines and carboxylic acid esters
with LDA cf. *36*, 740; with KOBu-*t* s. G. Cainelli et al., Synthesis *1994*, 805-8.

Potassium tert-*butoxide/oxygen* KOBu-t/O$_2$
2-Pyridones from α,β-ethyleneketones
via 6-hydroxy-2-piperidones cf. *41*, 636; 3-cyano-2-pyridones in one pot from α,β-ethyleneoxo compds. with KOBu-*t* under O$_2$ s. M.A. Ciufolini et al., Tetrahedron Letters *36*, 3307-10 (1995).

Organolithium compds. RLi
Synthesis of cyclic 2-ethylenealcohols from bicyclic 2,3-epoxyethers

363. [structure: MeO, Bu-substituted oxabicyclic compound] + BuLi ⟶ [Bu, Bu-substituted cyclopentenol with OH]

2.5 eqs. 1.6 *M* n-Butyllithium in hexanes added dropwise to a stirred soln. of *syn*-4-butyl-4-methoxy-1-methyl-6-oxabicyclo[3.1.0]hexane in dry THF at -78° under argon, the mixture allowed to warm to room temp., and stirred for 1 h before quenching with pH 7 phosphate buffer → 2,3-dibutyl-1-methylcyclopent-2-enol. Y >95%. Reaction proceeds via a carbenoid derived from the metalated oxirane, followed by insertion of the alkyllithium and subsequent elimination of Li-methoxide (an addition-elimination mechanism being precluded since the corresponding diol monoethers failed to give the allylic alcohol). Monocyclic analogs simply gave 3-ene-1,2-diol 1-monoethers. F.e. incl. cyclohexene analog s. L. Dechoux et al., Chem. Commun. *1996*, 549-50.

n-*Butyllithium* BuLi
α-Ketotrithioorthocarboxylic acid esters from carboxylic acid esters COOR → COC(SR)$_3$
Synthesis with addition of one C-atom

364. PhCOOMe + [LiC(SMe)$_3$] ⟶ PhCOC(SMe)$_3$

1.1 eqs. 2.5 *M* BuLi in hexane added dropwise during 5 min to a soln. of tris(methylthio)methane in anhydrous THF at -95° under N$_2$, stirred for 2 h, 0.8 eq. methyl benzoate in the same solvent added dropwise during 5 min, the resulting homogeneous soln. stirred at -95° for 5 min, and quenched directly with 1:1 ether/water → 2,2,2-tris(methylthio)-1-phenylethanone. Y 85%. With ca. 2-2.4 eqs. tris(methylthio)methane and BuLi at -78°, the corresponding **α-ketomercaptals** were obtained. F.e. and improved yields with added N-(methylthio)phthalimide s. R. Fochi et al., J. Org. Chem. *60*, 6017-24 (1995).

n-*Butyllithium/diethyl phosphorochloridate/potassium* tert-*butoxide*
(E)-α,β-Ethylenephosphonic acid esters from oxo compds.
via β-phosphoryloxyphosphonic acid esters

C═C—PO(OR)$_2$ ←

365.

One-pot conversion. 1.1 eqs. 1.6 M n-BuLi in hexane added under N$_2$ to a stirred soln. of diethyl methylphosphonate in dry THF at -78°, p-methoxybenzaldehyde added, stirred for 1 h, 1.1 eqs. diethyl chlorophosphate added, the mixture allowed to stand at room temp. for 3 h, recooled to -78°, a soln. of 1.2 eqs. K-*tert*-butoxide in THF added, stirred for 10 min, and worked up → product. Y 87% (single stereoisomer). F.e.s. C.-W. Lee et al., Synth. Commun. *25*, 2013-7 (1995).

sec-*Butyllithium/potassium* tert-*butoxide*
Synthesis of (E)-enolethers from acetals

s-*BuLi/KOBu*-t
CHC(OR)$_2$ → C═C(OR)R'

366.

Startg. acetal and *2.5 eqs.* sublimed *t*-BuOK added sequentially to a stirred soln. of *2.5 eqs.* s-BuLi in THF at -95°, 1 eq. pivaldehyde added after 3 h, kept at -50° for 1 h, and quenched with aq. THF → (E)-2-methoxy-4,4-dimethyl-1-phenylpent-1-en-3-ol. Y 72%. F.e. **and 2-acetylenealcohols** with 4 eqs. base, also acidic hydrolysis to ketones s. P. Venturello et al., J. Chem. Soc. Perkin Trans. I *1995*, 2757-60.

Lithium diisopropylamide/trimethylsilyl or triphenyltin chloride
α,β-**Ethylenephosphonic acid esters**
via α-silylation of phosphonic acid esters

CO → C═C–PO(OR)$_2$ ←

s. *42*, 725; (Z)-1-alkoxy-1,3-diene-2-phosphonic acid esters from formic acid esters s. Tetrahedron Letters *37*, 2951-4 (1996); α,β-ethylene-α-stannylphosphonic acid esters via α,α-bisstannylation (cf. *42*, 725s*47*) s. Synth. Commun. *25*, 1921-32 (1995).

Lithium diisopropylamide/hexamethylphosphoramide
Cyclopropanes from cyclic glycol sulfates and active methylene groups

i-*Pr$_2$NLi/(Me$_2$N)$_3$PO* ▽

367.

1-Aminocyclopropanecarbonitriles. A soln. of (dibenzylamino)acetonitrile in THF added to a soln. of 2.2 eqs. LDA/HMPA in the same solvent at -70° under N$_2$, 1.2 eqs. startg. cyclic sulfate added, stirred for 1.5 h, and treated with satd. NH$_4$Cl → 1-(dibenzylamino)cyclopropane carbonitrile. Y 45-70%. F.e. and method s. H.-P. Husson et al., Bull. Soc. Chim. France *131*, 391-6 (1994).

Lithium N-isopropylcyclohexylamide
α-Chloro-β-lactones from α-chlorocarboxylic acid phenyl esters and oxo compds.

368.

A soln. of startg. phenyl ester in THF at -70° added slowly over 30 min to 1 eq. Li-N-isopropylcyclohexylamide in the same solvent at -78°, stirred for 20 min below -70°, a precooled soln. of cyclohexanone (0.83 eq.) in THF added over 20 min at the same temp., allowed to warm to 0° within 2 h, diluted with ether, and quenched with aq. NaOH → product. Y 83%. Interestingly, elimination of LiOPh is more facile than LiCl, so that the anticipated Darzens condensation does not take place. The procedure is generally applicable to the preparation of α,β,β-trisubst. and α,β-disubst. derivs., and both enolizable and non-enolizable aldehydes may be used as well as ketones. F.e.s. C. Wedler et al., Angew. Chem. Intern. Ed. 34, 2028-9 (1995).

Potassium cyanide s. under NiCl$_2$(PPh$_3$)$_2$ KCN
Cesium fluoride   CsF
α-Alkylation of β-dicarbonyl compds. with mesylates   H → R

369.

Inversion of configuration. A mixture of 3 eqs. each of diethyl malonate and CsF in DMF stirred under N$_2$ for 1 h, a soln. of ethyl (S)-2-(methanesulfonyloxy)propanoate in DMF added, and stirred at 45° for 5 h → diethyl (S)-2-(ethoxycarbonyl)-3-methylbutanedioate. Y 68% (e.e. 99%). Under these almost neutral conditions, acid- and base-sensitive functions (e.g. ester, THP-ethers) remained unaffected. F.e. incl. α-alkylation of α-cyanocarboxylic acid esters s. T. Sato, J. Otera, J. Org. Chem. 60, 2627-9 (1995).

β-Picoline
3-Chromenes from phenols
and α,β-ethylenealdehydes cf. 27, 785; from *electron-deficient* phenols and α,β-ethyleneacetals s. J.T. North et al., J. Org. Chem. 60, 3397-400 (1995).

Magnesium/lithium bromide/dichloro[(S)-2-(dimethylamino)-1-(diphenylphosphino)-3-phenylpropane]palladium(II)
Asym. synthesis of diaryls by palladium-catalyzed cross-coupling   Ar—Ar'

370.

ca. 2 eqs. 1.8 M Phenylmagnesium bromide in ether added at -30° to equivalent amounts of 1-[2,6-bis[[(trifluoromethyl)sulfonyl]oxy]phenyl]naphthalene and LiBr in toluene containing ca. 5 mol% PdCl$_2$[(S)-Phephos], stirred for 48 h, and hydrolyzed with 10% HCl → (S)-product. Y 78% (e.e. 100% after recrystallization). The high enantioselectivity is partly a consequence of kinetic resolution on formation of by-product, 1-[2,6-(diphenyl)phenyl]naphthalene, by displacement of the second triflate group. F.e. and elaboration of the product by subsequent displacement of the remaining triflate group s. T. Hayashi et al., J. Am. Chem. Soc. 117, 9101-2 (1995).

Magnesium/trimethylsilyl chloride Mg/Me_3SiCl
γ-Lactones from α,β-ethylenecarboxylic acid esters and aldehydes
with SmI_2 cf. *41*, 723; β-aryl-γ-lactones with Mg/Me_3SiCl s. I. Nishiguchi et al., J. Org. Chem. *60*, 458-60 (1995).

Magnesium/N-methylmethoxylamine hydrochloride   ←
Synthesis of ketones from carboxylic acid esters   COOR → COR'
via hydroxamic acid esters
One-pot procedure

371.

2 M Phenylmagnesium chloride in THF added over 2 h to a slurry of N,O-dimethylhydroxylamine hydrochloride and startg. ester in the same solvent at -5° under N_2 maintaining the temp. between -2° and -5°, after 1 h at -5° the mixture warmed to 25° over 1 h, aged for 8 h, then quenched with 1 N HCl → 17β-benzoyl-4-aza-5α-androst-1-en-3-one. Y 87%. Formation of tert. alcohol by-products was kept to a minimum under these conditions. F.e. and with isolation of the intermediate hydroxamic acid esters s. J.M. Williams et al., Tetrahedron Letters *36*, 5461-4 (1995).

Magnesium/dichlorobis(triphenylphosphine)nickel(II) $Mg/NiCl_2(PPh_3)_2$
Regio- and stereo-specific synthesis of ethylene derivs. from alkoxy-2-ethylenes   OR → R'
Internal Lewis bases as directing group

372.

Under nickel catalysis, Grignard compds. (notably those lacking a β-hydrogen) couple with otherwise unreactive *sec.* allyl ethers, an enhancement of regio- and stereo-selectivity being effected by placement of a suitable Lewis basic residue [a phosphino group] in the substrate. E: A soln. of startg. allyl ether, 5 eqs. MeMgBr, and 5 mol% $(Ph_3P)_2NiCl_2$ in THF allowed to react for 12 h → product. Y 75% (<1% regioisomer; *cis:trans* >49:1; d.s. >49:1). Reaction is thought to involve *anti*-insertion of Ni(0) to form a metal-allyl system (stabilized by association with the appended diphenylphosphino group), followed by reductive *syn*-elimination. F.e. and *cis→trans*-rearrangement s. M.T. Didiuk et al., J. Am. Chem. Soc. *117*, 7273-4 (1995).

Zinc/cuprous cyanide/lithium chloride/trimethylsilyl chloride/hexamethylphosphoramide   ←
2-Cycloalkenone-2-carboxylic from α,β-acetylenecarboxylic acid esters
2-cyclopentenone-2-carboxylic acid esters s. *46*, 721; 2-cyclohexenone analogs with added $CuCN·LiCl/Me_3SiCl$ s. M.T. Crimmins et al., Tetrahedron Letters *36*, 7061-4 (1995).

Zinc/tetracarbonyl(cyclopentadienyl)vanadium/trimethylsilyl chloride   ←
1,3-Dioxolanes from three aldehyde molecules via pinacol coupling

373.   3 $C_5H_{11}CHO$ →

A mixture of 1.5 eqs. activated zinc (Aldrich), 3 mol% $CpV(CO)_4$, and 1 eq. trimethylsilyl chloride in DME stirred at room temp. for 10 min under argon, 1 eq. startg. aldehyde added, and stirring continued at room temp. for 3 h → product. Y 84%. *anti*-Isomers were favoured. There was no reaction in the absence of the vanadium catalyst or chlorosilane, and the system Zn/TiCl$_4$ failed. F.e.s. T. Hirao et al., J. Org. Chem. *61*, 366-7 (1996).

Grignard compds. s. under Ti(OPr-i)$_4$	RMgHal
Zinc cyanide s. under Pd(PPh$_3$)$_4$	Zn(CN)$_2$
Aluminum trifluoromethanesulfonimide	Al(NTf$_2$)$_3$
Regiospecific Friedel-Crafts acylation	ArH → ArCOR

374. MeO–⟨⟩ + Ac$_2$O —Al(NTf$_2$)$_3$→ MeO–⟨⟩–Ac

***p*-Acylation of phenolethers.** Anisole and 2 eqs. acetic anhydride added to a soln. of 20 mol% Al(NTf$_2$)$_3$ in nitromethane at room temp., and the mixture stirred for 10 min → *p*-methoxyacetophenone. Y 99%. The method is superior to other routes in that conditions are mild, only a catalytic amount of the Lewis acid is required, and carboxylic acid by-products are not a problem. F.e. and metal triflimides (e.g. (*i*-PrO)$_2$Ti(NTf$_2$)$_2$ and Yb(NTf$_2$)$_3$) s. K. Mikami et al., Synlett *1996*, 171-2.

Alumina or Bentonite or Silica gel ←
Heterogeneous Knoevenagel condensation CO → C=C
with Al$_2$O$_3$ s. *38*, 756; α-cyano-α,β-ethylenecarboxylic acid thioamides s. D. Villemin, B. Martin, Synth. Commun. *23*, 2259-63 (1993); with mexican bentonite under an IR lamp without solvent cf. F. Delgado et al., ibid. *25*, 753-9 (1995); trifluoromethyl α-alkylidene-β-ketocarboxylic acid esters with silica gel cf. K. Shibata et al., Chem. Lett. *1996*, 179.

Aluminosilicate FSM-16 ←
***meso*-Tetraarylporphines from ar. aldehydes** ○
with montmorillonite cf. *23*, 427s48; with *mesoporous* aluminosilicate FSM-16 s. T. Shinoda et al., Chem. Commun. *1995*, 1801-2; with high-valent transition metals (TiCl$_4$, VOCl$_3$, VO(OEt)Cl$_2$, VO(OPr-*i*)Cl$_2$ and Mn(OAc)$_3$) s. E.F. Llama et al., J. Chem. Soc. Perkin Trans. I *1995*, 2611-3.

Zeolite ←
Pechmann-Duisberg coumarin synthesis
with H$_2$SO$_4$ cf. *1*, 591; under heterogeneous catalysis with zeolite H-BEA or Amberlyst-15 s. E.A. Gunnewegh et al., Rec. Trav. Chim. Pays-Bas *115*, 226-30 (1996).

Envirocat EPZG ←
1,1-Nitroethylene derivs. from aldehydes CHO → CH=CNO$_2$
with anion exchange resin cf. *19*, 844; under heterogeneous catalysis with Envirocat EPZG in the absence of solvent s. B.P. Bandgar et al., Synlett *1996*, 149-50.

Boron fluoride BF$_3$
Asym. ring opening of cyclic ketones via aldol condensation
intramolecularly cf. *48*, 751; (E)-ethylenecarboxylic acid esters from aldehydes by intermolecular variant s. H. Suemune et al., Tetrahedron Letters *36*, 7259-62 (1995).

Boron fluoride-dimethyl sulfide BF$_3$-Me$_2$S
Friedel-Crafts acylation ArH → ArCOR
s. *4*, 725; of polycyclic arenes at low temp., regioselectivity, s. A.S. Kiselyov, R.G. Harvey, Tetrahedron Letters *36*, 4005-8 (1995).

Aluminum chloride AlCl$_3$
Replacement of acoxy groups by aryl OAc → Ar
of benzylic acoxy groups s. *19*, 839; of allylic acoxy groups, (Z)-(2-cyanoallyl)arenes and (E)-(2-carbalkoxyallyl)arenes, s. D. Basavaiah et al., Synlett *1996*, 393-5.

Scandium(III) triflate Sc(OTf)$_3$
Chromans from phenols and 2-ethylenealcohols ○
with BF$_3$ cf. *27*, 790a; with a little (0.1 mol%!) Sc(OTf)$_3$ s. H. Yamamoto et al., Bull. Chem. Soc. Japan *68*, 3569-71 (1995).

Ytterbium(III) triflate $Yb(OTf)_3$
Quinolines from N-arylaldimines and enolethers
with $H[Co(CO)_4]$ cf. *26*, 723; with $Yb(OTf)_3$, also 4-siloxy-1,2,3,4-tetrahydroquinolines from enoxysilanes, s. K. Takaki et al., Synthesis *1995*, 801-4.

Scandium(III) or ytterbium(III) triflate/lithium perchlorate $Sc(OTf)_3$ or $Yb(OTf)_3/LiClO_4$
Friedel-Crafts acylation   ArH → ArCOR
with $Sc(OTf)_3$ or $Yb(OTf)_3$ cf. *49*, 763; acceleration with added $LiClO_4$ s. Chem. Commun. *1996*, 183-4.

Samarium diiodide s.a. under Benzotriazole SmI_2
Samarium diiodide/trifluoroacetic acid SmI_2/CF_3COOH
Reductive coupling of acetals $2\ C(OR)_2 \rightarrow C(OR)C(OR)$
with $LAH/TiCl_4$ cf. *34*, 723; pinacol and benzyl ethers s. A. Studer, D.P. Curran, Synlett *1996*, 255-7.

Piperidinium acetate/zinc bromide/diisobutylaluminum hydride   ←
Polymer-based synthesis of cyclic 5'-ene-1,3-diols from ethylenealdehydes   ←

375.

via stereospecific **Knoevenagel condensation-intramolecular ene reaction**. Startg. malonate (prepared from Merrifield resin, 1,3-propanediol and methyl malonyl chloride) allowed to react with 3 eqs. startg. aldehyde in the presence of 0.1-0.2 eq. piperidinium acetate in methylene chloride at 20° for 1-3 h, treated with 1.1 eqs. $ZnBr_2$, allowed to react for 3 days, and the ene adduct treated with 4 eqs. *i*-Bu_2AlH in toluene at 0° for 12 h → product. Y 61% based on concentration of free hydroxyl groups in the startg. polymer (97.2:2.8 diastereoisomeric mixture; *trans*:*cis* >99:1). Condensation of α-subst. aldehydes required added Na_2SO_4 and elevated temp. (40°). Using this simple, polymer-based strategy, chromatographic separation of the product is not necessary and isomerization of the double bond in the intermediate 2-alkylidene-1,3-dicarbonyl compds. does not occur. F.e. incl cyclopentane analog, generation of a small combinatorial library, and cleavage via transesterification [with 1 eq. $Ti(OEt)_4$] s. L.F. Tietze, A. Steinmetz, Angew. Chem. Intern. Ed. *35*, 651-2 (1996).

Benzotriazole/samarium diiodide BtH/SmI_2
Pyrrolidines from 3-ethyleneamines and aldehydes
Regio- and stereo-specific radical ring closure

376.

A mixture of 1.08 eqs. benzotriazole, startg. 3-ethyleneamine, 1 eq. *n*-butanal, and 4 Å molecular sieves in dry benzene stirred at room temp. for 12 h, worked up, the residue dissolved in dry deoxygenated THF, added dropwise to a soln. of 2.16 eqs. SmI_2 in the same solvent at -10°, stirred for 30 min, warmed to room temp. over 40 min, and quenched with satd. K_2CO_3 → product. Y 67% (d.r. 73:27). The method is applicable to ethyleneamines having an activated double bond. Reaction occurs via 5-*exo-trig* radical cyclization giving moderate diastereoselectivity. F.e.s. J.M. Aurrecoechea, A. Fernández-Acebes, Synlett *1996*, 39-42.

Trifluoroacetic acid CF_3COOH
Polymer-based Pictet-Spengler ring closure s. *8*, 823s*51*

Silica gel/microwaves (s.a. under Al_2O_3)   ←
Supported Knoevenagel condensation under microwave irradiation   CO → C=C
with montmorillonite cf. *46*, 713s*50*; with *neutral* SiO_2 without solvent s. F. Langa et al., Tetrahedron Letters *37*, 1113-6 (1996); (Z)-5-alkylidene-2-imino-4-imidazoles without support or solvent cf. D. Villemin, B. Martin, Synth. Commun. *25*, 3135-40 (1995).

Trimethylsilyl cyanide/titanium tetrachloride $Me_3SiCN/TiCl_4$
α-Cyanohydroximinochlorides from 1,1-nitroethylene derivs. s. *51*, 206   C(CN)C(Cl)=NOH
Titanium tetraisopropoxide/ethylmagnesium bromide $Ti(OPr-i)_4/EtMgBr$
***tert*-Aminocyclopropanes from N,N-disubst. carboxylic acid amides** ▽

377.

A suspension of 2.5 eqs. ethylmagnesium bromide in THF at -78° under N₂ treated with 1 eq. Ti(OPr-*i*)₄ in THF via a steel cannula, after stirring for 2 min startg. dibenzylamide in THF added, the mixture allowed to warm to room temp., then refluxed for 24 h → product. Y 63%. The method has wide potential, and gives rise to previously inaccessible aminocyclopropanes. F.e.s. V. Chaplinski, A. de Meijere, Angew. Chem. Intern. Ed. *35*, 413-4 (1996).

Titanium tetraisopropoxide/cyclohexylmagnesium chloride $Ti(OPr-i)_4/c-C_6H_{11}MgCl$
Cyclopropanols from carboxylic acid esters and terminal ethylene derivs.

378.

Modified Kulinkovich reaction. The Kulinkovich cyclopropanol synthesis (*45*, 476) can be performed more simply and efficiently by using terminal ethylenes and *cyclohexylmagnesium chloride* in place of 2 eqs. alkylmagnesium bromide (one of which is sacrificed as the corresponding alkane). E: 4.5 eqs. Cyclohexylmagnesium chloride (commercial) in THF added slowly over 1 h to 1 eq. startg. ester and 1.5 eqs. startg. alkene in the same solvent in the presence of 1 eq. Ti(OPr-*i*)₄ → product. Y 64%. Di- and tri-subst. alkenes, as well as bromo- and siloxy-groups in the olefin, and double bonds in the ester function were tolerated. Reaction is driven by formation of cyclohexene which cannot participate further in the reaction. F.e. and hydroxyalkyl derivs. from lactones s. J. Lee et al., J. Am. Chem. Soc. *118*, 4198-9 (1996).

Hafnium(IV) triflate $Hf(OTf)_4$
***o*-Acylation of phenols with carboxylic acids**   ArH → ArCOR

379.

Equivalent amounts of 1-naphthol and acetic acid in 20:3 toluene/nitromethane added to 20 mol% Hf(OTf)₄ at room temp., stirred for 6 h at 100°, cooled to room temp., and quenched with water → 1-hydroxy-2-acetonaphthone. Y 81%. The method avoids the use of [more expensive] acid anhydrides or chlorides. Only a trace of product was formed with conventional Lewis acids (AlCl₃, BF₃-etherate, TiCl₄, SnCl₄). F.e.s. S. Kobayashi et al., Tetrahedron Letters *37*, 4183-6 (1996).

Titanium tetrachloride (s.a. under Me₃SiCN) $TiCl_4$
***meso*-Tetraarylporphines from ar. aldehydes** s. *23*, 427s*51*   ○

Trimethylphosphine/1,1'-(azodicarbonyl)dipiperidine $R_3P/RN=NR$
Triphenylphosphine/diethyl azodicarboxylate
Mitsunobu C-alkylation   H → R
of 1,1-disulfones with PPh₃/DEAD cf. *49*, 941; of methyl phenylsulfonylacetate with Me₃P/1,1'-(azodicarbonyl)dipiperidine s. J. Yu et al., Synlett *1995*, 1127-8; of α-cyanosulfones with added imidazole, also cyclic nitriles from diols, s. Tetrahedron Letters *36*, 5691-4 (1995); Mitsunobu *o*-benzylation of phenols with inversion of configuration s. S. Fukumoto et al., J. Chem. Soc. Perkin Trans. I *1996*, 1021-6.

Diethyl phosphorochloridate s. under BuLi (EtO)$_2$POCl
Tetracarbonyl(cyclopentadienyl)vanadium s. under Zn CpV(CO)$_4$
p-Toluenesulfonic acid TsOH
Pictet-Spengler ring closure with asym. induction s. *8*, 823s*51* ○
Pyridinium tosylate/triisobutylaluminum C$_5$H$_6$N$^+$TsO$^-$/i-Bu$_3$Al
Regiospecific reductive ketal-Claisen rearrangement s. *51*, 349 ←
Tricarbonyl(η^6-cycloheptatriene)tungsten/2,2′-bipyridyl/sodium salt
Regio- and stereo-specific C-α-allylation with allyl carbonates H → C–C≡C
with Mo(CO)$_6$/2,2′-bipy cf. *42*, 750; retention of configuration with W(CO)$_3$(η^6-cycloheptatriene) s. J. Lehmann, G.C. Lloyd-Jones, Tetrahedron *51*, 8863-74 (1995); syntheses **via molybdenum π-allyl complexes** with Mo(CO)$_6$/BnEt$_3$NCl, e.g. alkoxy-2-ethylenes (cf. *38*, 768), s. P. Kocovský et al., Tetrahedron Letters *36*, 6351-4 (1995).
Hydrogen chloride HCl
Elaboration of the Ugi 4-component condensation with a convertible isonitrile ←

380.

If 1-isocyanocyclohexene is used as the isonitrile in the Ugi 4-component condensation (cf. *17*, 809), the resulting α-acylaminocarboxamides can be readily converted to other carboxylic acid derivs., in certain instances without isolation of the N-cyclohexenyl amide. **E: α-Acylaminocarboxylic acid esters.** A 1 M soln. of benzaldehyde and 1.25 eqs. each of acetic acid and *p*-methoxybenzylamine in methanol stirred for 10 min, the mixture added in one portion to 1 eq. 1-isocyanocyclohexene, stirred at room temp. until reaction complete by TLC (12 h), 10 eqs. acetyl chloride added (for generation of anhydrous HCl), and warmed to 55° for 3 h → product. Y 79%. F.e. incl. the free acids, N-unsubst. amides, and thiolic acid ester derivs. s. T.A. Keating, R.W. Armstrong, J. Am. Chem. Soc. *117*, 7842-3 (1995); polymer-based adaptation s. Tetrahedron Letters *37*, 1149-52 (1996).

Bis(dimethylglyoximato)cobalt(III) dichloride complex ←
Regiospecific C-α-allylation of β-dicarbonyl compds. H → C–C≡C
with acoxy-2-ethylenes/CoCl$_2$ cf. *22*, 782s*49*; improved regioselectivity with 2-ethylenealcohols/bis(dimethylglyoximato)cobalt(III) dichloride complex s. M. Mukhopadhyay, J. Iqbal, Tetrahedron Letters *36*, 6761-4 (1995).

Bis[1,2-bis(diphenylphosphino)butane]nickel(0)/tetra-n-butylammonium hexafluorophosphate
C-α-Allylation s. *51*, 152

Dichlorobis(triphenylphosphine)nickel(II) s.a. under Mg NiCl$_2$(PPh$_3$)$_2$
Dichlorobis(triphenylphosphine)nickel(II)/triphenylphosphine/zinc/potassium cyanide ←
Nickel-catalyzed coupling with aryl mesylates

381.

Aryl mesylates are cheap alternatives to aryl triflates in nickel-catalyzed coupling. **E: Ar. nitriles.** A mixture of startg. aryl mesylate, 10 mol% NiCl$_2$(PPh$_3$)$_2$, 20 mol% PPh$_3$, 1 eq. Zn,. and 1.5 eqs. KCN dried at 25° under reduced pressure for 1 h in a Schlenk tube, DMF added under argon via syringe, stirred at room temp. for 5 min then at 80° for 12 h, cooled to 25°, treated with chloroform and 10% aq. HCl, and stirring continued for 20 min → product. Y 93%. The procedure is convenient and does not require pre-treatment of the Ni(II) complex with Zn prior to addition of reactants. F.e. and reactions, incl. prepn. of **alkylarenes and diaryls** from the corresponding organomagnesium or -zinc halides, s. V. Percec et al., J. Org. Chem. *60*, 6895-903 (1995); **sym. diaryls** and f. reactions s. ibid. 1066-9, 176-85; cf. *51*, 453.

Palladous acetate/1,3-bis(diphenylphosphino)propane/triethylamine/trioctylsilane ←
Ar. aldehydes from aryl triflates via reductive carbonylation $OSO_2CF_3 \rightarrow CHO$

382. $MeO-\langle{}\rangle-OSO_2CF_3 \xrightarrow{CO} MeO-\langle{}\rangle-CHO$

CO bubbled into a soln. of *p*-methoxyphenyl triflate, 2 mol% Pd(OAc)$_2$, and 2 mol% dppp in DMF at 70° over 20 min, 2.5 eqs. Et$_3$N added dropwise via a syringe followed by the addition of 2 eqs. tri-*n*-octylsilane, and the mixture stirred at 70° for 20 h → *p*-methoxybenzaldehyde. Y 70% (72% conversion). The method is simple, efficient, and general (for triflates with electron-donating or electron-withdrawing functions) and startg. ms. are readily available. Sensitive functional groups such as esters, cyclic acetals and phenolethers remained unaffected. F.e., also α,β-ethylenealdehydes from enol triflates, s. H. Kotsuki et al., Synthesis *1996*, 470-2.

Palladous acetate/tris(2,6-dimethoxyphenyl)phosphine/tri-n-butylstannyl acetate ←
4-Alk-1-ynyl-2(5H)-furanones from terminal acetylene derivs. s. *51*, 333   O

Tetrakis(triphenylphosphine)palladium(0) $Pd(PPh_3)_4$
Alkynyl-N-heterocyclics from triflyloxy-N-heterocyclics $OSO_2CF_3 \rightarrow C\equiv CR$
s. *41*, 736s*43*; oligopyridine derivs. with Pd(PPh$_3$)$_4$ s. F.M. Romero, R. Ziessel, Tetrahedron Letters *35*, 9203-6 (1994); 3-alkynylpyridazines with Pd(PPh$_3$)$_2$Cl$_2$/CuI/*i*-Pr$_2$NH s. D. Toussaint et al., Heterocycles *38*, 1273-86 (1994); *o*-silylethynyl-N-heterocyclics with Pd$_2$(dba)$_3$/(*o*-Tol)$_3$P/*i*-Pr$_2$NEt s. T. Okita, M. Isobe, Tetrahedron *51*, 3737-44 (1995).

Tetrakis(triphenylphosphine)palladium(0)/zinc cyanide $Pd(PPh_3)_4/Zn(CN)_2$
Ar. nitriles from aryl triflates   ArOTf → ArCN
s. *45*, 447s*48*; with Pd(PPh$_3$)$_4$/Zn(CN)$_2$ cf. H.G. Selnick et al., Synth. Commun. *25*, 3255-61 (1995).

Bis(dibenzylideneacetone)palladium/chiral 2,6-dimethyl-9-phenyl-9-
  phosphabicyclo[3.3.1]nonane ←
Tris(dibenzylideneacetone)dipalladium/chiral bis(o-acylaminophosphines) ←
Bis(allylpalladium chloride)/chiral ligands ←
Asym. C-α-allylation   H → C–C=C
s. *48*, 772s*50*; with chiral sulfoximines as ligand s. C. Bolm et al., Tetrahedron Letters *37*, 3985-8 (1996); with a chiral *tridentate* bis(phosphino)pyridine as ligand cf. X. Zhang et al., ibid. 4475-8; with monodentate chiral 2,6-dimethyl-9-phenyl-9-phosphabicyclo[3.3.1]nonane s. Y. Hamada et al., ibid. 7565-8; with C$_2$-symmetric *tert*-aminophosphines based on 1,1'-binaphthyl s. H. Kubota, K. Koga, ibid. *35*, 6689-92 (1994); with (S,S)-1,2-bis(*p*-tolylsulfinyl)benzene as ligand cf. M. Shibasaki et al., ibid. *36*, 8035-8 (1995); with C$_2$-symmetric 2-aminoethers or 2-hydroxythioethers s. B. Koning et al., Rec. Trav. Chim. Pays-Bas *115*, 49-55 (1996); with chiral macrocyclic diphosphine ligands s. M. Widhalm et al., J. Organometal. Chem. *523*, 167-78 (1996); with chiral bis(phosphinites) based on (R,R)-α,α,α',α'-tetraphenyl-2,2-dimethyl-1,3-dioxolane-4,5-dimethanol cf. D. Seebach et al., Helv. Chim. Acta *78*, 1636-50 (1995); with carbohydrate-based bis(phosphinites) cf. T.V. RajanBabu et al., Synlett *1996*, 745-6; with chiral dihydrobenzazaphosphole-borane complexes as ligand s. G. Brenchley, M. Wills, Tetrahedron *51*, 10581-92 (1995); with Pd$_2$(dba)$_3$/chiral bis(*o*-acylaminophosphines) s. *51*, 188; with 1,1'-bis(phosphino)-2,2'-bis(Δ^2-oxazolidon-2-yl)ferrocenes s. I. Ikeda et al., Tetrahedron Letters *37*, 4545-8 (1996).

*Bis(π-allylpalladium chloride)/(R,S)-1,2-bis[o-(diphenylphosphino)benzoylamino]
cyclohexane/sodium hydride* ←
**Synthesis of acoxy-2-ethylenes from α,β-ethyleneacylals
with asymmetrization** $CH(OAc)_2 \rightarrow CH(OAc)R$

383.

Unfavourable 1,2-addition of stabilized nucleophiles to α,β-ethyleneoxo compds. can be effected indirectly and enantioselectively via asymmetrization of the corresponding α,β-ethyleneacylals. E: A mixture of dimethyl methylmalonate and 0.85 eq. NaH (60% dispersion) in THF stirred at room temp. until evolution of H_2 ceased, the soln. cannulated into 0.5 eq. startg. acylal and a preformed catalyst (generated by mixing ca. 1 mol% [(π-allyl)PdCl]$_2$ and ca. 2.5 mol% chiral diamine) in THF, stirred at room temp. for 5 h, and poured into aq. $NaHSO_4$ → product. Y 75% (95% e.e.). Interestingly, the chiral ligand not only provided asym. induction, but also improved the yield and regioselectivity relative to an achiral ligand. Inferior results were obtained with other counterions. F.e.s. B.M. Trost et al., J. Am. Chem. Soc. *117*, 7247-8 (1995).

Chiral dichloropalladium phosphine complex s. under Mg ←

Nitrogen ↑   CC ⇅ N

Sodium hydride/n-butyllithium   *NaH/BuLi*
n-Butyllithium   *BuLi*
Ketones from carboxylic acid amides   CON< → COR
s. *14*, 930; from pseudoephedrin-based amides, chiral α-(carbalkoxyamino)ketones, s. A.G. Myers, T. Yoon, Tetrahedron Letters *36*, 9429-32 (1995); β,δ-diketocarboxylic acid esters from N-acyl-aziridines with NaH/n-BuLi s. B. Lygo, ibid. *35*, 5073-4 (1994).

n-Butyllithium/zinc bromide   *BuLi/ZnBr$_2$*
**Regiospecific synthesis of ketones
from oxo compds. and 1-subst. benzotriazoles**   COR → CHCOR

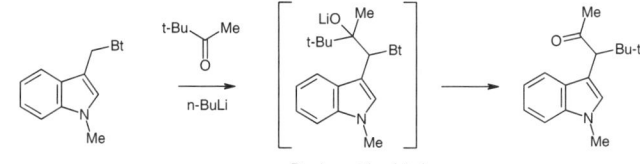

384.

via alkyl group migration. A soln. of startg. benzotriazole in THF treated with *n*-BuLi at -78°, *tert*-butyl methyl ketone added, and heated with a ca. 3-fold excess of $ZnBr_2$ at 65° for 3 h → product. Y 87%. Reaction is generally applicable to 1-alkyl-, 1-allyl, 1-α-alkoxy- and 1-α-arylthio-benzotriazoles for regiospecific insertion into aliphatic or ar. ketones or aldehydes, as well as into cyclic ketones **with ring expansion**. A cleaner alternative to insertion of diazo compds. is thus at hand. F.e. and details (in Supporting Information) s. A.R. Katritzky et al., J. Am. Chem. Soc. *117*, 12015-6 (1995); synthesis of ketones **from 1-α-alkoxybenzotriazoles** s. J. Org. Chem. *60*, 7619-24 (1995); α,β-acetyleneketones s. ibid. 7612-8; **vinyl ketones** s. ibid. 7589-96; **enolethers** with Grignard compds. s. ibid. 7605-11.

n-*Butyllithium/zinc bromide/magnesium* BuLi/ZnBr$_2$/Mg
Syntheses via 2-alkoxy-3-ethyleneepoxides

385.

3-**Ene-1,2-diol 2-monoethers**. A soln. of N-(α-ethoxyallyl)benzotriazole in THF stirred at -78° under argon with ca. 1 eq. 2 *M* BuLi for 10 min, 0.88 eq. cyclopentanone added, stirring continued at -78° for 3 h then at 20° for 12 h, 1 eq. ZnBr$_2$ in THF injected into the soln., stirred for 5 h, the suspension treated with ca. 2 eqs. 3 *M* ethereal methylmagnesium iodide at 20° for 10 h, and quenched with water → 1-(1-ethoxy-1-methylallyl)cyclopentan-1-ol. Y 71%. This is the first example of a 3-ene-1,2-diol deriv. having two tertiary oxy-functions. F.e., reactions, and isolation of the intermediate s. A.R. Katritzky, J. Jiang, J. Org. Chem. *60*, 7597-604 (1995); 2-alkoxy-2-cyclopentenones from carboxylic acid esters s. ibid. 7605-11.

n-*Butyllithium/sulfuric acid* BuLi/H$_2$SO$_4$
Isoxazoles from ketoximes and hydroxamic acid esters

386.

2 eqs. 2.5 *M* n-BuLi in hexanes added dropwise over 5 min to a soln. of acetone oxime in THF at 0°, after 30 min 0.83 eq. N-methoxy-N-methylphenylacetamide in THF added dropwise over 20 min, after a further 30 min the mixture poured into 4:1 THF/water containing concd. H$_2$SO$_4$, and refluxed for 1 h → 3-methyl-5-(phenylmethyl)isoxazole. Y 94.3%. The procedure is superior to that based on O-ethyl-N,N-dibutylamide carbocation salts (cf. *26*, 792s*38*), and particularly useful for the preparation of **5-alkylisoxazoles**. F.e. and regio- and stereo-selectivity s. T.J. Nitz et al., J. Org. Chem. *59*, 5828-32 (1994).

Lithium diisopropylamide/trimethylsilyl chloride i-Pr$_2$NLi/Me$_3$SiCl
β-*prim*-**Aminocarboxylic acid esters** H → C–NH$_2$
from carboxylic acid esters and 1-α-(phosphoranylideneamino)benzotriazoles

387.

β-*prim*-**Amino-α-arylcarboxylic acid esters**. 1 eq. 1.5 M LDA in hexane added at -40° under argon to startg. ester in dry THF, stirred for 1 h, 1 eq. startg. phosphine imine in the same solvent added in one portion, stirred overnight, allowed to warm slowly to room temp., worked up, the resulting oil dissolved in methanol, hydrolyzed with water for 15 h at room temp., solvent removed, the residue dissolved in ether, treated sequentially with 2 eqs. methanol and 2 eqs. Me$_3$SiCl under argon, and stirred for 10 min → methyl 3-amino-2-phenylpropionate hydrochloride. Y 78%. Reaction is not applicable to α-alkyl derivs. F.e.s. A.R. Katritzky et al., Khim. Geterotsikl. Soedin. 1995, 1023-5 (Eng.).

1,8-Diazabicyclo[5.4.0]undec-7-ene or 1,1,3,3-Tetramethylguanidine ←
Pyrroles from 1,1-nitroethylene derivs. ○
s. *44*, 736; benzyl pyrrole-2-carboxylates s. D.H. Burns et al., Synth. Commun. *25*, 379-87 (1995); **also isoindole ring** from ar. nitro compds. s. N. Ono et al., J. Chem. Soc. Perkin Trans. I *1996*, 417-23; *in situ*-generation of substrates **from 2-acoxynitro compds.**, 2-cyanopyrroles, s. M. Adamczyk, R.E. Reddy, Tetrahedron *52*, 14689-700 (1996).

Potassium tetracyanocuprate(I) $K_3[Cu(CN)_4]$
Cuprous cyanide/tert-butyl nitrite $CuCN/t\text{-}BuONO$
Ar. nitriles from amines $NH_2 \rightarrow CN$
with CuCN cf. *3*, 666s*5*; with $K_3[Cu(CN)_4]$ s. N. Yonezawa et al., Synth. Commun. *26*, 1575-8 (1996); hindered compds. with CuCN/*t*-BuONO via aprotic diazotization s. A.G. Giumanini et al., Tetrahedron *52*, 7137-48 (1996).

Chiral copper(II) 2,2'-bis(Δ²-oxazolines) ←
Cuprous triflate/chiral 1,2-diamines ←
Asym. cyclopropanation with diazo compds. s. *23*, 819s*51* ▽

Cupric trifluoromethanesulfonate $Cu(OTf)_2$
Aziridine-2-carboxylic acid esters from aldimines s. *50*, 388

Magnesium or Organolithium compds. *Mg or RLi*
Ketones from hydroxamic acid esters $CON(OMe)R \rightarrow COR'$
s. *37*, 806; **polymer-based synthesis** s. T.Q. Dinh, R.W. Armstrong, Tetrahedron Letters *37*, 1161-4 (1996); peptidyl pentafluoroethyl ketones s. ibid. *33*, 3265-8 (1992); α,β-acetyleneketones s. I. Delamarche, P. Mosset, ibid. *34*, 2465-8 (1993); s.a. J.J. De Voss et al., J. Med. Chem. *37*, 665-73 (1994); α-chloroketones s. L.F. Frey et al., Synlett *1996*, 225-6.

Zeolite ←
β-Ketocarboxylic acid esters from aldehydes $CHO \rightarrow COCH_2COOR$
with Al_2O_3 cf. *45*, 462s*49*; with zeolites s. H.R. Sonawane et al., Synlett *1996*, 369-70.

Methylrhenium trioxide $MeReO_3$
Aziridine-2-carboxylic acid esters from aldimines

388. Ph\\⟶N-C₆H₁₃-n   N₂⟶CO₂Et   ⟶   Ph\\N(C₆H₁₃-n)⟶CO₂Et

A mixture of startg. ar. imine and ethyl diazoacetate (30 mmole scale) treated under anaerobic conditions with 3 mol% $MeReO_3$ until elimination of N_2 ceased → *trans*-product. Y 93%. F.e., **also glycidic acid esters from oxo compds.**, s. Z. Zhu, J.H. Espenson, J. Org. Chem. *60*, 7090-1 (1995); *cis*-isomers with $Cu(OTf)_2$ cf. K.G. Rasmussen, K.A. Jørgensen, Chem. Commun. *1995*, 1401-2.

meso-Tetra-p-tolylporphyrinatoiron(II) ←
Cyclopropanes from ethylene derivs. and diazo compds. ▽
with a cationic iron(II) carbonyl complex cf. *23*, 819s*49*; from terminal and 1,1-disubst. alkenes with *meso*-tetra-*p*-tolylporphyrinatoiron(II), stereo- and chemo-selectivity, s. L.K. Woo et al., J. Am. Chem. Soc. *117*, 9194-9 (1995).

Bis(triethoxysilyl)bis(triphenylphosphine)ruthenium(0) $Ru[Si(OEt)_3]_2(PPh_3)_2$
Stereospecific cyclopropanation of styrenes s. *48*, 784s*51*
Dichlorotris(triphenylphosphine)ruthenium(II) $RuCl_2(PPh_3)_3$
Indoles from ar. amines ○
and glycols cf. *41*, 732; and *triethanolamine* s. S.C. Shim et al., Synth. Commun. *26*, 1349-53 (1996).

Rhodium(II) acetate $Rh(OAc)_2$
Cyclopropane ring from diazo compds. and ethylene derivs. ▽
s. *23*, 819s*41*; glycal derivs. s. J.O. Hoberg, D.J. Claffey, Tetrahedron Letters *37*, 2533-6 (1996).

Indolo[2,3-a]carbazole ring

Ar = 3,4-(MeO)$_2$C$_6$H$_3$

from 2,2'-biindoles and α-diazoketones. A soln. of startg. diazo compd. and 2,2'-biindole in degassed pinacolone heated with 0.1 eq. Rh$_2$(OAc)$_4$ at 120° until reaction complete → product. Y 62%. The choice of solvent was critical for solubilization of the substrate and compatibility with the carbenoid chemistry. N-t-Butyl, N-PMB and N-benzyl groups were tolerated. F.e.s. J.L. Wood et al., J. Am. Chem. Soc. *117*, 10413-4 (1995); f. method **from two indole molecules** cf. M.M. Faul et al., Synthesis *1995*, 1511-6; s.a. M. Ohkubo et al., Tetrahedron *52*, 8099-112 (1996).

Rhodium(II) acetate/boron fluoride Rh(OAc)$_2$/BF$_3$
(E)-γ,δ-Ethylene-α-hydroxyketones from α-diazoketones and 2-ethylenealcohols ←
via asym. [3.3]-sigmatropic rearrangement-1,2-allyl group migration

One-pot procedure. A soln. of startg. α-diazoketone and (S)-but-1-en-3-ol in benzene heated in the presence of a little Rh$_2$(OAc)$_4$, BF$_3$-etherate added, and worked up after 2 h at 25° → (−)-product. Y 75% (e.e. 92%). F.e.s. J.L. Wood et al., J. Am. Chem. Soc. *117*, 10413-4 (1995).

Rhodium(II) acetate/camphor-derived 1,3-oxathiane ←
Asym. synthesis of epoxides from aldehydes and diazo compds.
with chiral thioethers s. *50*, 480; with a camphor-derived 1,3-oxathiane s. Tetrahedron:Asym. *6*, 2557-64 (1995).

Chiral rhodium(II) N-(arenesulfonyl)prolinates ←
Asym. cyclopropanation with diazo compds.
s. *23*, 819s*46,48-50*; solvent effect s. M.P. Doyle et al., Tetrahedron Letters *37*, 4129-32 (1996); substituent effects s. H.M.L. Davies et al., ibid. 4133-6; chiral cyclopropanecarboxylic acid esters with chiral Cu(II)-bis(Δ^2-oxazolines) (cf. *23*, 819s*50*) s. A.V. Bedekar, P.G. Andersson, ibid. 4073-6; with CuOTf/(1S,2S)-N,N'-di(mesitylmethyl)-1,2-diphenyl-1,2-ethanediamine cf. S. Kanemasa et al., ibid. *35*, 7985-8 (1994).

Palladium-carbon or Palladous acetate Pd-C or Pd(OAc)$_2$
Arylation of ethylene derivs. with diazonium salts C=CH → C=CAr
s. *38*, 793s*48*; 3-aryl-2,3-dihydrothiophene 1,1-dioxides s. S. Sengupta, S. Bhattacharyya, Synth. Commun. *26*, 231-6 (1996); cinnamic acid esters with Pd-C s. M. Beller, K. Kühlein, Synlett *1995*, 441-2; in aq. media cf. Tetrahedron Letters *36*, 4475-8 (1995).

Dichloro(p-cymene)osmium dimer ←
Cyclopropanes from ethylene derivs. and diazo compds.
with [(TTP)Os]$_2$ cf. *48*, 784; with dichloro(p-cymene)osmium dimer s. A. Demonceau et al., Tetrahedron Letters *37*, 1025-6 (1996); from styrenes stereoselectively with Ru-catalysts, e.g. Ru[Si(OEt)$_3$]$_2$(PPh$_3$)$_2$, s. ibid. *36*, 3519-22 (1995).

Via intermediates   *v.i.*
Trisubst. ethylene derivs. from diazo compds. and vinylmagnesium bromides   C=CHR
via generation of ruthenium π-allyl from carbene complexes

391.

A soln. of acetato(cyclopentadienyl)(triphenylphosphine)ruthenium(II) in toluene treated with 1 eq. diphenyldiazomethane, stirred at room temp. for 5 min, solvent replaced by acetone, ca. 1.3 eqs. Et$_3$N·HCl added, and stirring continued for 30 min → intermediate carbene complex (Y 62%), suspended in benzene, treated with 2 eqs. 0.75 M vinylmagnesium bromide in THF, and stirred for 45 min at room temp. → intermediate π-allyl complex (Y 63%; *exo:endo* 1.9:1), in benzene treated with acetic acid at room temp. until cleavage complete → product (Y 100%). F.e.s. T. Braun et al., J. Am. Chem. Soc. *117*, 7291-2 (1995).

Halogen ↑   CC ⇅ Hal

Lithium   *Li*
Ketones from carboxylic acids and chlorides   COOH → COR
s. *16*, 853; rapid procedure under ultrasonication s. J.L. Luche et al., Synlett *1995*, 459-60.

Potassium hydride/n-butyllithium   *KH/BuLi*
Synthesis of 1-acetylene-1-thioethers from mercaptans   S–C≡C–R
via 2,2-dichlorovinyl thioethers

392.

One-pot procedure. *n*-Octanethiol in THF added over 10 min to a well-stirred suspension of 1.5 eqs. oil-free KH in anhydrous THF under argon, the mixture cooled to -50° after H$_2$ evolution had ceased (15-120 min), 1.1 eqs. trichloroethylene in THF added dropwise over 5 min, followed by the addition of a little anhydrous methanol, allowed to warm to 20°, stirred until gas evolution ceased (ca. 1 h), re-cooled to -70°, treated over 15 min with 2.2 eqs. 2.5 M *n*-BuLi in hexanes, after 30 min the mixture warmed to -40° over 30 min, 3 eqs. methyl iodide in HMPA added dropwise over 10 min, stirred at 20° for 1 h, quenched slowly with a little methanol, then poured into satd. aq. NH$_4$Cl → product. Y 98%. The method is simple, efficient, versatile (being applicable to prim., sec., tert. and ar. mercaptans) and high-yielding. F.e. incl. ethynyl derivs. with methanol as electrophile **and silylethynyl thioethers** with Me$_3$SiCl, s. P. Nebois et al., J. Org. Chem. *60*, 7690-2 (1995).

Sodium hydroxide/benzyltriethylammonium chloride   *NaOH/BnEt$_3$NCl*
Cyclopropanes from 1,2-dihalides   ▽
cyanocyclopropanes with LiNH$_2$ cf. *27*, 841; 1-aryl-1-cyanocyclopropanes with NaOH and BnEt$_3$NCl as phase transfer catalyst s. M. Fedorynski, A. Jonczyk, Org. Prep. Proc. Intern. *27*, 355-9 (1995).

Sodium hydroxide/2-benzylidene-1,3-bis(triethylammonio)propane dichloride   ←
1,1-Dichlorocyclopropanes from ethylene derivs.
with added BnEt$_3$NCl cf. *27*, 833; with multi-site, water-soluble, 2-benzylidene-1,3-bis-(triethylammonio)propane dichloride s. T. Balakrishnan, J.P. Jayachandran, Synth. Commun. *25*, 3821-30 (1995).

Potassium hydroxide/potassium carbonate/(2S)-2-hydroxymethyl-1-methyl- ←
1-(diphenylmethyleneamino)pyrrolidinium iodide
Asym. 1-alkylation of azomethines H → R
s. *27*, 843s*47*; with KOH/K$_2$CO$_3$ and a little (2S)-2-hydroxymethyl-1-methyl-1-(diphenylmethylene amino)pyrrolidinium iodide s. J.J. Eddine, M. Cherqaoui, Tetrahedron:Asym. *6*, 1225-8 (1995).

Potassium tert-*butoxide* KOBu-t
Synthesis of α-*prim*-aminocarboxylic acids ←
via C-alkylation of α-[N-[bis(alkylthio)methylene]amino]carboxylic acid esters
s. *31*, 804; asym. synthesis with 8-phenylmenthyl esters s. C. Alvarez-Ibarra et al., J. Org. Chem. *60*, 7934-40 (1995); chiral N-protected α-deuterio-α-aminocarboxylic acids s. Y. Elemes, U. Ragnarsson, J. Chem. Soc. Perkin Trans. I *1996*, 537-40.

n-*Butyllithium* BuLi
Ketones from 1-α-alkoxybenzotriazoles s. *51*, 384 Hal → COR
Functionalized 2-vinylcyclopropanecarbonyl from α,β-ethylenecarbonyl compds. ▽
Asym. synthesis with addition of three C-atoms
via intramolecular 1,4-addition-alkylation

393.

*Tri*subst, *tri*functionalized cyclopropanes are obtainable by asym. cyclopropanation with a chiral 3-chloroallylphosphonic diamide. E: A soln. of startg. chiral *trans*-3-chloroallylphosphonic diamide in THF treated with *n*-BuLi at -78°, and the resulting anion mixed with 2-methyl-2-cyclopentenone → *endo,endo*-product. Y 90%. The *exo,endo*-product predominated with the corresponding *cis*-allylphosphonodiamide as substrate. Reaction is generally applicable to enones, enoates, and α,β-ethylene-lactones and -lactams, affording enantiomerically *pure* (or highly enriched) cyclopropanes which are amenable to further manipulation. F.e.s. S. Hanessian et al., J. Am. Chem. Soc. *117*, 10393-4 (1995).

n-*Butyllithium/N,N,N',N'-tetramethylethylenediamine/trimethyl borate/palladous acetate/* ←
triphenylphosphine
Styrenes from ar. bromides and trisylhydrazones C═C(Ar)
via *in situ*-generated boronic acid esters
Shapiro-Suzuki coupling

394.

Ar = 2,4,6-(i-Pr)$_3$C$_6$H$_2$

A suspension of 2-methylcyclohexanone (2,4,6-triisopropylphenylsulfonyl)hydrazone in 1:1 hexanes/TMEDA stirred at -78° for 10 min, treated with 3 eqs. 2.5 *M* n-BuLi in hexanes for 10 min, warmed to 0° (N$_2$ evolution), the mixture re-cooled to -78° after ca. 15 min, 2 eqs. B(OMe)$_3$ added, stirred at 0° for 1 h, hexanes removed *in vacuo*, toluene, 2 *M* Na$_2$CO$_3$, 4 mol% Pd(OAc)$_2$, 8 mol% PPh$_3$ and 1 eq. 1-bromonaphthalene added, and the mixture refluxed under N$_2$ for 14 h → 6-methyl-1-(1-naphthyl)cyclohex-1-ene. Y 55%. F.e.s. M.S. Passafaro, B.A. Keay, Tetrahedron Letters *37*, 429-32 (1996).

tert-*Butyllithium/zinc bromide* t-BuLi/ZnBr$_2$
Asym. C-α-alkylation of N-protected 3-glycyl-4-imidazolidones s. *51*, 398 H → R
Lithium diisopropylamide i-Pr$_2$NLi
Asym. α-alkylation of hydrazones
s. *31*, 812s*48*; of lactone hydrazones and conversion to **chiral lactones** s. D. Enders et al., Synthesis

1995, 947-51; chiral α-(organothio)aldehyde hydrazones s. Tetrahedron *50*, 3349-62 (1994); chiral protected α-hydroxyaldehyde hydrazones s. Synlett *1994*, 792-4.

Asym. α-alkylation of 3-acyl-2-oxazolidones H → R
s. *44*, 776s*46,50*; *47*, 784; of indan-fused 3-acyl-2-oxazolidones s. A. Sudo, K. Saigo, Tetrahedron:Asym. *6*, 2153-6 (1995); asym. α-cyanomethylation s. S. Azam et al., J. Chem. Soc. Perkin Trans. I *1996*, 621-7; asym. α-bromodifluoromethylation s. Y. Kobayashi et al., Tetrahedron Letters *36*, 3711-4 (1995); asym. alkylation of a polymer-based 3-acyl-2-oxazolidone s. S.M. Allin, S.J. Shuttleworth, ibid. *37*, 8023-6 (1996); asym. α-(carbalkoxydifluoromethylation) s. ibid. *35*, 7399-400 (1994).

Asym. Michael addition-alkylation with α-aminonitriles as acyl carbanion equivalents ←
Chiral γ-diketones from α,β-ethyleneketones

395.

A soln. of (4S,5S,R/S)-(+)-[N-(2,2-dimethyl-4-phenyl-1,3-dioxan-5-yl)-N-methylamino]phenylacetonitrile in THF treated with a soln. of 1.1 eqs. LDA in the same solvent at -78°, 1 eq. 2-cyclohexenone added after 1 h, stirred for 2 h, 1:1 HMPA/THF added via syringe during 30 min, treated with 1.1 eqs. benzyl bromide, allowed to warm to 0° overnight, and hydrolyzed with satd. NH₄Cl → crude (S,S,S,R,R,S)-adduct, in THF treated with 2 eqs. 2 N aq. AgNO₃, and stirred overnight → (S,R)-product. Y 70% (e.e. ≥ 98%). F.e.s. D. Enders et al., Synthesis *1995*, 659-66.

Geospecific synthesis of α-hydroxyketene mercaptals $C(OH)C=C(SR)_2$
from aldehydes, dithioacetic acid esters and halides

396.

Methyl dithioacetate in THF treated with LDA at -78° until deprotonation complete, 1 eq. acetaldehyde added, the mixture allowed to react for 1 h, 1.5 eqs. benzyl bromide added, stirred at -78° for 2 h, then quenched with NH₄Cl/H₂O → product. Y 67% (Z/E 99/1). F.e.s. S. Tchertchian, Y. Vallée, Tetrahedron Letters *36*, 6225-6 (1995).

Lithium diisopropylamide/sodium bis(trimethylsilyl)amide   i-$Pr_2NLi/(Me_3Si)_2NNa$
3-Ethylene-1-vinyllactones from cyclic α,β-ethyleneketones
via bicyclic β'-stannyl-γ-vinyl-γ-lactols

397.

2-Methyl-2-cyclohexenone treated with 2 eqs. LDA and 1 eq. tri-*n*-butyltin hydride at -78°, followed by the addition of (E)-1,4-dibromobut-2-ene, allowed to warm to room temp., 1.5 eqs. NaN(SiMe₃)₂ added, and the crude vinyl ether subjected to column chromatography → intermediate γ-lactol (Y 71%), in benzene treated with Pb(OAc)₄, and refluxed for 10-5 min → product (Y 84%; single isomer). The first step proceeds via Michael addition of LiSnBu₃, followed by α-alkylation-ring closure, the bulky tributylstannyl group dictating the stereoselectivity. F.e.s. D.W. Landrey et al., Synlett *1995*, 543-4; stereospecific synthesis of γ-acoxy-δ,ε-ethyleneketones from ketones via γ-vinyl-γ-lactols s. J. Org. Chem. *60*, 2668-9 (1995).

Lithium diisopropylamide/lithium chloride   i-Pr_2NLi/LiCl
Asym. synthesis of α-aminocarboxylic acids via C-alkylation of chiral glycinamides   H → R

398.

A wide range of (D)- or (L)-α-amino acids is now available from glycinamides based on inexpensive (S,S)- or (R,R)-pseudoephedrine, which is readily removable for reuse. **E**: A soln. of (S,S)-pseudoephedrine glycinamide in THF added to a slurry of 6 eqs. LiCl and *1.95 eqs.* LDA in the same solvent at -78°, warmed to 0° for 20 min (to generate the (Z)-enolate), ethyl iodide added, and worked up after 1.5 h → intermediate α-alkylated amide (Y 76%; d.e. ≥ 99% after recrystallization), in pure water refluxed for 10 h (without added base) → product (Y 88%; e.e. ≥99%). Many of the formed amides are crystalline compds. so that purification can be readily achieved. Alkaline hydrolysis (NaOH) in water or aq. methanol is recommended if the resulting free amino acid requires subsequent *in situ* N-protection (e.g. with Boc_2O or FmocCl). Slightly less than 2 eqs. LDA is essential in order to avoid deleterious side reactions. F.e. incl. cyclic analogs s. A.G. Myers et al., J. Am. Chem. Soc. *117*, 8488-9 (1995); asym. synthesis of N-protected α-aminocarboxylic acids via asym. C-α-alkylation of N-protected 3-glycyl-4-imidazolidones (with *t*-BuLi/$ZnBr_2$) cf. D. Seebach et al., Helv. Chim. Acta *78*, 1185-206 (1995).

Lithium bis(trimethylsilyl)amide $(Me_3Si)_2$NLi
**Asym. synthesis of carboxylic acids
via C-α-alkylation of 2-acyl-1,3-dithiane 1-oxides**

399.

Chiral α-arylcarboxylic acids. 1.1 eqs. LiN$(SiMe_3)_2$ in THF added to a stirred soln. of (1R,2R)-2-ethyl-*anti*-2-[2-(1-naphthyl)acetyl]-1,3-dithiane 1-oxide in the same solvent at -78°, after 20 min, 1.5 eqs. MeI added, allowed to warm to 25° over 17 h, and poured into satd. aq. NH_4Cl → (1R,2R)-2-ethyl-*anti*-2-[2S-(1-naphthyl)propanoyl]-1,3-dithiane 1-oxide (Y 80%; single diastereomer), in acetone added to a stirred soln. of 8 eqs. NBS in 97:3 acetone/water at 0°, and stirred for 30 min → (2S)-2-(1-naphthyl)hexane-3,4-dione (Y 81%), in methanol treated dropwise with 2 eqs. aq. $NaIO_4$ while stirring at room temp., and worked up after 12 h → (2S)-2-(1-naphthyl)propanoic acid (Y 68%; e.e. 87%). Remarkably, cleavage of the auxiliary takes place with retention of chirality. F.e.s. P.C. Bulman Page et al., J. Chem. Soc. Perkin Trans. I *1995*, 2673-6.

Chiral lithium amide/lithium chloride   ←
Asym. benzylic metalation of tricarbonylchromium-complexed benzyl ethers

400.

A soln. of ca. 1 eq. LiCl in THF added via cannula to a soln. of the chiral dilithium diamide [prepared by treating 1.1 eqs. of the corresponding diamine in THF at -78° with 2.2 eqs. 1.6 *M* BuLi in hexanes, warming to room temp. with stirring, then re-cooling to -78°], followed by addition of (benzyl methyl ether)tricarbonylchromium(0) in THF via cannula over ca. 2 min, stirring continued for ca. 20 min, methanol added, and the mixture allowed to warm to room temp. → product. Y 96% (e.e. 97%). F.e.s. S.E. Gibson et al., Chem. Commun. *1996*, 839-40; **asym. *o*-lithiation** of tricarbonylchromium-complexed benzaldehyde via *in situ*-protection as chiral lithium 1-aminoalkoxides s. A. Alexakis et al., Tetrahedron:Asym. *6*, 2135-8 (1995); of complexed chiral ar. aldehyde hydrazones cf. E.P. Kündig et al., Helv. Chim. Acta *75*, 2657-60 (1992); of complexed chiral ketal analogs s. J. Aubé et al., J. Org. Chem. *57*, 3563-70 (1992); **asym. 1,4-addition-alkylation** of tricarbonylchromium-complexed 2-aryl-Δ^2-oxazolines in the presence of *chiral diethers* s. D. Amurrio et al., J. Org. Chem. *61*, 2258-9 (1996).

Potassium bis(trimethylsilyl)amide $(Me_3Si)_2NK$
Asym. α-alkylation of methyl (S)-N-benzyl-N-methylalaninate H → R
via borane complexation

401.

A cooled soln. of startg. borane-amino ester adduct [prepared by stirring methyl (S)-N-benzyl-N-methylalaninate with 0.95 eq. BH$_3$-SMe$_2$ in *hexane* at 20° for 24 h, and collecting the solid] in toluene added to a soln. of 1.17 eqs. KN(SiMe$_3$)$_2$ in toluene/THF at -78°, the mixture stirred at -23° for 0.25 h then re-cooled to -78°, 1.48 eqs. benzyl bromide added via syringe, the mixture stirred at -78° for 2 h, then allowed to warm to 0° over 2 h, and hydrolyzed with satd. aq. NH$_4$Cl → (S)-product. Y 78% (e.e. 82%, corrected for 90% optical purity of startg. ester). The α-chirality of the substrate is temporarily 'stored' in the form of an asymmetric quaternary nitrogen centre which effectively directs the enolate alkylation enantioselectively (cf. *50*, 502). F.e. and with LDA/HMPA for unactivated halides s. C. Mioskowski et al., Angew. Chem. Intern. Ed. *35*, 430-2 (1996).

Potassium carbonate-alumina s. under ZnCl$_2$-SiO$_2$ K_2CO_3-Al_2O_3
Triethylamine Et_3N
2-Azetidinones from azomethines with asym. induction
s. *7*, 836s*46-9*; chiral cis-2-azetidinones based on 3-carene s. B.M. Bhawal et al., Tetrahedron 52, 3741-56 (1996); chiral *cis*-3-amino-derivs. from a chiral sultam-based ketene s. ibid. 5579-84; chiral 3-aryloxy-derivs. s. Y. Hashimoto et al., Tetrahedron Letters *36*, 8821-4 (1995); chiral 4-carbalkoxy-derivs. s. C. Palomo et al., ibid. *33*, 4823-6 (1992).
2-Azetidinones from enolizable aldimines

402.

4-Alkyl-2-azetidinones can now be obtained stereospecifically by the classical [2+2]-route (*7*, 836) **from N-[bis(trimethylsilyl)methyl]aldimines** which, unlike simple enolizable aldimines, are stable and do not undergo competitive deprotonation. **E: Chiral 3-amino-2-azetidinones**. Triethylamine and a soln. of 2 eqs. startg. acyl chloride in dry chloroform added dropwise at 0° to a soln. of startg. crude N-[bis(trimethylsilyl)methyl]aldimine in the same solvent containing 4 Å molecular sieves under N$_2$, and refluxed for ca. 20 h → product. Y 55% (*cis:trans* >98:2). The silyl groups stabilize the adjacent electron-deficient carbanionic site of the aldimine. The corresponding N-benzyl- and N-(4-methoxyphenyl)-aldimine afford enamides or hydrolysis products. F.e. and cleavage of the protective group s. J.M. Aizpurua et al., Angew. Chem. Intern. Ed. *35*, 1239-41 (1996).

1,8-Diazabicyclo[5.4.0]undec-7-ene/lithium iodide DBU/LiI
C-Alkylation H → R
s. *31*, 819s*33*; of α-subst. β-diketones with added LiI s. K. Fuji et al., Synthesis *1995*, 1069-70; with hydrotalcite cf. C. Cativiela et al., Synth. Commun. *25*, 1745-50 (1995).

Copper/cupric perchlorate $Cu/Cu(ClO_4)_2$
Ar. allylation   H → C–C=C
o-allylation of phenols with Ag_2O cf. *22*, 844; with $Cu/Cu(ClO_4)_2$, also o-allylation of anisole, s. J.B. Baruah, Tetrahedron Letters *36*, 8509-12 (1995); ar. monoallylation with $ZnCl_2$-SiO_2/K_2CO_3-Al_2O_3 cf. M. Kodomari et al., Chem. Commun. *1995*, 1895-6.

Dilithium trichlorocuprate(I) s. under Mg Li_2CuCl_3
Cuprous chloride s.a. under Cp_2ZrBu_2   CuCl
Cuprous chloride/triethylamine/aniline/hydrogen chloride   ←
β-Diketones from terminal acetylene derivs. and carboxylic acid chlorides $COCH_2CO$
via α,β-acetyleneketones and β-amino-α,β-ethyleneketones

403.

One-pot conversion. 1.75 eqs. Et_3N added to 0.15 eq. CuCl in benzene, stirred for 10 min under N_2, 1.5 eqs. dimethylethynylcarbinol methyl ether added, heated at 55-60° for 30 min, 1.5 eqs. 2-methylhexanoyl chloride added, left for 1 h, washed with dil. HCl and water, the benzene layer concentrated, diluted with methanol, 1.5 eqs. aniline added, stirred at 60° for 2 h, and hydrolyzed with 25% HCl → 2,6-dimethyl-2-methoxydecane-3,5-dione. Y 80%. F.e. and with isolation of the intermediates s. I.E. Sokolov et al., Izv. Akad. Nauk Ser. Khim. *1995*, 710-4 (Russ.).

Cuprous iodide s. under Zn, Pd-C and $Pd(OAc)_2$ CuI
Cupric chloride/alumina $CuCl_2/Al_2O_3$
Diarylmethanes from arenes $ArCH_2Ar'$
with montmorillonite cf. *43*, 703s*46*; with $CuCl_2$-Al_2O_3 s. M. Kodomari et al., Nippon Kagaku Kaishi *1994*, 1137-9 (Jap.); with $ZnCl_2$/montmorillonite cf. J.J. Vanden Eynde et al., Tetrahedron Letters *36*, 3133-6 (1995).

Magnesium   Mg
Carboxylic acid derivs. from Grignard compds.   RMgHal → RC(O,S)-
Amides, thioamides and thiolic acid esters s. *51*, 297

Magnesium/dilithium trichlorocuprate(I) Mg/Li_2CuCl_3
Dimerization of halides   2 RHal → R–R
with $Mg/BH_3/AgNO_3$ cf. *28*, 814; of mixed dihalides with Mg/Li_2CuCl_3, selectivity, s. D.K. Johnson et al., Tetrahedron Letters *36*, 8565-8 (1995).

Magnesium/1,2-dibromotetrafluoroethane $Mg/CBrF_2CBrF_2$
Grignard synthesis with phenolethers   ArOR → Ar–Ar'
diaryls s. *24*, 839; *44*, 702; 7-arylindolines with added $CBrF_2CBrF_2$ s. R.H. Hutchings, A.I. Meyers, J. Org. Chem. *61*, 1004-13 (1996).

Magnesium/N-chlorosuccinimide   Mg/NCS
4-Arylation of pyridines   H → Ar
s. *37*, 659s*38*; by Grignard addition to 3-pyridylcarboxamides in the presence of NCS cf. J. Mulzer et al., Tetrahedron *51*, 9531-42 (1995).

Magnesium/dichloro[1,2-bis(diphenylphosphino)ethane]nickel(II) $Mg/Ni(dppe)Cl_2$
Ketones from carboxylic acid chlorides   COCl → COR
with $Mg/Fe(acac)_3$ cf. *40*, 563; with $Ni(dppe)Cl_2$ s. C. Malanga et al., Tetrahedron Letters *36*, 9185-8 (1995).

Zinc (s.a. under $TiCl_4$)   Zn
β-ketoesters cf. *10*, 625; β,γ-ethyleneketones from β,γ-ethylenebromides under ultrasonication s. B.C. Ranu et al., Tetrahedron Letters *37*, 1109-112 (1996).

β-Keto- from α-halogeno-carboxylic acid esters   Hal → COR
and 3-acyl-2-oxazolidones cf. *48*, 808; and 1-acylpyrazoles, chiral γ-carbalkoxyamino-β-ketoesters s. J. Heterocyc. Chem. *32*, 723-5 (1995).

Zinc or Zinc,copper Zn or Zn,Cu
[2+2]-Cycloaddition with dichloroketene
2,2-dichlorocyclobutanone ring cf. 34, 811; chiral γ-siloxy- and γ-carbalkoxyamino-α,α-dichloro-β-lactones with asym. induction s. C. Palomo et al., Chem. Commun. 1995, 1735-6; cyclobutenediones from acetylene derivs. via 4,4-dichlorocyclobutenones with Zn under ultrasonication s. M.S.A. Parker, C.J. Rizzo, Synth. Commun. 25, 2781-9 (1995).

Zinc/cuprous iodide Zn/CuI
Aldehydes from halides via radical carbonylation   Hal → CHO
with Bu_3SnH/AIBN cf. 45, 488; from iodides with Zn/CuI or Sm/CuI s. Synlett 1995, 1249-51.

Zinc/cobaltous bromide $Zn/CoBr_2$
Sym. ketones from organozinc halides   2 RZnHal → R_2CO

404.   2 $Cl(CH_2)_4ZnI$ $\xrightarrow{CO}$ $Cl(CH_2)_4CO(CH_2)_4Cl$

Sym. functionalized ketones. A soln. of 4-chlorobutylzinc iodide (prepared from 4-iodochlorobutane and 2 eqs. Zn in THF) added dropwise to 0.75 eq. $CoBr_2$ in N-methyl-2-pyrrolidone at 0° while CO was bubbled slowly through the mixture, stirred for 3 h at room temp. with continued CO bubbling and for 2 h without, poured into hexane, and stirred for a further 2 h → bis(4-chlorobutyl) ketone. Y 56%. Reaction is generally applicable to alkyl-, benzyl- and arylzinc halides, and tolerates ester and chloride substitution (not cyano). Surprisingly, there was no homo-coupling. F.e. incl. **cyclic ketones** from bis(organozinc halides) s. A. Devasagayaraj, P. Knochel, Tetrahedron Letters 36, 8411-4 (1995).

Dialkylzinc s. under $Ni(acac)_2$ R_2Zn
Diethylzinc Et_2Zn
Simmons-Smith cyclopropanation with asym. induction
s. 41, 797s47; 47, 806s49; of glycals s. J.O. Hoberg, J.J. Bozell, Tetrahedron Letters 36, 6831-4 (1995); stereospecific cyclopropanation of racemic 2-ethylene-5-hydroxysilanes s. P. Mohr, ibid. 7221-4.

Diethylzinc/chiral 1,2-di(sulfonylamino)cyclohexanes or 1,3,2-dioxaborolane-4,5-dicarboxamides or (aR)-N,N,N',N'-tetraethyl-1,1'-bi-2-naphthol-3,3'-dicarboxamide   ←
Asym. Simmons-Smith cyclopropanation
s. 47, 806s47-9; of allyl alcohols with chiral 1,2-di(sulfonylamino)cyclohexanes s. S.E. Denmark et al., Tetrahedron Letters 36, 2215-8, 2219-22 (1995); of 3-hydroxyenestannanes s. S. Kobayashi et al., ibid. 35, 7045-8 (1994); with chiral 1,3,2-dioxaborolane-4,5-dicarboxamides cf. A.B. Charette, H. Juteau, J. Am. Chem. Soc. 116, 2651-2 (1994); s.a. J. Org. Chem. 60, 1081-3 (1995); with added (aR)-N,N,N',N'-tetraethyl-1,1'-bi-2-naphthol-3,3'-dicarboxamide s. T. Katsuki et al., Chem. Lett. 1995, 1113-4.

Zinc chloride s.a. under Cp_2ZrBu_2 and $Pd(PPh_3)_4$ $ZnCl_2$
Zinc chloride/montmorillonite
Diarylmethanes from arenes s. 43, 703s51 $ArCH_2Hal → ArCH_2Ar'$
Zinc chloride-silica/potassium carbonate-alumina   ←
Ar. monoallylation with β,γ-ethylenechlorides s. 22, 844s51   H → C–C=C
Diisobutylaluminum hydride/tetrakis(triphenylphosphine)nickel(0) i-$Bu_2AlH/Ni(PPh_3)_4$
(E)-Allylarenes from benzyl chlorides and acetylene derivs.   Cl → C=CH

405. $\equiv\!\!-\!\!\diagdown_{C_5H_{11}\text{-}n}$ $\xrightarrow{i\text{-}Bu_2AlH}$ $[i\text{-}Bu_2Al\!-\!\!\diagup\!\!=\!\!\diagdown C_5H_{11}\text{-}n]$ → $t\text{-}Bu\!-\!C_6H_4\!-\!CH_2\!-\!\diagup\!\!=\!\!\diagdown C_5H_{11}\text{-}n$

via hydroalumination. 1.2 eqs. i-Bu_2AlH added dropwise under argon to a soln. of startg. alkyne in dry hexane at room temp., heated with stirring at 50°, cooled to room temp. when hydroalumination complete (6-8 h), 50% of the solvent removed *in vacuo*, a soln. of startg. benzyl chloride and 5 mol% freshly prepared $Ni(PPh_3)_4$ in dry THF (preliminarily stirred for 5-10 min) added via cannula with cooling in ice, allowed to warm to room temp., and stirred until reaction complete by GC (2-6 h) → product. Y 92% (E:Z 96:4). The electronic nature of the benzyl chloride is of little consequence, and a hydroxyl group in the alkyne can be accomodated (with 2 eqs. i-Bu$_2$AlH). Reaction **also** takes place **via hydrozirconation** (using Cp$_2$ZrHCl) with incorporation of a greater variety of functionality (e.g. an ester group in the halide). F.e. and from internal alkynes s. B.H. Lipshutz et al., Tetrahedron *52*, 7265-76 (1996).

Triethylborane/oxygen Et_3B/O_2
Synthesis of epoxides from 2-ethylenemonoperoxyketals via radical ring closure

406.

A 1 *M* soln. of Et$_3$B in hexane added dropwise to a soln. of startg. peroxyketal and 1 eq. methyl iodoacetate in benzene at 20°, and worked up when GC indicated completion of reaction → product. Y 72%. The key step in the conversion is the generation of an alkyl radical by iodine abstraction with methyl radical, itself produced via β-elimination of 1-methoxy-1-methylethoxy radical from the peroxyketal. The unsaturation may be electron-rich or electron-poor. F.e. and with AIBN as initiator s. F. Ramon et al., J. Org. Chem. *61*, 2071-4 (1996).

Trimethyl borate s. under BuLi $(MeO)_3B$
Montmorillonite s. under ZnCl$_2$ ←
Hydrotalcite ←
α-Alkylation of β-diketones s. *31*, 819s*51* H → R
*Samarium/*tert-*butyldimethylsilyl chloride* Sm/t-$BuMe_2SiCl$
Samarium-mediated Simmons-Smith cyclopropanation
with Sm,Hg cf. *42*, 822s*47,50*; improved procedure for cyclopropanation of 2-ethylene- and 2-allene-alcohols with Sm/*t*-BuMe$_2$SiCl s. J. Org. Chem. *61*, 2210-4 (1996).

Scandium(III) or Hafnium(IV) triflate $Sc(OTf)_3$ or $Hf(OTf)_4$
o-**Acylation of phenols via Fries rearrangement** ArH → ArCOR

407.

A soln. of 1-naphthol, 5 mol% Sc(OTf)$_3$, and 1.1 eqs. acetyl chloride in 6.7:1 toluene/nitromethane stirred at 100° for 6 h, and quenched with water → 1-(2-hydroxynaphthyl) methyl ketone. Y 93%. The method is mild and superior to the standard Fries rearrangement since the product can be obtained directly from the phenol without isolation of the ester. F.e.s. S. Kobayashi et al., Synlett *1995*, 1153-4; from the intermediate s. Chem. Commun. *1995*, 1527-8; using Hf(OTf)$_4$ with or without isolation of the intermediate, cf. Tetrahedron Letters *37*, 2053-6 (1996); Fries rearrangement with AlCl$_3$ under microwave irradiation (cf. *4*, 676) s. V. Sridar, V.S.S. Rao, Indian J. Chem. *33B*, 184-5 (1994); with ZrCl$_4$ at room temp. cf. D.C. Harrowven, R.F. Dainty, Tetrahedron Letters *37*, 7659-60 (1996).

Samarium diiodide SmI_2
Cyclopentanones from cyclobutanones under neutral conditions

408.

1 eq. Diiodomethane and 3-phenylcyclobutanone added to a soln. of 2.1 eqs. 0.1 *M* SmI$_2$ in THF at room temp., stirred for 15 h, and worked up → product. Y 61%. Extended reaction times were necessary to ensure complete conversion of the cyclobutanone. F.e. and regioselectivity, also with isolation of the intermediate iodohydrin, s. S. Fukuzawa, T. Tsuchimoto, Tetrahedron Letters *36*, 5937-8 (1995).

Samarium diiodide/hexamethylphosphoramide $SmI_2/(Me_2N)_3PO$
Samarium-mediated syntheses with α-bromocarboxylic acid esters ←
β-ketoesters cf. *50*, 515; sym. succinic acid esters with added HMPA s. É. Balaux, R. Ruel, Tetrahedron Letters *37*, 801-4 (1996).

N-Chlorosuccinimide s. under Mg NCS
Di-n-butylzirconocene/cuprous chloride/N,N′-dimethyl-N,N′-propyleneurea ←
Benzene ring from *o*-dihalides and zirconacyclopentadienes ○

409.

Naphthalenes. 2.1 eqs. CuCl, *3 eqs. DMPU*, and 1 eq. *o*-diiodobenzene added at room temp. to a soln. of startg. zirconacyclopentadiene (prepared from Cp_2ZrBu_2 and 2 eqs. 3-hexyne in THF), and stirred at 50° for 2 h → product. Y 70%. *o*-Bromoiodides may also be used, but *o*-dibromides were unreactive. So, too, were α-silyl-substituted zirconacyclopentadienes. Significantly, there was no reaction in the absence of DMPU, and the proportion of the latter (2.5-3 eqs. being optimum) was critical. F.e. incl. an anthracene and benzo[*b*]thiophene deriv. s. T. Takahashi et al., J. Am. Chem. Soc. *118*, 5154-5 (1996).

Di-n-butylzirconocene/zinc chloride/tetrakis(triphenylphosphine)palladium(0) ←
Vinylzirconocene halides from α,β-ethylenehalides ←

410.

Cross-coupling. A soln. of startg. vinyl chloride treated with 1 eq. Cp_2ZrBu_2 [in ether] at room temp., $ZnCl_2$ and a little $Pd(PPh_3)_4$ added, and allowed to react with iodobenzene → product. Y 78%. There was no isomerization of the intermediate to the more stable *trans*-1,2-disubst. vinylzirconocene (as takes place on hydrozirconation of terminal alkynes). Vinyl bromides reacted similarly, but addition of aryl halides failed. F.e. and electrophiles, also isolation of the intermediates, s. T. Takahashi et al., J. Am. Chem. Soc. *117*, 11039-40 (1995).

Hafnium(IV) triflate s.a. under Sc(OTf)₃ $Hf(OTf)_4$
Hafnium(IV) triflate/lithium perchlorate $Hf(OTf)_4/LiClO_4$
Friedel-Crafts reactions ←
acylation s. *50*, 466; alkylation s. Bull. Chem. Soc. Japan *68*, 2053-60 (1995).

Titanium tetrachloride $TiCl_4$
Ene reaction with thionium ions
s. *40*, 575; **with selenonium ions** generated from 1,1-chloroselenides with $TiCl_4$, α-arylseleno-γ,δ-ethylenecarboxylic acid esters, s. C.C. Silveira et al., Synthesis *1995*, 1305-10.

Titanium tetrachloride/zinc/N,N,N′,N′-tetramethylethylenediamine $TiCl_4/Zn/TMEDA$
Alkoxycyclopropanes from carboxylic acid esters ▽

411.

A soln. of startg. azulene in THF stirred for 3 h at 25° with 2.2 eqs. *methylene bromide*, 4 eqs. $TiCl_4$, 9 eqs. Zn, and 8 eqs. TMEDA, and poured on ice/2 *N* NaOH → 2-(1-methoxycyclopropyl)-1-methylazulene. Y 80%. F.e.s. R.-A. Fallahpour, H.-J. Hansen, Helv. Chim. Acta *77*, 2297-302 (1994).

Hexamethyldistannane s. under Pd(PPh₃)₄ $(Me_3Sn)_2$

Tri-n-butylstibine Bu_3Sb
α,β-Ethylenecarboxylic acid esters from oxo compds.   CO → C=C–COOR
and α-bromoesters cf. *41*, 812; alkylidenemalonic from α,α-dibromomalonic acid esters and ketones s. A.P. Davis, K.M. Bhattarai, Tetrahedron *51*, 8033-42 (1995).

Lithium perchlorate s. under Hf(OTf)$_4$ $LiClO_4$
Cobaltous bromide s. under Zn $CoBr_2$
Nickel(II) acetoacetonate/dialkylzinc $Ni(acac)_2/R_2Zn$
Nickel-catalyzed cross-coupling of ethylene-*prim*-iodides with dialkylzinc compds.   I → R

412.

The presence of a remote double bond in a prim. iodide permits selective Ni-catalyzed cross-coupling reactions between sp^3 C centres, intramolecular complexation of the double bond on the nickel centre favouring reductive elimination rather than transmetalation to organozinc compd. **E:** THF, N-methyl-2-pyrrolidone, and startg. alkenyl iodide added successively by syringe to 7.5 mol% Ni(acac)$_2$ at -40° under argon, the mixture cooled to -78°, a soln. of 2 eqs. bis(5-acetoxypentyl)zinc in THF added slowly (final ratio of THF:NMP 3:1), the mixture allowed to warm to -35°, stirred for 15 h at this temp., and excess of dialkylzinc quenched with satd. aq. NH$_4$Cl → product. Y 78%. Bulky dialkylzincs or poorly coordinating alkene groups, however, favour transmetalation. F.e.s. P. Knochel et al., Angew. Chem. Intern. Ed. *34*, 2723-5 (1995).

Tetrakis(triphenylphosphine)nickel(0) s. under i-*Bu*$_2$*AlH* $Ni(PPh_3)_4$
Dichloro[1,2-bis(diphenylphosphino)ethane]nickel(II) s. under Mg $Ni(dppe)Cl_2$
Palladium-carbon/triphenylphosphine/cuprous iodide/potassium carbonate   ←
Arylacetylenes from ar. halides   C≡CH → C≡CAr
with Et$_3$N as base cf. *46*, 798; with K$_2$CO$_3$ in aq. medium, also hetarylacetylenes, and from triflates, s. L. Bleicher, N.D.P. Cosford, Synlett *1995*, 1115-6; **tolans** with Pd(OAc)$_2$ and a little hexadecyltrimethylammonium bromide in an aq. organic emulsion s. D.V. Davydov, I.P. Beletskaya, Izv. Akad. Nauk Ser. Khim. *1995*, 995 (Russ.); **sym. tolans** with acetylene gas (using PdCl$_2$(PPh$_3$)$_2$) s. M. Pal, N.G. Kundu, J. Chem. Soc. Perkin Trans. I *1996*, 449-51.

Palladous acetate s.a. under BuLi $Pd(OAc)_2$
Palladous acetate/sodium hydrogen carbonate $Pd(OAc)_2/NaHCO_3$
Cross-coupling of terminal acetylene derivs. with unsatd. iodonium salts   C≡CH → C≡CR
1,3-diynes with CuCN/*n*-BuLi cf. *47*, 797; also arylacetylenes and 1,3-enynes with Pd(OAc)$_2$/NaHCO$_3$ in aq. medium under mild conditions s. S.-K. Kang et al., Chem. Commun. *1996*, 835-6; arylacetylenes with PdCl$_2$(PPh$_3$)$_2$/CuI/Bu$_3$N/K$_2$CO$_3$ s. N.A. Bumagin et al., Izv. Akad. Nauk Ser. Khim. *1995*, 789 (Russ.).

Palladous acetate/potassium carbonate $Pd(OAc)_2/K_2CO_3$
Carboxylic acid esters from halides by carbonylation   Hal → COOR
β,γ-ethylenecarboxylic acid esters with Na$_2$PdCl$_4$ cf. *12*, 867s48; with Pd(OAc)$_2$/K$_2$CO$_3$ at atm. pressure under heterogeneous conditions s. J. Kiji et al., Bull. Chem. Soc. Japan *69*, 1029-31 (1996); α,β-acetylenecarboxylic acid esters with PdBr$_2$/LiBr/Et$_3$N cf. O.N. Temkin et al., Mendeleev Commun. *1995*, 3-4.

*Palladous acetate/cuprous iodide/potassium carbonate/hexadecyl-
   trimethylammonium bromide*   ←
Tolans from ar. iodides s. *46*, 798s51   ArC≡CH → ArC≡CAr'
Palladous acetate/triphenylphosphine/triethylamine $Pd(OAc)_2/Ph_3P/Et_3N$
Effect of high pressure on catalyst turnover in palladium-catalyzed coupling   ←

413.

On homogeneous coupling in the presence of palladium phosphine complexes, high pressure stabilizes the coordination between metal and phosphine ligand thereby increasing the life-time of the catalyst. **E**: A soln. of iodobenzene, 3 eqs. 2,3-dihydrofuran, 3 eqs. Et$_3$N, and 1:2 Pd(OAc)$_2$/ PPh$_3$ (0.01 mol%) in 1:1 THF/acetonitrile heated at 60° for 36 h under 8 kbar → product. Y 71% (turnover number ≥ 10,000). The turnover number was 280 at 1 atm. after 12 h (Y 2.8%). F.e. and temperature dependence s. S. Hillers et al., J. Am. Chem. Soc. *118*, 2087-8 (1996).

Palladous acetate/tris(m-sulfophenyl)phosphine trisodium salt/triethylamine ←
1,3-Diynes from α,β-acetylenehalides and terminal acetylene derivs. C≡C–C≡C
with CuCl/EtNH$_2$ cf. *15*, 634; in aq. organic media with Pd(OAc)$_2$/water-soluble triarylphosphine s. J.P. Genêt et al., J. Org. Chem. *60*, 6829-39 (1995).

Benzofurans from *o*-iodophenols and terminal acetylene derivs. ○
with PdCl$_2$(PPh$_3$)$_2$/CuI/Et$_3$N cf. *47*, 839; in aq. organic media with Pd(OAc)$_2$/Et$_3$N and water-soluble triarylphosphine, **also N-acylindoles** (cf. *47*, 829), s. J.P. Genêt et al., J. Org. Chem. *60*, 6829-39 (1995).

Tetrakis(triphenylphosphine)palladium(0) s.a. under Cp$_2$ZrBu$_2$ Pd(PPh$_3$)$_4$
Tetrakis(triphenylphosphine)palladium(0)/potassium carbonate   Pd(PPh$_3$)$_4$/K$_2$CO$_3$
Benzocyclobutenes from ar. bromides and ethylene derivs. □

414.

A soln. of 1.2 eqs. startg. ethylene and 2,3-dimethylbromobenzene in DMF added to a mixture of 5 mol% Pd(PPh$_3$)$_4$ and 1.2 eqs. K$_2$CO$_3$ under N$_2$, and stirred at 105° for 12-24 h → 5,6-dimethyl-1,2,3,4,4a,8b-hexahydro-1,4-methanobiphenylene. Y 96%. The method is mild and practical. The reaction is prone to steric effects and subject to the electronic nature of the substituent on the ar. ring. F.e.s. M. Catellani, L. Ferioli, Synthesis *1996*, 769-72.

Tetrakis(triphenylphosphine)palladium(0)/potassium carbonate/tert-butyl isocyanide ←
Ring closures via intramolecular oxypalladation of 4-allenealcohols ○
s. *48*, 831; improved procedure with Pd(PPh$_3$)$_4$/K$_2$CO$_3$ and added *t*-BuNC, 2-vinyltetrahydrofurans and aryl 1-(tetrahydrofuran-2-yl)vinyl ketones via carbonylation s. Tetrahedron Letters *36*, 3805-8 (1995).

Tetrakis(triphenylphosphine)palladium(0)/lithium chloride/hexamethyldistannane ←
Palladium-catalyzed ar. cross-coupling of aryl triflates with ar. bromides   Ar–Ar'
via *in situ*-Stille coupling

415.

Arylpyridines. 1 eq. Hexamethyldistannane and anhydrous dioxane added sequentially to equivalent amounts of 4-bromoacetophenone and 2-pyridyl triflate, 5 mol% Pd(PPh$_3$)$_4$, and 3 eqs. LiCl under N$_2$ in a dry atmosphere, heated to reflux for 16 h under N$_2$, cooled, poured into satd. aq. KF and ethyl acetate, and vigorously stirred for 2 h → product. Y 68%. Reaction is made possible by the differing reactivity of the two electrophiles, the pyridyl triflate reacting initially with the distannane to give the corresponding pyridylstannane, which then undergoes Stille coupling with the halide in the presence of the same catalyst. The procedure is of potential where the stability of the stannane is an issue. F.e. and with hetaryl halides s. S.A. Hitchcock et al., Tetrahedron Letters *36*, 9085-8 (1995).

Tetrakis(triphenylphosphine)palladium(0)/zinc chloride/triethylamine ←
Ring closures of *o*-iodocarboxylic acids with terminal acetylene derivs. ○
3-alkylidenephthalides with added CuI cf. *48*, 839; **isocoumarins** with added ZnCl$_2$ s. H.-Y. Liao, C.-H. Cheng, J. Org. Chem. *60*, 3711-6 (1995).

trans-*Di(m-acetato)bis[o-(di-o-tolylphosphino)benzyl]dipalladium(II)/tetra*-n-*butylammonium bromide/sodium acetate*
Heck reaction with palladacyclics as catalyst ←

416.

Structurally defined and easy-to-handle palladacyclics surpass all previously known catalysts of the Heck reaction in terms of thermal stability and lifetime, and facilitate coupling with both ar. bromides and normally unreactive chlorides. E: 1.4 eqs. *n*-Butyl acrylate injected into a degassed soln. of 4-chlorobenzaldehyde and 1.1 eqs. Na-acetate in diethyleneglycol di-*n*-butyl ether under argon, heated to 100°, 0.1 mol% Pd catalyst in degassed DMA added via syringe, heated to 130°, and worked up after 24 h → product. Y 81% (conversion 90%; turnover no. 810). There was no loss of catalyst by Pd deposition (as may take place with conventional Heck catalysts), even at 140°; turnovers as high as 200,000 can be achieved for reactions with ar. bromides using as little as *0.005 mol%* catalyst, and no additional phosphine ligands are required. For reactions with ar. chlorides, addition of bromide ion is necessary to stabilize the catalyst. F.e.s. W.A. Herrmann et al., Angew. Chem. Intern. Ed. *34*, 1844-8 (1995); **diaryls** by Suzuki cross-coupling of ar. *chlorides* with 0.001 to 0.02 mol% palladacyclic catalyst s. M. Beller et al., ibid. 1848-9.

Palladous halide/lithium halide/triethylamine ←
α,β-Acetylene-carboxylic acid esters from -halides by carbonylation Hal → COOR
s. *12*, 867s*51*

Dichlorobis(triphenylphosphine)palladium(II)/potassium carbonate/ethanol ←
Regiospecific intramolecular carbopalladation of allenes ←
via palladium π-allyl complexes

417.

with subsequent nucleophile capture. 5 Mol% Cl$_2$Pd(PPh$_3$)$_2$ added under argon to a soln. of 5,5-bis(ethoxycarbonyl)-8-*n*-propyl-9-iodo-1,2,8(Z)-dodecatriene, 5 eqs. K$_2$CO$_3$, 10 eqs. ethanol, and 3 eqs. diethyl malonate in DMF, and heated at 120° for 23 h → 6,6-bis(ethoxycarbonyl)-3-[2',2'-bis(ethoxycarbonyl)ethyl]-1,2-di-*n*-propyl-1,3-cyclooctadiene. Y 51%. The intermediate Pd-allyl species can also be trapped by O- and N-nucleophiles, both inter- and intra-molecularly, the latter effecting **triple ring closure** in one operation. F.e. incl. 5- to 7-membered ring closures, also large-ring carbo- and hetero-cyclics via β-elimination, s. S. Ma, E. Negishi, J. Am. Chem. Soc. *117*, 6345-57 (1995); benzo-condensed heterocyclics by trapping with amines in the presence of Pd(OAc)$_2$/PPh$_3$/K$_2$CO$_3$ cf. R. Grigg et al., Chem. Commun. *1995*, 1903-4.

Water-soluble dichloropalladium phosphine complex/sodium carbonate/ ←
sodium dodecyl sulfate
Sym. diaryls from ar. halides 2 ArHal → Ar–Ar
with Pd,Hg/N$_2$H$_4$ cf. *34*, 825s*26*; in aq. organic microemulsion with a water-soluble Pd-phosphine complex/Na$_2$CO$_3$ and Na-dodecyl sulfate under H$_2$ s. D.B. Davydov, I.P. Beletskaya, Izv. Akad. Nauk Ser. Khim. *1995*, 1180-1 (Russ.).

Sulfur ↑ CC ⇅ S

Irradiation s. under (Bu$_3$Sn)$_2$ ⫽

Potassium hydroxide   KOH
Aziridines from azomethines and sulfonium salts   ⟶N⟵
s. *23*, 871s*38*; 1-sulfonyl-2-vinylaziridines from N-sulfonylimines with KOH s. A.-H. Li et al., Chem. Commun. *1996*, 491-2; s.a. J. Chem. Soc. Perkin Trans. I 1996, 2725-9; catalytic procedure from allyl halides and a little Me$_2$S cf. ibid. 867-9; 2-acyl-derivs. s. U.K. Nadir, A. Arora, Synth. Commun. *26*, 2355-61 (1996); N-sulfinylaziridines from N-sulfinylimines with asym. induction s. F.A. Davis et al., Tetrahedron:Asym. *6*, 1511-4 (1995).

α-Hydroxyketones from 1-sulfonyl-1-isonitriles and aldehydes via 2-alkoxy-Δ^3-oxazolines

418.

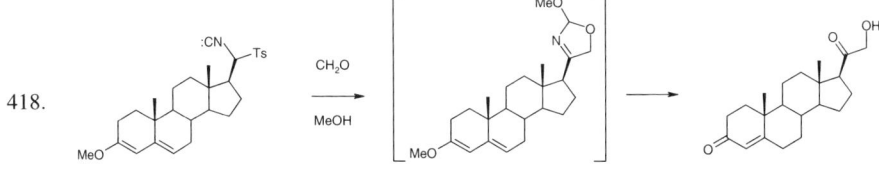

4.4 eqs. Powdered KOH added to a stirred mixture of 17β-(isocyanotosylmethyl)-3-methoxy-androsta-3,5-diene, 2.2 eqs. 37% aq. formaldehyde, and 5 eqs. methanol in THF at 0°, the cooling bath removed, and stirring continued for 1 h at room temp. → 3-methoxy-17β-(2-methoxy-3-oxazolin-4-yl)androsta-3,5-diene (Y 96%), in THF treated with 4 N H$_2$SO$_4$, and allowed to stand at room temp. for 40 h → 21-hydroxypregn-4-ene-3,20-dione (Y 93%). F.e. **and from 2-alkoxy-4-vinyl-Δ^3-oxazolines** (via reduction with NaBH$_4$ or 9-BBN-H) s. D. van Leusen et al., J. Org. Chem. *59*, 5650-7 (1994).

Sodium/alcohol   NaOR
3-Cyano-2-pyridone ring
4-alkylthio-derivs. cf. *32*, 847; 4-alkoxy-derivs. s. M.L. Purkayastha et al., Synthesis *1995*, 641-3.

Organolithium compds.   RLi
Synthesis of sec. amines from N,N-disubst. trifluoromethanesulfonylamines   NHC(R)

419.

A mixture of triflic acid N,N-dibenzylamide and *tert*-butyllithium in THF stirred at -60° for 3 h → product. Y 93%. The presence of an activated β-hydrogen and the use of a strong base of low nucleophilicity is essential for reaction. F.e. incl cyclic sec. amines, and bases, s. H. Handel et al., Tetrahedron Letters *36*, 6063-6 (1995).

Synthesis of ethylene derivs. from thioenolethers   C=C(SR) → C=CR'
with RMgBr/CuI cf. *29*, 854; β-alkoxy-α,β-ethyleneketones from α-ketoketene O,S-acetals s. H. Junjappa et al., Tetrahedron Letters *36*, 9377-80 (1995); (Z)-2-ethylene-1,1,1-trifluorides with RLi, also from 1-(trifluoromethyl)enolethers, s. J.-P. Bégué et al., ibid. *37*, 171-4 (1996).

n-*Butyllithium*   BuLi
α-Ketomercaptals from carboxylic acid esters s. *51*, 364   COOR → COCH(SR)$_2$
Enolethers from α-alkoxysulfones and sulfones   CHSO$_2$R → C=C(OR)

420.

1.05 eqs. 1.6 M n-BuLi in hexane added dropwise under argon at 0° to a degassed soln. of startg. sulfone in THF, warmed to room temp., stirred for 30 min, a degassed soln. of 1.5 eqs. lithiated *tert*-butoxymethyl phenyl sulfone (also prepared in THF by metalation with *n*-BuLi at -78°) added slowly via cannula with stirring over 1 h, and stirring continued at room temp. for 2 h → 1-*tert*-

butoxyhex-1-ene. Y 71% (60:40 E/Z). F.e., **also thioenolethers from α-(alkylthio)sulfones**, s. M. Julia et al., Bull. Soc. Chim. France *131*, 965-72 (1994).

n-*Butyllithium/methylene chloride* $BuLi/CH_2Cl_2$
3,3-Dibromalcohols from cyclic glycol sulfates
Synthesis with addition of one C-atom

421.

1.25 eqs. Precooled 1.55 M n-BuLi in hexanes added to a soln. of 1.96 eqs. methylene chloride in 1:1 ether/THF at -100°, stirred for 15 min, a precooled soln. of 1.49 eqs. methylene bromide in THF added, stirred for 45 min, a precooled soln. of startg. cyclic sulfate in THF added, stirred at -100° for 2 h and at -95° for 2 h, allowed to warm to -78° overnight, 5 drops of concd. H_2SO_4 and 0.98 eq. water added, stirred for 2 h at 20°, and quenched with aq. pH 7 buffer → 1,1-dibromo-4,4-dimethylpentan-3-ol. Y 88%. F.e.s. H.C. Stiasny, Synthesis *1996*, 259-64.

Lithium diisopropylamide   i-Pr_2NLi
2-Azetidinones from azomethines and thiolic acid esters s. *48*, 854s*51*

Piperidine   ←
Synthesis of (E)-2-ethylenealcohols from aldehydes   CHCHO → C(OH)C=C
and functionalized methyl sulfones via vinylogous Pummerer rearrangement
(E)-α,β-ethylene-γ-hydroxycarboxylic acid esters cf. *38*, 858; (E)-α,β-ethylene-γ-hydroxyphosphonic acid esters s. J. Nokami et al., Chem. Lett. *1995*, 1025-6; (E)-α,β-ethylene-γ-hydroxysulfoxides s. J.M. Llera et al., Tetrahedron Letters *36*, 4889-92 (1995).

Aluminum bromide $AlBr_3$
2-Azetidinones from azomethines
and 2-pyridyl thiolates with $TiCl_4$ cf. *48*, 854s*50*; with 0.5 eq. $AlBr_3$ or $EtAlCl_2$, stereoselectivity, s. R. Annunziata et al., Tetrahedron *52*, 2583-90 (1996); 3-fluoro-2-azetidinones from phenyl esters with LDA cf. T. Ishihara et al., ibid. 255-62.

Samarium diiodide SmI_2
Reductive coupling of sulfones with oxo compds.   ←
s. *50*, 531; glycol monobenzyl ethers from α-alkoxysulfones s. T. Skrydstrup et al., Chem. Commun. *1996*, 515-6.

Tetramethylguanidine   ←
Pyrroles from 1,1-nitroethylene derivs.   ○
and ethyl isocyanoacetate cf. *44*, 736; and thioiminoesters with tetramethylguanidine s. M. Yokoyama et al., Bull. Chem. Soc. Japan *68*, 2735-8 (1995); **from α,β-ethylenesulfones** cf. G. Haake et al., Tetrahedron Letters *35*, 9703-4 (1994).

Di-tert-*butyl peroxide* $(t\text{-}BuO)_2$
Radical C-allylation with allyl sulfones
    ←

422.

A soln. of startg. alkyl allyl sulfone and *3-6 eqs.* allyl tolyl sulfone in degassed chlorobenzene heated under reflux with 1 drop of di-*tert*-butyl peroxide until TLC indicated completion of reaction (2-8 h) → product. Y 60%. Excess of allyl tolyl sulfone (which in effect serves as the principle allylating agent) acts as a relay which prevents loss of product by a subsequent radical addition. F.e. and **radical 1,2-addition-allylation** to electron-deficient alkenes s. B. Quiclet-Sire, S.Z. Zard, J. Am. Chem. Soc. *118*, 1209-10 (1996).

Titanocene dichloride/organolithium compds./triethyl phosphite
Vinylcyclopropanes from 1,3-bis(arylthio)ethylenes
Titanocene α,β-ethylenecarbene complexes as intermediates

423.

2 eqs. *tert*-BuLi in pentane added at -78° to a suspension of Cp₂TiCl₂ in THF, after 15 min 0.5 eq. 4-phenyl-1,3-bis(phenylthio)-1-butene in THF and 1 eq. (EtO)₃P added sequentially, stirred for 15 min, the cooling bath removed, and stirring continued for 3 h → 1,1-dimethyl-2-(3-phenyl-1-propenyl)cyclopropane. Y 80%. Yields were enhanced in the presence of a phosphite or phosphine ligand which stabilizes the intermediate titanocene olefin complex. F.e. **and from cyclic α,β-ethylenemercaptals** s. T. Takeda et al., Tetrahedron Letters *36*, 8835-8 (1995).

Titanium tetrachloride/triphenylphosphine $2TiCl_4·PPh_3$
1,2,3,4-Tetrahydroquinolines from ethylene derivs.
and arylaminomethyl thioethers s. *38*, 760s51

Hexabutyldistannane/acetone/irradiation $(Bu_3Sn)_2/Me_2CO/⫽⫽$
Oxo compds. from halides via O-benzyloximes   Hal → C(=NOR)R'
Radical conversion under mild conditions

424.

Aldehydes. A soln. of startg. iodide (0.3 *M*), 1.2 eqs. hexabutyldistannane, 2 eqs. O-benzyl-1-(phenylsulfonyl)formaldoxime, and 5 eqs. acetone (sensitizer) in benzene irradiated at 300 nm for 4 h → intermediate alkoxime (Y 94%), treated with 30% formaldehyde soln. in THF (1:3) containing a little HCl → product (Y 90%). **Ketones** were obtained similarly **from O-benzyl-1-sulfonyloximes**. Bromides were also effective but reaction times were longer and yields lower. Reaction can be performed with complex molecules where more conventional routes would be inappropriate (acetals, esters, alcohols, and carbamates being unaffected). F.e. (high yield) incl. highly functionalized C-acylglycoside derivs., **also stereospecific radical ring closure-acylation**, s. S. Kim et al., J. Am. Chem. Soc. *118*, 5138-9 (1996).

Stannic chloride $SnCl_4$
1,2,3,4-Tetrahydroquinolines from ethylene derivs.
and arylaminomethyl ethers cf. *38*, 760; and arylaminomethyl sulfones or arylaminoacetonitriles with SnCl₄ s. U. Beifuss et al., Synlett *1996*, 34-6; from arylaminomethyl thioethers with 2 TiCl₄/PPh₃ cf. Chem. Commun. *1995*, 2137-8.

Triethyl phosphite s. under Cp₂TiCl₂ $(EtO)_3P$

Remaining Elements ↑   CC ⇅ Rem

Without additional reagents   w.a.r.
Dötz reaction
s. *43*, 860; *p*-aminophenols from aminocarbene complexes s. W.D. Wulff et al., J. Org. Chem. *60*, 4566-75 (1995); prepn. of chromium alkoxycarbene complexes from dialkylzincs s. H. Stadtmüller, P. Knochel, Organometallics *14*, 3163-6 (1995).

5H-Azepin-2(1H)-ones from α,β-ethyleneazomethines
and chromium α,β-acetylene(alkoxy)carbene complexes

425.

A soln. of startg. α,β-ethyleneazomethine and chromium carbene complex in hexane allowed to react at 20° for 3 h → intermediate η¹-azepine chromium complex (Y 90%), in THF or benzene heated at 50° until demetalation complete → product (Y 93%). Reaction involves nucleophilic addition of the imine nitrogen to the carbene carbon, rather than consecutive [2+1]-cycloaddition and [3,3]-rearrangement, the driving force being an unprecedented **1,2-migration of the pentacarbonylchromium group**. F.e.s. J. Barluenga et al., J. Am. Chem. Soc. *118*, 695-6 (1996).

Irradiation (s.a. under MeAl(OAr)₂ and 9,10-Dicyanoanthracene)
Cyclopropanes from iron carbene complexes
s. *47*, 527; from a (siloxymethyl)iron complex cf. H. Du et al., Synth. Commun. 26, 1371-7 (1996); dihalogenocyclopropanes from porphyrinatoiron dichlorocarbene complexes s. C.J. Ziegler, K.S. Suslick, J. Am. Chem. Soc. *118*, 5306-7 (1996); **asym. cyclopropanation** with chiral complexes s. R.D. Theys, M.M. Hossain, Tetrahedron Letters *36*, 5113-6 (1995).

Bridged 4-imidazolidones
from chromium N-[ω-(alkylideneamino)alkyl]aminocarbene complexes

426.

A soln. of pentacarbonyl[[[3-(p-anisylideneamino)propyl]amino](phenyl)carbene]chromium(0) in dry, degassed, CO-satd. ether pressurized to 2.5 atm. CO, irradiated for 24 h in a Pyrex testtube with a 450 W Hg-lamp, and the resulting crude complex demetalated by air oxidation under direct sunlight → 8-(p-anisyl)-7-phenyl-1,5-diazabicyclo[3.2.1]octan-6-one. Y 85% (single diastereomer). The expected bridged β-lactams were not formed. F.e. and prepn. of the startg. m. s. B. Alcaide et al., J. Am. Chem. Soc. *117*, 5604-5 (1995).

Electrolysis
Alkoxy-3-ethylenes $\quad$ C(OR)SR' → C(OR)C–C≡C
from 2-ethylenesilanes and monothioacetals
with $SnCl_4$ cf. *37*, 892; under electrolysis s. J. Yoshida et al., Tetrahedron Letters *37*, 3157-60 (1996).

Dihalogenomethylene compds. from oxo compds. $\quad$ CO → C=CHal$_2$
dichloromethylene compds. via Horner synthesis cf. *47*, 872; diiodomethylene compds. (with $KN(SiMe_3)_2$) s. B. Bonnet et al., Synthesis *1993*, 1071-4; with $[(Me_2N)_3PCCl_3]^+BF_4^-$ under electrolysis s. P. Jubault et al., Bull. Soc. Chim. France *131*, 1001-6 (1994); difluoromethylene compds. s. ibid. *132*, 850-6 (1995).

Microwaves s. under SiO_2
Lithium/4,4'-di-tert-butylbiphenyl
Generation of alkyllithiums from alkyl aryl selenides

427.

Sec. alcohols from aldehydes. Startg. selenide added to 3 eqs. Li-4,4'-*tert*-butylbiphenylide in THF *at -78°*, the mixture allowed to react for 30 min, then quenched with benzaldehyde →

product. Y 76%. The method is general for the preparation of *prim-, sec-* and *tert*-alkyllithiums which may not be accessible by other routes. F.e. and electrophiles, also reactions with benzyl selenides, **cyclopentanes from 5-ethyleneselenides**, and comparison with Li-naphthalenide, s. A. Krief et al., Tetrahedron Letters *36*, 8111-4 (1995); **reductive lithiation of selenoacetals** with trapping of the intermediate 1,1-lithioselenides s. ibid. 8115-8.

Sodium hydride   *NaH*
Horner synthesis   CO → C=C
s. *23*, 879; of functionalized α,β-ethylenecarboxylic acid thioamides s. S. Le Roy-Gourvennec, S. Masson, Synthesis *1995*, 1393-6; of (E,E)-1-trifluoromethyl-1,3-dienes s. V. Martin et al., Synth. Commun. *25*, 3519-28 (1995); of enoates with Triton B s. K. Ando, Tetrahedron Letters *36*, 4105-8 (1995); of conjugated pentaenecarboxylic acids s. A.A. Souto et al., ibid *35*, 5907-10 (1994).

Potassium hydroxide/ammonium ceric nitrate   *KOH/(NH$_4$)$_2$Ce(NO$_3$)$_6$*
1,2-Addition of nitro compds. to electron-rich ethylene derivs.   ←
via oxidative generation of α-nitroalkyl radicals

428.

(E)-α,β-Ethyleneketones from enoxysilanes. A soln. of 1-nitro-4-phenylbutane in methanol treated with 1.3 eqs. KOH at room temp., the mixture added to a soln. of 1.9 eqs. CAN and 0.85 eq. α-trimethylsiloxystyrene in the same solvent at -78°, and allowed to react to completion → crude 3-nitro-1,6-diphenylhexan-1-one (Y high), in methanol treated with Et$_3$N at room temp. → 1,6-diphenyl-2-hexen-1-one (Y 95%). Such C-alkylation of nitronate ion is unprecedented. F.e. incl addition to enolethers and styrenes, **also double ring closure of nitroethylene derivs.** by an intramolecular variant, s. N. Arai, K. Narasaka, Chem. Lett. *1995*, 987-8.

Potassium hydroxide/18-crown-6 polyether   *KOH/crown*
Heterogeneous Wittig synthesis   CHO → CH=C
with NaOH cf. *29*, 868; (E)- and (Z)-stilbenes with KOH/18-crown-6 under solid-liq. phase transfer catalysis s. G. Bellucci et al., Tetrahedron Letters *37*, 4225-8 (1996).

Potassium tert-*butoxide*   *KOBu-t*
Wittig synthesis of thioenol derivs.   CO → C=C
of dithioenolesters s. *23*, 879s39; of (E)-vinyl dithiocarbamates s. Z.-Z. Huang, L.-L. Wu, Synth. Commun. *26*, 509-14 (1996).

Potassium tert-*butoxide/lithium bromide*   *KOBu-t/LiBr*
Aldol-type condensation with enoxysilanes via alkali-metal enolates   C(OH)C–CO
Effect of counterion on regioselectivity

429.

A soln. of KOBu-*t* in THF added under argon to a soln. of 1 eq. startg. enoxysilane in the same solvent at -15°, stirred for 45 min, 5 eqs. LiBr added, stirring continued for 20 min, cooled to -78°, 1 eq. benzaldehyde in THF added, stirred for 1 h, and quenched with water → 2-(hydroxy phenylmethyl)-2-methylcyclohexan-1-one. Y 72% (*threo:erythro* 2.2:1). The corresponding K-enolate, however, gave the regioisomeric aldol adduct (with KOBu-*t* alone), which was also obtained under the same conditions from the regioisomeric enoxysilane. The results suggest that reaction takes place under thermodynamic control with K-enolates and under kinetic control with Li-enolates after a short reaction time. F.e.s. P. Duhamel et al., J. Org. Chem. *61*, 2232-5 (1996).

n-*Butyllithium* *BuLi*
3-Alkylidenecyclobutanols from oxo compds.

430.

via Wittig synthesis. 1.02 eqs. 1.6 M n-BuLi in hexane added via syringe to a soln. of methyl triphenylphosphonium iodide in dry *toluene* at 0°, stirred for 20 min, 1 eq. epichlorhydrin in toluene added slowly via syringe, the mixture stirred at 0° for 30 min, a further 2.04 eqs. 1.6 M n-BuLi in hexane added dropwise, cooled to -40° when the colour changed from yellow to reddish orange, 1.5 eqs. benzaldehyde in toluene added dropwise via syringe, stirring continued at -40° for a further 1 h, warmed to room temp., then quenched with water → 3-benzylidenecyclobutanol. Y 62%. The choice of base and solvent was found to be critical for the reaction, the formation of isomeric (2-alkylidenecyclopropyl)carbinols being negligible under these conditions; yields were better with aromatic aldehydes, whereas ketones gave a low yield. F.e.s. K. Okuma et al., Tetrahedron Letters *36*, 5591-4 (1995).

Lithium diisopropylamide i-*Pr₂NLi*
Peterson olefination CO → C=C
s. *28*, 856s*31*; arylthio as stereo-directing group, (Z)-γ-arylthio-α,β-ethylenecarboxylic acid esters, s. N.Y. Grigorieva et al., Izv. Akad. Nauk Ser. Khim. *1995*, 509-16 (Russ.).
Horner synthesis of 2-functionalized 1,3-dienes
s. *14*, 877s*46*; of 2-alkoxy-1,3-dienes s. K. Fettes et al., J. Chem. Soc. Perkin Trans. I *1995*, 2123-7; f. method from prop-2-ynyltriphenylphosphonium bromide, chiral derivs., cf. J. Barluenga et al., Chem. Commun. *1995*, 1785-6.
α-Allenecarboxylic acid esters from ketenes C=C=O → C=C=C–COOR
by Horner synthesis with NaH cf. *28*, 851; *22*, 864; with LDA (the ketenes being generated from carboxylic acid esters with *n*-BuLi/ZnCl₂ or SnCl₂) s. K. Tanaka et al., Synlett *1995*, 933-4; **from α,β-ethylenecarboxylic acid esters** by 1,4-addition-Horner synthesis s. Tetrahedron Letters *36*, 9513-4 (1995).

Potassium bis(trimethylsilyl)amide/18-crown-6 polyether *KN(SiMe₃)₂/crown*
Horner synthesis CO → C=C
of (Z)-enoates s. *39*, 851; of (Z)-α,β-ethylenesulfoxides s. J. Motoyoshiya et al., Synthesis *1996*, 637-40; prepn. of KN(SiMe₃)₂ s. J. Åhman, P. Somfai, Synth. Commun. *25*, 2301-3 (1995).
Asym. Horner synthesis
with chiral 2,2'-binaphthyl esters cf. *48*, 866; with chiral 1-(α-phosphonylacyl)benzopyrano[4,3-*c*]isoxazolidines using KN(SiMe₃)₂/18-crown-6 as base s. A. Abiko, S. Masamune, Tetrahedron Letters *37*, 1077-80 (1996); **kinetic resolution of aldehydes** with chiral α-phosphonylesters s. T. Rein et al., ibid. *36*, 2303-6 (1995); Angew. Chem. Intern. Ed. *33*, 556-8 (1994).

Cesium fluoride *CsF*
Epoxides from oxo compds. and α-silylsulfonium salts
with KOBu-*t* cf. *33*, 854; with CsF s. K. Hioki et al., Synthesis *1995*, 649-50.
Generation of azomethium ylids from N-α-silylazomethines
s. *38*, 907; 1,2,3,5,6,11b-hexahydroindolizino[8,7-*b*]indoles s. G. Poissonnet et al., J. Org. Chem. *61*, 2273-82 (1996); 1,3-dipolar cycloaddition with *in situ*-generated (1,3-dithiolane-2-ylidene)ammonium methylids s. C.W.G. Fishwick et al., Tetrahedron Letters *36*, 9409-12 (1995); γ-lactams s. ibid. *37*, 3915-8 (1996); with cyclic [thio]imidate methylids s. ibid. 5163-6; with isothiourea analogs s. S. Kohra, Y. Tominaga, Heterocycles *38*, 1217-20 (1994).

*Lithium bromide s. under KOBu-*t *LiBr*
Sodium iodide s. under BiCl₃ *NaI*
Triton B ←
Horner synthesis of α,β-ethylenecarboxylic acid esters s. *23*, 879s*51*   CO → C=CHCOOR

Triethylamine   *Et_3N*
Stereospecific synthesis of 3-ethylenealcohols   CO → C(OH)C–C=C
from 2-ethylene(trifluoro)silanes
with CsF cf. *43*, 843; 1,7-diene-4,5-diols from α-diketones with Et_3N, and preferential allylation of the less enolized keto group, s. R. Gewald et al., Synthesis *1996*, 111-5; coupling with *o*-amino- or *o*-hydroxy-ketones and with β-diketones s. H. Sakurai et al., Chem. Lett. *1995*, 281-2.

Triethylamine or ethyldiisopropylamine/lithium chloride or bromide   *$R_3N/LiHal$*
Horner synthesis   CO → C=C
s. *39*, 854s*48,49*; of α-diazo-β,γ-ethylenecarboxylic acid esters s. H.G. Viehe et al., Synthesis *1995*, 920-2; **polymer-based Horner synthesis** of (E)-α,β-ethylenecarboxylic acid amides s. C.R. Johnson, B. Zhang, Tetrahedron Letters *36*, 9253-6 (1995).

Ethanolamine   *$H_2NCH_2CH_2OH$*
Asym. allylboration   CO → C(OH)C–C=C
with B-allyldiisocampheylborane s. *33*, 865s*46,48*; chiral *anti*-3-ene-1,2-diols from B-(3-silylallyl)-derivs. s. A.G.M. Barrett, J.W. Malecha, J. Chem. Soc. Perkin Trans. I *1994*, 1901-5; from tartrate-based 3-silylallylboronates cf. J.A. Hunt, W.R. Roush, Tetrahedron Letters *36*, 501-4 (1995); from B-allenyl-1,3,2-dioxaborinane/Ipc_2BH s. H.C. Brown, G. Narla, J. Org. Chem. *60*, 4686-7 (1995); with B-allylditerpenylboranes cf. ibid. *56*, 401-4 (1991); *57*, 6608-14, 6614-7 (1992); chiral 4-hydroxy-2-methyleneethers s. T.A.J. van der Heide et al., Tetrahedron Letters *34*, 4655-6 (1993); chiral 2-amino-3-ethylenealcohols s. A.G.M. Barrett, M.A. Seefeld, Tetrahedron *49*, 7857-70 (1993); chiral 3-alkoxy-4-hydroxy-1,5-enynes s. P. Ganesh, K.M. Nicholas, J. Org. Chem. *58*, 5587-8 (1993); chiral 5-ethylene-3-hydroxy epoxides s. W.R. Roush et al., ibid. *56*, 1636-48 (1991); chiral 3-ethylenealcohols with a quaternary carbon centre s. Y. Yamamoto et al., Synlett *1996*, 883-4.

Chiral diamines   ←
3-Ethylenealcohols from aldehydes   CHO → CH(OH)C–C=C
Asym. synthesis with addition of three C-atoms
with $Sn(OTf)_2$/chiral diamines cf. *41*, 892; allylation with diallyltin dibromide without $Sn(OTf)_2$ s. S. Kobayashi, K. Nishio, Tetrahedron Letters *36*, 6729-32 (1995).

1,8-Diazabicyclo[5.4.0]undec-7-ene/sodium borate   *$DBU/NaBO_3$*
Synthesis of sec. alcohols   CH=$NNSO_2R$ → CH(OH)R′
from aldehyde trisylhydrazones and trialkylboranes

431.   PhCH(=NNHTris) BBu_3 → [PhCHBu(BBu_2)] → PhCHBu(OH)

Sec. benzylalcohols. 1 eq. 1 *M* tri-*n*-butylborane in THF added under argon to a soln. of startg. trisylhydrazone in the same solvent via syringe, treated with 1 eq. DBU, stirred at room temp. until reaction complete by TLC (2 h), and oxidized with $NaBO_3 \cdot 4H_2O$ → product. Y 90%. The non-nucleophilic base prevents deborinylation and subsequent protonolysis to alkanes (except for hydrazones with an electron-withdrawing group in the *ortho* position). The method is general and unaffected by steric hindrance. F.e. and oxidation with peracetic acid s. G.W. Kabalka et al., Synth. Commun. *26*, 999-1006 (1996); **from** the corresponding **1,1-dichlorides** with *t*-BuLi cf. Tetrahedron Letters *36*, 8545-8 (1995).

Lithium alkyl(cyano)cuprates   *RCu(CN)Li*
Synthesis of ethylene derivs. from enetellurides   C=CTeR → C=CR′
and Grignard compds. cf. *37*, 868; and Li-alkyl(cyano)cuprates, (Z)-isomers, s. A. Chieffi, J.V. Comasseto, Synlett *1995*, 671-4; with bromomagnesium diorgano(cyano)cuprates cf. Tetrahedron Letters 35, 4063-6 (1994); cross-coupling **via higher-order lithium vinylcuprates** s. ibid. 1145-8; details s. J. Org. Chem. *61*, 6975-89 (1996).

Cuprous thiophene-2-carboxylate   *CuOCOR*
Copper(I)-mediated cross-coupling with stannanes under mild conditions   ←

432.

Cuprous thiophene-2-carboxylate effects rapid and efficient cross-coupling of aryl-, hetaryl- or vinyl-stannanes with vinyl iodides and certain aryl iodides at or below room temp., thereby providing a useful alternative to the conventional Pd-catalyzed (Stille) coupling. E: A soln. of 2-(tri-*n*-butylstannyl)-4,5-dimethoxybenzaldehyde and (E)-2-(2-iodovinyl)-4-bromothiophene in N-methyl-2-pyrrolidone treated with *1.5 eqs.* Cu(I)-thiophene-2-carboxylate at 23° for *5 min* → product. Y 78%. Reaction takes place **with retention of configuration** (at both vinylstannane and vinyl iodide residue) and leaves carbonyl groups, alkyl iodides, benzyl bromides, and most (het)aryl bromides and iodides unaffected. Copper(I) carboxylates were more effective than Cu(I) halides which were required in larger excess to drive the reversible reaction to completion. F.e.s. G.D. Allred, L.S. Liebeskind, J. Am. Chem. Soc. *118*, 2748-9 (1996); cross-coupling of α-functionalized stannanes with CuCN cf. J.R. Falck et al., ibid. *117*, 5973-82 (1995).

Silver acetate s. under PdCl$_2$ *AgOAc*
Cuprous cyanide s. under Zn   *CuCN*
Cuprous chloride   *CuCl*
Sym. 1,3-dienes from enestannanes   2 C=C(Sn≤) → C=C–C=C
with Pd(II) cf. *43*, 847; with CuCl, with retention of configuration, s. E. Piers et al., Tetrahedron Letters *37*, 1173-6 (1996).

Silver trifluoromethanesulfonate/(S)-2,2′-bis(diphenylphosphino)-1,1′-binaphthyl   ←
3-Ethylenealcohols from 2-ethylenestannanes and aldehydes   CH(OH)C–C=C
Asym. synthesis under silver(I) catalysis

433.

A novel chiral Ag(I) phosphine complex is now available for catalytic asym. allylation of aldehydes with allylstannanes. E: A soln. of ca. 5 mol% AgOTf and ca. 5 mol% (S)-BINAP in dry THF stirred under argon at 20° for 10 min with exclusion of direct light, a soln. of benzaldehyde in THF and allyltributyltin added dropwise in this order at -20°, stirred for 8 h, and treated with 1 *N* HCl and solid KF at room temp. for 30 min → (S)-product. Y 88% (e.e. 96%). Other chiral Ag(I)-phosphine complexes were less effective. Reaction is generally applicable to aromatic and α,β-unsatd. aldehydes, but yields and enantioselectivity were significantly lower with aliphatic aldehydes. Methallylstannanes reacted similarly **without double bond shift**. F.e.s. H. Yamamoto et al., J. Am. Chem. Soc. *118*, 4723-4 (1996).

Zinc s.a. under NiCl$_2$(dppf)   *Zn*
Zinc/cuprous cyanide/lithium chloride/boron fluoride   *Zn/CuCN/LiCl/BF$_3$*
Terminal ethylene derivs. from aldehydes   CHO → CH=CH$_2$
Synthesis with addition of one C-atom

434.

A soln. of Knochel's dioxyborylmethylzinc iodide in THF added dropwise at -30° to a soln. of 1.2 eqs. CuCN and 2.4 eqs. LiCl in THF under N$_2$, slowly warmed to 0°, cooled to -78°, 0.5 eq. startg. aldehyde and 2 eqs. BF$_3$-etherate added successively, stirred for 30 min at -78°, warmed to room temp. over 2 h, and treated with satd. NH$_4$Cl (pH 8) for 1 h at room temp. → 1-vinylnaphthalene. Y 84%. This boron-Wittig methylenation leaves ester and (notably) keto groups unaffected. F.e. and method for aliphatic aldehyde adducts, also conversion of the intermediate to the corresponding **glycols** with H$_2$O$_2$/OH⁻ s. M. Sakai, N. Miyaura et al., Tetrahedron *52*, 915-24 (1996).

Dialkylzinc R_2Zn
Generation of dialkylzinc compds. from trialkylboranes ←
s. *49*, 877; **1,3-dizinc compds.** and their Cu(I)-catalyzed cross-coupling s. H. Eick, P. Knochel, Angew. Chem. Intern. Ed. *35*, 218-20 (1996).

Zinc triflate/chiral bis(Δ^2-oxazolines)/triethylborane ←
Lewis acid-catalyzed asym. radical 1,4-addition-allylation ←

435.

A soln. of N-acryloyl-2-oxazolidone in 3:2 methylene chloride/pentane treated with 5 eqs. *tert*-butyl iodide, 5 eqs. allyltributylstannane, 1 eq. Zn(OTf)$_2$, and 1.2 eqs. (R,R)-bis(Δ^2-oxazoline) *at -78°* in the presence of Et$_3$B as initiator → product. Y 92% (e.e. 90%). The Lewis acid activates the alkene towards addition of the nucleophilic alkyl radical, and the subsequent addition-fragmentation of the resulting radical with allyltributylstannane determines the stereoselectivity. The enantioselectivity is critically dependent on the chiral ligand, the solvent, the Lewis acid, and the initiating radical. F.e.s. J.H. Wu et al., J. Am. Chem. Soc. *117*, 11029-30 (1995).

Magnesium bromide/triethylborane/oxygen $MgBr_2/Et_3B/O_2$
Lewis acid-catalyzed asym. radical cross-coupling ←

436.

The combination of chiral 2-oxazolidone and Lewis acid is highly effective in securing outstanding levels of selectivity in acyclic radical substitution. **E:** 2 eqs. Allyltributylstannane added under N$_2$ at -78° to a soln. of startg. 2-oxazolidone and 1 eq. MgBr$_2$·etherate in methylene chloride, treated with 2 eqs. 1 *M* Et$_3$B in hexane followed via syringe over 2 min by O$_2$, and stirring continued at -78° for 2 h → product. Y 94% (RS:RR ≥ 100:1). The ratio was 1:1.8 in the absence of Lewis acid. The latter chelates the two carbonyls, thereby stabilizing one radical rotomer in readiness for face-selective attack, the bulky benzhydryl group on the ring providing a cooperative effect. Sc(OTf)$_3$ was equally effective. F.e.s. M.P. Sibi, J. Ji, Angew. Chem. Intern. Ed. *35*, 190-2 (1996); chiral α-allyl-β-siloxy-esters with Ln(fod)$_3$/AIBN under irradiation s. H. Nagano et al., J. Chem. Soc. Perkin Trans. I *1996*, 389-94; chiral α-allyl-β-alkoxycarboxylic acid esters s. Y. Guindon et al., Synlett *1995*, 449-51; chiral 4-ethylenealcohols from 1,2-iodohydrins with methylaluminum bis(2,6-di-*tert*-butyl-4-methylphenoxide) and AIBN under irradiation s. N. Moufid, P. Renaud, Helv. Chim. Acta *78*, 1001-5, 1006-12 (1995).

Zinc chloride $ZnCl_2$
1,5-Dienes from 2-ethylenesilanes and siloxy-2-ethylenes Si∈→C-C=C
Regiospecific conversion

437.

Siloxy-2-ethylenes serve *as electrophile* in Lewis acid-catalyzed coupling with allylsilanes. **E:** Solns. of startg. siloxy-2-ethylene and 1.1 eqs. startg. allylsilane in dichloromethane added sequentially to *10 mol%* dry ZnCl$_2$ in the same solvent at room temp. under argon, stirred at 25° for 42 h, and quenched with a few drops of Et$_3$N → product. Y 83% (92% major regioisomer). The allylsilane attacks predominantly at the least substituted position of the siloxy-2-ethylene (except where an adjacent phenylthio group is present in the electrophile) with exclusive double bond shift (in the allylsilane). F.e.s. T. Yokozawa et al., Tetrahedron Letters *36*, 5243-6 (1995).

2,3-Dihydro-4(1H)-pyridones from aldimines via asym. Diels-Alder reaction ○
from chiral imines s. *47*, 888s*50*; from chiral Cr(CO)$_3$-complexed imines, also chiral 2,3-dihydro-4-pyrones from aldehydes (cf. *38*, 886), s. C. Baldoli et al., Synlett *1996*, 258-60.

Chiral 3-sulfonyl-1,3,2-oxazaborolidin-5-ones/tetra-n-butylammonium fluoride ←
β-Alkoxycarbonyl compds. from cyclic acetals via asym. aldol-type condensation C
chiral β-alkoxycarboxylic acids with TiCl$_4$/CF$_3$COOH cf. *40*, 635; improved asym. ring opening of cyclic acetals based on (S)-1-(2,6-dichlorophenyl)propane-1,3-diol (with allylsilanes and enoxysilanes), and conversion to the corresponding chiral sec. alcohols (cf. *38*, 902), s. M.B. Andrus, S.D. Lepore, Tetrahedron Letters *36*, 9149-52 (1995); chiral β-alkoxycarboxylic acid esters and thioesters, and β-alkoxyketones with chiral 3-sulfonyl-1,3,2-oxazaborolidin-5-ones/ Bu$_4$NF s. T. Harada et al., Synlett *1996*, 43-5.

Zinc bromide s. under RCp$_2$ZrCl   ZnBr$_2$
Triethylaluminum s. under Ni(acac)$_2$ Et$_3$Al
Methylaluminum bis(2,6-di-tert-butyl-4-methylphenoxide)/azodiisobutyronitrile/
irradiation ←
Asym. radical cross-coupling of 1,2-iodohydrins s. *51*, 436   ←
Trimethylsilyl tetrakis(triflyloxy)borate   Me$_3$SiB(OTf)$_4$
Asym. synthesis of 3-ethylene-*tert*-alcohols   CO → C(OR)C–C=C
from ketones and 2-ethylenesilanes via *tert*-alkoxy-3-ethylenes

438.

Synth.Meth. *42*, 876s*48* has been applied to ketones in a new route to chiral 3-ethylene-*tert*-alcohols. E: A soln. of ethyl methyl ketone and startg. chiral alkoxysilane in methylene chloride treated with 0.1 eq. trimethylsilyl tetrakis(triflyloxy)borate (or 0.1 eq. each of Me$_3$SiOTf and TfOH) at -78°, allyltrimethylsilane added, and worked up when reaction complete → intermediate ether (Y 82%; 89% based on conversion; 89:11 mixture of diastereomers), treated with Na in liq. NH$_3$ in the presence of methanol → product (Y 76%). There was no loss of chirality in the second step. It is thought that reaction proceeds via an oxazolidinium ion as the amide group in the alkoxysilane appears critical. Ar. ketones did not react, and selectivity and yield were reduced with enones. However, β-alkoxyketones, ketoesters, and γ,δ-ethyleneketones may be used as substrate. F.e.s. L.F. Tietze et al., J. Am. Chem. Soc. *117*, 5851-2 (1995).

Alumina   Al$_2$O$_3$
Stereospecific Wittig synthesis under mild conditions   CO → C=C
in the presence of an inorganic solid support

439.

(E)-α,β-Ethylenecarboxylic acid esters. A mixture of benzaldehyde and 1.1 eqs. startg. P-ylid stirred *without solvent* at 25° for 10 min under N$_2$, 10 eqs. (by weight) of alumina (activated at 200° for 4 h *in vacuo*) added, and the solid mixture stirred for 1.5 h → ethyl cinnamate. Y 88% (E:Z 93:7). Reaction is fast and generally applicable to aliphatic and [het]ar. aldehydes. F.e.s. D.D. Dhavale et al., J. Chem. Res. (S) *1995*, 414-5; with silica gel in non-polar media cf. V.J. Patil, U. Mävers, Tetrahedron Letters *37*, 1281-4 (1996); with SiO$_2$ under microwave irradiation without solvent cf. C. Xu et al., Synth. Commun. *25*, 2229-33 (1995); *trans*-chalcones s. Org. Prep. Proc. Intern. *27*, 559-61 (1995).

Montmorillonite ←
Lewis acid-catalyzed reactions of acoxy-2-ethylenes with Si-nucleophiles ←
2,3-unsatd. C-glycosides from O^3-acylglycals with SnCl$_4$ cf. *37*, 871s*49*; 2,3-unsatd. C-allyl glycosides with montmorillonite K 10 s. K. Toshima et al., Chem. Commun. *1996*, 1379-80; α-isomers with DDQ cf. Chem. Lett. *1993*, 2013-6.

*9-Bromo-9-borabicyclo[3.3.1]nonane/2,6-di-*tert-*butylpyridine*
β-Hydroxyketones from enoxysilanes and oxo compds. $CO \rightarrow C(OH)C-CO$
with $Bu_2BOSO_2CF_3$ cf. *30*, 621s*36*; *syn*-products **via (Z)-enol borinates** from (Z)- or (E)-enoxy silanes with 9-bromo-9-borabicyclo[3.3.1]nonane/2,6-di-*tert*-butylpyridine s. D.A. Evans et al., Tetrahedron Letters *36*, 9245-8 (1995); *erythro*-products with $Ph_4P^+HF_2^-$ cf. A. Bohsako et al., Synlett *1995*, 1033-4.

Lewis acids ($ZnCl_2$, BF_3, $TiCl_4$ or $SnCl_4$)
3-Ethylenealcohols from oxo compds. and 2-ethylene-silanes or -stannanes $C(OH)C-C\equiv C$
s. *36*, 879s*46-9*; from 2-ethylenestannanes with reversal of diastereoselectivity by $ZnCl_2$ s. Y. Nishigaichi, A. Takuwa, Chem. Lett. *1994*, 1429-32; effect of O-functionalization on reactivity of allylstannanes s. J.A. Marshall et al., J. Org. Chem. *61*, 2904-7 (1996); *anti*-3-ethylenealcohols, incl. 3-ene-1,2-diol 2-monoethers, with $InCl_3$ cf. ibid. *60*, 1920-1 (1995); *trans*-4-hydroxyenesilanes from 2-ethylene-1,1-silylstannanes, also 4-alkoxy-derivs. from acetals, s. M. Lautens et al., Tetrahedron *52*, 7221-34 (1996); 1(E),3(E),7-trien-5-ols from (π-acylallyl-)tricarbonyliron-lactone complexes s. S.V. Ley, L.R. Cox, Chem. Commun. *1996*, 657-8; 5-β-hydroxyisoxazoles (with $TiCl_4$) s. M. Yamamoto et al., Synth. Commun. *26*, 2177-87 (1996); 2-methylene-1,4-chlorhydrins s. M. Taddei et al., Synlett 1993, 119-21; from aldehydes with Me_3SiOTf/Me_2S **via 1,1-siloxysulfonium salts** cf. S. Kim, S.H. Kim, Tetrahedron Letters *36*, 3723-4 (1995).

Boron fluoride (s.a. under Zn) BF_3
Regiospecific synthesis of β-aminoketones $COC-CH_2N\!<$
**from α-silylketones with asym. induction
via Mannich reaction with (Z)-2'-silylenoxysilanes**

440.

A soln. of startg. chiral α-silylketone (prepared by asym. α-silylation of the corresponding SAMP hydrazone) in THF added dropwise over 5 min to a vigorously stirred soln. of 1.1 eqs. LDA in 4:1 THF/HMPA at -95°, treated with 1.2 eqs. Me_3SiCl in one portion, and quenched with satd. aq. $NaHCO_3$ → crude enoxysilane, in methylene chloride added dropwise to a soln. of 1.15 eqs. dibenzylmethoxymethylamine in the same solvent at -95°, treated with 1.15 eqs. BF_3-etherate, allowed to warm to -70°, and worked up after 3 h → intermediate (S,R)-aminosilane (Y 95%; d.e. ≥ 96%), in THF *at -70°* treated sequentially with 50 eqs. solid NH_4F and 1.2 eqs. 1 *N* Bu_4NF in THF, allowed to warm to room temp., and pH 6 buffer added → (R)-product (Y 97%; e.e. 93%). Desilylation with Bu_4NF in THF at room temp. resulted in complete racemization. F.e.s. D. Enders et al., Angew. Chem. Intern. Ed. *35*, 981-4 (1996).

Synthesis of selenides from 2-hydroxyselenides via seleniranium ions   ←
Retention of configuration

441.

Chiral ketones. A soln. of startg. chiral hydroxyselenide (e.e. >99.7%) and 4 eqs. acetophenone trimethylsilyl enolether in methylene chloride treated with 2 eqs. BF_3-etherate at 0° for 8 h → intermediate ketoselenide (Y 84%; e.e. >99.7%), in toluene refluxed with Bu_3SnH/AIBN for 4 h → product (Y 82%; e.e. >99.7%). The hindered Se-2,4,6-tri-*tert*-butylphenyl group serves to prevent undesirable selenophilic attack of the C-nucleophile and to **suppress racemization** at the chiral centre in the episelenonium ion. F.e. and Si-nucleophiles s. A. Toshimitsu et al., J. Am. Chem. Soc. *118*, 2756-7 (1996).

Asym. synthesis of *syn*-3-ene-1,2-diols from aldehydes   C(OH)C(OSi≤)C═C
via 2-monoethers cf. *48*, 889; via 2-monosilyl ethers s. J. Org. Chem. *60*, 2662-3 (1995).
Stereospecific 1,2-addition to α-subst. β-alkoxyaldehydes   ←
under non-chelating conditions

442.

The first examples of *anti-Felkin* carbophilic addition are reported, wherein dominant stereocontrol emanates from a remote β-oxy function under conditions which preclude chelate organisation. E: **Aldol-type condensation**. A soln. of startg. *syn*-aldehyde and enoxysilane in toluene treated with 1 eq. BF$_3$-etherate at -78° until reaction complete → product. Y 93% (94% anti-Felkin). The 'non-reinforcing' *syn*-disposition of the α-alkyl and β-hetero functions in the substrate is essential; in the corresponding *anti*-substrate, Felkin addition predominates under conditions where the two substituents are mutually reinforcing. Similar results were obtained with allylsilanes as well as Li- and boron-enolates (not Grignard compds.). F.e.s. D.A. Evans et al., J. Am. Chem. Soc. *117*, 6619-20 (1995).

Samarium/mercuric chloride   Sm/HgCl$_2$
Generation of non-stabilized carbonyl ylids from two 1,1-siloxyiodide molecules   ←

443.

Stereospecific 1,3-dipolar cycloaddition. A suspension of 2.1 eqs. Sm metal and 0.4 eq. HgCl$_2$ in toluene stirred for 20 min, 1 eq. startg. alkene added to the white suspension, cooled to -78°, a soln. of 4 eqs. startg. siloxyiodide in toluene added, allowed to warm to room temp., stirred for a further 1 h, and poured into satd. aq. NaHCO$_3$ → product. Y 65% (>95/<5 mixture of diastereomers). The procedure is notably applicable to generating *alkyl-substituted* carbonyl ylids. F.e. and dipolarophiles, also with SmI$_2$/Sm(0), and sym. epoxides with SmI$_2$/tetramethylurea, s. M. Hojo et al., J. Am. Chem. Soc. *118*, 3533-4 (1996).

Indium(III) chloride   InCl$_3$
***anti*-3-Ethylenealcohols from oxo compds. and 2-ethylenestannanes**   C(OH)C-C═C
s. *36*, 879s*51*

Ammonium ceric nitrate s. under KOH   (NH$_4$)$_2$Ce(NO$_3$)$_4$

Lanthanide(III) β-diketonates/azodiisobutyronitrile   Ln(III)/AIBN
Asym. radical cross-coupling   ←
with α-halogeno-β-siloxycarboxylic acid esters s. *51*, 436

Lanthanide(III) or Scandium(III) triflates   Ln(OTf)$_3$ or Sc(OTf)$_3$
3-Ethyleneamines from 2-ethylenestannanes and aldimines s. *40*, 633s*51*   CH(NHR)C-C═C

Ytterbium(III) triflate   Yb(OTf)$_3$
3-Component synthesis of β-aminocarboxylic acid esters   ←
from O-silyl O-alkyl keteneacetals, aldehydes and aminosilanes with LiClO$_4$ cf. *50*, 551; from amines with Yb(OTf)$_3$, Er(OTf)$_3$ or Tm(OTf)$_3$ and 4Å molecular sieves s. S. Kobayashi et al., Tetrahedron Letters *36*, 5773-6 (1995); asym. induction with chiral amines s. P.G. Cozzi et al., ibid. *37*, 1691-4 (1996); **from azomethines** with Yb(OTf)$_3$ or Sc(OTf)$_3$ (cf. *33*, 879s*51*), also β-aminothiolic acid esters, cf. Synlett *1995*, 233-4.

Yttrium(III) triflate/(R,R)-N,N'-bis(trifluoromethanesulfonyl)-1,2-diphenylethylenediamine/
sodium hydride
2,3-Dihydro-4-pyrones from aldehydes ←
via asym. hetero-Diels-Alder reaction s. *43*, 851s*51* O

Nafion-supported scandium(III)   *Sc(III)*
3-Ethylenealcohols from oxo compds. and 2-ethylenestannanes   C(OH)C–C=C
in aq. media with Sc(OTf)$_3$ cf. *49*, 909s*50*; with readily removable Nafion-supported scandium(III), also in organic solvents, s. S. Kobayashi, S. Nagayama, J. Org. Chem. *61*, 2256-7 (1996).

Azodiisobutyronitrile   RN=NR
Stereospecific radical cross-coupling of 2-ethylenestannanes ←
with halides or selenides s. *46*, 832s*47-9*; *syn*- and *anti*-γ,δ-ethylene-β'-oxysulfoxides s. P. Renaud, T. Bourquard, Synlett *1995*, 1021-3; synthesis of enesilanes s. Chimia *48*, 366-9 (1994); coupling of 2-ethylene[bis(disilylamino)]halogenostannanes, improved work-up, s. E. Fouquet et al., Chem. Commun. *1995*, 2387-8.

γ,δ-Ethyleneketones from diazomethyl ketones   CH=N$_2$ → CH$_2$CH$_2$CH=CH$_2$
Synthesis with addition of three C-atoms

444. n-C$_5$H$_{11}$–CO–CH=N$_2$ + CH$_2$=CH–CH$_2$–SnBu$_3$ ⟶ n-C$_5$H$_{11}$–CO–CH$_2$–CH$_2$–CH=CH$_2$

A soln. of 1-diazoheptan-2-one, 2 eqs. allyltri-*n*-butylstannane, and a little AIBN in benzene heated under reflux for 3 h, cooled to room temp., diluted with ether, and shaken with satd. aq. KF → dec-1-en-5-one. Y 68%. F.e.s. H.-S. Dang, B.P. Roberts, J. Chem. Soc. Perkin Trans. I *1996*, 769-75.

9,10-Dicyanoanthracene/biphenyl/irradiation ←
Co-sensitized 1,4-addition with 1,1-silylurethans under mild conditions ←

445. C$_6$H$_{11}$–N(SiMe$_3$)–CO–OMe + (MeO$_2$C)CH=CH(CO$_2$Me) ⟶ C$_6$H$_{11}$–N(CH(CO$_2$Me)CH$_2$CO$_2$Me)–CO–OMe

A stirred, water-cooled mixture of startg. 1,1-silylurethan, 2 eqs. dimethyl maleate, 0.3 eq. biphenyl, and 0.1 eq. 9,10-dicyanoanthracene in degassed 2:1 acetonitrile/water irradiated for 3.5 h with a 450 W Xenon lamp through a filter (>345 nm), a further 0.1 eq. 9,10-dicyanoanthracene added, and irradiation continued for 3.5 h → product. Y 62%. The two sensitizers work in concert to generate a biphenyl radical cation which oxidizes the substrate under diffusion control. The method is quite general with acceptor-substituted alkenes (with very reactive alkenes, the alkene was added continuously to prevent oligo- and poly-merization); however, electron-rich alkenes, such as cyclohexene and 1-hexene, were unreactive. F.e.s. E. Meggers et al., Angew. Chem. Intern. Ed. *34*, 2137-9 (1995).

(L)-Bis(2,4-dimethyl-3-pentyl) tartrate ←
Asym. synthesis of 2-α-hydroxy-1,3-dienes   CH(OH)C(=CH$_2$)CH=CH$_2$
from cyclic acetals and 1-trimethylsilyl-2,3-butadiene cf. *47*, 901; from aldehydes and diisopropyl 2,3-butadien-1-ylboronate with (L)-bis(2,4-dimethyl-3-pentyl) tartrate (cf. *38*, 890) s. H.C. Brown et al., J. Org. Chem. *61*, 100-4 (1996).

2,3-Dichloro-5,6-dicyanoquinone   *DDQ*
2,3-Unsatd. C-α-allyl glycosides from O^3-acylglycals s. *37*, 871s*51* ←

Silica gel or Silica gel/microwaves ←
Stereospecific Wittig synthesis s. *51*, 439   CO → C=C

Dimethyltitanocene   *Cp$_2$TiMe$_2$*
Tebbe-type methylenation with dimethyltitanocene   CO → C=CH$_2$
s. *46*, 870; enoxysilanes, cyclic ketene acetals, thioenolethers, eneselenides, enesilanes, enamines, and bis(methylenation) of dicarboxylic acid anhydrides and imides s. N.A. Petasis, S.-P. Lu,

Tetrahedron Letters *36*, 2393-6 (1995); 3-methylenecyclobutenes s. ibid. 6001-4; isopropenyl ethers s. T. Le Diguarher et al., Synth. Commun. *25*, 1633-9 (1995); 2-functionalized enesilanes with bis(trimethylsilylmethyl)titanocene/bis(trimethylsilyl)ethyne s. N.A. Petasis et al., Tetrahedron Letters *36*, 3619-22 (1995); 2-(trifluoromethyl)enesilanes s. J.-P. Begue, M.H. Rock, J. Organometal. Chem. *489*, C7-C8 (1995); 4-alkylidene-1,3-dioxolanes and -1,3-dioxanes s. *51*, 345.

Titanium silicate or Sulfated zirconia ←
Heterogeneous aldol-type condensation under mild conditions   CHO → CH(OH)C–CO

446.   PhCHO + [structure with OSiMe$_3$ and OMe] → [Ph-CH(OH)-C(Me)-C(O)OMe structure]

Titanium silicate has been used for the first time in a non-oxidative conversion. E: Titanium silicate molecular sieve (TS-1) added to a mixture of startg. silyl enol ether and 1 eq. benzaldehyde in *dry* THF, stirred at 25-27° until TLC indicated completion of reaction, and worked up in the normal way → product. Y 85%. Rather than display a Lewis acid effect, the catalyst activates the enolic oxygen of the O-Si bond prior to double bond shift. F.e., stereoselectivity with enoxysilanes, comparison with other zeolites, and effect of added water s. R. Kumar et al., Chem. Commun. *1996*, 129-30; with sulfated zirconia cf. K.V. Srinivasan et al., Synlett *1996*, 239-40.

Titanium tetraisopropoxide/(R)-1,1'-bi-2-naphthol   ←
2,3-Dihydro-4-pyrones from aldehydes via asym. hetero-Diels-Alder reaction   O
with chiral 3-sulfonyl-1,3,2-oxazaborolidin-5-ones cf. *43*, 851s*50*; improved enantioselectivity with Ti(OPr-*i*)$_4$/(R)-BINOL s. G.E. Keck et al., J. Org. Chem. *60*, 5998-9 (1995); with Y(OTf)$_3$/ (R,R)-N,N'-bis(trifluoromethanesulfonyl)-1,2-diphenylethylenediamine/NaH s. K. Mikami et al., Synlett *1995*, 975-7.

Titanium tetraisopropoxide/racemic 1,1'-bi-2-naphthol/diisopropyl D-*tartrate*   ←
Asym. synthesis of 3-ethylenealcohols from aldehydes   CHO → CH(OH)C–C═C
Modified Keck procedure

447.   Ph-CHO + Bu$_3$Sn-CH$_2$-CH=CH$_2$ → Ph-CH(OH)-CH$_2$-CH=CH$_2$

In asym. synthesis of 3-ethylenealcohols from aldehydes and allylstannanes by the Keck procedure (cf. *49*, 898), high enantioselectivity can be achieved in the presence of an *inexpensive* combination of Ti(OPr-*i*)$_4$/chiral diol and Ti(OPr-*i*)$_4$/(rac)-BINOL in place of the expensive Ti(OPr-*i*)$_4$/(S)- or (R)-BINOL system. E: 20 Mol% *racemic* BINOL, 30 mol% diisopropyl D-tartrate (D-DIPT) and 30 mol% Ti(OPr-*i*)$_4$ heated under reflux in methylene chloride in the presence of 4 Å molecular sieves for 1 h, benzaldehyde added at room temp., cooled to -78°, treated with 1.1 eqs. allyltributyltin, and kept at -23° for 70 h prior to the usual work-up → (S)-product. Y 63% (e.e. 91%). The true catalyst is considered to be a mixture of dimeric complexes, Ti$_2$(OPr-*i*)$_4$[(R)-BINOL][D-DIPT] and Ti$_2$(OPr-*i*)$_4$[(S)-BINOL][D-DIPT], the former being less active. F.e.s. J.W. Faller et al., J. Am. Chem. Soc. *118*, 1217-8 (1996).

Zirconium tetraisopropoxide/(S)-1,1'-bi-2-naphthol   ←
with Cl$_2$Ti(OPr-*i*)$_2$/(S)-1,1'-bi-2-naphthol cf. *49*, 898; with Zr(OPr-*i*)$_4$, notably for reaction of ar. and *hindered* aldehydes, s. E. Tagliavini et al., Tetrahedron Letters *36*, 7897-900 (1995).

Bis[2-(o-hydroxyphenyl)-Δ^2-oxazolinato]-titanium(IV) or -zirconium(IV) bistriflate   ←
β-Aminocarboxylic acid esters   C═N → C(NH)C—COOR
from azomethines and O-silyl O-alkyl keteneacetals
with TiCl$_4$ cf. *33*, 879; with bis[2-(*o*-hydroxyphenyl)-Δ^2-oxazolinato]-titanium(IV) or zirconium(IV) bistriflate as Lewis acid, and f. reactions (incl. catalytic allylation, hydrocyanation, and [hetero] Diels-Alder reactions) s. P.G. Cozzi, C. Floriani, J. Chem. Soc. Perkin Trans. I *1995*, 2557-63.

Trimethylsilyl chloride/ammonium fluoride   *Me$_3$SiCl/NH$_4$F*
3-Ethyleneamines from 2-ethylenestannanes and aldimines   CH(NHR)C–C═C
with TiCl$_4$ cf. *40*, 633; with Me$_3$SiCl/NH$_4$F (from allyltributylstannane) s. D.-K. Wang et al.,

Tetrahedron Letters 36, 8649-52 (1995); with lanthanide(III) triflates or Sc(OTf)₃, also with asym. induction, s. C. Bellucci et al., ibid. 7289-92; with bis(η³-allylpalladium chloride), retention of aldehydes, cf. H. Nakamura et al., Chem. Commun. *1996*, 1459-60; N-homoallylanilines from *in situ*-generated N-arylformimines s. H.-J. Ha et al., J. Chem. Soc. Perkin Trans. I *1995*, 2631-4.

Organozirconocene chlorides/zinc bromide/benzaldehyde $RCp_2ZrCl/ZnBr_2/PhCHO$
Ketones from aldehydes via Oppenauer-type oxidation $CHO \rightarrow COR$
One-pot procedure

448.

α,β-Ethyleneketones. A 0.8 M soln. of hydrocinnamaldehyde and 1.1 eqs. 2-cyclopentylvinyl zirconocene chloride in THF treated with 0.2 eq. ZnBr₂ at 0° for 15 min, 1 eq. benzaldehyde added, and stirred at 25° for 3 h → product. Y 91%. Reaction takes place with retention of configuration. F.e.s. B. Zheng, M. Srebnik, J. Org. Chem. *60*, 3278-9 (1995).

Titanium tetrachloride (s.a. under BF₃) $TiCl_4$
Asym. ring opening of cyclic acetals with Si-nucleophiles   ○
Improved conversion s. *40*, 635s*51*

Stannous chloride/acetyl chloride $SnCl_2/AcCl$
3-Methylenetetrahydrofurans from trimethylenemethane equivalents   ○
and aldehydes with Pd(OAc)₂/Bu₃SnOAc cf. *42*, 908; **from acetals** with 2-(trimethylsilyloxymethyl)-allyltrimethylsilane and SnCl₂/AcCl as catalyst s. T. Oriyama et al., Tetrahedron Letters *36*, 5581-4 (1995).

Stannic chloride s. under BF₃ $SnCl_4$
Tetraphenylphosphonium hydrogen fluoride $Ph_4P^+HF_2^-$
***erythro*-β-Hydroxyketones from enoxysilanes and aldehydes** s. *30*, 621s*51*   C(OH)C–CO
2-Acetylenealcohols from silylacetylenes and oxo compds. s. *30*, 624s*51*   C(OH)C≡C
Bismuth trichloride/sodium iodide $BiCl_3/NaI$
β-Diketones from enoxysilanes   C(OSi≤)=C → COC–CO
Preferential C-acylation

449.

A suspension of 1.1 eqs. acetyl chloride, 0.05 eq. BiCl₃, and 0.15 eq. NaI in 9:1 dichloromethane/ether stirred at room temp. for 5 min under argon, propiophenone trimethylsilyl enol ether in the same solvent added in one portion with rapid stirring, and quenched with satd. NaHCO₃ after 40 min at room temp. → 1-phenyl-2-methyl-1,3-butanedione. Y 93% (by GC). This is the first *catalytic* C-acylation of enoxysilanes. For less reactive acyl chlorides, the use of ZnI₂ instead of NaI increased yields and shortened reaction times. F.e.s. C. Le Roux et al., J. Org. Chem. *61*, 3885-7 (1996).

Bis(fluorosulfonyl)imine $HN(SO_2F)_2$
3-Ethylenealcohols from oxo compds.   CO → C(OH)C–C=C
Synthesis with addition of three C-atoms

450.

Hydrocinnamaldehyde treated with 1.2 eqs. allyltrimethylsilane in methylene chloride in the presence of 5 mol% HN(SO₂F)₂ at 0° for 5 min under N₂ → product. Y 89%. The aldehyde reacted preferentially in the presence of cyclopentanone, α-tetralone or acetophenone; with other ketones, however, reaction took place at lower temp. (-40°). This highly efficient method illustrates the first example of this transformation **under Brönsted acid catalysis**. F.e.s. G. Kaur et al., Chem. Commun. *1996*, 581-2.

Trimethylsilyl triflate/dimethyl sulfide Me_3SiOTf/Me_2S
3-Ethylenealcohols from aldehydes via 1,1-siloxysulfonium salts $CH(OH)CH_2CH=CH_2$
s. *46*, 879s51
Ammonium fluoride s. *under* Me_3SiCl NH_4F
*Tetra-*n-*butylammonium fluoride* Bu_4NF
Synthesis of alcohols from oxo compds. and silanes $CO \rightarrow C(OH)R$
s. *30*, 624s*46,48*; 3-alkylidene-4-α-hydroxy-2-azetidinones s. K. Suda et al., Chem. Pharm. Bull. *44*, 466-8 (1996); 2,2-difluoroalcohols with KF s. T. Hagiwara, T. Fuchikami, Synlett *1995*, 717-8; 1-trifluoromethylbenzocyclobutenols s. D.P. Becker, D.L. Flynn, ibid. *1996*, 57-9; 2-acetylene-alcohols with $Ph_4P^+HF_2^-$ s. A. Bohsako et al., ibid. *1995*, 1033-4; β-hydroxy-α-methyleneketones s. K. Matsumoto et al., Chem. Lett. *1994*, 1211-4.

Nickel acetoacetonate/triphenylphosphine/triethylaluminum $Ni(acac)_2/Ph_3P/Et_3Al$
Regio- and stereo-specific cross-coupling of 2-ethyleneamines $N< \rightarrow R$

451.

with boronic acids. The amino function of allylamines is a better leaving group than allylic acoxy or hydroxyl in Ni-catalyzed cross-coupling with aryl, vinyl and methyl-boronic acids (or esters), which proceeds with *inversion* of configuration at the displaced carbon atom. E: 20 Mol% 1 *M* Et$_3$Al in hexane added slowly to 10 mol% Ni(acac)$_2$ and 40 mol% PPh$_3$ in toluene, stirred at 60° for 20 min, transferred via cannula to a mixture of startg. allylamine and 1.1 eqs. phenylboronic acid in toluene at 60°, and heated at reflux for 10 h → product. Y 57%. In general, bulky phosphine ligands promote reaction at the less-substituted allylic terminus, while bidentate ligands, notably 1,1′-binaphthyl-2,2′-bis(diphenylphosphinite), favour the more substituted position. Vinylboronic acids react with retention of geometry, (E)-isomers coupling more efficiently. For methylation, however, methylboronic esters, e.g. 2-methyl-1,3,2-benzodioxaborole, are preferred. F.e.s. B.M. Trost, M.D. Spagnol, J. Chem. Soc. Perkin Trans. I *1995*, 2083-96.

*Dichlorobis(triphenylphosphine)nickel(II)/triphenylphosphine/*n-*butyllithium* $\leftarrow$
Allylarenes from benzyl chlorides and enalanes $Cl \rightarrow C=C$

452.

Retention of configuration. 0.5 *mol%* Ni(0) is a more rapid and cheaper catalyst than Pd(0) for coupling benzyl chlorides with enalanes. E: **Ubiquinones**. 1.5 eqs. 2 *M* Me$_3$Al in hexane added at 0° to 0.25 eq. Cp$_2$ZrCl$_2$ under argon, stirred under reduced pressure to remove the solvent, dichloromethane added, warmed to room temp. over 30 min, startg. alkyne added, stirred at 0° for 30 min to complete carboalumination, the crude enalane (Zr-free) taken up in THF, the dark blue soln. obtained by adding 0.67 eq. benzyl chloride to a soln. of 5 mol% Ni(0) (prepared by stirring 5 mol% (PPh$_3$)$_2$NiCl$_2$, 10 mol% PPh$_3$, and 5 mol% 0.49 *M* n-BuLi [in hexane] in THF at room temp. under argon for 30 min) transferred via cannula to the enalane soln. at room temp., dil. with ether after <*15 min*, and quenched dropwise at 0° with 1 *M* HCl → product. Y 87%. Both electron-rich and -poor benzyl chlorides reacted rapidly, and good functional group compatibility is anticipated (e.g. esters were unaffected). F.e.s. B.H. Lipshutz et al., J. Am. Chem. Soc. *118*, 5512-3 (1996).

Dichloro[1,1'-bis(diphenylphosphino)ferrocene]nickel(II)/n-butyllithium or zinc/potassium phosphate ←
Diaryls from aryl mesylates and arylboronic acids $ArB(OH)_2 \rightarrow Ar-Ar'$
Nickel-catalyzed Suzuki coupling under mild conditions

453.   MeO$_2$C—⟨ ⟩—OMs + (OH)$_2$B—⟨ ⟩   →   MeO$_2$C—⟨ ⟩—⟨ ⟩

Freshly distilled THF added via syringe to a mixture of methyl 4-mesyloxybenzoate, 1.1 eqs. phenylboronic acid, 10 mol% NiCl$_2$(dppf), 1.7 eqs. Zn-powder, and 3 eqs. K_3PO_4 in a dry Schlenk tube under N$_2$, and stirred at 67° for 24 h → product. Y 48% by GC (67% in dioxane at 95°). Reaction is cheaper than the Pd-catalyzed conversion with aryl triflates, and there was no homocoupling of the [less reactive] mesylate (as took place with the corresponding triflate under Ni-catalysis). A number of functional groups were tolerated (ketones, esters, phenolethers). F.e.s. V. Percec et al., J. Org. Chem. *60*, 1060-5 (1995); **from ar. chlorides** with NiCl$_2$(dppf)/*n*-BuLi/K$_3$PO$_4$ cf. S. Saito et al., Tetrahedron Letters *37*, 2993-6 (1996).

Palladium-carbon/cuprous iodide/triphenylarsine ←
Stille coupling with α,β-unsatd. halides or triflates s. *34*, 862s*51*   Sn≼ → R
Palladous acetate/triisopropyl phosphite $Pd(OAc)_2/(i-PrO)_3P$
2-Cyclopentenone ring from α,β-ethylenesulfones   ○
via 3-sulfonyl-1-methylenecyclopentane ring

454.

Asym. induction. A soln. of 3(S)-*tert*-butyldimethylsilyloxy-1-(phenylsulfonyl)cyclohexene, ca. 5 mol% Pd(OAc)$_2$, and 0.3 eq. triisopropyl phosphite in toluene treated with 2-[(trimethylsilyl)-methyl]allyl pivalate at 80°, further amounts of the latter ester added after 24 and 48 h, and worked up when TLC indicated completion of reaction → 2-methylene-4(S)-(*tert*-butyldimethyl-siloxy)-7a(S)-(phenylsulfonyl)-1,2,3a(R),4,5,6,7,7a-octahydroindene (Y 83%), in 1:1 methanol/methylene chloride subjected to O$_3$-bubbling at -78° until the blue colour persisted, purged with N$_2$, treated with Me$_2$S, warmed to room temp., Et$_3$N added, and stirred overnight → 7(S)-(*tert*-butyldimethylsiloxy)-1,4,5,6,7,7a(S)-hexahydroinden-2-one (Y 94%). Details s. B.M. Trost et al., J. Am. Chem. Soc. *117*, 9662-70 (1995).

Bis(dibenzylideneacetone)palladium(0) $Pd(dba)_2$
4-Aryl-2-ethylenesilanes from 1,3-dienes, ar. iodides and silylstannanes $C(Ar)C=C-C(Si≼)$

455.   Bu$_3$SnSiMe$_3$ + ⟩=⟨ + I—⟨ ⟩   →   Me$_3$Si—⟨ ⟩—⟨ ⟩

A soln. of 1.6 *M* 1,3-butadiene in toluene, 1.6 mol% Pd(dba)$_2$, and 0.33 eq. each of tributyl(trimethylsilyl)stannane and iodobenzene in toluene purged with argon, and heated with stirring in a glass-lined autoclave at 100° for 4 h → product (E:Z 88:12). Y 42% (containing 16% regioisomer). The selectivity is not as high as in the 3-component reaction with acyl chlorides and disilanes (cf. *49*, 934). F.e.s. Y. Obora et al., J. Am. Chem. Soc. *117*, 9814-21 (1995).

Bis(dibenzylideneacetone)palladium(0)/triphenylstibine/lithium chloride ←
Allylarenes from ar. halides and allylstannanes   Hal → C-C=C
with (PPh$_3$)$_4$Pd cf. *33*, 883; regiospecific synthesis of 2-ethylenesilanes from 2-ethylene-4-silylstannanes with Pd(dba)$_2$/Ph$_3$Sb/LiCl, also coupling with alkyl or aryl triflates, s. Y. Tsuji et al., J. Org. Chem. *60*, 4647-9 (1995).

Palladous chloride/silver acetate/triethylamine $PdCl_2/AgOAc/Et_3N$
Heck-type arylation of electron-deficient ethylene derivs. with diaryl tellurides $C=C(Ar)$

456.

(E)-Stilbenes. Dry methanol, 2 eqs. Et$_3$N, and styrene added to 5 mol% PdCl$_2$, 2 eqs. AgOAc, and 0.5 eq. bis(p-methoxyphenyl) telluride, and the heterogeneous mixture stirred at 25° for 20 h → (E)-p-methoxystilbene. Y 99%. Vinylation was less effective. F.e.s. Y. Nishibayashi et al., J. Organometal. Chem. *507*, 197-200 (1996).

Palladous chloride/1,1'-bis(diphenylphosphino)ferrocene/tetra-n-butylammonium fluoride/ triethylamine   ←
α,β-Acetyleneketones from halides by carbonylation   Hal → COC≡C
with terminal alkynes s. *37*, 828; **also from silylacetylenes** in one pot with added Bu$_4$NF s. S. Cacchi et al., Synlett *1995*, 823-4.

Bis(η^3-allylpalladium chloride)   ←
3-Ethyleneamines from 2-ethylenestannanes and aldimines s. *40*, 633s51   CH(NHR)C-C=C
Bis(η^3-allylpalladium chloride)/2,4-bis[(4,6-dimethyl-1,3,2-dioxaphosphorinan-2-yl)oxy]pentane   ←
Regiospecific transition metal-catalyzed high-pressure cycloaddition   O

457.

under kinetic control. High pressure has been shown to reverse the regioselectivity of transition metal-catalyzed cycloaddition by comparison with the conversion at atm. pressure. **E: [3+2]-Cycloaddition with trimethylenemethane equivalents**. A soln. of coumarin, startg. trimethylenemethane precursor, 2.5 mol% (η^3-C$_3$H$_5$PdCl)$_2$, and 5 mol% 2,4-bis[(4,6-dimethyl-1,3,2-dioxaphosphorinan-2-yl)oxy]pentane in 7:3 toluene/benzene heated at 70° under 15 kbar → product. Y 77% (ca. 60% ethylidene-isomer). With 5 mol% Pd(PPh$_3$)$_4$ and 5 mol% PPh$_3$ at 65° under atm. pressure, the thermodynamically more stable methylene-isomer predominated (cf. *36*, 894). F.e. and effect on [6+4]-cycloaddition s. B.M. Trost et al., J. Am. Chem. Soc. *117*, 3284-5 (1995).

Palladium phosphine complexes   ←
Suzuki coupling of boronic acids with halides or triflates   B(OH)$_2$ → R
s. *37*, 902s46-50; 4-biphenylyl glycosides s. H. Müller, C. Tschierske, Chem. Commun. *1995*, 645-6; aryl-subst. phenylalaninates s. M.J. Burk et al., J. Am. Chem. Soc. *116*, 10847-8 (1994); aryl-subst. cyclic tryptophan analogs s. D.E. Zembower, M.M. Ames, Synthesis *1994*, 1433-6; 4-arylpyridines s. A. Godard et al., J. Organometal. Chem. *517*, 25-36 (1996); phenylpyrroles s. C.K. Chang et al., J. Org. Chem. *60*, 7030-2 (1995); 2-arylporphyrins s. X. Zhou et al., ibid. *61*, 3590-3 (1996); 4-arylcoumarins s. J. Chem. Soc. Perkin Trans. I *1996*, 2591-7; 2-aryl-Δ^2-carbapenems s. N. Yasuda et al., Tetrahedron Letters *34*, 3211-4 (1993); **with triarylboroxines** s. Z.Z. Song, H.N.C. Wong, J. Org. Chem. *59*, 33-41 (1994); Angew. Chem. Intern. Ed. *32*, 432-4 (1993); s.a. Y. Satoh, C. Shi, Synthesis *1994*, 1146-8.
Stille coupling with halides   Sn≤ → R
s. *34*, 862s46-49; incorporation of ^{11}CH$_3$ groups s. Y. Andersson et al., Acta Chem. Scand. *49*, 683-8 (1995); coupling with α,β-unsatd. halides (or triflates) with *inexpensive* Pd-C/CuI/Ph$_3$As cf. G.P. Roth et al., Tetrahedron Letters *36*, 2191-4 (1995).
Stille coupling of enestannanes with halides
s. *39*, 887s46-49; 2-α-hydroxy-1,3(4)-dienes s. W. Adam, P. Klug, J. Org. Chem. *59*, 2695-9 (1994); 3-subst. cyclobutenones s. L.S. Liebeskind et al., ibid. 7917-20; (Z)-α,β-ethyleneketones (with Pd(PPh$_3$)$_4$/CuI) s. T. Takeda et al., Tetrahedron *51*, 2515-24 (1995); C$_1$-vinylglycals s. A.

Abas et al., Synlett *1995*, 1264-6; chiral 2-sulfinyl-1,3-dienes s. R.S. Paley et al., Tetrahedron Letters *36*, 3605-8 (1995); 2-subst. penems s. M.A. Armitage et al., ibid. 775-6.

Stille coupling of hetarylstannanes with halides   Sn≤ → CR
s. *34*, 862s*46,48,49*; with 5-stannylisoxazoles s. Acta Chem. Scand. *49*, 53-6 (1995); with N-protected 3-stannylpyrroles s. J. Wang, A.I. Scott, Tetrahedron Letters *37*, 3247-50 (1996); with 1-carboxy-2-stannylindoles s. R.L. Hudkins et al., J. Org. Chem. *60*, 6218-20 (1995); prepn. of oligopyridyls s. J.A. Zoltewitz et al., Tetrahedron *51*, 11393-400 (1995); s.a. D.J. Cárdenas, J.-P. Sauvage, Synlett *1996*, 916-8; polypyridylarenes s. M. Fujita et al., Tetrahedron Letters *36*, 5247-50 (1995).

Bis(benzonitrile)dichloropalladium(II)/triethyl phosphite/catechol/triethylamine   ←
Cross-coupling with pentacoordinated enesilicates   ←
s. *44*, 889; with enesilicates *in situ*-generated from ene(trialkoxy)silanes and catecholborane/ Et_3N cf. Bull. Soc. Chim. France *132*, 499-508 (1995).

trans-Di(m-acetato)bis[o-(di-o-tolylphosphino)benzyl]dipalladium(II)/potassium carbonate   ←
Diaryls by Suzuki cross-coupling s. *51*, 416 $ArB(OH)_2$ → Ar–Ar′

Carbon ↑   CC ↕ C

Without additional reagents   w.a.r.
C-Alkylation with acyl peroxides   H → R
in carboxylic acids cf. *10*, 637; perfluoroalkylation of 1,4-naphthoquinones and 9,10-anthraquinones in Freon 113 s. M. Matsui et al., Bull. Chem. Soc. Japan *68*, 1042-51 (1995); of coumarins s. Synlett *1991*, 113-4.

N-Heterocyclic ketones from stannanes   Sn≤ → COR

458.

2-Acylpyrimidines. A soln. of 1.2 eqs. startg. α-ketoacyl chloride in dry benzene added dropwise to a stirred soln. of startg. stannylpyrimidine in the same solvent under N_2, and stirring continued at room temp. for 15 min → 2-benzoyl-4,6-dimethylpyrimidine. Y 97%. Yields were generally much lower by the standard method with acyl chlorides (cf. *34*, 862s*48,49*). F.e. incl. **4-acylpyrimidines**, also carbethoxy derivs. with EtOCOCOCl, s. Y. Yamamoto et al., Heterocycles *41*, 1275-90 (1995).

Pyrroles from acetylene derivs. via 1,3-dipolar cycloaddition
with 5-hydroxyoxazolium betaines s. *19*, 911; *22*, 617s*31*; from transition metal α,β-acetylene (alkoxy)carbene complexes s. S.C. Shin et al., Synth. Commun. *25*, 2043-50 (1995); with 5-aminothiazolium salts (cf. *26*, 919s*30*), 2-organothiopyrroles, cf. F. Berrée, G. Morel, Tetrahedron *51*, 7019-34 (1995).

Decarboxylative 1,3-dipolar cycloaddition via non-stabilized azomethinium ylids
s. *44*, 897s*46*; *trans*-3-aryl-4-nitropyrrolidines s. M. Nyerges et al., Tetrahedron *51*, 6783-8 (1995).

Eneheterocyclics from ketenes
s. *25*, 638; 3-diarylmethylene-2,3-dihydrofurans s. N.M. Igidov et al., Izv. Akad. Nauk Ser. Khim. *1995*, 331-4 (Russ.).

4-Dimethylaminopyridine   DMAP
o-Acetonylphenols from p-hemiquinols via Carroll rearrangement ←

459.

2 Mol% DMAP added to a stirred mixture of startg. hemiquinol and 1.1-1.4 eqs. diketene in methylene chloride at room temp. → product. Y 72%. If the hemiquinol is *o,o'*-disubstituted, the corresponding 6-acetonyl-2,4-cyclohexadienone may be formed in high yield. However, no reaction took place with sterically congested substrates. F.e.s. K.L. Sorgi et al., Tetrahedron Letters *36*, 3597-600 (1995).

Magnesium chloride/triethylamine or ethyldiisopropylamine $MgCl_2/R_3N$
β-Ketocarboxylic acid esters from malonic acid monoesters   C–COOH → C–COR
with *i*-PrMgBr cf. *14*, 888; α-acylamino-β-ketocarboxylic acid esters with $MgCl_2/Et_3N$ or *i*-$PrNEt_2$ s. D.J. Krysan, Tetrahedron Letters *37*, 3303-6 (1996).

Boron fluoride BF_3
(E)-α,β-Ethylenealdehydes from aldehydes   CHO → CH=CHCHO
Synthesis with addition of two C-atoms

460.

Cinnamaldehydes. 1 eq. BF_3-etherate added via syringe to benzaldehyde in methylene chloride at -78°, treated dropwise with 1 eq. 2-(tri-*n*-butylstannyl)vinyl ethyl ether (*cis*- or *trans*-isomer) in the same solvent, stirred at -78° for 1 h, quenched with 1:1 aq. methanol, and allowed to warm to room temp. → product. Y 86%. Other Lewis acids failed. No reaction was observed with ketones or satd. aldehydes, thereby affording scope for selective conversion. F.e. incl. **2,4-dienals** s. J.A. Cabezas, A.C. Oehlschlager, Tetrahedron Letters *36*, 5127-30 (1995).

Samarium triiodide SmI_3
C-Acyl exchange   CAc → CAc′
with Mg/EtOH cf. *19*, 915; β-diketones with **51**,SmI_3, and sym. β-diketones by double acyl exchange, s. W. Hao et al., Synth. Commun. *26*, 2421-7 (1996).

Acetic anhydride Ac_2O
Pyrroles from acetylene derivs. and α-acylaminocarboxylic acids
s. *30*, 631; polymer-based synthesis s. A.M.M. Mjalli et al., Tetrahedron Letters *37*, 2943-6 (1996).

Trifluoroacetic acid/triethylamine/2,3-dichloro-5,6-dicyanoquinone ←
Porphyrins from tripyrranedicarboxylic acid *tert*-butyl esters and pyrrole-2,5-dicarboxaldehydes

461.

One-pot conversion. A soln. of startg. tripyrranedicarboxylate treated with TFA under N_2 for 10 min, diluted 20 times with dichloromethane, 1 eq. startg. dialdehyde added to the resulting diacid, stirred at room temp. for 2 h, neutralized with Et₃N, and treated with 1 eq. DDQ for 1 h → product. Y 72-83%. The synthesis can be conducted on a gram scale. F.e.s. Y. Lin, T.D. Lash, Tetrahedron Letters *36*, 9441-4 (1995); from the diacids cf. A. Boudif, M. Momenteau, J. Chem. Soc. Perkin Trans. I *1996*, 1235-42; **sym. porphyrins** from 5-aminomethylpyrrole-2-carboxylic acid esters s. L. Cheng, J. Ma, Org. Prep. Proc. Intern. *27*, 224-8 (1995).

Stannic chloride $SnCl_4$
3-Hydroxytetrahydrofuran-2-carboxylic acid esters from β-benzyloxyoxo compds.
Stereospecific ring closure

462.

Ethyl diazoacetate and 0.25 eq. $SnCl_4$ added sequentially to a stirred soln. of 0.5 eq. startg. aldehyde (0.125 *M*) in methylene chloride at -78°, and worked up after 30 min → product. Y 84% (single diastereomer). There was no formation of β-ketoester (cf. *45*, 462) by using $SnCl_4$ as Lewis acid. Although reaction took place with other O-protective groups (e.g. MeOCH₂, Et₃Si, Bn), O-(*p*-methoxybenzyl) proved the most effective. Furthermore, ring closure appears unaffected by α,α-disubstitution of the aldehyde. F.e. and with $ZrCl_4$ (for ketone derivs.) s. S.R. Angle et al., J. Am. Chem. Soc. *117*, 8041-2 (1995).

Nitrosyl fluoroborate $NOBF_4$
Sym. diarylmethanes from benzyl ethers $2\ ArCH_2OR \rightarrow Ar_2CH_2$

463.

2-(Methoxymethyl)-3,5,6-trimethyl-1,4-dimethoxybenzene added under argon to a slurry of 1 mol% $NOBF_4$ in dichloromethane at ca. 0° [dry box], warmed to room temp., stirred for 4 h, Zn dust added, and stirring continued for a further 5 min → product. Y 94%. Reaction is initiated by a catalytic amount of one-electron oxidant (NO⁺, aromatic cation radical, or anodically). F.e.s. R. Rathore, J.K. Kochi, J. Org. Chem. *60*, 7479-90 (1995).

Palladous acetate/ethyldiisopropylamine/benzyltriethylammonium chloride
1-Acylindoles from *o*-iodoacylamines via 1-acyl-2-hydroxyindolines

464.

A soln. of startg. iodoacylamine, 1.5 eqs. *i*-Pr₂NEt, 1 eq. BnEt₃NCl, 1.5-2 eqs. vinylene carbonate, and 3 mol% Pd(OAc)₂ in DMF heated at 75° for ca. 24 h → intermediate indoline (Y 72%), in ethanol stirred with 5 mol% TsOH at room temp. for 13 h → product (Y 81%). Yields were low from the corresponding bromides. F.e.s. K. Samizu, K. Ogasawara, Heterocycles *41*, 1627-9 (1995).

Palladous acetate/tri-n-*butylphosphine/potassium carbonate*
Regiospecific synthesis of 2-ethylenealcohols from 4-vinyl-1,3-dioxolan-2-ones
with Grignard compds. and CuCN cf. *46*, 918; 2,5-dienols **with asym. induction** s. S.-K. Kang et al., Tetrahedron:Asymm. *5*, 21-2 (1994); chiral *syn*-2-vinyl-1,3-diols from aldehydes with Pd(PhCN)₂Cl₂/SnCl₂ s. Synth. Commun. *25*, 1359-66 (1995); (E)-4-aryl-2-ethylenealcohols from ar. iodides with Pd(OAc)₂/Bu₃P/K₂CO₃ s. Tetrahedron Letters *36*, 8047-50 (1995).

Palladous acetate/triphenylphosphine/triethylamine/potassium bromide/tetra-n-butylammonium fluoride
**2-Vinyl-2,5-dihydrofurans
from 4-alk-1-ynyl-1,3-dioxolan-2-ones and electron-deficient ethylene derivs.**

465.

Startg. cyclic carbonate treated with 3 eqs. methyl acrylate in the presence of 5 mol% Pd(OAc)$_2$, 10 mol% Ph$_3$P, 2 eqs. Et$_3$N, 2 eqs. KBr and 1 eq. Bu$_4$NF at 75° for 20 h [in DMF] → product. Y 75% (42% after 50 h without Bu$_4$NF). A little water was necessary for the reaction to take place but an excess had a negative effect. Reaction is believed to take place **via zwitterionic allenylpalladium species**. F.e., also reaction of acrylamides under modified conditions, s. P.H. Dixneuf et al., Chem. Commun. *1996*, 919-20; **α-allene-δ-hydroxycarboxylic acid esters** by trapping the intermediate with CO in methanol using Pd(dba)$_2$/Bu$_3$P, also biscarbonylation, s. Synlett *1996*, 218-20.

Elimination

Hydrogen ↑

Phenyl iodosotrifluoroacetate/boron fluoride   PhI(OCOCF$_3$)$_2$/BF$_3$
Intramolecular oxidative non-phenolic coupling
with VOF$_3$/BF$_3$ cf. *29*, 910s*46,49*; with PhI(OCOCF$_3$)$_2$/BF$_3$ s. Y. Kita et al., Chem. Commun. *1996*, 1481-2.

2,3-Dichloro-5,6-dicyanoquinone   DDQ
2,5-Cyclohexadienone from 2-cyclohexenone ring
with SeO$_2$ cf. *12*, 878; 9-acoxyeudesma-1,4-dien-3-ones with DDQ s. Y. Li et al., Synth. Commun. *26*, 551-7 (1996).

Manganese(III) acetate/cupric acetate   Mn(OAc)$_3$/Cu(OAc)$_2$
**Cyclic β,γ-ethyleneketones from ethyleneketones
Oxidative regiospecific radical ring closure**

466.

The scope of Mn(III)-based oxidative ring closures (cf. *46*, 930) has been extended to cyclization of *simple* unsaturated ketones. E: A 0.1 *M* soln. of startg. ketone in acetic acid heated at 80° for 1.5 h with 2.5 eqs. Mn(OAc)$_3$·2H$_2$O and 1 eq. Cu(OAc)$_2$·H$_2$O → product. Y 75%. Higher yields are obtained for cyclic (notably bridged bicyclic) ketones which cannot enolize, so that subsequent oxidation is not possible. The formation of ketoalkyl radicals by SET from the Mn(III)-enolate appears to be the rate-limiting step. F.es. B.B. Snider, B.M. Cole, J. Org. Chem. *60*, 5376-7 (1995); 8-vinylbicyclo[3.3.1]nonane-2,9-diones s. Tetrahedron *51*, 12983-94 (1995).

Oxygen ↑   CC ↑ O

*Lithium/4,4′-di-*tert-*butylbiphenyl*
Cyclopropanes from cyclic 1,3-diol sulfates   ←
             ▽

467.

5-(4-*tert*-Butylbenzyl)-1,3,2-dioxathiane 2,2-dioxide in THF added during 30 min at 0° under argon to a suspension of 14 eqs. Li-powder and 10 mol% 4,4′-di-*tert*-butylbiphenyl in the same solvent, stirred at the same temp. for 10 min, and hydrolyzed with water → (4-*tert*-butylbenzyl)cyclopropane. Y 85%. F.e. incl. fused and spiro-analogs s. D. Guijarro, M. Yus, Tetrahedron *51*, 11445-56 (1995).

Sodium hydride/diethyl phosphorochloridate/potassium tert-*butoxide*   ←
α,β-Acetylenephosphonic from β-ketophosphonic acid esters   COCH$_2$ → C≡C
with Tf$_2$O/*i*-Pr$_2$NEt cf. *44*, 929s*48*; via α,β-ethylene-β-phosphoryloxyphosphonic acid esters in one pot with NaH/(EtO)$_2$P(O)Cl/KOBu-*t* s. D.Y. Oh et al., Synth. Commun. *26*, 1563-7 (1996).

n-*Butyllithium*   BuLi
Cyclic α,β-ethyleneketones from cyclic 2,3-epoxyalcohols with 1,2-alkyl migration   ←
2,2′-Dioxycarbenes as intermediates

468.

3 eqs.1.6 M *n*-BuLi in hexane added dropwise at -78° under argon to a stirred soln. of 2,5-di methyl-6-oxabicyclo[3.1.0]hexan-2-ol in anhydrous THF, allowed to warm to room temp., stirred for 1 h, and quenched with water → 2,3-dimethylcyclopent-2-enone. Y 74%. The favourable *cis*-disposition of the oxy-functionalities in the substrate facilitates regiospecific deprotonation leading to a carbenoid which then undergoes 1,2-alkyl shift prior to elimination of Li$_2$O. Reaction is also applicable to cyclohexene oxides, but unconstrained acyclic analogs underwent classical [non-carbenoid] β-elimination to the allyl alcohol. The migratory aptitude of the organo groups was in the order Me > *n*-Bu > aryl > *t*-Bu, a mixture of isomers normally being formed from substrates where the alkyl groups were different. F.e.s. E. Doris et al., J. Am. Chem. Soc. *117*, 12700-4 (1995).

Lithium telluride   Li$_2$Te
Ethylene derivs. from 1,2-dimesylates   C=C
s. *49*, 938; **from cyclic glycol sulfates** or 1,3-dioxolane-2-thiones with retention of configuration s. B. Chao et al., Tetrahedron Letters *36*, 7209-12 (1995).

Potassium fluoride/alumina   KF/Al$_2$O$_3$
Benzofurans from *o*-alkoxyoxo compds.   O
with HCl cf. *23*, 927; 2-arylbenzofurans with KF/Al$_2$O$_3$ s. D. Hellwinkel, K. Göke, Synthesis *1995*, 1135-41.

Alkylmagnesium halides s. *under* Ti(OPr-i)$_4$ *and* ClTi(OPr-i)$_3$ RMgHal
Titanium/chlorosilanes   Ti/ClSi≤
Activation of commercial titanium with chlorosilanes   ←

469.

Commercial titanium, which is passivated by a tightly bound surface layer of titanium oxide, can be activated by chlorosilanes which effectively remove the oxide as soluble $TiCl_3$. The system displays an exceptional template effect, permitting intramolecular McMurray-type formation of **macrocyclic ethylene derivs.** *without* high dilution. **E:** A suspension of 64 eqs. commercial Ti-powder in DME and 64 eqs. Me$_3$SiCl refluxed for 68 h, 1,26-bis(4-benzoylphenyl)hexacosane added at once, and refluxed for 6 h → 1,2-diphenyl(2,26)[36]-paracyclophan-1-ene. Y 90%. Interestingly, aliphatic oxo compds. are relatively inert, while substrates possessing both keto and enone functionality react preferentially at the latter. F.e. and *catalytic* McMurray-type coupling with TiCl$_3$ (10 mol%) and Zn-dust in the presence of Me$_3$SiCl s. A. Fürstner, A. Hupperts, J. Am. Chem. Soc. *117*, 4468-75 (1995).

Titanium tetraisopropoxide/isopropylmagnesium chloride $Ti(OPr\text{-}i)_4/i\text{-}PrMgCl$
Syntheses via titanium-mediated nucleophilic intramolecular acyl substitution ←

470.

Titanium η^2-olefin complexes, generated *in situ* from unsaturated carbonic and carboxylic acid esters, play a central role in nucleophilic intramolecular acyl substitution with the intermediate formation of an organotitanium(IV) species for further manipulation as required. **E: γ-Lactones**. 2.6 eqs. *i*-PrMgCl in ether added dropwise with stirring at -50° to 1.3 eqs. Ti(OPr-*i*)$_4$ and startg. ethylenecarbonate in the same solvent, stirred for 1 h at -45 to -40°, the generated Ti(IV) complex treated with 3 *N* HCl, and stirred for 30 min at room temp. → product. Y 92%. F.e.s. F. Sato et al., J. Am. Chem. Soc. *118*, 2208-16 (1996); α-alkylidenelactones s. Tetrahedron Letters *36*, 6075-8 (1995).

Trimethylsilyl iodide/hexamethyldisilazane $Me_3SiI/(Me_3Si)_2NH$
Enolethers from mixed acetals $CHC(OR)(OR') \rightarrow C{=}C(OR)$
chiral α-vinyloxycarboxylic acid esters with Me$_3$SiOTf/Et$_3$N cf. *50*, 569; (Z)-3-alkoxy-2-ethylenestannanes with Me$_3$SiI/(Me$_3$Si)$_2$NH s. Y. Yamamoto et al., Tetrahedron Letters *37*, 3195-8 (1996).

Chlorotitanium triisopropoxide/n-butylmagnesium chloride $ClTi(OPr\text{-}i)_3/BuMgCl$
Condensed cyclopropanols from ethylenecarboxylic acid esters
Stereospecific intramolecular Kulinkovich reaction

471.

A soln. of startg. enoate, 5 eqs. ethereal BuMgCl, and 0.5 eq. ClTi(OPr-*i*)$_3$ in ether allowed to react at room temp. for 1-2 h → product. Y 55%. The procedure is simple, inexpensive, and suitable for preparing cyclopropanes fused to 5-, 6- and 7-membered rings. A limitation, however, is that the alkene group may only be mono-substituted, and the yield is poor if an allylic function is present. F.e.s. J.K. Cha et al., J. Am. Chem. Soc. *118*, 291-2 (1996).

Titanium trichloride/zinc $TiCl_3/Zn$
Pyrrole ring from *o*-acylaminoketones
indoles with Ti/C$_8$ cf. *47*, 940; isolated pyrroles from β-acylamino-α,β-ethyleneketones, also with TiCl$_3$/Zn, s. J. Org. Chem. *60*, 6637-41 (1995).

Titanium tetrachloride $TiCl_4$
2-Subst. *p*-quinones from *p*-subst. phenols via 4-subst. 4-peroxy-2,5-cyclohexadienones ←

472.

with 1,2-alkyl migration. 4 eqs. 3.3 M *tert*-Butyl hydroperoxide in dry benzene added dropwise with stirring at room temp. over 2 h to a soln. of 4-methylphenol and 3 mol% $RuCl_2(PPh_3)_3$ in ethyl acetate, stirred for 3 h, and excess of the hydroperoxide removed with a soln. of $NaHSO_3$ → intermediate peroxide (Y 85%), in dry methylene chloride added dropwise with stirring during 30 min at -78° to 1.2 eqs. $TiCl_4$ in the same solvent, and stirred for a further 1 h at room temp. → 2-methyl-*p*-benzoquinone (Y 82%). F.e. and Diels-Alder reaction with *in situ*-generated quinones, also with 1,2-aryl migration, s. S. Murahashi et al., J. Am. Chem. Soc. *118*, 2509-10 (1996).

Triphenyl phosphate/quinoline   ←
Dehydration of sec. alcohols   CHCH(OH) → C≡CH
with $[(PhO)_3PMe]I/HMPA$ cf. *28*, 916; cycloalkenes with $(PhO)_3PO$/quinoline or $(PhO)_2P(O)Cl$ s. H. Quast, T. Dietz, Synthesis *1995*, 1300-4.

Diethyl phosphorochloridate s. under NaH $ClPO(OEt)_2$

Phosphoric acid/silica or trifluoromethanesulfonic acid/silica   ←
Isocyclics from alcohols   ○
with H_3PO_4 cf. *1*, 745; under *heterogeneous* conditions with H_3PO_4/SiO_2, CF_3SO_3H/SiO_2 or $MeSO_3H/SiO_2$ s. P.J. Kropp et al., J. Org. Chem. *60*, 4146-52 (1995).

Phosphorus oxide chloride $POCl_3$
2-Aminoindoles from N-arylcarboxylic acid hydrazides
Kost reaction

473.

A mixture of startg. hydrazide and 3 eqs. $POCl_3$ in dioxane stirred until reaction complete → product. Y 94% (as the hydrochloride). F.e.s. V.P. Zhestkov et al., Khim. Geterotsikl. Soedin. *1995*, 1502-6 (Russ.).

Bischler-Napieralski ring closure
1,2,3,4-tetrahydroisoquinolines s. *17*, 942; chiral 1,2,3,4-tetrahydroisoquinoline-3-carboxylic acids by polymer-based synthesis and adaptation for multiple synthesis s. W.D.F. Meutermans, P.F. Alewood, Tetrahedron Letters *36*, 7709-12 (1995).

Trifluoromethanesulfonic anhydride/4-dimethylaminopyridine $Tf_2O/DMAP$
Isocarbostyril ring by Bischler-Napieralski-type ring closure
3,4-dihydro-derivs. with $POCl_3/SnCl_4$ cf. *33*, 694; also unsatd. ring with Tf_2O/4-DMAP *under mild conditions* s. M.G. Banwell et al., Chem. Commun. *1995*, 2551-3.

p-*Toluenesulfonic acid*   TsOH
1-Acylindoles from 1-acyl-2-hydroxyindolines s. *51*, 464   C(OH)CH → C=C

Lithium perchlorate/trifluoroacetic acid $LiClO_4/CF_3COOH$
Intramolecular ionic Diels-Alder reaction
with *in situ*-generated heteroatom-stabilized allyl cations

474.

A 0.02 M soln. of startg. cyclohexenol in ether added slowly via syringe pump during 1 h to 5 M $LiClO_4$ in ether containing 10 mol% trifluoroacetic acid, and worked up after 30 min at room temp. → *exo*-adduct. Y 87%. The highly polar medium further stabilizes the intermediate allyl cation, and the anhydrous perchlorate effectively sequesters water, thereby preventing cleavage to the corresponding cyclohexenone. A concerted mechanism is invoked. F.e.s. P.A. Grieco et al., J. Am. Chem. Soc. *118*, 2095-6 (1996).

Nitrogen ↑

Without additional reagents   w.a.r.
Fused pyridine ring via intramolecular Diels-Alder reaction with 1,2,4-triazines   ←
s. *40*, 669; pyrid[3,4-*b*]indoles from thiourea-tethered 1,2,4-triazines s. W.-H. Fan et al., Tetrahedron Letters *36*, 6591-4 (1995).

Sodium/ethylene glycol $Na/HOCH_2CH_2OH$
tert-Butyllithium   t-*BuLi*
Ethylene derivs. from sulfonylhydrazones $CHC(=NNHSO_2Ar) \rightarrow C{=}CH$
s. *23*, 935; *27*, 949; deuterio- and tritio-ethylene derivs. from trisylhydrazones, stereoselectivity, s. M. Saljoughian et al., Tetrahedron Letters *37*, 2923-6 (1996); **exocyclic enolethers** with $Na/HOCH_2CH_2OH$ s. S. Chandrasekhar et al., Chem. Lett. *1996*, 211-2.

Lithium diisopropylamide   i-Pr_2NLi
Catalyzed Shapiro reaction with N-aziridin-1-ylimines   C=C

475.

cis-**Ethylene derivs.** A soln. of 6-undecanone phenylaziridinylhydrazone in ether treated with *0.3 eq.* LDA at -20° for 1 h and at 0° for 3 h → 5-undecene. Y 84% (*cis/trans* 99.4:0.6). Reaction proceeds with high regioselectivity and *cis*-stereoselectivity, and is adaptable on the 30 mmol scale with as little as 0.05 eq. base. Lithium 2,2,6,6-tetramethylpiperidide was equally effective, but $LiNEt_2$ and $LiN(SiMe_3)_2$ gave less satisfactory results. F.e. incl. **chiral ethylene derivs.** s. K. Maruoka et al., J. Am. Chem. Soc. *118*, 2289-90 (1996); uncatalyzed formation of **cyclic enolethers** from 1-(aziridin-1-ylimino)ethers s. S. Kim et al., Chem. Commun. *1996*, 909-10; from epoxy-derivs. cf. Tetrahedron Letters *36*, 4845 (1995).

Ethyldiisopropylamine   i-Pr_2NEt
Hofmann elimination of a polmer-based acrylate support s. *51*, 172   ←
Copper   Cu
Cyclic tert. amines from diazo-*tert*-amines   ○
3-piperidones with $Rh(OAc)_4$ cf. *48*, 955; 2-morpholones with Cu or $Cu(acac)_2$ s. F.G. West, B.N. Naidu, J. Org. Chem. *59*, 6051-6 (1994).

Chiral copper(II) semicorrin complexes   ←
Chiral copper(I) bis(Δ^2-oxazoline) complex   ←
Asym. double ring closure of ethylenediazo compds.   ▽
with chiral Cu(II) Schiff base complexes cf. *46*, 954s*48*; with chiral Cu(II) semicorrin complexes

s. C. Piqué et al., Synlett *1995*, 491-2; with a chiral CuOTf-bis(Δ^2-oxazoline) complex cf. E.J. Corey et al., Tetrahedron Letters *36*, 8745-8 (1995).

Silver hexafluoroantimonate/chiral bis(Δ^2-oxazolines) ←
Asym. insertion of diazo compds. into carbon-hydrogen bonds O
Optimization of catalyst by high throughput parallel screening

476.

By parallel screening of 96 catalytic combinations *on a microtitre plate*, it was revealed that AgSbF$_6$ in the presence of a chiral bis (Δ^2-oxazoline) is the optimum system for asym. intramolecular carbene insertion. **E**: A soln. of startg. diazo compd. in THF added to 10 mol% each of AgSbF$_6$ and chiral bis(Δ^2-oxazoline) in the same solvent, stirred at 25° for 24 h, 1.5 eqs. solid DDQ added, and stirred for a further 1 h → product. Y 44% (2.7:1 mixture of diastereomers). Use of a silver salt is unprecedented, and considered more suitable than the more familiar copper(I) or Rh catalysts. F.e. and description of the screening procedure s. K. Burgess et al., Angew. Chem. Intern. Ed. *35*, 220-2 (1996).

Diisobutylaluminum hydride i-Bu$_2$AlH
Reductive retrodiene scission by elimination of 1,2,4-triazoline-3,5-diones C
with LAH cf. 27, 959; steroidal 5,7-dienes with *i*-Bu$_2$AlH s. H. Takayama et al., J. Chem. Soc. Perkin Trans. I *1995*, 2679-80.

Samarium diiodide SmI$_2$
Pyrrolidines by intramolecular radical 1,4-addition s. *51*, 376 O

Tetrathiafulvalene ←
Tetrathiafulvalene-catalyzed radical ring closure-nucleophilic substitution
s. *50*, 98; spiro-O-heterocyclics by double ring closure s. Chem. Commun. *1996*, 737-8; stereospecific application to tetracyclic N-/O-heterocyclics s. R.J. Fletcher et al., ibid. 739-40.

Rhodium(II) trifluoroacetate Rh(OCOCF$_3$)$_2$
(Z)-α,β-Ethylenecarbonyl from α-diazocarbonyl compds. CHCN$_2$ → C=CH
(Z)-enoates cf. *37*, 946; also (Z)-enones s. D.F. Taber et al., J. Org. Chem. *61*, 2908-10 (1996).

Dirhodium(II) tetrakis(methyl 2-pyrrolidone-5(S)-carboxylate) ←
Dirhodium(II) tetrakis(methyl 2-oxazolidone-4(S)-carboxylate) ←
Dirhodium(II) tetrakis(methyl 1-acyl-2-imidazolidone-4(S)-carboxylate ←
Asym. intramolecular carbene insertion with α-diazocarbonyl compds. O
s. *47*, 955s*48,49,50*; chiral pyrrolizidines with dirhodium(II) tetrakis(methyl 1-acyl-2-imidazolidone-4(S)-carboxylates s. M.P. Doyle et al., Tetrahedron Letters *37*, 1371-4 (1996); γ-lactones from *sec-* or *tert*-alkyl diazoacetates s. ibid. *36*, 7579-83, 4745-8 (1995); also with dirhodium(II) tetrakis(methyl 2-pyrrolidone-5(S)-carboxylate) s. J. Am. Chem. Soc. *118*, 8837-46 (1996); chiral 2-azetidinones with dirhodium(II) tetrakis(methyl 2-oxazolidone-4(S)-carboxylate) s. Synlett *1995*, 1075-6; benzo-condensed O- and S-heterocyclics s. M.A. McKervey et al., J. Chem. Soc. Perkin Trans. I *1995*, 1373-9.

Halogen ↑   CC ↑ Hal

Irradiation s. under (Triphenyltin)cobaloxime ∰
Electrolysis/nickel(II) cyclamin complexes ←
Indirect electrocatalytic ring closures of unsatd. halides O
s. *41*, 942s*46,48,49*; 2-alkoxytetrahydropyrans from O-homoallyl α-bromoacetals s. K. Fukumoto

et al., J. Org. Chem. *61*, 677-84 (1996); s.a. Chem. Pharm. Bull. *43*, 362-4 (1995); benzo-condensed O-heterocyclics by ring closures of ar. ethylenehalides s. S. Olivero et al., Tetrahedron Letters *36*, 4429-32 (1995).

Microwaves s. under KOBu-t ←

Potassium tert-*butoxide* KOBu-t
Bicyclic N-sulfonylenamines from ω-(sulfonylamino)-α,β-acetyleneiodonium salts ○
Stereospecific intramolecular nucleophilic substitution-carbene insertion

477.

Synth.Meth. *46*, 504 has been adapted *intramolecularly* by using a suitably activated nitrogen nucleophile [an imide anion], thereby securing a useful biheterocyclization stereospecifically. **E:** A soln. of startg. stannylacetylene treated with cyano(phenyl)iodonium triflate according to *46*, 458s*48*, and the resulting iodonium salt (Y 86% by ^{1}H NMR) treated immediately with *t*-BuOK or LiN(SiMe$_3$)$_2$ in THF → product. Y 73%. The fused nitrogen ring may be 5-, 6- or 7-membered, and the carbene can insert into C-H bonds positioned on prim., sec. or tert. carbon atoms. An N,N'-ditosylurea may also serve as nucleophile. F.e. and details (in Supporting Information) s. K.S. Feldman et al., J. Am. Chem. Soc. *117*, 7544-5 (1995).

Potassium tert-*butoxide/tetra-*n-*butylammonium bromide/microwaves* ←
Ketene acetals from α-bromacetals C=C(OR)$_2$
with KOH/Bu$_4$NBr cf. *3*, 749s*48*; with KOBu-*t* under microwave irradiation in the absence of solvent, also ketene mercaptals from α-chloromercaptals s. A. Díaz-Ortiz et al., Tetrahedron Letters *37*, 1695-8 (1996).

n-*Butyllithium* BuLi
Allyl enolethers from O-silyl O-allyl α-iodoacetals s. *51*, 196 ←
Intramolecular Barbier reaction with ar. iodides ○
s. *46*, 956; 3-hydroxyindolines s. Heterocycles *41*, 2279-87 (1995).

sec-*Butyllithium/(-)-sparteine* ←
N-Protected 2-arylpyrrolidines from N-benzyl-3-chloramines via asym. deprotonation

478.

A soln. of N-benzyl-N-carbo-*tert*-butoxy-3-chloropropylamine *in toluene* added to 1.5 eqs. (-)-sparteine and 1.2 M *s*-BuLi in cyclohexane at -78°, stirred for 5 h, and quenched with water and ether → Boc-(S)-2-phenylpyrrolidine. Y 72% (e.e. 96%). Reaction was shown to occur via asym. deprotonation followed by cyclization, rather than by non-stereospecific deprotonation followed by asym. ring closure. F.e.s. S. Wu et al., J. Am. Chem. Soc. *118*, 715-21 (1996).

Lithium bis(trimethylsilyl)amide LiN(SiMe$_3$)$_2$
Cyclodec-3-ene-1,5-diynes from 1,10-dibromo-2,8-diynes

479.

2.2 eqs. *Li-bis(trimethylsilyl)amide* added slowly (3 h) to a 2.5 x 10^{-2} mol. dm^{-3} soln. of startg.

dibromide in THF containing 22 eqs. HMPA at -45°, worked up with ether, and treated with 3 eqs. Co$_2$(CO)$_8$ → product. Y 92%. F.e.s. G.B. Jones et al., Chem. Commun. *1995*, 1791-2.

1,8-Diazabicyclo[5.4.0]undec-7-ene DBU
Cyclopropanes from iodides ▽
with NaOH cf. *26*, 967s*30*; *cis*-trifluoromethanesulfonylcyclopropanes with DBU s. A. Mahadevan, P.L. Fuchs, J. Am. Chem. Soc. *117*, 3272-3 (1995).

Ethylmagnesium bromide/diisopropylamine EtMgBr/*i*-Pr$_2$NH
Dihalogenomethylene compds. from 2-acoxy-1,1,1-trihalides C=C(Hal)$_2$
with Al/PbBr$_2$ cf. *44*, 955; (Z)-β-bromo-β-fluorostyrenes with EtMgBr/*i*-Pr$_2$NH, and (E)-β-bromo-α,β-difluorostyrenes by debromofluorination s. M. Kuroboshi et al., Tetrahedron Letters *36*, 6271-4 (1995).

Ethylmagnesium bromide/dichloro[1,2-bis(diphenylphosphino)ethane]nickel(II) ←
Ethylene derivs. from 1,2-dibromides under mild conditions C(Br)C(Br) → C=C

480.

2.1 eqs. EtMgBr in anhydrous THF added slowly to a soln. of startg. dibromide and a little Ni(dppe)Cl$_2$ in the same solvent at 0° (instantaneous reaction) → product. Y 96% (1:1 mixture of stereoisomers). The reaction is very fast (a few seconds to 15 min), there was no alkene isomerization, and esters, carbonyl compds., and THP ethers were unaffected. F.e.s. C. Malanga et al., Tetrahedron Letters *36*, 9189-92 (1995).

Chiral methylaluminum 1,1'-bi-2-naphthoxide s. under Bu$_3$SnH ←
Indium In
Ring expansion of cyclic ketones via 1,3-acyl group migration in aq. medium ◯

481.

Medium-ring γ-alkylidene-δ-ketocarboxylic acid esters. 2 eqs. In powder added to a mixture of startg. cyclic ketone in 3:1 aq. 0.1 *N* HCl/methanol, stirred vigorously at room temp. for 10 h, quenched with 1 *N* HCl, worked up, and the resulting isomeric α-vinyl derivs. stirred overnight with 2 eqs. DBU in THF at room temp. → product. Y 50%. This **Barbier-type ring expansion** was also applied to the formation of 8-, 9-, 10- and 14-membered cyclic ketones, incl. a bicyclic deriv. F.e. and stereospecificity s. C.-J. Li et al., J. Am. Chem. Soc. *118*, 4216-7 (1996); 3-methylenecyclopentanol-5-carbonyl compds. by intramolecular Barbier-type reaction s. Tetrahedron Letters *37*, 471-4 (1996).

Samarium diiodide SmI$_2$
Hydroxyketones from acoxyiodides via O→C-acyl group migration ←
chiral δ-hydroxyketones s. *49*, 955; chiral γ-hydroxyketones s. J. Org. Chem. *59*, 3445-52 (1994).

Samarium diiodide/hexamethylphosphoramide SmI$_2$/(Me$_2$N)$_3$PO
Regio- and stereo-specific intramolecular nucleophilic acyl substitution- ◯
radical ring closure

482.

Bicyclic tert. alcohols. 4-5 eqs. Methylene iodide added to a vigorously stirred soln. of 4-5 eqs. Sm in dry THF, stirring continued for 2.5 h at room temp., 5 eqs. HMPA added, stirred for 15 min, cooled to 0°, 1 eq. 0.05 M startg. iodoester in dry THF added dropwise over 2 h at 0°, warmed to room temp., stirred for a further 30-45 min, and quenched with satd. aq. NaHCO₃ → product. Y 70% (single diastereomer). Reaction was markedly enhanced by using an activated alkene group as radical acceptor, and extended to silyl-activated acetylene derivs. Highly strained *trans:anti:cis* **linear triquinanes, angular triquinanes, and spiroisocyclics** were obtained by the same procedure, and lactones could be used instead of esters (with ring opening to hydroxyalkyl derivs.). F.e.s. G.A. Molander, C.R. Harris, J. Am. Chem. Soc. *118*, 4059-71 (1996).

Bicyclo[n.m.0]alkan-1-ols from dihalogenolactones
Stereospecific intramolecular nucleophilic acylation-Barbier cyclization ←

483.

Synth.Meth. *49*, 956 has been extended to dihalogenolactones, thereby providing *two* sites for tandem generation of organosamarium species which react sequentially in a novel bicyclization. E: 4.5 eqs. Methylene iodide added to 5 eqs. Sm in THF at room temp., stirred vigorously for 2.5 h, HMPA added, stirring continued for 15 min, cooled to 0°, a 0.03-0.04 M soln. of dihydro-3,3-bis(3-iodopropyl)furan-2(5H)-one in THF added dropwise over 2-2.5 h, allowed to warm to room temp., stirred for 2 h, and quenched with satd. aq. NaHCO₃ → *cis*-5-(2-hydroxyethyl)bi cyclo[3.3.0]octan-1-ol. Y 83%. A level of regioselectivity can be achieved by using chloroiodides in place of diiodides, and steric hindrance around the carbonyl group is tolerated. F.e. and tricyclo[m.n.0.0]alkanes (incl. angular triquinanes), also by stereospecific ring expansion-cyclization of α,ω-bis(halogenalkyl)lactones, s. G.A. Molander, C.R. Harris, J. Am. Chem. Soc. *117*, 3705-16 (1995).

Tris(trimethylsilyl)silane/tri-n-butyltin chloride/azodiisobutyronitrile ←
Regio- and stereo-specific intramolecular radical addition ○
to 3-(α,β-ethyleneacyl)-2-oxazolidones
Organotin halides as Lewis acid

484.

A soln. of 1.5 eqs. (Me₃Si)₃SiH and a little AIBN in benzene added via syringe pump over 20 h to a soln. of startg. 2-oxazolidone and *1 eq.* Bu₃SnCl in the same solvent at 80° under N₂ → product. Y 87% (d.e. >97%). Lewis acidic organotin halides control the rotamer population in such a way as to favour 5-*exo*-cyclization over both 6-*endo*-cyclization and direct replacement of halogen by hydrogen. However, reductive dehalogenation did take place exclusively in the absence of the tin halide, and was favoured by stronger Lewis acids (MgBr₂, Et₂AlCl) or Bu₃SnH/AIBN. Raising the temperature also favoured 5-*exo*-cyclization. F.e. and with Bu₃SnH/Bu₃SnCl s. M.P. Sibi, J. Ji, J. Am. Chem. Soc. *118*, 3063-4 (1996).

Chlorobis(cyclopentadienyl)titanium(III) (Cp₂TiCl)₂
Glycals from O²-acylglycosyl bromides s. *42*, 957s*51* C(OAc)C(Hal) → C=C

Tri-n-butyltin hydride/triethylborane/chiral methylaluminum 1,1'-bi-2-naphthoxide ←
Asym. intramolecular 1,4-addition of vinyl radicals ○
to a chiral enoate with achiral MeAl(OAr)₂ as Lewis acid cf. *50*, 579; also with i-Bu₃Al s. Synlett *1995*, 1045-7; chiral 2-methylenecycloalkaneacetic acid derivs. with a chiral methylaluminum 1,1'-bi-2-naphthoxide as catalyst s. Chem. Commun. *1996*, 579-80.

Tri-n-butyltin hydride/azodiisobutyronitrile Bu₃SnH/RN=NR
Regio- and stereo-specific radical ring closures of ethylenehalides ○
s. *29*, 970s*46-50*; condensed chlorocyclopropanes s. Y. Tanabe et al., Chem. Lett. *1994*, 1757-60; furanolignans s. C. Roy et al., J. Chem. Soc. Perkin Trans. I *1995*, 927-9; azasilabicyclics s. S.

Mignani et al., Synlett *1996*, 890-92; 4-α-arylthio-2-azetidinones **with asym. induction** s. H. Ishibashi et al., ibid. *1995*, 915-7; chiral *trans*-5-subst. 2-hydroxymethylpyrrolidines s. S. Shibuya et al., J. Chem. Soc. Perkin Trans. I *1996*, 465-73.

Intramolecular radical 1,4-addition   ○
s. *42*, 960s*48-50*; of α-amino-α,β-ethylenenitriles s. C.-C. Yang, J.-M. Fang, J. Chem. Soc. Perkin Trans. I *1995*, 879-87; 3-phosphonylmethyl-2,3-dideoxyribosides **with asym. induction** s. S. Shibuya et al., Synlett *1995*, 1280-2; polyoxycyclopentanes from unsatd. aldonolactones s. I. Lundt et al., ibid. 918-20; 2-oxabicyclo[2.2.1]heptanes s. D.E. Shaw et al., Chem. Commun. *1994*, 2447-8; chiral indolizidines s. E. Lee et al., Tetrahedron Letters *37*, 1445-6 (1996); chiral hexahydroquinolizinones and tetrahydroindolizinones s. P. Mangeney et al., ibid. 1599-602; pyroglutamates s. Tetrahedron *52*, 6739-58 (1996).

Ring expansion of 2-(ω-halogenoalkyl)cyclobutanones   ○
s. *43*, 960s*49*; fused cycloheptanones s. Tetrahedron *50*, 12579-92 (1994); also fused cyclooctanones s. Tetrahedron Letters *36*, 2729-32 (1995); bridged tricyclic ketones s. ibid. *35*, 5563-6 (1994).

Methylenecycloheptane from 2-γ-bromo(methylene)cyclobutane ring   ←
via transannular radical ring opening

485.

A soln. of 1.5 eqs. Bu_3SnH and a little AIBN in benzene added over 4 h to a refluxing soln. of startg. bromide in the same solvent, refluxing continued for an additional 1 h, the mixture cooled to room temp., and worked-up with DBU → *cis*-fused product. Y 94% (92% purity). The method has distinct advantages over ring expansion of the corresponding ketones, since formation of the intermediate cyclopentylcarbinyl radical is ca. 9 times faster than formation of an alkoxyl radical, thereby reducing the amount of direct reduction product; furthermore, ring opening of the carbinyl radical is ca. 4 times slower than that of the alkoxyl radical, so that fragmentation of the former is more selective. F.e.s. W. Zhang, P. Dowd, Tetrahedron Letters *36*, 8539-42 (1995); also methylenecyclooctane ring from the corresponding 1-iodomethylbicyclo[n.2.0]alkanes cf. G.L. Lange, C. Gottardo, J. Org. Chem. *60*, 2183-7 (1995); transannular chloromethylcyclopentane ring opening cf. S. Sarkar, S. Ghosh, Tetrahedron *50*, 921-30 (1994).

Tri-n-butyltin chloride s. under $(Me_3Si)_3SiH$ Bu_3SnCl

Chromous acetate/disodium ethylenediaminetetraacetate $Cr(OAc)_2/Na_2EDTA$
Glycals from O^2-acylglycosyl halides $C(OAc)C(Hal) → C≡C$
with $Zn/Ag-C_8$ cf. *42*, 957; with $Cr(OAc)_2/Na_2EDTA$ via glycosylchromium(III) compds. s. G. Kovács et al., Tetrahedron Letters *37*, 1293-6 (1996); with $(Cp_2TiCl)_2$ cf. C.L. Cavallaro, J. Schwartz, J. Org. Chem. *60*, 7055-7 (1995); acid- and base-labile derivs. s. Tetrahedron Letters *37*, 4357-60 (1996).

(Triphenyltin)cobaloxime/irradiation   ←
Generation of alkyl radicals from bromides   ←
with a diiodocobalt(III) Schiff base complex/Zn cf. *48*, 968; improved procedure with (triphenyltin)cobaloxime under *non-reducing* conditions s. M. Tada, K. Kaneko, J. Org. Chem. *60*, 6635-6 (1995).

Bis(tri-n-butylphosphine)(1,5-cyclooctadiene)nickel(0) $(Bu_3P)_2Ni(cod)$
Benzocyclobutenes by reductive ring closure   ○
trans-1,2-diiodobenzocyclobutenes with NaI/KI cf. *14*, 968; *trans*-1,2-dibromo-derivs. with

(Bu₃P)₂Ni(cod) s. R. Boese et al., J. Org. Chem. *61*, 2549-52 (1996).
Dichloro[1,2-bis(diphenylphosphino)ethane]nickel(II) s. under EtMgBr $Ni(dppe)Cl_2$

Pallådous acetate/triphenylphosphine/tetra-n-butylammonium chloride   ←
Macrocyclization by polymer-supported intramolecular Heck arylation   ○

486.

Macrocyclic β-aryl-α,β-ethylenecarboxylic acid amides. A mixture of startg. resin-supported amide in 9:1:1 DMF/water/acetonitrile containing a little Pd(OAc)₂ (0.01 *M*), PPh₃ (0.02 *M*) and Bu₄NCl (0.02 *M*) allowed to react at room temp. overnight, the resin filtered, washed with DMF, dried, and cleaved with 50:47:3 trifluoroacetic acid/methylene chloride/anisole at room temp. for 30 min → product. Y 78% (predominantly *trans*). The resin support creates a 'pseudodilution' effect so that intermolecular condensation is minimized (as in high dilution techniques). This is the first instance of a Pd-catalyzed carbon-carbon macrocyclization on a solid support. F.e. and application to the **combinatorial synthesis of macrocyclic libraries** s. M. Hiroshige et al., J. Am. Chem. Soc. *117*, 11590-1 (1995); **peptoid isocarbostyrils** with Pd(PPh₃)₄ s. D.A. Goff, R.N. Zuckermann, J. Org. Chem. *60*, 5748-9 (1995).

Pallådous acetate/triphenylphosphine/tetraethylammonium chloride/piperidine/formic acid   ←
Reductive intramolecular Heck arylation   ○
of iodoarylethylenes s. *47*, 974; of iodoarylacetylenes s. R. Grigg et al., Tetrahedron *52*, 11479-502 (1996).

Sulfur ↑   CC ↑↑ S

Electrolysis/triethylamine $\frac{z}{}/Et_3N$
Pyrroles from β-amino-α,β-ethylenecarboxylic acid thioamides   ○
3-aminopyrroles with Br₂/Et₃N cf. *48*, 976; by electrolysis s. J. Prakt. Chem. *337*, 310-12 (1995).

Electrolysis/tetraethylammonium chloride $\frac{z}{}/Et_4NCl$
Cyclic ketones from 2-hydroxythioethers with ring expansion   ○
with TlOAc cf. *39*, 965; under electrolysis with Et₄NCl, also acyclic ketones with aryl or benzyl migration, s. Y. Sawaki et al., Tetrahedron *52*, 4303-10 (1996); cyclopentanones from (1-arylthiocyclobutyl)carbinols with SnCl₄ cf. L. Fitjer et al., Tetrahedron Letters *36*, 4985-8 (1995).

Lithium/naphthalene $LiC_{10}H_7$
Acetylene derivs. from α-hydroxymercaptals   C(SR)=C(SR) → C≡C
via ene-1,2-di(thioethers)

487.

A soln. of startg. hydroxymercaptal (readily prepared from octanal and phenylbis (phenylthio)methane), 2 eqs. methanesulfonyl chloride, and 3 eqs. Et₃N in benzene (or toluene) refluxed for 12-15 h → intermediate enedithioether (Y 88%), in THF treated with 6 eqs. Li-naphthalenide at -100° for 1 h → product (Y 69%). This is part of a multistep route to unsym. acetylenes **from aldehydes**, and features the novel formation of a triple bond from an alkenyl sulfide. F.e. incl. sym. acetylene derivs., **also ethylene derivs.** from α-acoxymercaptals via 1,2-di(thioethers) and 1,2-di(sulfones) s. T. Sato et al., Synlett *1995*, 628-30.

Aluminum amalgam   *Al,Hg*
Allenes from β,γ-acetylene-β′-ketosulfones   C≡C-C(SO$_2$R) → CH=C=C
via retro-ene elimination of sulfur dioxide

488.

Startg. acetylene treated with Al-amalgan in 10% aq. THF at room temp. for 30 min → product. Y 91%. The method is simple, versatile and high-yielding. F.e. and deuteriated products s. J.E. Baldwin et al., Tetrahedron Letters *36*, 7925-8 (1995); **from β-ketosulfoxides** via β,γ-ethylene-β-triflyloxysulfoxides (with LDA/*n*-BuLi) cf. T. Satoh et al., Tetrahedron *51*, 9327-38 (1995).

Samarium diiodide/hexamethylphosphoramide   ←
Ethylene derivs. from β-hydroxysulfones   C(OH)C(SO$_2$R) → C=C

489.

Trisubst. ethylene derivs. A soln. of startg. β-hydroxysulfone in THF containing 1-5 mol% HMPA treated with SmI$_2$ at 0° until reaction complete → product. Y 66% (E/Z ca. 2/1). The standard Julia route with Na/Hg failed. Such olefins may **also** be obtained more readily (at ≤ -78°!) **from β-acoxysulfones** so that a selective conversion can be achieved in the presence of a β-hydroxysulfonyl group. F.e. and improved prepn. of the startg. m. from ketones s. I.E. Markó et al., Tetrahedron Letters *37*, 2089-92 (1996); without HMPA s. A.S. Kende, J.S. Mendoza, ibid. *31*, 7105-8 (1990); (E)-isomers from β-acoxysulfones with Mg/EtOH/HgCl$_2$ cf. G.H. Lee et al., ibid. *36*, 5607-8 (1995).

Dibromodifluoromethane/potassium hydroxide-alumina   *CBr$_2$F$_2$/KOH-Al$_2$O$_3$*
Ethylene derivs. from sulfones   CHSO$_2$CH → C=C
s. *29*, 973s*49*; 3-ene-1,5-diynes s. Tetrahedron Letters *37*, 1049-52 (1996); 1,3(E),5-trienes s. Chem. Commun. *1995*, 1297-9.

Tri-n-butyltin hydride/azodiisobutyronitrile   *Bu$_3$SnH/RN=NR*
Ethylene derivs. from thiiranes

490.

A soln. of 1.5 eqs. Bu$_3$SnH in benzene added dropwise over 30 min to a refluxing soln. of 1,2-epithio-3-phenoxypropene in the same solvent containing a little AIBN → 3-phenoxypropene. Y 93%. The procedure is generally applicable to alkyl-, α-aryloxy- and α-alkoxy-thiiranes, although certain electron-withdrawing functions (e.g. *p*-nitrophenyl) inhibit reaction. (E)-Isomers predominated with 2,3-disubst. thiiranes. It was not possible to intercept the intermediate β-thioalkyl radical. F.e.s. E. Turos et al., J. Org. Chem. *60*, 470-2 (1995).

Triphenyltin hydride/azodiisobutyronitrile   *R$_3$SnH/RN=NR*
Radical ring closures of arylthioacetylenes
s. *43*, 972; 1-alkylidene-2-spiroindan ring of fredericamycin A analogs s. D.L.J. Clive et al., Tetrahedron *52*, 6085-116 (1996).

Stannic chloride $SnCl_4$
Cyclopentanones from (1-arylthiocyclobutyl)carbinols s. *39*, 965s*51*   ○

Remaining Elements ↑   CC ↑ Rem

*Lithium/4,4'-di-*tert*-butylbiphenyl*   ←
Cyclopentanes from 5-ethyleneselenides s. *51*, 427   ○

Methyllithium   *MeLi*
Cyclic 3-ethylenealcohols from 2,5-bridged 3-stannyltetrahydrofurans s. *51*, 28   ○

n-*Butyllithium*   *BuLi*
**Ring closures of unsatd. cyclic 1,1-aminostannanes
via α-aminoalkyllithium compds. with retention of configuration**   ○

491.

α-Aminoalkyllithium compds. are sufficiently stable *in hexane/ether at room temp.* to support anionic cyclization with retention of α-chirality. **E:** A soln. of startg. chiral aminostannane (e.e. 94%) in 10:1 hexane/ether treated with 2 eqs. *n*-BuLi at -78° to room temp., re-cooled to -78°, and quenched with methanol → (+)-pseudoheliotridane. Y 87% (e.e. 94%). Chelation of lithium to the alkene residue, as well as the poorly donating nature of the solvent, are considered responsible for the configurational stability of the intermediate. F.e. and trapping with electrophiles other than proton s. I. Coldham et al., J. Am. Chem. Soc. *118*, 5322-3 (1996).

Potassium bis(trimethylsilyl)amide $KN(SiMe_3)_2$
α-Oxo-N-heterocyclics by intramolecular Horner-Emmons synthesis
s. *25*, 693; isocarbostyrils with $KN(SiMe_3)_2$ s. A. Couture et al., Tetrahedron *52*, 4433-48 (1996).

Potassium fluoride   KF
1,3-Diols from β-hydroxyacylsilanes via stereospecific 1,2-Si→C-aryl migration   ←

492.

syn,syn-**1,3-Diols**. Startg. acylsilane (96% *syn*) in DMSO treated with KF at room temp. for 5 h in the presence of a little water → product. Y 92% (86% *syn,syn*-isomer). *anti*-Acylsilanes led predominantly to *anti,syn*-products. The presence of an α-alkyl group was essential for improved stereoselectivity, the α-unsubst. analogs leading to a 50:50 mixture of *syn/anti* isomers. F.e. and with *n*-Bu₄NF (in THF) s. K. Oshima et al., Tetrahedron Letters *36*, 5555-8 (1995).

Cuprous chloride   CuCl
Cyclic 1,3-dienes from bis(enestannanes)   ○

493.

A soln. of startg. bis(enestannane) in DMF added rapidly with stirring to ca. 5 eqs. CuCl slurried in the same solvent at ca. 60°, and worked up after 15 min → product. Y 67%. The procedure appears generally applicable to the formation of 4- to 8-membered rings, although somewhat more dilute conditions are required for the larger-ring systems. F.e.s. E. Piers, M.A. Romero, J. Am. Chem. Soc. *118*, 1215-6 (1996).

Ammonium ceric nitrate $(NH_4)_2Ce(NO_3)_6$
α,β-Ethyleneketones from enoxysilanes CHC≡C(OSi≤) → C≡C-CO
with diallyl carbonates/Pd(OAc)$_2$/dppe cf. *39*, 969; with CAN s. P.A. Evans et al., Tetrahedron Letters *36*, 3985-8 (1995).

Titanium tetrachloride $TiCl_4$
Stereospecific ring closures of 2-alkylidene-1,3-di(silanes)
2,3-*trans*-Disubst. pyrrolidines

494.

1 eq. 1 M TiCl$_4$ in toluene added dropwise with vigorous stirring to a soln. of startg. imine in methylene chloride at -78°, allowed to warm gradually to room temp. over 2-3 h, stirring continued for 2 h, the mixture transferred via cannula with vigorous stirring into satd. aq. KHCO$_3$ at 0°, and stirred for 30 min at room temp. → *trans*-2-isopropyl-3-[(3-trimethylsilyl)isopropenyl]pyrrolidine. Y 98% (*trans:cis* >50:1). Other initiators were less effective. F.e. and double ring closures s. T. Kercher, T. Livinghouse, J. Am. Chem. Soc. *118*, 4200-1 (1996).

Stannic chloride $SnCl_4$
Intramolecular 1,4-addition with 2-ethylenesilanes
with BF$_3$ cf. *38*, 911s*49*; **with asym. induction** (using SnCl$_4$), chiral α-(2-vinylcycloalkyl)malonic acid derivs., s. L.F. Tietze, C. Schünke, Angew. Chem. Intern. Ed. *34*, 1731-3 (1995).

Thionyl chloride/triethylamine $SOCl_2/Et_3N$
Cyclopropanes from 3-hydroxystannanes ▽
with 2-fluoro-1-methylpyridinium tosylate cf. *39*, 966s*43*; *trans*-2-vinylcyclopropanecarboxylic acid esters with SOCl$_2$ s. A. Krief et al., Synlett *1995*, 121-2.

Carbon ↑   CC ↑ C

Without additional reagents w.a.r.
Retrodiene scission C
s. *17*, 198s*46,48,49*; N-protected 2-subst. Δ^3-pyrrolines s. C. Cinquin et al., Tetrahedron *52*, 6943-52 (1996); 3-ene-1,5-diynes by *anionic* retrodiene scission with KH s. M.E. Bunnage, K.C. Nicolaou, Angew. Chem. Intern. Ed. *35*, 1110-2 (1996).

Irradiation ⇝
Photochemical bisdecarbonylation of cyclic α-diketones
ethylene derivs. cf. *32*, 985; benzene ring (phthalic acid esters and benzofulvenes) s. A. Thomas et al., Tetrahedron *52*, 2481-8 (1996).

Potassium hydride KH
Anionic retrodiene scission - 3-Ene-1,5-diynes s. *17*, 198s*51*

Potassium hydroxide KOH
in situ-Generation of α-alkoxy-α-allenecarboxylic acid esters ←
from α-alkoxy-β,γ-acetylenemalonic acid esters
and subsequent Myers-type cycloaromatization

495.

A soln. of startg. malonate in 24:1 ethanol/water treated with 0.5 eqs. KOH at 25° → product. Y 78%. D-Labelling studies suggest that reaction involves self-quenching by fairly fast disproportionation of toluene biradicals producing zwitterionic species. F.e.s. M. Shibuya et al., Tetrahedron Letters *37*, 865-8 (1996).

3,6-Di(2'-pyridyl)-s-tetrazine
Naphthalenes by elimination of etheno bridges
with PhN₃ as trapping agent cf. *26*, 987; with 3,6-di(2'-pyridyl)-*s*-tetrazine, also anthracenes, s. H. Heaney et al., Tetrahedron *51*, 7755-76 (1995).

Chloroaluminumtitanium complex
Cyclic enolethers from acoxyethylenes
Tebbe methylenation-ring closing metathesis

496.

A soln. of startg. ester in THF treated with 4 eqs. Tebbe reagent at 25° for 30 min, and refluxed for 10 h → product. Y 71%. The procedure is generally applicable to 6- and 7-membered cyclic enolethers, and is tolerant of trisubst. olefin groups. F.e. and with Cp₂TiMe₂ s. K.C. Nicolaou et al., J. Am. Chem. Soc. *118*, 1565-6 (1996).

Molybdenum carbene complex
Ring closing metathesis
s. *48*, 988; fused N-heterocyclics s. S.F. Martin et al., Tetrahedron *52*, 7251-64 (1996); **with kinetic resolution** using a chiral Mo-carbene complex s. O. Fujimura, R.H. Grubbs, J. Am. Chem. Soc. *118*, 2499-500 (1996).

trans-*Dichlorobis(2,6-dibromophenoxy)oxotungsten(VI)/tetraethyllead*
Tetrachlorobis(2,6-diphenylphenoxy)tungsten(VI)/tetrabutyllead
Ring closing metathesis

497.

Ring closing metathesis **of non-conjugated dienes** can now be performed efficiently with a user-friendly oxotungsten complex - as an alternative to the expensive Ru-carbene complex (cf. *49*, 985) and Mo-carbene complex (cf. *48*, 988). E: A soln. of (+)-citronellene and 2 mol% trans-dichlorobis(2,6-dibromophenoxy)oxotungsten(VI) (recrystallized) in 1,2,4-trichlorobenzene treated with 4 mol% Et₄Pb, heated at 90° (oil bath) for 2 h, cooled, filtered, the crude product distilled out under N_2, and heated at reflux under air for 30 min (to remove dissolved isobutylene) 5 (S)-3-methylcyclopentene. Y 70%. Reaction proceeds with retention of chirality, and esters, amides, allyl ethers, and siloxy groups were unaffected. Cyclohexenes, 2,5-dihydrofurans and D^3-pyrrolines were also prepared, but 1,9-decadiene and 1,13-tetradecadiene gave polymeric products. F.e. and with the crude catalyst s. W.A. Nugent et al., J. Am. Chem. Soc. *117*, 8992-8 (1995); macrocyclic carbohydrate derivs. with tetrachlorobis(2,6-diphenylphenoxy)tungsten(VI)/ Bu₄Pb s. G. Descotes et al., Tetrahedron *52*, 10903-20 (1996).

Ruthenium carbene complex
Ring closing metathesis
s. *49*, 985s*50*; macrocyclic [rigidified] α-aminoacids and peptides s. R.H. Grubbs et al., J. Am. Chem. Soc. *118*, 9606-14; hydrazulenes, details, s. Tetrahedron *51*, 13003-14 (1995); azasugars s. Tetrahedron Letters *36*, 1621-4 (1995).

Formation of Electron Pair on Nitrogen

Elimination ⇈

Oxygen ↑   EIN ⇈ O

Benzyltriethylammonium tetrathiomolybdate $(BnEt_3N)_2MoS_4$
N-Deoxygenation $\geqslant NO \rightarrow \geqslant N$
with Mb(IV) oxide complex/PPh$_3$ cf. *37*, 994; of aldehyde nitrones and N-oxides with (BnEt$_3$N)$_2$MoS$_4$ s. P. Ilankumaran, S. Chandrasekaran, Tetrahedron Letters *36*, 4881-2 (1995).

Triphenylphosphine PPh_3
N-Deoxygenation
s. *14*, 988; of N-arylaldimine N-oxides s. S. Sivasubramanian et al., Org. Prep. Proc. Intern. *27*, 221-4 (1995); of 3-nitropyridine N-oxides with PCl$_3$ (cf. *12*, 952) s. R. Nesiet et al., J. Org. Chem. *57*, 3713-6 (1992).

Carbon ↑   EIN ⇈ C

Sodium hydride telluride   *NaHTe*
N-Dequaternization $\geqslant N^+R \rightarrow \geqslant N$
s. *41*, 993; of phase transfer catalysts s. W. Li, X.-J. Zhou, Synth. Commun. *25*, 3635-9 (1995).

Ethyldiisopropylamine   i-Pr_2NEt
N-Dequaternization by Hofmann elimination of a polymer support s. *51*, 172

Formation of Electron Pair on Sulfur

Elimination ⇈

Oxygen   EIS ⇈ O

Titanium dioxide/irradiation TiO_2/⚡
Titanium tetrachloride or titanocene dichloride/samarium $TiCl_4$ or Cp_2TiCl_2/Sm
Tellurium tetrachloride/sodium iodide $TeCl_4$/NaI
Thioethers from sulfoxides $\geqslant SO \rightarrow \geqslant S$
with TiCl$_4$/NaI cf. *41*, 995s46; with TiCl$_4$/Sm s. J.Q. Wang, Y.M. Zhang, Synth. Commun. *25*, 3545-7 (1995); with Cp$_2$TiCl/Sm cf. Y. Zhang et al., ibid. 1825-30; photoreduction on TiO$_2$ [semiconductor particles], **also from sulfones**, s. N. Somasundaram et al., Chem. Commun. *1994*, 1473; with TeCl$_4$/NaI cf. R.H. Khan, R.C. Rastogi, Indian J. Chem. *33B*, 293-4 (1994).

Formation of Electron Pair on Remaining Elements

Elimination ⇑

Halogen ↑ ElRem ⇑ Hal

Tris(trimethylsilyl)phosphine $(Me_3Si)_3P$
Phosphines from phosphine dihalides $\geqslant PHal_2 \rightarrow\, \geqslant PHal$
with hydrocarbons cf. *43*, 954; tris(perfluoroalkyl)phosphines from their difluorides with $(Me_3Si)_3P$ s. J.J. Kampa et al., Angew. Chem. Intern. Ed. *34*, 1241-4 (1995).

Heteropolar Bond

Uptake ⇓

Addition to Nitrogen Het ⇓ N

Without additional reagents *w.a.r.*
Quaternary ammonium salts from tert. amines $\geqslant N \rightarrow\, \geqslant N^+R$
s. *25*, 694; selective conversion with retention of phosphino groups via W-complexation s. I.V. Komarov et al., Tetrahedron *51*, 11271-80 (1995).
N-Quaternization with chloroformic acid esters ←
nitrilium salts s. *48*, 995; pyridinium tetraphenylborates s. J.A. King, Jr., G.L. Bryant, Jr., Synth. Commun. *24*, 1923-35 (1994).

Addition to Sulfur Het ⇓ S

Without additional reagents *w.a.r.*
Sulfonium salts from thioethers $>S \rightarrow\, >S^+R$
s. *1*, 790; sulfonium perchlorates in acetone s. V.K. Aggarwal et al., Tetrahedron Letters *35*, 8659-60 (1994).

Resolutions Res

Chromatography ←
Optical resolution on chiral stationary phases ←
s. *5*, 666s*49*; of alkyl nitrates by gas chromatography s. M. Schneider, K. Ballschmiter, Chem. Eur. J. *2*, 539-44 (1996); of Δ^2-isoxazolines by column chromatography s. C. Ticozzi, A. Zanarotti, Tetrahedron Letters *35*, 7421-4 (1994).
Optical resolution via chiral derivatization ←
s. *17*, 393s*49*; resolution **of amines** via chromatographic separation of diastereoisomeric α-(*tert*-butoxycarbonylamino)carboxamides s. C.K. Miao et al., Org. Prep. Proc. Intern. *24*, 87-90 (1992); via flash chromatography of diastereoisomeric α-aminocarboxylic acid esters cf. G. Curotto et al., Tetrahedron:Asym. *6*, 849-52 (1995); of 2-acetylenealcohols via diastereoisomeric 4-(alk-2-ynyloxy)-6,6-dimethyl-3-oxabicyclo[3.1.0]hexan-2-ones s. J.P. Genet et al., Synthesis *1995*, 165-7.

n-*Butyllithium/polymer-based α-hydroxycarboxylic acid esters* ←
Asym. protonation of enolates ←
s. *39*, 993s*49*; of achiral enoxysilanes with a polymer-based α-hydroxyester s. F. Cavelier et al., Tetrahedron Letters *35*, 2891-4 (1994).

Chiral amines ←
Optical resolution via salt formation
s. *5*, 666s*49*; resolution **of alcohols** via fractional crystallization of diastereoisomeric phthalic acid monoester salts with (S)-α-methylbenzylamine s. A. Reyes, E. Juaristi, Synth. Commun. *25*, 1053-8 (1995); of N-acylarylglycines with (R)-2-aminobut-1-yl benzyl ethers as base s. J. Touet et al., Tetrahedron *51*, 1709 (1995); of 1,1'-bi-2-naphthol with N-benzylcinchonidinium chloride s. D. Cai et al., Tetrahedron Letters *36*, 7991-4 (1995).

Determination of enantiomeric purity ←
s. *5*, 666s*49*; of chiral dihydroxybiaryls by NMR assay of 1,3,2-diazaphospholidine S-sulfide derivs. s. A. Alexakis et al., Tetrahedron Letters *35*, 5125-8 (1994); of chiral glycidols by direct HPLC s. J. Chen, W. Shum, ibid. *34*, 7663-6 (1993); of cyclic phosphoric acid diesters via diastereoselective salt formation with chiral amines s. R. Hulst et al., Rec. Trav. Chim. Pays-Bas *114*, 220-4 (1995).

Determination of abs. configuration ←
s. *5*, 666s*49*; of constituent amino acids in peptides by an advanced Marfey's method s. K. Harada et al., Tetrahedron Letters *36*, 1515-8 (1995); **of alcohols and amines** by NMR assay of chiral aryl(methoxy)acetic acid derivs. s. J.M. Seco et al., ibid. *35*, 2921-4 (1994); of long chain sec. alcohols with a chiral 9-anthryl(methoxy)acetic acid as auxiliary s. T. Kusumi et al., ibid. 4397-400; s.a. ibid. *37*, 4541-4 (1996); of amines and alcohols by ^{19}F NMR assay of 2-fluoroacetic acid derivs. s. M. Barrelle, S. Hamman, J. Chem. Res. (S) *1995*, 316-7; of benzylcarbinamines by CD measurements s. H.E. Smith, J.R. Neergaard, J. Am. Chem. Soc. *119*, 116-24 (1997); of amines with poly((4-carboxyphenyl)acetylene) as probe s. E. Yashima et al., ibid. *117*, 11596-7 (1995); of cyclic sec. amines via ^{1}H NMR assay of Mosher amides s. T.R. Hoye, M.K. Renner, J. Org. Chem. *61*, 2056-64 (1996); of trisubst. allenes with permethylated β-cyclodextrin as solvating agent s. G. Uccello-Barretta et al., ibid. *60*, 2227-31 (1995); **of carboxylic acids** via NMR assay of (S)-methyl mandelate esters s. E. Tyrrell et al., Tetrahedron *52*, 9841-52 (1996); of halogenated furanones s. A.D. Wright et al., Helv. Chim. Acta *78*, 758-64 (1995); **of cyanohydrins** via ^{1}H NMR assay of tricyclic γ-lactol derivs. s. C.R. Noe et al., Liebigs Ann. *1995*, 1353-60.

Lipase ←
Kinetic resolution by asym. hydrolysis ←
Chiral alcohols and carboxylic acids s. *28*, 13s*51*; *51*, 6
Kinetic resolution by asym. O-acylation with enolesters s. *44*, 214s*51* ←
Deracemization of α-(arylthio)thiolic acid esters via asym. hydrolysis s. *51*, 94 ←

Lipase/silica ←
Dynamic kinetic resolution of 1,1-hydroxythioethers via asym. O-acylation s. *51*, 233 ←

Lipase/triphenylphosphine/diethyl azodicarboxylate ←
Deracemization of sec. alcohols
via asym. enzymatic O-acylation-Mitsunobu reaction s. *51*, 103

Aminoesterase ←
Kinetic resolution of α-aminocarboxylic acid esters s. *48*, 8s*51* ←

Subtilisin ←
Kinetic resolution via N-carbalkoxylation s. *51*, 146 ←

Cyanohydrin lyase ←
Kinetic resolution of ketone cyanohydrins s. *22*, 693s*51* ←

Horseradish peroxidase
Kinetic resolution of α-hydroperoxycarboxylic acid esters via enzymatic reduction

498.

Startg. peroxide, 1 eq. guaiacol, and a little horseradish peroxidase allowed to react in pH 6 phosphate buffer at 15-20° until 50% conversion had been attained → (R)-hydroperoxide (e.e. 97%) and (S)-alcohol (e.e. 97%). F.e.s. W. Adam et al., Tetrahedron:Asym. *6*, 1047-50 (1995); **resolution of hydroperoxides** cf. J. Am. Chem. Soc. *117*, 11898-901 (1995).

Microorganisms
Deracemization of sec. alcohols via microbial oxidation-reduction

499.

A sterilized nutrient broth (consisting of glucose, batotryptone and yeast extract in water) inoculated with *Bacillus stearothermophilus*, incubated for 48 h at 39° with stirring, centrifuged, the cells suspended in phosphate buffer at pH 7.2, startg. racemic alcohol in DMSO added, stirred at 39° for 5 h, centrifuged, the supernatant added to centrifuged cells of *Yarrowia lipolytica* Y10 [cultured in water containing glucose, $(NH_4)_2SO_4$, KH_2PO_4, $CaCl_2$, $MgSO_4$, inositol, H_3BO_3, $ZnSO_4$, $MnCl_2$, $FeCl_2$, $CuSO_4$, tiamine, biotin, Ca-pantothenate, pyridoxane and nicotinic acid for 48 h at 28°], and stirring continued for 24 h at 28° → (R)-product. Y 91% (e.e. 100%) and 9% ketone. F.e. and conditions s. A. Medici et al., Tetrahedron:Asym. *6*, 3047-53 (1995); deracemization of sec. benzylalcohols with the oxidoreductase system of *Geotrichum candidum* cf. K. Nakamura et al., Tetrahedron Letters *36*, 6263-6 (1995); kinetic resolution of 1-arylethanols (cf. *50*, 601) s. ibid. *35*, 4375-6 (1994).

(S)-3,5-Dihydro-3,3,5,5-tetramethyl-4H-dinaphth[2,1-c:1',2'-e]azepine-N-oxyl/sodium hypochlorite/potassium bromide
Ketones from sec. alcohols with kinetic resolution

500.

A chiral N-oxyl catalyst is now available for the first non-enzymatic kinetic resolution of sec. alcohols by asym. oxidation. E: A soln. of racemic methyl(phenyl)carbinol in methylene chloride/water stirred rapidly for 30 min at 0° with 1 mol% (S)-3,5-dihydro-3,3,5,5-tetramethyl-4*H*-dinaphth[2,1-*c*:1',2'-*e*]azepine-N-oxyl, 0.1 eq. KBr, and 0.6-0.7 eq. 0.35 *M* NaOCl (pH 8.6) → acetophenone and unreacted (R)-methyl(phenyl)carbinol (e.e. 98% at 87% conversion). Reaction is effective with activated (benzylic and propargylic) sec. alcohols, and partial resolution is also possible (e.e. 19%) with prim. alcohols possessing an adjacent chiral centre. F.e.s. S.D. Rychnovsky et al., J. Org. Chem. *61*, 1194-5 (1996).

Titanocene dichloride Cp_2TiCl_2
Kinetic resolution by asym. N-acylation with 3-acyl-2-oxazolidones s. *51*, 160

Potassium osmate/potassium hexacyanoferrate/6-(9-O-dihydroquinidine)(3-[(S)-1-(1-anthryl)-2,2-dimethylprop-1-yl]pyridazine/potassium carbonate/tert-butanol
Kinetic resolution of ethylene derivs. via asym. dihydroxylation
s. *49*, 997; of sec. *p*-methoxybenzoyloxy-2-ethylenes with a 9-O-dihydroquinidine anthrylpyridazine as chiral ligand (cf. *50*, 72) s. E.J. Corey et al., J. Am. Chem. Soc. *117*, 10817-24 (1995).

Reviews

This is a collection of reviews in the field of synthetic organic chemistry published mainly in 1996. The layout is to aid access via the Supplementary Reference Index, each entry being indexed in the Subject Index, e.g.
Uronic acids
-, synthesis and reactions, review **18**, 299s51

2, 368 **Reactions of hydrazines and hydroxylamines** with α,β-unsatd. and β-dicarbonyl compds., K.N. Zelenin, Org. Prep. Proc. Intern. 27, 519-40 (1995).

3, 708 **Oxidative dehydrogenation** of lower alkanes on vanadium oxide-based catalysts, E.A. Mamedov, V. Cortés Corberán, Appl. Catal. A 127, 1-40 (1995).

5, 666 **Chiral discrimination** by modified cyclodextrins, C.J. Easton, S.F. Lincoln, Chem. Soc. Rev. 25, 163-70 (1996).

13, 831 Synthesis and reactions of **intermediates in the Wittig, Peterson, and related reactions**, T. Kawashima, R. Okazaki, Synlett 1996, 600-8.

18, 299 **Uronic acids:** synthesis and reactions, V.A. Timoshchuk, Russ. Chem. Rev. 64, 675-703 (1995).

19, 265 **Acyl glucuronides**, P.J. Hayball, Chirality 7, 1-9 (1995).

20, 271 **Diazo transfer** reagents, F.W. Bollinger, L.D. Tuma, Synlett 1996, 407-13.

20, 647 **Heterogeneous catalytic oxidation** with molecular oxygen and sulfur, O.V. Krylov, V.A. Matyshak, Russ. Chem. Rev. 64, 167-86 (1995).

21, 786 **Cubanes** and cage-related molecules, A. Bashir-Hashemi et al., Chem. Ind. 1995, 551-5; cage compds. and molecular clips derived from glycoluril, R.P. Sijbesma, R.J.M. Nolte, Topics Curr. Chem. 175, 25-56 (1995); polycyclic cage compds., A.P. Marchand, Aldrichimica Acta 28, 95-104 (1995).

23, 139 **Epoxides**, catalytic reactions with CO_2, Coord. Chem. Rev. 153, 155-74 (1996).

23, 757 **Carboxylation and aminomethylation of alkanes and arenes** via transition metal-activation of C-H bonds, Y. Fujiwara et al., Synlett 1996, 591-9.

26, 97 **Nucleoside 5'-phosphates labelled with radioactive phosphorus** isotopes, synthesis, Yu.S. Skoblov et al., Russ. Chem. Rev. 64, 799-807 (1995).

27, 162 **Iodine(III) and thallium(III) reagents**, oxidative rearrangements with -, O. Prakash, Aldrichimica Acta 28, 63-71 (1995).

28, 13 **2-Amino-1-phenylethanols**: biocatalytic syntheses, L.T. Kanerva, Acta Chem. Scand. 50, 234-42 (1996); **artificial enzymes**, Y. Murakami et al., Chem. Rev. 96, 721-58 (1996); **chiral α-subst. α-amino- and α-hydroxycarboxylic acids**: enzymatic synthesis, H.E. Schoemaker et al., Acta Chem. Scand. 50, 225-33 (1996).

28, 511 Synthesis of **radiopharmaceuticals via organotin intermediates**, H. Ali, J.E. van Lier, Synthesis 1996, 423-44.

28, 543 **Thiophenes** from hydrocarbons and H_2S, catalytic routes, M.A. Ryashentseva, Russ. Chem. Rev. 63, 437-47 (1994).

28, 765 **N-Macroheterocyclics** by Mannich reaction, A.V. Bordunov et al., Synlett 1996, 933-48.

29, 744 **N-Heterocyclics, 5- and 6-membered** from iminoesters, V.I. Kelarev, V.N. Koshelev, Russ. Chem. Rev. 64, 317-48 (1995).

29, 964   N,O-Bis(trimethylsilyl)acetamide as mild silyl transfer agent, A.M. El-Khawaga, H.M.R. Hofmann, J. Prakt. Chem. *337*, 332-4 (1995).

30, 242   **Enisothiocyanates**, N.A. Nedolya et al., Sulfur Reports *17*, 183-395 (1996).

30, 348   **Sulfenyl chlorides** in organic synthesis, I.V. Koval, Russ. Chem. Rev. *64*, 731-51 (1995).

30, 561   **Low-valent titanium** in organic chemistry, A. Fürstner, B. Bogdanovic, Angew. Chem. Intern. Ed. *35*, 2442-69 (1996).

32, 47   **Stereospecific reduction of arenes**, T.J. Donohoe et al., Tetrahedron:Asym. *7*, 317-44 (1996); **Birch reduction**, A.J. Birch, Pure Appl. Chem. *68*, 553-6 (1996).

32, 867   **Difluoromethylene compds.**, synthesis, M.J. Tozer, T.F. Herpin Tetrahedron *52*, 8619-83 (1996).

32, 974   **Sultines**, O.B. Bondarenko et al., Russ. Chem. Rev. *65*, 147-66 (1996).

33, 100   **Silyl triflates**, preparation and use as silylating agents, W. Uhlig, Chem. Ber. *129*, 733-9 (1996).

33, 315   **Smiles rearrangement** of *o*-aminophenyl phenyl ethers, T.N. Gerasimova, E.F. Kolchina, Russ. Chem. Rev. *64*, 133-40 (1995).

33, 659   **Organomercury compds.**, C.E. Holloway et al., J. Organometal. Chem. *495*, 1-31 (1995).

33, 786   **Oxiranyl and aziridinyl anions**, T. Satoh, Chem. Rev. *96*, 3303-25 (1996).

34, 55   **Hydrogenolysis, Pd-catalyzed** of allylic and propargylic compounds with various hydrides, J. Tsuji, T. Mandai, Synthesis *1996*, 1-24.

34, 630   **1,3-Dipolar cycloaddition** to allenes, G. Broggini, G. Zecchi, Gazz. Chim. Ital. *126*, 479-88 (1996).

35, 634   **Asym. synthesis of heterocyclics** from α-aminocarboxylic acids, F.J. Sardina, H. Rapoport, Chem. Rev. *96*, 1825-72 (1996).

36, 430   **Centropolyindans** by cyclodehydration, D. Kuck, Synlett *1996*, 949-65.

36, 652   **Michael addition, stereoselective** of enolates to α,β-unsatd. carbonyl compds., A. Bernardi, Gazz. Chim. Ital. *125*, 539-47 (1995).

36, 775   **Umpolung in organometallic chemistry** - from carbanionic nucleophiles to metallic electrophiles, J.J. Eisch, J. Organometal. Chem. *500*, 101-15 (1995).

36, 898   **Pyridinium ylids** as nucleophile, V.P. Litvinov, Russ. J. Org. Chem. *30*, 1658-83 (1994); **cycloaddition** with pyridinium ylids and oxidopyridiniums, W. Sliwa, Heterocycles *43*, 2005-29 (1996).

37, 623   **Diastereofacial nucleophilic addition** to unsym. subst. trigonal carbons, Tetrahedron *52*, 5263-301 (1996).

37, 688   **Enamines from 2-ethyleneamines**, asym. Rh(I)-catalysis, S. Otsuka, Acta Chem. Scand. *50*, 353-60 (1996).

38, 39   **Selective hydrogenation** over ruthenium catalysts, P. Kluson, L. Cerveny, Appl. Catal. A *128*, 13-31 (1995).

38, 470   **Trichloroacetonitrile**, chemistry, S.M. Sherif, A.W. Erian, Heterocycles *43*, 1083-118 (1996).

38, 661   **7-Azabicyclo[2.2.1]-heptanes**, -hepta-2,5-dienes and -hept-2-enes, chemistry, Z. Chen, M.L. Trudell, Chem. Rev. *96*, 1179-94 (1996).

39, 200   **Carbon-silicon bond oxidation**, G.R. Jones, Y. Landais, Tetrahedron *52*, 7599-662 (1996); **hydroxyl groups from silanes**, I. Fleming, Chemtracts *9*, 1-64 (1996).

39, 225   **Oxidation of prim. and sec. alcohols** with stable N-oxide radicals, A.E.J. de Nooy et al., Synthesis *1996*, 1153-74.

39, 292   **Photochemical reactions of pyrroles**, M. D'Auria, Heterocycles *43*, 1305-34 (Part 1); 1529-58 (Part 2) (1996); **photoamination** for heterocyclic synthesis, M. Yasuda et al., ibid. 2513-22.

39, 458 **Electrophilic N-fluoro compds.**, G.S. Lal et al., Chem. Rev. *96*, 1737-56 (1996).

39, 593 **Baylis-Hillman reaction**, D. Basavaiah et al., Tetrahedron *52*, 8001-62 (1996).

40, 384 **Quinone methids** as intermediates, P. Wan et al., Can. J. Chem. *74*, 465-75 (1996).

40, 567 **Aqueous Barbier-Grignard type reactions** - scope, mechanism, and synthetic applications, C.-J. Li, Tetrahedron *52*, 5643-68 (1996).

41, 54 **1,2,3,4-Tetrahydroquinolines**, recent syntheses, A.R. Katritzky et al., Tetrahedron *52*, 15031-70 (1996).

41, 173 **Oligosaccharide synthesis**, strategies, G.-J. Boons, Tetrahedron *52*, 1095-121 (1996); oligosaccharide synthesis using glycosidases, C. Bucke, J. Chem. Technol. Biotechnol. *67*, 217-20 (1996); oligosaccharides and glycoconjugates from glycals, S.J. Danishefsky, M.T. Bilodeau, Angew. Chem. Intern. Ed. *35*, 1380-419 (1996); **septanose synthesis**, Z. Pakulski, Pol. J. Chem. *70*, 667-707 (1996); **higher carbon sugars** by 'tail-to-tail' coupling of monosaccharides, S. Jarosz, Pol. J. Chem. *70*, 141-57 (1996); carbohydrates and carbohydrate mimetics by chemoenzymatic routes, H.J.M. Gijsen et al., Chem. Rev. *96*, 443-73 (1996).

41, 175 **Kinetic resolution of 2-methylalkanoic acids** by asym. lipase-catalyzed esterification with long-chain alcohols, H. Edlund et al., Acta Chem. Scand. *50*, 666-71 (1996).

41, 305 **Azasugars** and multistep cascade rearrangement via hetero-Diels-Alder cycloaddition with nitroso compds., J. Streith, A. Defoin, Synlett *1996*, 189-200.

41, 463 **Fluorinated peroxides**, H. Sawada, Chem. Rev. *96*, 1779-808 (1996); **fluoroalkyl radicals**, chemistry, W.R. Dolbier, Jr., ibid. 1557-84; **fluorinated carbenes**, D.L.S. Brahms, W.P. Dailey, ibid. 1585-632; **fluorinated carbanions**, W.B. Farnham, ibid. 1633-40; **fluorinated ylids** and related compds., D.J. Burton et al., ibid. 1641-716.

41, 660 **Electrophilic perfluoroalkylating agents**, T. Umemoto, Chem. Rev. *96*, 1757-78 (1996).

41, 678 **Reactions in micellar media**, S. Tascioglu, Tetrahedron *52*, 11113-52 (1996).

42, 45 **Asym. catalysis with metal-BINAP complexes**, industrial applications, H. Kumobayashi, Rec. Trav. Chim. Pays-Bas *115*, 201-10 (1996); s.a. S. Akutagawa, Appl. Catal. A *128*, 171-207 (1995); **asym. Noyori-type hydrogenation**, R. Noyori, Acta Chem. Scand. *50*, 380-90 (1996).

42, 209 **Silyl-1,3-dienes** and derivs. in organic synthesis, M.D. Stadnichuk, T.I. Voropaeva, Russ. Chem. Rev. *64*, 25-46 (1995).

42, 595 **Diphenylphosphinyl group**, stereocontrol in organic synthesis using the -, J. Clayden, S. Warren, Angew. Chem. Intern. Ed. *35*, 241-70 (1996).

42, 616 **Polyfunctionalized dialkylzincs**, prepn. and use, P. Knochel, Chemtracts *8*, 205-21 (1995).

42, 703 **[4+4]-Cycloaddition** and its strategic application in natural product synthesis, S.McN. Sieburth, N.T. Cunard, Tetrahedron *52*, 6251-82 (1996).

42, 771 **Trichloroethylene** in organic synthesis, R.V. Kaberdin, V.I. Potkin, Russ. Chem. Rev. *63*, 641-59 (1994).

42, 814 **Fluorine-containing heterocyclics, synthesis,** Russ. J. Org. Chem. *30*, 1792-842 (1994).

42, 852 **Group(VI) metal carbene complexes** in synthesis, J. Barluenga, Pure Appl. Chem. *68*, 543-52 (1996).

42, 865 **Polyfluoro-alkanes and -alkenes, cyclic** and bicyclic, J.C. Tatlow, J. Fluorine Chem. *75*, 7-34 (1995); **polyfluorocarbocations**, C.G. Krespan, V.A. Petrov, Chem. Rev. *96*, 3269-301 (1996).

43, 151 **2,3-Unsatd. carbohydrates**, B. Fraser-Reid, Acc. Chem. Res. *29*, 57-66 (1996).

43, 425 **Isoureas** - synthesis and applications, A.A. Bakibaev, V.V. Shtrykova, Russ. Chem. Rev. *64*, 929-38 (1995).

43, 540 **Organotellurium compds.**, reactivity, I.D. Sadekov, V.I. Minkin, Russ. Chem. Rev. *64*, 491-522 (1995).

43, 700	**Chiral C-tetrasubst. α-aminocarboxylic acids**, B. Kapeteini et al., Chim. Oggi *14*, 9-12 (1996).
43, 769	**anti-Bredt polycyclic ethylene derivs.**, W.T. Borden, Synlett *1996*, 711-9; hindered polycyclic alkenes, B.M. Lerman, Russ. Chem. Rev. *64*, 1-24 (1995).
43, 833	**Cathodic deprotonation** of acids, V.A. Petrosyan, Russ. Chem. Bull. *44*, 1353-64 (1995).
43, 943	**Cascade processes of metallo carbenoids**, A. Padwa, M.D. Weingarten, Chem. Rev. *96*, 223-69 (1996).
43, 989	**Methylenecyclobutane chemistry**, M.G. Vinogradov, A.V. Zinenkov, Russ. Chem. Rev. *65*, 131-45 (1996).
44, 7	**Bis-protected hydroxylamines** in synthesis, J.L. Romine, Org. Prep. Proc. Intern. *28*, 249-88 (1996).
44, 65	**Silicon in organic chemistry**, reactivity and ground-state effects, J.M. White, Australian J. Chem. *48*, 1227-51 (1995).
44, 469	**Mercaptans and dialkyl thioethers** from alcohols and H_2S under heterogeneous catalysis, A.V. Mashkina, Russ. Chem. Rev. *64*, 1131-47 (1995).
44, 568	**Asym. ene reaction with carbonyl compds.**, K. Mikami, Pure Appl. Chem. *68*, 639-44 (1996).
44, 577	**Trifluoromethyl group** in organic synthesis, A.S. Kiselyov, L. Strekowski, Org. Prep. Proc. Intern. *28*, 291-318 (1996).
44, 651	**Microwave-assisted chemistry**, C.R. Strauss, R.W. Trainor, Australian J. Chem. *48*, 1665-92 (1995).
44, 837	**Phosphorus ylids, C-element-subst.**, O.I. Kolodiazhnyi, Tetrahedron *52*, 1855-929 (1996).
44, 936	**Isoindoles**, V.A. Kovtunenko, Z.V. Voitenko, Russ. Chem. Rev. *63*, 997-1018 (1994).
45, 6	**α-Acylaminocarboxylic acids**, synthesis and applications, A.P. Mikhalkin, Russ. Chem. Soc. *64*, 259-75 (1995).
45, 308	**Enediynes, enyne allenes and related compounds**, J.W. Grissom et al., Tetrahedron *52*, 6453-518 (1996); **cascade cyclization via 1,4-diradicals** generated from enediynes, enyne-allenes, and enyne-ketenes, K.K. Wang, Chem. Rev. *96*, 207-22 (1996); **1,5-cyclization of alkyl propargyl 1,4-diradicals**, W.C. Agosta, P. Margaretha, Acc. Chem. Res. *29*, 179-82 (1996).
45, 414	**Tandem reactions** in organic synthesis: novel strategies for natural product elaboration and the development of new synthetic methodology, P.J. Parsons et al., Chem. Rev. *96*, 195-206 (1996); **cascade reactions**, Tetrahedron Symposia-in-Print No.62, Tetrahedron *52*, 11385-664 (1996); **domino reactions**, L.F. Tietze, Chem. Rev. *96*, 115-36 (1996); **tandem [4+2]/[3+2] cycloadditions** of nitroalkenes, S.E. Denmark, A. Thorarensen, Chem. Rev. *96*, 137-65 (1996).
45, 555	**Sequential palladium-catalyzed reactions**, M. Catellani, Russ. Chem. Bull. *44*, 397-405 (1995); **cross-coupling**, Pd- and/or Cu-mediated, R. Rossi et al., Org. Prep. Proc. Intern. *27*, 127-60 (1995).
46, 3	**Protecting group strategies**, M. Schelhaas, H. Waldmann, Angew. Chem. Intern. Ed. *35*, 2056-83 (1996); **preferential O-desilylation**, T.D. Nelson, R.D. Crouch, Synthesis *1996*, 1031-69.
46, 160	**Mitsunobu reactions**, progress, D.L. Hughes, Org. Prep. Proc. Intern. *28*, 127-64 (1996).
46, 350	**Peptide synthesis** with aminoacid halides, L.A. Carpino et al., Acc. Chem. Res. *29*, 268-74 (1996).
46, 484	**α-Amino- and α-acylamino-alkyl radicals**, generation and stereoselective reactions, P. Renaud, L. Giraud, Synthesis *1996*, 913-26.
46, 540	**Triethylamine trishydrofluoride** in synthesis, G. Haufe, J. Prakt. Chem. *338*, 99-113 (1996).

46, 631 **Transition metal-mediated cycloaddition** M. Lautens et al., Chem. Rev. *96*, 49-92 (1996).

46, 679 **Ring closure via carbopalladation**, E. Negishi et al., Chem. Rev. *96*, 365-93 (1996); **carbocyclization**, transition metal-catalyzed, I. Ojima et al., ibid. 635-62; **radical and transition metal-catalyzed cascade** reactions for preparing complex polycyclics from acyclics, M. Malacria, ibid. 289-306; **palladium-catalyzed cascade reactions**, A. Heumann, M. Réglier, Tetrahedron *52*, 9289-346 (1996).

46, 720 **Carbon dioxide**, heterogeneous catalytic reactions, O.V. Krylov, A.Kh. Mamedov, Russ. Chem. Rev. *64*, 877-900 (1995).

46, 811 **Arene-catalyzed lithiation**, M. Yus, Chem. Soc. Rev. *25*, 155-61 (1996).

46, 921 Synthesis of **β-carbolines**, B.E. Love, Org. Prep. Proc. Intern. *28*, 1-64 (1996).

46, 930 **Heterogeneous oxidative radical reactions**, M. Yu Sinev et al., Russ. Chem. Rev. *64*, 349-64 (1995); oxidative radial ring closure based on Mn(III), B.B. Snider, Chem. Rev. *96*, 339-63 (1996).

46, 954 **Asym. intramolecular cyclopropanation**, A.M.P. Koskinen, H. Hassila, Acta Chem. Scand. *50*, 323-7 (1996).

47, 22 **Carbonylation** of ar. nitro compds., F. Ragaini, S. Cenini, Chim. Ind. (Milan) *78*, 421-7 (1996); **selective catalytic reduction of ar. nitro compds.** to ar. amines, isocyanates, carbamates and ureas with CO, A.M. Tafesh, J. Weiguny, Chem. Rev. *96*, 2035-52 (1996).

47, 49 **Hydrogenation, selective** of unsatd. ketones and acetylenealcohols, E.M. Sulman, Russ. Chem. Rev. *63*, 923-36 (1994).

47, 304 **Peptide thioamides**, T. Hoeg-Jensen, Phosphorus Sulfur Silicon, *108*, 257-78 (1996).

47, 444 **Bis(trichloromethyl) carbonate** in organic synthesis, L. Cotarca et al., Synthesis *1996*, 553-76.

47, 542 **Asym. hydrosilylation** with palladium 2-(diphenylphosphino)-1,1'-binaphthyl complexes, T. Hayashi, Acta Chem. Scand. *50*, 259-66 (1996).

47, 623 **Ketene and ketene acetals**, H.P.S. Pflaum, Chim. Oggi *14*, 13-7 (1996).

47, 646 **Asym. synthesis**, I.W. Davis, P.J. Reider, Chem. Ind. *1996*, 412-5; s.a. D. Enders, W. Bettray, Pure Appl. Chem. *68*, 569-80 (1996); chirotechnology: designing economic chiral syntheses, R.A. Sheldon, J. Chem. Tech. Biotechnol. *67*, 1-14 (1996); **asym. catalysis**, non-linear effects, H.B. Kagan et al., Acta Chem. Scand. *50*, 345-52 (1996); asym. catalysis with chiral aziridines, D. Tanner et al., ibid. 361-8; chiral pyridinamides as ligands in asym. catalysis, C. Moberg et al., ibid. 195-202 (1996); design of chiral ligands for asym. catalysis from C_2-symmetric semicorrins and bisoxazolines to non-symmetric phosphinooxazolines, A. Pfaltz, ibid. 189-94; (S)-2-methoxymethyl pyrrolidine as chiral auxiliary, D. Enders, M. Klatt, Synthesis *1996*, 1403-18; chiral zinc mercaptides as catalyst, R.M. Kellogg, R.P. Hof, J. Chem. Soc. Perkin Trans. I *1996*, 1651-7; 1,2-aminoalcohols and their heterocyclic derivs. as chiral auxiliaries in asym. synthesis, D.J. Ager et al., Chem. Rev. *96*, 835-75 (1996); chiral aminoacid-derived compds. as stoichiometric auxiliaries in asym. synthesis, A. Studer, Synthesis *1996*, 793-815; *trans*-4-hydroxy-L-proline as chiral starting block, P. Remuzon, Tetrahedron *52*, 13803-35 (1996); chiral aminals in asym. synthesis, A. Alexakis et al., Pure Appl. Chem. *68*, 531-4 (1996).

47, 750 **Propargyl compds.**, palladium-catalyzed reactions, J. Tsuji, T. Mandai, Angew. Chem. Intern. Ed. *34*, 2589-612 (1995).

47, 954 **Vinyl ethers** of aminoalcohols and their derivs., B.F. Kukharev et al., Russ. Chem. Rev. *64*, 523-40 (1995).

47, 955 **Asym. synthesis of lactones and lactams** with chiral dirhodium carboxamidates, M.P. Doyle, Aldrichimica Acta, *29*, 3-11 (1996).

48, 106 **Schenck reaction** - diastereoselective oxyfunctionalization with singlet oxygen, M. Prein, W. Adam, Angew. Chem. Intern. Ed. *35*, 477-94 (1996).

48, 108 **Microbial epoxide hydrolases**, K. Faber et al., Acta Chem. Scand. *50*, 249-58 (1996).

48, 134 **Biomimetic catalytic oxidation** with manganese complexes, Rec. Trav. Chim. Pays-Bas *115*, 385-96 (1996).

48, 210 **Elemental fluorine** in non-organofluorine chemistry, S. Rozen, Acc. Chem. Res. *29*, 243-8 (1996).

48, 291 **Catalysis with low-valent ruthenium complexes** as redox Lewis acid and base catalysts, S.-I. Murahashi, T. Naota, Bull. Chem. Soc. Japan *69*, 1805-24 (1996).

48, 488 **Polyvalent organoiodine compds.**, P.J. Stang, V.V. Zhdankin, Chem. Rev. *96*, 1123-78 (1996).

48, 607 **Aldolase-catalyzed carbohydrate synthesis**, C.-H. Wong, Acta Chem. Scand. *50*, 211-8 (1996); **monosaccharides** from non-carbohydrate sources, T. Hudlicky et al., Chem. Rev. *96*, 1195-220 (1996)

48, 625 Ligand-mediated **addition of organometallic reagents to azomethines**, S.C. Denmark, O.J.-C. Nicaise, Chem. Commun. *1996*, 999-1004.

48, 640 **Covalent fullerene chemistry**, F. Diederich, C. Thilgen, Science *271*, 317-23 (1996); **heterocycle-containing [60]fullerene derivs.**, synthesis, S. Eguchi et al., Fullerene Sci. Technol. *4*, 303-27 (1996).

48, 767 Enantioselective synthesis through **enzymatic desymmetrization**, E. Schoffers et al., Tetrahedron *52*, 3769-826 (1996); **desymmetrization of epoxides**, D.M. Hodgson et al., ibid. 14361-84.

48, 772 **Asym. allylation** under transition metal catalysis, B.M. Trost, D.L. Van Vranken, Chem. Rev. *96*, 395-422 (1996); **catalytic asym. synthesis** via palladium p-allyl complexes, J.M.J. Williams, Synlett *1996*, 705-10; **asym. synthesis** catalyzed by palladium and copper, J.-E. Bäckvall, Acta Chem. Scand. *50*, 661-5 (1996).

48, 791 **1,1-Diorganometallics** (sp^3-geminate), synthesis and reactivity, I. Marek, J.-F. Normant, Chem. Rev. *96*, 3241-67 (1996).

48, 794 Regiospecific, diastereoselective and enantioselective **lithiation**, syntheses via -, P. Beak et al., Acc. Chem. Res. *29*, 552-60 (1996).

48, 818 **Samarium diiodide**, sequencing reactions with -, G.A. Molander, C.R. Harris, Chem. Rev. *96*, 307-38 (1996); **radical-ion ring closure**, S. Hintz et al., Topics Curr. Chem. *177*, 77-124 (1996).

48, 822 **Tandem radical reactions of carbon monoxide**, isonitriles, and other reagent equivalents of the geminal radical acceptor/radical precursor synthon, I. Ryu, N. Sonoda, Chem. Rev. *96*, 177-94 (1996).

48, 850 **Chelation assistance** in activation of mercaptals and acetals, T.-Y. Luh, Synlett *1996*, 201-8.

48, 866 **Asym. Horner-type synthesis**, T. Rein, O. Reiser, Acta Chem. Scand. *50*, 369-79 (1996).

48, 889 **Chiral allyl- and allene-stannanes** as reagents for asym. synthesis, J.A. Marshall, Chem. Rev. *96*, 31-47 (1996).

49, 100 **Polyfluorooxaziridines**, synthesis and reactivity, Chem. Rev. *96*, 1809-23 (1996).

49, 305 Recent advances in the use of **immobilized lipases** directed toward the asym. syntheses of complex molecules, H. Akita, Biocatal. Biotransform. *13*, 1-156 (1996).

49, 366 **Pyrroloindoles**, S.A. Samsonia et al., Russ. Chem. Rev. *63*, 815-32 (1994).

49, 372 **Inorganic supports** in organic synthesis, J.H. Clark, D.J. Macquarrie, Chem. Soc. Rev. *25*, 303-10 (1996); organic reactions **at** well-defined **oxide surfaces**, M.A. Barteau, Chem. Rev. *96*, 1413-30 (1996); **silicon dioxide** surfaces for organic reactions, V.A. Basiuk, Russ. Chem. Rev. *64*, 1003-19 (1995).

49, 499 Organic synthesis utilizing **2-phenylthiocyclobutyl ketones**, T. Takeda, T. Fujiwara, Synlett *1996*, 481-92.

49, 697 **Alkyne-Co$_2$(CO)$_6$ complexes**, development of highly stereo- and regio-selective reactions with -, C. Mukai, M. Hanaoka, Synlett *1996*, 11-7.

49, 882 **Water** as solvent, A. Lubineau, Chem. Ind. *1996*, 123-6; Lewis acid-catalyzed synthesis in aq. media, J.B.F.N. Engberts et al., Rec. Trav. Chim. Pays-Bas *115*, 457-64 (1996).

49, 939 **Zirconocene and Titanocene**, unusual reactions, A. Ohff et al., Synlett *1996*, 111-8.

50, 15 **Chiral ethylene-1,2-bis(η^5-4,5,6,7-tetrahydro-1-indenyl)-titanium and -zirconium complexes** as catalysts for enantioselective C-C and C-H bond formation, A.H. Hoveyda, J.P. Morken, Angew. Chem. Intern. Ed. *35*, 1262-84 (1996).

50, 116 **Carbonic acid esters**, A.-A.G. Shaikh, S. Sivaram, Chem. Rev. *96*, 951-76 (1996); dimethyl carbonate in synthesis, Y. Ono, Pure Appl. Chem. *68*, 367-76 (1996).

50, 120 **Organometallic oxides** as catalysts, W.A. Herrmann, J. Organometal. Chem. *500*, 149-73 (1995).

50, 155 **Stereospecific Diels-Alder reactions** with 2-amino-1,3-dienes, D. Enders, O. Meyer, Liebigs Ann. *1996*, 1023-35.

50, 157 **Chiral β-aminocarboxylic acids** and a-subst. derivs., asym. synthesis, G. Cardillo, C. Tomasini, Chem. Soc. Rev. *25*, 117-28 (1996).

50, 252 **Asym. Pummerer-type reactions** induced by O-silylated ketene acetals, Y. Kita, N. Shibata, Synlett *1996*, 289-96.

50, 260 **Radical substitution** in the formation of bonds to heteroatoms, C.H. Schiesser, L.M. Wild, Tetrahedron *52*, 13265-314 (1996).

50, 271 **Selective fluorination** with hypofluorites, S. Rozen, Chem. Rev. *96*, 1717-36 (1996).

50, 343 **Chlorobis(cyclopentadienyl)hydridozirconium** as reagent, J.-P. Majoral et al., Chem. Ber. *129*, 879-86 (1996); **organo(chloro)zirconocene complexes**, synthetic applications, P. Wipf, H. Jahn, Tetrahedron *52*, 12853-910 (1996).

50, 389 **Asym. protonation of enolates** and enols, C. Fehr, Angew. Chem. Intern. Ed. *35*, 2566-87 (1996).

50, 396 **Tandem Diels-Alder cycloadditions**, J.D. Winkler, Chem. Rev. *96*, 167-76 (1996).

50, 400 **Dispiroketals**: a new functional group for organic synthesis, S.V. Ley et al., Cont. Org. Synth. *2*, 365-92 (1995).

50, 419 **Organosilicon hydrides/carbon monoxide** as potent reactant combination in developing new transition metal-catalyzed reactions, N. Chatani, S. Murai, Synlett *1996*, 414-24.

50, 433 **Asym. [3.3]-sigmatropic rearrangement**, D. Enders et al., Tetrahedron:Asym. *7*, 1847-82 (1996).

50, 438 **2,5-Disubst. pyrrolidines**, stereospecific synthesis, M. Pichon, B. Figadère, Tetrahedron:Asym. *7*, 927-64 (1996).

50, 555 **Organic chemistry on solid supports**, J.S. Früchtel, G. Jung, Angew. Chem. Intern. Ed. *35*, 17-42 (1996); the emerging art of **solid-state synthesis**, B. Parkinson, Science *270*, 1157-8 (1995); survey of recent solid-phase reactions, P.H.H. Hermkens et al., Tetrahedron *52*, 4527-54 (1996); **combinatorial chemistry** (6 papers), Acc. Chem. Res. *29*, Issue No.3, 114-70 (1996); combinatorial synthesis of small organic molecules, F. Balkenhohl et al., Angew. Chem. Intern. Ed. *35*, 2288-337 (1996); synthesis and application of small molecule libraries, L.A. Thompson, J.A. Ellman, Chem. Rev. *96*, 555-600 (1996); combinatorial chemistry - a rational approach to chemical diversity, X. Williard, A. Tartar et al., Eur. J. Med. Chem. *31*, 87-98 (1996); compound libraries for lead discovery, H. Matter, D. Lassen, Chim. Oggi *14*, 9-15 (1996); combinatorial synthesis and screening of biological and peptide libraries, M. Rinnova, M. Lebl, Collection Czech. Chem. Commun. *61*, 171-231 (1996); split synthesis and solid-phase synthetic methodology, J.J. Baldwin, I. Henderson, Med. Res. Rev. 16, 391-405 (1996); **liquid-phase combinatorial synthesis** with fluorous reagents and media, D.P. Curran, Chemtracts, *9*, 75-88 (1996).

51, 31   **Supercritical fluids** and medium, T. Clifford, K. Bartle, Chem. Ind. *1996*, 449-52.

51, 317 **Asym. 1,3-dipolar cycloaddition**, metal-catalyzed, K.V. Gothelf, K.A. Jørgensen, Acta Chem. Scand. *50*, 652-60 (1996); **chiral isoxazolidines** via asym. 1,3-dipolar cycloaddition of nitrones to alkenes, M. Frederickson, Tetrahedron *53*, 403-65 (1997).

51, 446 **Sulfated zirconia**-based strong solid-acid catalysts: recent progress, X. Song, A. Sayari, Catal. Rev. Sci. Eng. *38*, 329-412 (1995).

Index to Volume 51

As in previous volumes, reactions are indexed from both the starting material and product aspects, e.g. '**Azides** startg. m.f. amines' and '**Amines** from azides'. Nomenclature for complex functions can be located under the 'special s.' sub-entry, e.g. '**Carboxylic acids** special s. aminocarboxylic acids' or by consulting the Formula Index of Complex Functional Groups (Volume *48*, p. 471).

Hydrogenated and functionalized ring systems are indexed by the conventional reversal, e.g. '**Pyridines, aryl-**', the only important exception to the rule being alkylideneisocyclics which are indexed as such, e.g. '**Alkylidenecyclopentanes**'.

As from Volume *51*, '**Epoxides**' will be used in place of 'Oxido compds.'; '**Thiiranes**' in place of 'Sulfido compds.'; '**Diels-Alder reaction**' in place of 'Diene synthesis'; and '**Benzo[*b*]thiophenes**' in place of 'Thianaphthenes'.

References to abstracts in this volume are in the format **51**, 234. An entry such as '**2-Pyridones, 3-cyano- 41**, 636s**51**' refers to the indexing of a supplementary reference, which must be followed up via the Supplementary References Index (p. 287).

Absolute asym. synthesis
s. Synthesis, asym., absolute
Acetaldehyde
– as reagent **51**, 73
Acetals (s.a. Ketals)
–, activation, chelation-controlled, review **48**, 850s**51**
– from
  alkoxysilanes **41**, 221s**51**
– special s.
  azidoacetals
  formals
  halogenacetals
– startg. m. f.
  2-acetylenealcohols **51**, 366
  (E)-enolethers, synthesis **51**, 366
  ketones, – **51**, 366
  mercaptals **51**, 226
  monothioacetals **51**, 226
Acetals, cyclic
– special s.
  carbohydrate O,O-alkylidene derivs.
  1,3-dioxolanes
– startg. m. f.
  α-alkoxynitriles, asym. synthesis **51**, 277
  β-hydroxycarbonyl compds., – – **40**, 635s**51**
Acetic acid esters
– special s.
  4-chromanoneacetic acid esters
Acetic anhydride
– as reactant **51**, 55
Acetone
– as sensitizer **51**, 424
Acetonitrile
– as reactant **51**, 296
o-**Acetonylphenols**
– from
  p-hemiquinols **51**, 459
Acetoxy... s. Acoxy...
Acetyl... s.a. Acyl...
O-Acetylation
– under super-Lewis acid catalysis **51**, 81
Acetyl chloride
– as reactant **51**, 449
– as reagent **51**, 296
β,γ-Acetylene-α-acylaminomalonic acid esters
– startg. m. f.
  oxazole-4-carboxylic acid esters **51**, 116
2-Acetylenealcohol O-derivs.
–, hydrogenolysis, Pd-catalyzed, review **34**, 55s**51**
–, reactions, –, – **47**, 750s**51**
Acetylenealcohols
–, hydrogenation, selective, review **47**, 49s**51**
2-Acetylenealcohols
– from
  acetals **51**, 366
  acetylene derivs., terminal **51**, 273
  α,β-acetyleneiodides **51**, 278
  oxo compds. **51**, 273, 278
  silylacetylenes **30**, 624s**51**
–, resolution **17**, 393s**51**
– startg. m. f.
  (E)-α,β-ethyleneoxo compds. **51**, 33
3-Acetylenealcohols 40, 567s**51**
– from
  β,γ-acetylenehalides, synthesis **51**, 270

acoxy-2-acetylenes, – **51**, 289
allenesilanes, – **51**, 270
erythro-**3-Acetylenealcohols, terminal**
– from
  iodoallenes **51**, 267
α,β-Acetylene(alkoxy)carbene complexes
– special s.
  chromium α,β-acetylene(alkoxy)carbene complexes
Acetylenealuminates
– special s.
  sodium bis(alkynyl)diethylaluminates
2-Acetyleneamines
– special s.
  N-propargylanilines
– startg. m. f.
  pyrroles **51**, 337
  Δ²-pyrrolines **51**, 337
2-Acetylenecarbonic acid esters
–, reactions, Pd-catalyzed, review **47**, 750s**51**
2-Acetylenecarbonic acid esters, cyclic
– special s.
  1,3-dioxolan-2-ones, 4-alk-1-ynyl-
α,β-Acetylenecarboxylic acid esters
– by carbonylation **12**, 867s**51**
Acetylene derivs.
–, allylmagnesiation, Mn-catalyzed, regiostereospecific **51**, 303
– from
  aldehydes **51**, 487
  ene-1,2-di(thioethers) **51**, 487
  α-hydroxymercaptals **51**, 487
–, germylformylation **46**, 672s**51**
–, hydroboration **51**, 238
–, hydrosilylation **48**, 543s**51**
–, hydrozincation, regiostereospecific **51**, 195
–, silylformylation, intramolecular, regiostereospecific **51**, 332
– special s.
  acoxyacetylenes
  alkoxyacetylenes
  arylacetylenes
  diyn...
  enyn...
  2(5H)-furanones, 4-alk-1-ynyl-propargyl...
  tolans
– startg. m. f.
  2-allenealcohols, synthesis, asym. **51**, 268
  (E)-allylarenes **51**, 405
  1,3-dienes (from 2 different molecules) **51**, 314
  α-diketones **51**, 107
  enazides **51**, 130
  (E)-eneselenides **50**, 343s**51**
  (Z)-ethylene derivs. **16**, 72s**51**
  (E)-α,β-ethyleneiodides **51**, 195
  α,β-ethylene-β-(organothio)aldehydes **51**, 331
  ketones **38**, 111s**51**
–, vinylzirconation, stereospecific **51**, 314
Acetylene derivs., terminal
– startg. m. f.
  2-acetylenealcohols **51**, 273
  carboxylic acid esters, C-cleavage **51**, 107
  β-diketones **51**, 403
  1,3-diynes, sym. **51**, 358

2(5H)-furanones, 4-alk-1-ynyl- **51**, 333
furan-3-ylacetic acid esters **51**, 333
α-methylenephosphonic – – **51**, 239
Acetylenedicobalt hexacarbonyl complexes
–, reactions, regiostereospecific, review **49**, 697s**51**
Acetyleneepoxides
–, radical ring closure **51**, 318
α,β-Acetylenehalides
– special s.
  α,β-acetyleneiodides
β,γ-Acetylenehalides
– special s.
  1,10-dihalogeno-2,8-diynes
– startg. m. f.
  3-acetylenealcohols **51**, 270
  2-acetylenesilanes **51**, 270
  2-allenealcohols **51**, 270
  allenesilanes **51**, 270
α,β-Acetylene-γ-hydroxycarboxylic acid amides 19, 734s**51**
α,β-Acetylene-γ-hydroxycarboxylic acid esters
– startg. m. f.
  2(5H)-furanones, 4-alk-1-ynyl- **51**, 333
  furan-3-ylacetic acid esters **51**, 333
α,β-Acetylene-δ-hydroxycarboxylic – –
– startg. m. f.
  2-pyrones, 5,6-dihydro-, 3-γ-keto- **51**, 329
α,β-Acetyleneiodides
– startg. m. f.
  2-acetylenealcohols **51**, 278
α,β-Acetyleneiodonium salts
– special s.
  sulfonylamino-α,β-acetyleneiodonium salts
Acetyleneketones
–, hydrogenation, selective, review **47**, 49s**51**
α,β-Acetyleneketones
– from
  silylacetylenes **37**, 828s**51**
– startg. m. f.
  β-diketones **51**, 403
o-**Acetyleneketones**
– startg. m. f.
  1-naphthols **51**, 338
β,γ-Acetylene-β'-ketosulfones
– startg. m. f.
  allenes **51**, 488
β,γ-Acetylenemalonic acid esters
– special s.
  alkoxy-β,γ-acetylenemalonic acid esters
2-Acetylenesilanes
– from
  β,γ-acetylenehalides **51**, 270
– startg. m. f.
  2-allenealcohols **51**, 270
Acetylenestannanes 51, 241
β,γ-Acetylenesulfones
– special s.
  β,γ-acetylene-β'-ketosulfones
1-Acetylene-1-thioethers
– from
  mercaptans **51**, 392
– special s.
  silylethynyl thioethers
– startg. m. f.
  α,β-ethylene-N-sulfonylthioiminoesters **51**, 310

Acids, solid-supported (s.a. Solid acids)
– special s.
  Envirocat EPZG
Acoxyacetylenes
– startg. m. f.
  (E)-α-alkylidene-o-hydroxyketones
    51, 348
Acoxy-2-acetylenes
– startg. m. f.
  3-acetylenealcohols, synthesis 51, 289
Acoxy compds. (s.a. Carboxylic acid esters)
– from
  alcohols, inversion of configuration
    51, 80
– special s.
  diol esters
N-Acoxydicarboxylic acid imides
–, S-acylation with – 51, 223
γ-Acoxy-δ,ε-ethyleneketones
– from
  ketones 51, 397
Acoxyethylenes (s.a. Enolesters)
– startg. m. f.
  enolethers, cyclic via metathesis 51, 496
Acoxy-2-ethylenes (s.a. C-Acoxylation, allylic)
– from
  α,β-ethyleneacylals, synthesis, asym.
    51, 383
– startg. m. f.
  β,γ-ethylenesulfones, asym. conversion
    37, 527s51
Acoxy-3-ethylenes
– startg. m. f.
  cyclopropanols, 2-β-hydroxy- 51, 316
  γ-lactones 51, 470
(E)-Acoxy-3-ethylenes
– from
  3-ethyleneepoxides, synthesis, regiostereospecific 51, 290
3'-Acoxy-2-ethylenesilanes
– special s.
  3-acoxy-2-methylenesilanes
1,2-Acoxyhalides
– from
  epoxides 10, 412s51
α-Acoxyketones
–, reduction, asym. 29, 36s51
– special s.
  α-formoxyketones
– startg. m. f.
  ketones 51, 34
Acoxylation
– special s.
  trifluoroacetoxylation
C-Acoxylation, allylic, asym. (s.a. Kharasch acoxylation, asym.)
α-Acoxymercaptals
– startg. m. f.
  ethylene derivs. 51, 487
3-Acoxy-2-methylenesilanes
– special s.
  2-[(trimethylsilyl)methyl]allyl pivalate
α-Acoxynitriles
– from
  acylcyanides 51, 287
α-Acoxyphosphonic acid esters
–, hydrolysis, asym. 28, 13s51
β-Acoxysulfones
– startg. m. f.
  ethylene derivs. 51, 489

o-Acoxysulfoxides
–, hydrolysis, asym. 28, 13s51
1,1-Acoxythioethers
– from
  aldehydes, asym. conversion 51, 233
Acrylic acid aryl esters
– startg. m. f.
  1-indanones, 7-hydroxy- 51, 346
Activation
– of
  N-acylsulfonamides 51, 156
  carbon dioxide by superoxide 51, 169
Activation, chelation-controlled
– of mercaptals and acetals, review
    48, 850s51
Acylals
– from
  aldehydes 51, 55
– special s.
  ethyleneacylals
Acylamidrazones
– special s.
  formylamidrazones
Acylamines (s.a. Carboxylic acid amides)
–, 3-component synthesis 51, 296
– from
  acyl hydroxamates 51, 9
  alkoxylamines 51, 9
  hydroxamic acid esters 51, 9
  hydroxamic acids 51, 9
– special s.
  diacylamines
  halogenacylamines
  nitroacylamines
Acylamines, cyclic
– from
  azidoketals 51, 193
Acylamines, 2-functionalized
– via 1,2,3-oxathiazolidine 2,2-dioxides, 3-acyl- 51, 230
2-Acylaminoalcohols
– startg. m. f.
  acylamines, 2-functionalized 51, 230
–, chiral 42, 621s51
α-Acylaminoalkyl radicals
–, synthesis, stereospecific with –, review
    46, 484s51
α-Acylaminocarboxylic acid amides
– from
  aziridines, 2-imino- 51, 123
α-Acylaminocarboxylic acids
–, synthesis and reactions, review 45, 6s51
1-Acylamino-1,3-dienes
–, Diels-Alder reaction, asym. with –
    49, 622s51
α-Acylamino-β-ketocarboxylic acid esters 14, 888s51
α-Acylaminoketones
– special s.
  α-(trifluoroacetylamino)ketones
α-Acylaminomalonic acid esters
– special s.
  acetylene-α-acylaminomalonic acid esters
α-Acylaminophosphine oxides
  49, 560s51
o-Acylaminophosphines
– special s.
  bis(o-acylaminophosphines)
α-Acylaminophosphonic acid esters
  49, 560s51

Acylation
– special s.
  formylation
C-Acylation (s.a. Friedel-Crafts acylation)
C-Acylation, nucleophilic, intramolecular-Barbier cyclization, stereospecific 51, 483
o-Acylation
– of phenols 51, 379, 407
p-Acylation
– of phenolethers 51, 374
N-Acylation
– of
  2-oxazolidones 51, 167
  sultams 51, 167
– special s.
  N-formylation
  N-trifluoroacetylation
– with
  enolesters 47, 273s51
  N-acylsulfonic acid amides, polymer-based 51, 156
N-Acylation, preferential
– with 2-oxazolidones, 3-acyl- 51, 160
O-Acylation
– special s.
  O-acetylation
O-Acylation, asym.
– of 1,1-hydroxythioethers with dynamic kinetic resolution 51, 233
–, –, enzymatic-Mitsunobu reaction
–, deracemization of alcohols by – 51, 103
–, –, phosphine-catalyzed
– with carboxylic acid anhydrides 51, 83
–, Lewis acid-catalyzed
    51, 78
S-Acylation
– with N-acoxydicarboxylic acid imides
    51, 223
Acyl carbanion equivalents, chiral
–, α-aminonitriles as – 51, 395
S-Acylcoenzyme A derivs., lipophilic
    51, 223
Acylcyanides
– startg. m. f.
  α-acoxynitriles 51, 287
Acyl glucuronides
–, review 19, 265s51
C-Acylglycoside derivs. 51, 424
Acyl glycosides
– from
  glycosyl trichloroacetimidates
    44, 211s51
– special s.
  acyl glucuronides
1,3-Acyl group migration
–, ring expansion, Barbier-type via –
    51, 481
O→C-Acyl group migration 51, 348
Acyl halides s. Carboxylic acid halides
Acyl hydroxamates
– startg. m. f.
  acylamines 51, 9
N-Acyllactams, bicyclic, dipole-aligned
–, aldol condensation, asym. with –
    51, 274
Acylsilanes
– from
  imidazoles, 1-acyl-
    51, 244
– special s.
  hydroxyacylsilanes

Acyl substitution, nucleophilic, intramolecular
–, syntheses via – **51**, 316, 348, 470
– –, **nucleophilic-radical ring closure**
  51, 482
N-Acylsulfonic acid amides, polymer-based
–, N-acylation with – **51**, 156
Acyl thiohydroxamates, cyclic
– special s.
  2-pyridinethiones, 1-acoxy-
(Acylthio)thioethers
– from
  disulfides, cyclic **32**, 523s**51**
2-(Acylthio)thioethers
  5, 421s**51**
2-Adamantanones 51, 115
1,2-Addition s.a. CC⇓OC, CC⇓NC
–, **stereospecific**
– to α,β-ethylene-γ-(organothio)ketones
  51, 20
–, –, **non-chelating**
– to β-alkoxyaldehydes, α–subst. **51**, 442
1,2-Addition-alkylation
– of azo compds. **24**, 312s**51**
1,4-Addition (s.a. CC⇓CC, Michael addition, Radical 1,4-addition)
– with alkylzinc bromides **51**, 304
–, **intramolecular, asym.**
– with 2-ethylenesilanes **38**, 911s**51**
–, **photochemical, co-sensitized**
– with 1,1-silylurethans **51**, 445
1,4-Addition-alkylation, asym.
– of
  4-imidazolidones, 3-(α,β-ethyleneacyl)- **37**, 657s**51**
  Δ²-oxazolines, 2-aryl-, tricarbonyl chromium-complexed **51**, 400
–, **intramolecular, asym.**
  51, 393
1,4-Addition-allylation
– special s.
  radical 1,4-addition-allylation
1,4-Addition-Horner synthesis
  28, 851s**51**
1,6-Addition, nucleophilic
– to aryloxo compds. **51**, 300
Alcohols (s.a. Carbinols)
–, N-alkylation with – **51**, 150
– from
  boronic acid esters **51**, 98
  – acids **51**, 98
  ethylene derivs. via hydrozincation **51**, 60
  halides **51**, 91
  iodides **51**, 93
  oxo compds., reduction (HC⇓OC)
  51, 35
  – –, –, preferential and selective **51**, 23
  – –, (sec. alcohols), reduction, asym.
  51, 22, 24
  – –, (– –), synthesis (CC⇓OC) **51**, 284, 427
  – –, (– –), –, asym. **51**, 266
  – –, (– –), transfer-hydrogenation, asym.
  51, 26
  selenides (sec. alcohols), synthesis
  51, 427
  silanes, review **39**, 200s**51**
–, oxidation with N-oxide radicals, stable, review **39**, 225s**51**
– special s.
  acetylenealcohols

acylaminoalcohols
(alkylideneamino)alcohols
allenealcohols
aminoalcohols
2-arylalcohols
azidoalcohols
benzylalcohols
deuterioalcohols
diols
epoxyalcohols
ethylenealcohols
glycols
halogenhydrins
siloxyalcohols
(sulfonylamino)alcohols
tetraols
– startg. m. f.
  acoxy compds., inversion of configuration **51**, 80
  amines via Schmidt reaction **51**, 162
  amines, ar., sec. **51**, 151
  halides **51**, 207
  mercaptans (with H$_2$S), review
  44, 469s**51**
  oxo compds. (OC⇑H) **51**, 76
  – –, (ketones), kinetic resolution
  51, 500
  phosphoric acid esters, mixed **51**, 51
  thioethers (with H$_2$S), review
  44, 469s**51**
Alcohols, bicyclic
– from
  epoxides, fused, desymmetrization
  51, 343
Alcohols, cyclic
– from
  ethyleneoxo compds. with lactonization **51**, 313
Alcohols, prim.
– from
  carboxylic acid halides **51**, 35
  carboxylic acids **51**, 35
  ethylene derivs., terminal, synthesis, asym. **51**, 320
  imidazoles, 1-acyl- **37**, 57s**51**
– startg. m. f.
  carboxylic acids **15**, 261s**51**
Alcohols, sec.
–, deracemization **51**, 103, 499
– from
  aldehyde trisylhydrazones, synthesis
  51, 431
–, resolution, kinetic by oxidation **51**, 500
– special s.
  1,1,1,3,3,3-hexafluoro-2-propanol
Alcohols, sec., unsatd.
– from
  aldehydes, synthesis **51**, 288
Alcohols, tert., bicyclic
– from
  ethylene(halogeno)carboxylic acid esters **51**, 482
Alcohols, tert., cyclic
– from
  ketones, cyclic, diastereoselectivity, review **37**, 623s**51**
Alcohols, unsatd.
– from
  oxo compds., unsatd., hydrogenation, homogeneous **51**, 27
Aldehydes (s.a. Carbonyl compds., Hydroacylation, Oxo compds.)

– from
  alcohols s. OC⇑H and under Oxo compds.
  halides, with 1 extra C-atom **51**, 424
  nitriles **51**, 86
–, resolution, kinetic **48**, 866s**51**
– special s.
  acetaldehyde
  alkoxyaldehydes
  benzaldehyde
  dialdehydes
  ethylenealdehydes
  hydroxyaldehydes
  (organothio)aldehydes
  peptide aldehydes
  pivalaldehyde
  pyrimidinecarboxaldehydes
  pyrrole-2,5-dicarboxaldehydes
  silylaldehydes
– startg. m. f.
  3-acetylenealcohols, synthesis **51**, 289
  acetylene derivs. **51**, 487
  α-acoxynitriles **51**, 287
  1,1-acoxythioethers, asym. conversion
  51, 233
  acylals **51**, 55
  alcohols s. under Oxo compds.
  –, sec., unsatd., synthesis **51**, 288
  2-allenealcohols, synthesis, asym.
  51, 268
  α-aminophosphonic acids, asym. induction **51**, 240
  α-aminophosphonous –, – – **51**, 240
  1,3-diol esters (from 2 molecules)
  51, 308
  1,3-diols **51**, 298
  1,3-dioxolanes (from 3 molecules)
  51, 373
  diselenides, sym. **51**, 243
  3-ethylenealcohols, synthesis, asym., regiospecific **51**, 433
  –, –, – with 3 extra C-atoms **51**, 447
  –, –, regiostereospecific **51**, 272
  (E)-α,β-ethylenealdehydes, with 2 extra C-atoms **51**, 460
  ethylene derivs., terminal, with 1 extra C-atom **51**, 434
  syn-3-ethylene-2,1-halogenhydrins, synthesis, asym. **51**, 269
  α,β-ethyleneketones **51**, 448
  glycols, with 1 extra C-atom **51**, 434
  threo-glycols, sym. **51**, 283
  1,1-(homoallyloxy)silanols **51**, 354
  α-hydroxyketene mercaptals, synthesis
  51, 396
  α-hydroxyketones **51**, 418
  ketones **51**, 448
  1,2-oxasilacyclopentanes **51**, 298
  2H-1,3-oxazines, 3,6-dihydro-, 3-tosyl-
  51, 173
  pyrrolidines **51**, 376
  selenolic acid esters **51**, 256
  α-siloxynitriles **51**, 263
  tellurolic acid esters **51**, 256
  thiolic acid esters **51**, 256
Aldehydes, ar.
– from
  aryl triflates **51**, 382
  benzylalcohols **51**, 109
  methylarenes **51**, 74
Aldehydes, hindered
–, aldol condensation, regiostereospecific with – **51**, 282

Aldehyde trisylhydrazones
– startg. m. f.
 alcohols, sec., synthesis **51**, 431
Alder-ene reaction (s.a. Anti-Alder-ene reaction)
Aldimines (s.a. Azomethines)
– special s.
 N-(diphenylmethyl)aldimines
– startg. m. f.
 α-aminophosphonic acid esters, asym. conversion **51**, 236
 aziridine-2-carboxylic – – **51**, 388
 3-ethyleneamines, 1,3-asym. induction **51**, 295
Aldol condensation (s.a. Michael addition-aldol condensation)
Aldol condensation, asym.
– with
 N-acyllactams, bicyclic, dipole-aligned **51**, 274
 4H-1,3-benzoxazin-4-ones, 2,3-dihydro-, 3-acyl- **48**, 588s**51**
 carboxylic acid esters **51**, 275, 285
– –, **reductive, regiostereospecific**
–, equivalent **51**, 319
– with α,β-ethyleneketones **51**, 322
– –, **regiostereospecific**
– with aldehydes, hindered **51**, 282
– –, **stereospecific**
– using supersilylating agents **51**, 276
– with β-alkoxyaldehydes, α-subst. **51**, 442
Aldoses
– startg. m. f.
 glycosides **44**, 166s**51**
Aldoses, unprotected
– startg. m. f.
 N-glycosyl-N-heterocyclics **51**, 177
Alkali-metal enolates
–, aldol-type condensation, regiospecific via – **51**, 429
Alkanes s. Hydrocarbons
Alkene oxides s. Epoxides
Alkenes s. Ethylene derivs.
Alkoximes
– special s.
 O-allyloximes
 O-benzyloximes
 hydroxyalkoximes
 sulfonylalkoximes
– startg. m. f.
 ozonides **51**, 88
α-Alkoxy-β,γ-acetylenemalonic acid esters
– startg. m. f.
 α-alkoxy-α-allenecarboxylic acid esters **51**, 495
Alkoxyacetylenes
– startg. m. f.
 α-ketocarboxylic acid esters **51**, 67
β-Alkoxyaldehydes, α-subst.
–, aldol-type condensation, stereospecific with – **51**, 442
α-Alkoxy-α-allenecarboxylic acid esters 51, 495
α-Alkoxycarboxylic acid amides 48, 457s**51**

(Z)-1-Alkoxy-1,3-diene-2-phosphonic acid esters
42, 725s**51**
2-Alkoxy-1,3-dienes
– by Horner synthesis **14**, 877s**51**
2-Alkoxyenestannanes
– special s.
 2-(tri-n-butylstannyl)vinyl ethyl ether
2-Alkoxy-3-ethyleneepoxides
–, syntheses via – **51**, 385
β-Alkoxy-α,β-ethyleneketones
– from
 α-ketoketene O,S-acetals **29**, 854s**51**
Alkoxy-2-ethylenes
– startg. m. f.
 ethylene derivs., synthesis, regiostereospecific **51**, 372
tert-Alkoxy-3-ethylenes
– from
 2-ethylenesilanes, synthesis, asym. **51**, 438
**(Z)-3-Alkoxy-2-ethylenestannanes
50**, 569s**51**
Alkoxyglycol sulfates, cyclic
– startg. m. f.
 2-α-hydroxy-O-heterocyclics **51**, 113
3-Alkoxy-4-hydroxy-1,5-enynes, chiral 33, 865s**51**
Alkoxyamines
– special s.
 allyloxyamines
 O-ethylhydroxylamine
 N-methylmethoxylamine
– startg. m. f.
 acylamines **51**, 9
 amines **51**, 9
 hydroxyalkoximes **51**, 122
Alkoxyl radicals, unsatd.
–, ring closure, regiostereospecific **51**, 209
α-Alkoxynitriles
– from
 acetals, cyclic, asym. synthesis **51**, 277
 oxo compds., – – **51**, 277
β-Alkoxyoxo compds.
– special s.
 β-benzyloxyoxo compds.
(Alkoxy)polysilanes
–, rearrangement, skeletal **51**, 235
**Alkoxy-p-quinones
51**, 108
Alkoxysilanes (s.a. Siloxy..., O-Silylation, Silyl ethers)
– startg. m. f.
 acetals **41**, 221s**51**
1,1-Alkoxysilanols
– special s.
 1,1-(homoallyloxy)silanols
Alkoxystannanes
– special s.
 dialkoxystannanes
– startg. m. f.
 stannanes **51**, 241
α-Alkoxysulfones
– startg. m. f.
 enolethers **51**, 420
Alkylarenes
– from
 aryl mesylates **51**, 381
– special s.
 allylarenes
 methylarenes
– startg. m. f.

aryl ketones **51**, 76
Alkylation
– special s.
 allylation
– with (triphenylphosphoranylidene)-acetic acid esters **51**, 137
C-Alkylation
– of β-dicarbonyl compds. with mesylates, with inversion **51**, 369
C-Alkylation, asym.
– of
 1,3-dithiane 1-oxides, 2-acyl- **51**, 399
 glycinamides **51**, 398
 4-imidazolidones, 3-glycyl- **51**, 398
 methyl (S)-N-benzyl-N-methylalaninate **51**, 401
–, –, **polymer-based
44**, 776s**51**
C-β-Alkylation, stereospecific
– of ethylene derivs., electron-deficient **51**, 328
N-Alkylation
– special s.
 N-methylation
– with alcohols **51**, 150
–, **polymer-based 51**, 172
–, **reductive** (s.a. N-Cyclopropylation, reductive) **51**, 141
–, –, **polymer-based 46**, 317s**51**
O-Alkylation
– with (triphenylphosphoranylidene)-acetic acid esters **51**, 137
Alkyl hypohalites, prim.
– startg. m. f.
 carboxylic acid esters (from 2 molecules) **51**, 90
2-(Alkylideneamino)alcohols, chiral
– special s.
 2-[N-(salicylidene)amino]alcohols, chiral
Alkylidenecarbenes
–, ring closure, double via – **51**, 477
3-Alkylidenecyclobutanols 51, 430
(E)-Alkylidenecyclobutenes 51, 356
Alkylidenecyclopentanes
– special s.
 methylenecyclopentane...
**2-Alkylidene-3(4)-cyclopentenones
51**, 330
α-(2-Alkylidenecyclopent-1-yl)ketones 51, 305
2-Alkylidene-1,3-di(silanes)
–, ring closure, stereospecific with – **51**, 494
(E)-α-Alkylidene-o-hydroxyketones
– from
 acoxyacetylenes **51**, 348
γ-Alkylidene-δ-ketocarboxylic acid esters, medium-ring 51, 481
(E)-α-Alkylideneketones
– from
 α,β-ethyleneketones, synthesis **51**, 322
α-Alkylidenelactones 51, 470
α-Alkylidene-γ-lactones
– special s.
 α-methylene-γ-lactones
**Alkylidenemalonic acid esters
41**, 812s**51**
Alkylidenephosphoranes, metallo-subst.
–, review **44**, 837s**51**
**(Z)-3-Alkylidenephthalides
35**, 85s**51**; **41**, 500s**51**

Alkyl radicals
–, generation, Co-mediated **48**, 968s51
Alkyl silyl acetals
– as intermediates **51**, 36
– special s.
   O-allyl O-silyl α-halogenacetals
α-(Alkylthio)sulfones
– startg. m. f.
   thioenolethers **51**, 420
Alkylzinc bromides
–, 1,4-addition with – **51**, 304
Alkynes s. Acetylene derivs.
Alkynylsamarium(III) iodides
– as intermediates **51**, 278
2-Allenealcohols
– from
   acetylene derivs., synthesis, asym.
   51, 268
   β,γ-acetylenehalides **51**, 270
   2-acetylenesilanes **51**, 270
– special s.
   silyl-2-allenealcohols
2-Allenealcohols, chiral 49, 81s51
α-Allenecarboxylic acid esters
– special s.
   alkoxy-α-allenecarboxylic acid esters
– startg. m. f.
   (E)-α,β-ethylenecarboxylic acid esters, synthesis **50**, 411s51
α-Allene-δ-hydroxycarboxylic acid esters 51, 465
Allenes
–, C-α-allylation with – **51**, 131
–, carbopalladation, intramolecular with nucleophile capture **51**, 417
–, cycloaddition, 1,3-dipolar, review **34**, 630s51
– from
   β,γ-acetylene-β'-ketosulfones **51**, 488
   1,3-enynes, synthesis **51**, 334
   β-ketosulfoxides **51**, 488
– special s.
   enyne-allenes
   halogenallenes
   1,2,n-trien...
– startg. m. f.
   1,3-enynes **37**, 675s51
   (E)-2-ethylene-*tert*-amines **51**, 131
Allenesilanes
– from
   β,γ-acetylenehalides **51**, 270
– startg. m. f.
   3-acetylenealcohols **51**, 270
Allenestannanes
–, synthesis, asym. with –, review **48**, 889s51
Allenyldisilanes
–, [3+2]-cycloaddition with – **48**, 674s51
Allenyl ethers
– startg. m. f.
   2-ethyleneethers, synthesis **51**, 334
Allenylpalladium complexes, zwitterionic
– as intermediate **51**, 465
Allenyl thioethers
– startg. m. f.
   (E)-thioenolethers, synthesis **51**, 334
Allyl alcohol O-derivs.
–, allylation, Ni-catalyzed with – **51**, 152
Allyl alcohols s. 2-Ethylenealcohols
Allylarenes
– from
   benzyl halides **51**, 452
   enalanes **51**, 452
   styrenes **51**, 326
(E)-**Allylarenes**
– from
   acetylene derivs. **51**, 405
   benzyl halides **51**, 405
Allylation, Ni-catalyzed
– with allyl alcohol O-derivs. **51**, 152
–, asym., Pd-catalyzed
–, reversal of enantioselectivity **51**, 188
–, –, transition metal-catalyzed
–, review **48**, 772s51
C-Allylation (s.a. 1,4-Addition-allylation, C-Diallylation, Radical allylation)
–, intramolecular, stereospecific
–, 1,1-(homoallyloxy)silanols via –
   51, 354
C-α-Allylation
– with
   allenes **51**, 131
   2-ethylenealcohols **22**, 782s51
Allylation, ar. 22, 844s51
Allyl bromide
– as reagent **51**, 358
Allyl carbonates s. 2-Ethylenecarbonic acid esters
Allyl cations, heteroatom-stabilized
–, Diels-Alder reaction, ionic, intramolecular with – **51**, 474
Allyl enolethers
– from
   O-allyl O-silyl α-halogenacetals
   51, 196
   enoxysilanes **51**, 196
   2-ethylenealcohols **51**, 196
Allyl ethers s. Alkoxy-2-ethylenes, 2-Ethyleneethers
C-Allyl glycosides, 2,3-unsatd.
37, 871s51
Allyl group migration s. Rearrangement, [3.3]-sigmatropic-1,2-allyl group migration
Allyl halides s. β,γ-Ethylenehalides
Si-Allyl-1,1-hydroxysilanes
– startg. m. f.
   1,1-(homoallyloxy)silanols **51**, 354
α-Allyl-γ-lactones
– startg. m. f.
   2-cyclopentenones, 5-β-hydroxy-
   51, 344
Allyllithium compds.
– from
   allyl silyl ethers **51**, 362
Allylmagnesiation, Mn-catalyzed, regiostereospecific
– of acetylene derivs. **51**, 303
O-Allyloximes
– startg. m. f.
   isoxazolidines, 4-arylseleno- **51**, 250
Allyloxylamines
–, ring closures, selenylative **51**, 250
Allyl phosphates
– startg. m. f.
   3-ethylenealcohols, synthesis, regiostereospecific **51**, 272
Allylsilanes s. 2-Ethylenesilanes
Allyl silyl ethers
– startg. m. f.
   allyllithium compds. **51**, 362
O-Allyl O-silyl α-halogenacetals
– as intermediates **51**, 196

Allylstannanes s. 2-Ethylenestannanes
Allyl sulfones
–, radical C-allylation with – **51**, 422
Allyltri-n-butylstannane
– as reactant **51**, 444
Allyltrimethylsilane
– as reactant **51**, 450
Aluminates, organo-
– special s.
   acetylenealuminates
Aluminum alkoxides, organo-
– special s.
   vinylaluminum alkoxides
Aluminum amalgam 51, 488
– aroxides
   – tris(2,6-diphenylphenoxide) **51**, 300
– –, chiral, mixed-metal
– special s.
   lithium aluminum bis(1,1'-bi-2-naphthoxide)
– –, organo-
   methylaluminum bis(2,6-di-*tert*-butyl-4-methylphenoxide) **51**, 436
– chloride **51**, 125, 220, 407
– –/sodium chloride **51**, 346
– compds., organo- (s.a. Carboalumination)
   triethylaluminum **51**, 451
   triisobutylaluminum **51**, 345, 349
   trimethylaluminum (as reactant) **51**, 320
– special s.
   enalanes
– **enolates**
– as intermediates **51**, 345
– **halides, organo-**
   diethylaluminum chloride **51**, 326
   diisobutylaluminum chloride **51**, 23
– **hydrides, organo-**
   diisobutylaluminum hydride **51**, 375, 405
– special s.
   catecholalane
– **mercaptides, organo-**
   diisobutylaluminum mercaptides (as reactant) **51**, 256
– **oxide** (s.a. Nitrogen monoxide/alumina, Tetra-*n*-butylammonium fluoride/–)
   51, 58, 439
– **selenides, organo-**
   diisobutylaluminum selenides (as reactant) **51**, 256
– **tellurides, organo-**
   diisobutylaluminum tellurides (as reactant) **51**, 256
– **trifluoromethanesulfonimide**
   51, 78, 374
Amberlite IRA-400 tetrahydridoborate
(s.a. Copper/Amberlite IRA-400 tetrahydridoborate) **51**, 229, 234
Amidines
– special s.
   aminoamidines
   sulfonylamidines
Amidrazones
– special s.
   acylamidrazones
Aminals, chiral
– in asym. synthesis, review **47**, 646s51
Amination, photochemical s. Photoamination
α-Amination, asym.
– of carboxylic acids **51**, 182

Amines
–, N-alkylation with alcohols **51**, 150
– from
  alcohols, via Schmidt reaction **51**, 162
  alkoxylamines **51**, 9
  azides **51**, 11
  –, via Schmidt reaction **51**, 162
  N-nitramines **51**, 12
  nitroso compds. **30**, 10as**51**
  oximes **24**, 23s**51**
–, resolution **17**, 393s**51**
–, –, kinetic via N-carbalkoxylation **51**, 146
– special s.
  acetyleneamines
  acylamines
  diamines
  epoxyamines
  halogenamines
  nitramines
– startg. m. f.
  1,2-diamines (from 2 different molecules) **51**, 138
  guanidines **51**, 158
  nitro compds. **21**, 125s**51**
Amines, ar. (s.a. Anilines)
– from
  azides, ar. **51**, 163
  halides, ar. **51**, 171
  nitro compds., ar. (with CO), review **47**, 22s**51**
Amines, ar., prim.
– from
  o-nitrosulfonates **51**, 10
– startg. m. f.
  amines, ar., sec. **51**, 151
Amines, ar., sec.
– from
  alcohols **51**, 151
  amines, ar., prim. **51**, 151
– special s.
  diarylamines
Amines, ar., tert.
– special s.
  triarylamines
Amines, cyclic
– from
  phosphinyloxylamines **51**, 183
Amines, prim.
–, N-acylation, preferential **51**, 160
– from
  aziridines, N-phosphoryl-, synthesis **49**, 610s**51**
  ethylene derivs. **51**, 129
– startg. m. f.
  amines, sec. **51**, 141
  disilazanes **51**, 118
  isocyanates **51**, 133
  methylamines, sec. **28**, 346s**51**
  nitro compds. **51**, 45
Amines, sec.
– from
  amines, prim. **51**, 141
  azomethines **51**, 126
  –, asym. transfer hydrogenation **51**, 29
  –, synthesis, review (ligand effects) **48**, 625s**51**
  oxo compds. **51**, 141
  trifluoromethanesulfonic acid amides, N,N-disubst., synthesis **51**, 419
– special s.
  methylamines, sec.

– startg. m. f.
  azomethines **51**, 181
  nitrones **21**, 125s**51**
Amines, sec., cyclic, chiral **51**, 29
Amines, tert.
– from
  azomethines **51**, 126
– special s.
  ethyldiisopropylamine
Amines, tert., cyclic
– special s.
  N-methylmorpholine
Amines, tert., polycyclic, chiral
– special s.
  (–)-sparteine
2-Aminoalcohols
– special s.
  2-amino-1-phenylethanols
cis-**2-Aminoalcohols**
– from
  α-diketone monosiloximes, reduction, asym. **51**, 8
2-Aminoalcohols, chiral
– as reagent **51**, 26
– in asym. synthesis, review **47**, 646s**51**
2-*prim*-Aminoalcohols, chiral **46**, 601s**51**
2-Aminoalcohols, 3-functionalized
– from
  2,3-epoxyamines **51**, 120
Aminoalkyl enolethers
–, review **47**, 954s**51**
α-Aminoalkyllithium compds., chiral
– as intermediates **51**, 491
α-Aminoalkyl radicals
–, synthesis, stereospecific with –, review **46**, 484s**51**
α-*prim*-Aminoamidines
– from
  Δ¹-azirines, 2-amino- **51**, 124
1,1-Aminoazides
– special s.
  benzotriazoles, 1-α-azido-
α-Aminocarboxylic acid amides
– special s.
  glycinamides
α-Aminocarboxylic acid choline esters
– startg. m. f.
  peptides **51**, 145
α-Aminocarboxylic acid derivs.
– special s.
  1,1'-iminodicarboxylic acid derivs.
Aminocarboxylic acid esters
– special s.
  tetraaminocarboxylic acid esters
α-Aminocarboxylic acid esters
–, N-formylation **51**, 139
– special s.
  α-aminocarboxylic acid choline esters
– – –, chiral **51**, 15
– special s.
  prolinates
α-*tert*-Aminocarboxylic – –, chiral
– special s.
  methyl (S)-N-benzyl-N-methylalaninate
β-Aminocarboxylic acid esters
– from
  α-halogenocarboxylic acid esters, asym. induction **51**, 293
– special s.
  β-amino-α-arylcarboxylic acid esters

β-*prim*-Aminocarboxylic – –
– by Mannich-type reaction **51**, 387
α-Aminocarboxylic acids
– as chiral auxiliaries, stoichiometric, review **47**, 646s**51**
– from
  N-(diphenylmethyl)aldimines, conversion, asym. **51**, 294
  α-(diphenylmethylamino)nitriles, retention of configuration **51**, 294
  glycinamides, synthesis, asym. **51**, 398
– special s.
  α-amino-β-glycosyloxycarboxylic acids
  α-deuterio-α-aminocarboxylic –
– startg. m. f.
  heterocyclics, synthesis, asym., review **35**, 634s**51**
– –, α,α-disubst., chiral **16**, 384s**51**
– –, *a*-subst.
–, synthesis, asym., enzymatic, review **28**, 13s**51**
– –, tetrasubst., homochiral
–, synthesis, review **43**, 700s**51**
α-*prim*-Aminocarboxylic acids, chiral
–, synthesis, asym. **31**, 804s**51**; **50**, 380s**51**
β-Aminocarboxylic acids
–, synthesis, asym., review **50**, 157s**51**
o-**Aminodiaryl ethers**
–, Smiles rearrangement, review **33**, 315s**51**
2-Amino-1,3-dienes
–, Diels-Alder reaction, stereospecific with –, review **50**, 155s**51**
α-Aminodi(phosphonic acids) **36**, 608s**51**
2-Amino-3-ethylenealcohols, chiral **33**, 865s**51**
α-Amino-β-glycosyloxycarboxylic acids **51**, 5
β-Aminoketones
– from
  azomethinium salts **51**, 291
  ketimines **51**, 291
  α'-silylketones, asym. induction **51**, 440
Aminometalation-carbometalation, intramolecular
–, cycloisomerization, regiospecific via – **51**, 340
Aminomethylation, catalyzed
– of hydrocarbons, review **23**, 757s**51**
α-Aminonitriles
–, Michael addition-alkylation, asym. with – **51**, 395
– special s.
  α-(diphenylmethylamino)nitriles
α-Aminonitriles, chiral
– as acyl carbanion equivalents **51**, 395
N-Aminonucleosides, ¹⁵N-labelled **51**, 164
o-**Aminophenolethers**
– special s.
  o-aminodiaryl ethers
p-**Aminophenols**
– by Dötz reaction **43**, 860s**51**
2-Amino-1-phenylethanols
–, synthesis, biocatalytic, review **28**, 13s**51**
α-Aminophosphonic acid esters
– from
  aldimines, asym. conversion **51**, 236

α-Aminophosphonic acids
- from
  aldehydes, asym. induction **51**, 240
- special s.
  α-aminodi(phosphonic acids)
α-*prim*-Aminophosphonic acids
- from
  α-*sec*-aminophosphonous acids **51**, 240
α-Aminophosphonous acids
- from
  aldehydes, asym. induction **51**, 240
α-*prim*-Aminophosphonous acids
-, synthesis, polymer-based **49**, 556s**51**
α-*sec*-Aminophosphonous acids
- startg. m. f.
  α-*prim*-aminophosphonic acids **51**, 240
Aminoselenation, intramolecular, asym.
51, 258
Aminosilanes
- special s.
  N-(trimethylsilyl)diethylamine
- startg. m. f.
  disilazanes **51**, 118
1,1-Aminostannanes, cyclic, unsatd.
-, ring closures, stereospecific **51**, 491
Aminosugars
- special s.
  3-thioglucosamine...
Aminyl radicals
- from
  N-nitramines **51**, 12
Ammonia
- as reactant **51**, 168
Ammonia equivalent, chiral
42, 290s**51**
Ammonium cerium(IV) nitrate
51, 2, 128, 428
- as catalyst **51**, 15
- **hexafluorophosphate 51**, 33
- **hypophosphite 51**, 254
- **persulfate 51**, 250
- **salts, cyclic**
- special s.
  ketoammonium salts, cyclic
Aniline
- as reagent **51**, 403
Anilines (s.a. Amines, ar.)
-, *p*-bromination **51**, 201
Anthraquinones, 1-alkoxy-
- startg. m. f.
  oxo compds. **48**, 255s**51**
Anti-Alder-ene reaction 51, 329
Antimony(III) alkoxides
- ethoxide **51**, 187
Arbuzov-Michaelis rearrangement,
double 51, 260
Arenes (s.a. Benzene ring)
- from
  diarylmercury compds. **29**, 172s**51**
-, reduction, stereospecific, review
  32, 47s**51**
- special s.
  alkylarenes
  allylarenes
- startg. m. f.
  1,1-diarylethylenes **51**, 306
  diaryliodonium salts **51**, 204
Arenes, polycyclic
- startg. m. f.
  quinones, polycyclic **51**, 108
Arenesulfinic acids
- startg. m. f.
  diaryl sulfones **51**, 228

Arenesulfonic acid halides
- special s.
  o-nitrobenzenesulfonyl chloride
Arenetricarbonylchromium(0)
  complexes
-, 1,4-addition-alkylation, asym. **51**, 400
-, metalation, benzylic, asym. of benzyl
  ethers via – **51**, 400
-, *o*-metalation via – **51**, 400
Aromatization (s.a. Cycloaromatization)
Arylacetylenes
- from
  diaryliodonium salts **47**, 797s**51**
- startg. m. f.
  1,1-diarylethylenes **51**, 306
2-Arylalcohols 43, 562s**51**
N-Arylation, intramolecular 51, 171
-, -, Pd-catalyzed **51**, 190
Arylboronic acid esters
- from
  halides, ar. **51**, 259
Arylboronic acids
- startg. m. f.
  diaryls **51**, 453
  phenols **51**, 98
Arylcarboxylic acid esters
-, *o*-alkylation **49**, 679s**51**
α-**Arylcarboxylic acid esters**
- special s.
  amino-α-arylcarboxylic acid esters
N-Arylcarboxylic acid hydrazides
- startg. m. f.
  indoles, 2-amino- **51**, 473
α-**Arylcarboxylic acids, chiral 51**, 399
Arylcarboxylic acid salts
- special s.
  potassium benzoate
1-Arylenacylamines
-, hydrogenation, asym. **51**, 31
(E)-**4-Aryl-2-ethylenealcohols**
46, 918s**51**
γ-**Aryl-γ,δ-ethyleneketones 51**, 341
β-**Aryl-α,β-ethylenelactams,**
  macrocyclic 51, 486
(Z)-**1-Aryl-2-ethylenesilanes, chiral**
47, 542s**51**
4-Aryl-2-ethylenesilanes
- from
  1,3-dienes **51**, 455
C-Aryl glycosides 26, 725s**51**
Si→C-Aryl group migration 51, 492
Aryl ketones
- from
  alkylarenes **51**, 76
Aryl ketones, cyclic
- startg. m. f.
  benzylalcohols, cyclic, reduction, asym.
  51, 19
α-**Arylketones**
- special s.
  acetonylphenols
Aryl mesylates
- startg. m. f.
  alkylarenes **51**, 381
  diaryls **51**, 381, 453
  nitriles, ar. **51**, 381
Aryloxo compds.
-, 1,6-addition, nucleophilic to – **51**, 300
- from
  mandelic acids **23**, 296s**51**
(E)-**2-(Arylseleno)enoxysilanes**
43, 89s**51**

Arylseleno-O-heterocyclics
- from
  ethylenealcohols, asym. conversion
  51, 258
γ'-**Arylseleno-γ-lactolides 51**, 257
(**Arylseleno)lactones**
- from
  ethylenecarboxylic acids, asym.
  conversion **51**, 258
2-(Arylseleno)nitric acid esters
- as intermediate **29**, 601s**51**
Arylstannanes
- from
  halides, ar. **51**, 248
Aryltelluro group migration
-, radical ring closure with – **51**, 355
Aryl triflates
-, cross-coupling, ar. with halides, ar.
  51, 415
- startg. m. f.
  aldehydes, ar. **51**, 382
Ascorbic acid
- as reagent **51**, 257
Asymmetrization s. Desymmetrization
Asym. synthesis s. Induction, asym.,
  Synthesis, –
Automultiplication, asym., catalytic
51, 271
7-Azabicyclo[2.2.1]-heptanes, -hept-2-
  enes and -hepta-2,5-dienes
-, review **38**, 661s**51**
3-Azabicyclo[3.2.0]heptan-2-ones, chiral
38, 682s**51**
4-Azaenynes
- startg. m. f.
  Δ²-pyrroline ring, N-condensed **51**, 340
Aza-Payne rearrangement
51, 68, 120
Azasugars
-, synthesis, review **41**, 305s**51**
1,2-Azatitanacyclopent(a,e)nes
-, ring closures of azomethines, unsatd.
  via – **51**, 315
Aza-Wittig synthesis, intramolecular
-, benzothiazoles by – **51**, 231
-, -. polymer-based **51**, 184
5H-Azepin-2(1H)-ones
- from
  chromium α,β-acetylene(alkoxy)
  carbene complexes **51**, 425
Azetidines, 3-alkoxy- 22, 368s**51**
2-Azetidinones
- from
  aldimines, enolizable **51**, 402
  carboxylic acid halides **51**, 402
-, 3-amino-, chiral **7**, 836s**51**; **51**, 402
-, 3-fluoro- **48**, 854s**51**
-, N-unsubst. **51**, 402
3-Azetidinones, N-protected, chiral
35, 327s**51**
α-(**2-Azetidinon-1-yl)carboxylic acid**
  amides
- by Ugi reaction **51**, 360
Azides
- special s.
  aminoazides
  bridgehead azides
  carboxylic acid azides
  diazides
  enazides

Azi – Ben 254

(Azides)
– startg. m. f.
  amines 51, 11
  –, via Schmidt reaction 51, 162
  azomethines (from 2 molecules) 51, 163
  carbo-tert-butoxyamines 51, 161
Azides, ar.
– startg. m. f.
  amines, ar. 51, 163
Azide salts
– special s.
  guanidinium azides
α-Azidoacetals
– from
  enolethers 51, 127
Azidoalcohols
– startg. m. f.
  N-o-hydroxylactams 51, 159
β-Azido-α,β-ethylenecarboxylic acid
esters 51, 130
α-Azidohydroximinohalides
– from
  1,1-nitroethylene derivs. 51, 206
Azidoketals
–, Schmidt reaction with – 51, 193
Azines (C=N–N=C)
–, cleavage 34, 172s51
Aziridine-2-carboxylic acid esters
– from
  aldimines 51, 388
Aziridines
– from
  epoxides 51, 175
  phosphine imines 51, 175
–, chiral
– as reagent, review 47, 646s51
–, 2-imino-
– startg. m. f.
  α-acylaminocarboxylic acid amides
  51, 123
–, 1-metalated
–, review 33, 786s51
–, 1-phosphoryl-
– startg. m. f.
  amines, synthesis 49, 610s51
–, 1-sulfinyl-, chiral 23, 871s51
Aziridinium salts
– as intermediates 51, 138
Aziridinium salts, 2-α-siloxy-
– as intermediates 51, 120
Aziridin-2-ylcarbinols, N-sulfonyl-
– startg. m. f.
  2,3-epoxysulfonylamines 51, 68
N-Aziridin-1-ylimines
–, Shapiro reaction, catalyzed with –
  51, 475
1-(Aziridin-1-ylimino)ethers
– startg. m. f.
  enolethers, cyclic 51, 475
Δ¹-Azirines, 2-amino-
– startg. m. f.
  α-prim-aminoamidines 51, 124
2,2'-Azobis(3-ethylbenzothiazoline-6-
sulfonic acid)
– as reagent 51, 109
Azo compds.
–, 1,2-addition-alkylation 24, 312s51
Azodicarboxylic acid amides
– special s.
  N,N,N',N'-tetramethylazodicarboxamide
Azo N,N'-dioxides, sym.
– from

2-pyridinethiones, 1-acoxy- 51, 180
Azomethines (s.a. Imines)
– from
  amines, sec. 51, 181
  azides (2 molecules) 51, 163
  oxo compds. 51, 147
– special s.
  aldimines
  N-bis(trimethylsilyl)methylaldimines
  ethyleneazomethines
  ketimines
– startg. m. f.
  amines, sec. 51, 126
  –, –, asym. transfer hydrogenation
  51, 29
  –, –, synthesis, review (ligand effects)
  48, 625s51
  amines, tert. 51, 126
  β-aminocarboxylic acid esters, asym.
  induction 51, 293
  2-azetidinones (from aldimines,
  enolizable) 51, 402
  3-ethyleneamines, synthesis,
  regiostereospecific 51, 292
  N'-sulfonylamidines 51, 165
Azomethines, cyclic
– from
  diazides 51, 163
Azomethines, unsatd.
–, ring closures via 1,2-azatitana-
  cyclopent(a,e)nes 51, 315
Azomethinium salts
– startg. m. f.
  β-aminoketones 51, 291

Barbier cyclization (s.a. C-Acylation,
nucleophilic, intramolecular-Barbier
cyclization)
Barbier-Grignard-type syntheses
– in aq. media, review 40, 567s51
Barium 51, 292
Barium halides, organo-
– special s.
  2-ethylenebarium halides
Baylis-Hillman reaction
–, review 39, 593s51
Benzaldehyde
– as reagent 51, 448
Benzene ring (s.a. Arenes)
– by bisdecarbonylation 32, 985s51
– from
  o-dihalides 51, 409
  zirconacyclopentadienes 51, 409
Benzene ring, condensed
– from
  diynes 26, 725s51
Benzeneselenyl bromide
– as reactant 51, 250
Benz[e]indenes, 2,3-dihydro-
– by radical ring closure, double 51, 312
Benzocyclobutenes
– from
  halides, ar. 51, 414
1,4-Benzodiazepine-2,5-diones
–, polymer-based synthesis, multi-step
  51, 184

1,3,2-Benzodioxaaluminole
  s. Catecholalane
3(2H)-Benzofuranone-2-acetic acid
esters 51, 342
Benzofurans
– via styryl carbenes 51, 112
Benzopyran... s. Chrom[a,e]n...,
  Coumarin..., Isocoumarin...
Benzothiazoles
– from
  carboxylic acid halides 51, 231
  o-(phosphoranylideneamino)thioethers
  51, 231
Benzo[b]thiophenes
  (under Thianaphthenes in Vol. 1-50)
– via styryl carbenes 51, 112
Benzotriazole
– as reagent 51, 144, 376
Benzotriazole, N-(α-ethoxyallyl)-
– as reactant 51, 385
–, 1-formyl-
–, formylation with – 51, 157
–, 1-trifluoromethanesulfonyl-
– as reactant 51, 297
–, 1-trimethylsilyl-
–, 1,4-addition with – 51, 280
Benzotriazole-1-carboxamidinium
tosylate
– as reactant 51, 158
Benzotriazoles, 1-α-alkoxy-
– startg. m. f.
  ketones, synthesis 51, 384
Benzotriazoles, 1-α-azido-
– startg. m. f.
  eniminophosphoranes 51, 117
–, 1-cycloalkylideneamino-
– startg. m. f.
  nitriles with ring opening 51, 189
–, 1-α-(N-homoallylamino)-
– as intermediates 51, 376
–, 1-imidoyl-
– startg. m. f.
  [thio]iminoesters 51, 85
–, 1-α-lithiated
– startg. m. f.
  ketones, synthesis, regiospecific 51, 384
–, 1-α-(phosphoranylideneamino)-
– as intermediates 51, 117
– startg. m. f.
  β-prim-aminocarboxylic acid esters
  51, 387
2H-1,3-Benzoxazines, 3,4-dihydro-
46, 669s51
4H-1,3-Benzoxazin-4-ones, 2,3-dihydro-,
3-acyl-
–, aldol condensation, asym. with –
  48, 588s51
Benzoxazoles
– from
  o-hydroxyoximes 12, 229s51
Benzylalcohols
– startg. m. f.
  aldehydes, ar. 51, 109
Benzylalcohols, cyclic
– from
  aryl ketones, cyclic, reduction, asym.
  51, 19
–, sec., chiral 51, 24
Benzyl ethers
–, retention on hydrogenation, catalytic
  51, 32
– startg. m. f.

diarylmethanes, sym. **51**, 463
– –, tricarbonylchromium-complexed
–, metalation, benzylic, asym. **51**, 400
Benzyl halides
– startg. m. f.
allylarenes **51**, 405, 452
N-Benzyl-3-halogenamines, N-protected
– startg. m. f.
pyrrolidines, 2-aryl-, N-protected via
asym. deprotonation **51**, 478
Benzyllithium compds.
– from
benzyl silyl ethers **51**, 362
styrenes **51**, 299
N-Benzylnitrones, *in situ*-generated
–, cycloaddition, 1,3-dipolar,
stereospecific with – **51**, 352
O-Benzyloximes
– as intermediates **51**, 424
β-Benzyloxyoxo compds.
– startg. m. f.
furan-2-carboxylic acid esters,
tetrahydro-, 3-hydroxy- **51**, 462
Benzyl silyl ethers
– startg. m. f.
benzyllithium compds. **51**, 362
O-Benzyl-1-sulfonyloximes
– startg. m. f.
ketones **51**, 424
Benzyltriethylammonium chloride
– as reagent **51**, 464
– **tetrathiomolybdate**
– as reagent **51**, 163
Biaryls s. Diaryls
Bicyclo[n.1.0]alkanes, 1-siloxy-
– startg. m. f.
ethylenecarboxylic acids **51**, 97
Bicyclo[n.m.0]alkan-1-ols
– from
dihalogenolactones **51**, 483
**Bicyclo[2.2.1]hept-2-ene-2,3-
dicarboxylic acid esters, 5,6-*exo*-
epoxy-**
– startg. m. f.
bicyclo[3.1.0]hex-2-ene-1,2-
dicarboxylic acid 1-monoesters, 6-
formyl-, chiral **51**, 347
**Bicyclo[3.1.0]hex-2-ene-1,2-dicarboxylic
acid 1-monoesters, 6-formyl-, chiral
51**, 347
Bicyclo[2.2.2]octan-2-ones
–, synthesis, polymer-based **32**, 641s51
Bicyclo[3.3.0]oct-1-en-4-ols 51, 69
Bicyclo[2.1.0]pentan-2-ol ring
– startg. m. f.
cyclopent-3-enol ring **51**, 69
**Biginelli synthesis, polymer-based
51**, 361
2,2'-Biindoles
– startg. m. f.
indolo[2,3-*a*]carbazole ring **51**, 389
BINAP s. 2,2'-Bis(diphenylphosphino)-
1,1'-binaphthyl
1,1'-Bi-2-naphthol, chiral (s.a.
Lanthanum triisopropoxide/(R)-1,1'-bi-
2-naphthol, Metal-BINOL catalysts)
1,1'-Bi-2-naphthol, racemic, poisoned
–, induction, asym. with – **51**, 447
1,1'-Bi-2-naphthols, sym. 31, 719s51
1,1'-Binaphthyls, sym. 31, 719s51
Biphenyl/9,10-dicyanoanthracene
– as coupled sensitizers **51**, 445

4-Biphenylyl glycosides 37, 902s51
Birch reduction
–, review **32**, 47s51
Bis(*o*-acylaminophosphines), chiral
– as reagent **51**, 188
1,3-Bis(arylthio)ethylenes
– startg. m. f.
vinylcyclopropanes **51**, 423
**Bis[2,6-bis(1(R)-ethoxyethyl)phenyl]
diselenide**
– as reactant **51**, 258
Bis(catecholborane)
– as reactant **51**, 237
**Bischler-Napieralski ring closure,
polymer-based 17**, 942s51
Bis(decarbonylation)
–, carboxylic acids from α-hydroxymalonic acids **51**, 106
1,2-Bis(dicyclohexylphosphino)ethane
– as reagent **51**, 237
1,4-Bis(9-O-dihydroquinine)phthalazine
– as reagent **51**, 132
**2,4-Bis[(4,6-dimethyl-1,3,2-dioxa-
phosphorinan-2-yl)oxy] pentane**
– as reagent **51**, 457
**(R,S)-1,2-Bis[*o*-(diphenylphosphino)-
benzamido]cyclohexane**
– as reagent **51**, 383
**1,2-Bis[*o*-(diphenylphosphino)-
benzamido]-1,2-diphenylethane**
– as reagent **51**, 153
**(S)-2,2'-Bis(diphenylphosphino)-1,1'-
binaphthyl** (s.a. under Silver
trifluoromethanesulfonate)
1,4-Bis(diphenylphosphino)butane
– as reagent **51**, 334
1,2-Bis(diphenylphosphino)ethane
– as reagent **51**, 330
**(R,R)-2,6-Bis[1-(diphenylphosphino)-
ethyl]pyridine**
– as reagent **51**, 26
1,1'-Bis(diphenylphosphino)ferrocene
– as reagent **51**, 10
1,3-Bis(diphenylphosphino)propane
– as reagent **51**, 382
Bis(enestannanes)
– startg. m. f.
1,3-dienes, cyclic **51**, 493
Bis(α,β-ethyleneketones)
–, ring closure **51**, 305
Bis(fluorosulfonyl)imide
– as reagent **51**, 450
Bismuth(III) mandelate 51, 105
–(III) sulfate **51**, 226
– **tribromide 51**, 226
– **trichloride 51**, 226, 449
Bis(organozinc halides)
– startg. m. f.
ketones, cyclic **51**, 404
Bis(Δ²-oxazolines), chiral (s.a. under
Copper(I) trifluoromethane sulfonate)
– as ligand **51**, 435, 476
–, **C₂-symmetric**
– as ligands, review **47**, 646s51
Bis(tri-*n*-butyltin) oxide
– as reactant **51**, 248
– as reagent **51**, 313
Bis(trichloromethyl) carbonate
–, review **47**, 444s51
Bis(trifluoromethyl)dioxirane
– as reagent **51**, 75
N,O-Bis(trimethylsilyl)amide
– as silylating agent, review **29**, 964s51

N-[Bis(trimethylsilyl)methyl]aldimines
– startg. m. f.
2-azetidinones, N-unsubst. **51**, 402
Borane/dimethyl sulfide 51, 8, 21
Borane-amine complexes
–, α-alkylation, asym. of methyl
(S)-N-benzyl-N-methylalaninate via –
51, 401
Boranes
– special s.
eneboranes
Boranes, tert.
– special s.
trialkylboranes
triethylborane
Boric acid chloride esters, cyclic
– special s.
B-chlorocatecholborane
Boric acid esters
– special s.
trimethyl borate
Borinic acid esters
– special s.
B-methoxydiisopinocampheylborane
Borinic acid halides
– special s.
diisopinocampheylchloroborane
Borinyl triflates
– special s.
dicyclohexylborinyl triflate
diethylborinyl –
Boron enolates (s.a. Enol borinates)
Boron ester enolates, chiral
– as intermediates **51**, 275
Boron fluoride (s.a. Tetrabutylammonium
fluoride/boron fluoride) **51**, 100, 124,
159, 173, 269, 304, 354, 390,
434, 440, 441, 442, 460
Boronic acid esters
– special s.
arylboronic acid esters
di(boronic – –)
ethyleneboronic – –
– startg. m. f.
alcohols **51**, 98
Boronic acid esters, cyclic
– special s.
1,3,2-dioxaborolan...
Boronic acids (s.a. Suzuki...)
–, cross-coupling, regiostereospecific with
2-ethyleneamines **51**, 451
– special s.
arylboronic acids
– startg. m. f.
alcohols **51**, 98
Boron-Wittig methylenation 51, 434
Boroxines
– special s.
triarylboroxines
**Bridgehead azides
49**, 277s51
Bromination, ar., regiospecific
– under ultrasonication **51**, 200
p-**Bromination**
– of anilines **51**, 201
Bromine
– as reagent **51**, 240, 258
α-**Bromodifluoromethylation, asym.
47**, 784s51
Bromolactonization 51, 199

N-Bromosuccinimide (s.a. Triphenyl-
phosphine/N-bromosuccinimide)
- as reagent **51**, 50, 200, 215, 399
Buckminsterfullerene s. under Fullerenes
***tert*-Butoxyformic anhydride**
- startg. m. f.
  isocyanates **51**, 133
***tert*-Butyl hydroperoxide**
- as reactant **51**, 472
- as reagent **51**, 298
- –/1,8-diazabicyclo[5.4.0]undec-7-ene
- as reagent **51**, 62

Caesium s. Cesium
Cage compds., polycyclic
-, review **21**, 786s**51**
Camphorsulfonic acid
- as reagent **51**, 338
1,2-Carbalkoxy group migration
  51, 344
N-Carbalkoxylation
- special s.
  N-carbobenzoxylation
  N-carbo-*tert*-butoxylation
N-Carbalkoxylation, asym., enzymatic
- with carbonic acid esters **51**, 146
Carbamic acid esters s. N-(De)carb-
  alkoxylation, Urethans
Carbamyl halides
- special s.
  dimethylcarbamoyl chloride
Carbanion equivalents
- special s.
  acyl carbanion equivalents
Carbanions, fluorinated
-, review **41**, 463s**51**
Carbanions, nucleophilic
-, polarity reversal, review **36**, 775s**51**
Δ²-Carbapenems, 2-aryl- 37, 902s**51**
Carbene complexes
- special s.
  ethylenecarbene complexes
  transition metal carbene –
Carbenes
- special s.
  alkylidenecarbenes
  2,2'-dioxycarbenes
  vinylcarbenes
Carbenes, fluorinated
-, review **41**, 463s**51**
Carbenoids (s.a. Metallocarbenoids)
Carbinols (s.a. Alcohols)
- special s.
  aziridinylcarbinols
Carboalumination, asym., Zr-catalyzed
  51, 320
N-Carbobenzoxylation 41, 367s**51**
Carbo-*tert*-butoxyamines
- from
  azides **51**, 161
N-(Carbo-*tert*-butoxy)diamines
- startg. m. f.
  N-(carbofluorenylmethoxy)diamines
  with position switch **51**, 166
N-Carbo-*tert*-butoxylation, preferential
  42, 339s**51**

**Carbocyclization, transition metal-
catalyzed**
-, review **46**, 679s**51**
N-(Carbofluorenylmethoxy)diamines
- from
  N-(carbo-*tert*-butoxy)diamines with
  position switch **51**, 166
Carbohydrate O,O-alkylidene derivs.
  28, 141s**51**
Carbohydrates
- from
  substrates, non-carbohydrate, review
    48, 607s**51**
- special s.
  aldoses
  azasugars
  disaccharides
  glyc...
  ketosugars
  oligosaccharides
  septanoses
  thioglucos...
  uronic acids
-, synthesis, aldolase-catalyzed, review
    48, 607s**51**
-, –, chemoenzymatic, review **41**, 173s**51**
Carbohydrates, 2,3-unsatd.
-, review **43**, 151s**51**
β-Carbolines s. 9*H*-Pyrid[3,4-*b*]indoles
Carbolithiation, asym.
- of cinnamyl derivs. **51**, 299
Carbomagnesiation
- special s.
  allylmagnesiation
Carbometalation (s.a. Aminometalation-
  carbometalation)
- special s.
  carboalumination
  carbolithiation
  carbomagnesiation
  carbonickelation
  carbopalladation
  carbozirconation
Carbon dioxide
-, activation by superoxide **51**, 169
-, reactions, catalytic, heterogeneous,
  review **46**, 720s**51**
-, –, – with epoxides, review **23**, 139s**51**
Carbon dioxide, supercritical
-, hydrogenation, asym., homogeneous in
  – **51**, 31
Carbon disulfide
- startg. m. f.
  carboxylic acid thioamides **51**, 297
  2-ethylenedithiolcarbonic acid esters
    51, 221
  1,3-oxathiolane-2-thiones **51**, 218
  thiazolidine-2-thiones **51**, 232
Carbonic acid esters
-, N-carbalkoxylation, asym., enzymatic
  with – **51**, 146
-, review **50**, 116s**51**
- special s.
  acetylenecarbonic acid esters
  cyanohydrin carbonates
  dimethyl carbonate
  enol carbonates
  ethylenecarbonic acid esters
Carbonickelation, intramolecular
-, ring closures of enynones via – **51**, 305
Carbonium ions and salts
- special s.

  allyl cations
Carbon monoxide (s.a. Carbonylation)
-, reactions, catalytic with nitro compds.,
  ar., review **47**, 22s**51**
- startg. m. f.
  ketones, sym. **51**, 404
  2*H*-1,3-oxazin-2-ones, tetrahydro-
    51, 119
Carbon oxide sulfide
- startg. m. f.
  carboxylic acid amides **51**, 297
  thiolic acids/esters **51**, 297
Carbon tetrabromide
- as reagent **51**, 51
Carbon tetrachloride
- as reagent **51**, 209
Carbonylation (s.a. [4+1]-Cycloaddition,
  carbonylative, Cyclocarbonylation,
  Radical carbonylation, Thiolformylation)
-, aldehydes from triflates **51**, 382
- of 1,2-oxatitanacyclopentanes **51**, 315
Carbonylation, silylative
-, review **50**, 419s**51**
Carbonyl compds. (s.a. Aldehydes,
  Carboxylic acid..., Ketones, Oxo
  compds.)
- special s.
  dicarbonyl compds.
  (organothio)carbonyl –
Carbonyl ylids, non-stabilized
-, cycloaddition, 1,3-dipolar,
  stereospecific with – **51**, 443
- from
  1,1-siloxyiodides (2 molecules) **51**, 443
Carbopalladation
-, ring closures via –, review **46**, 679s**51**
–, intramolecular
- of allenes with nucleophile capture
    51, 417
Carbostyrils 31, 162s**51**
Carboxyl anion equivalent 51, 106
Carboxylation, catalyzed
- of hydrocarbons (CC⇓HC), review **23**,
  757s**51**
Carboxylic acid amides (s.a. Acylamines)
- from
  carboxylic acid azides **51**, 11, 163
  – – esters **51**, 187
  – acids **51**, 187
  Grignard compds. **51**, 297
  nitriles **51**, 58
- special s.
  acylaminocarboxylic acid amides
  (2-azetidinonyl)carboxylic – –
  dicarboxylic – –
  ethylenecarboxylic – –
  halogenocarboxylic – –
Carboxylic acid amides, N,N-disubst.
- startg. m. f.
  cyclopropanes, *tert*-amino- **51**, 377
Carboxylic acid anhydrides
-, O-acylation, asym., phosphine-
  catalyzed with – **51**, 83
- special s.
  dicarboxylic acid anhydrides
  p-nitrobenzoic anhydride
Carboxylic acid aryl esters
- special s.
  ethylenecarboxylic acid aryl esters
Carboxylic acid azides
- startg. m. f.

carboxylic acid amides **51**, 11, 163
Carboxylic acid choline esters
- special s.
  aminocarboxylic acid choline esters
Carboxylic acid esters (s.a. Acoxy...)
-, aldol condensation, asym. with -
  51, 275, 285
-, cleavage (HO⚡C)
- from
  acetylene derivs., terminal, C-cleavage
  51, 107
  alkyl hypohalites, prim. (2 molecules)
  51, 90
  carboxylic acid hydrazides **51**, 87
  - acids **51**, 78, 79
  α-cyanocarboxylic acid esters **51**, 43
- special s.
  acetic acid esters
  allenecarboxylic – –
  aminocarboxylic – –
  aziridine-2-carboxylic – –
  carboxylic acid aryl esters
  – –choline –
  – –vinyl –
  cyanocarboxylic acid esters
  diazocarboxylic – –
  epoxycarboxylic – –
  ethylenecarboxylic – –
  halogenocarboxylic – –
  hydroperoxycarboxylic – –
  hydroxycarboxylic – –
  ketocarboxylic – –
  phosphoranylidenecarboxylic – –
- startg. m. f.
  β-*prim*-aminocarboxylic acid esters,
  synthesis **51**, 387
  carboxylic acid amides **51**, 187
  – acids, asym. hydrolysis **51**, 6
  cyclopropanes, alkoxy- **51**, 411
  cyclopropanols **51**, 378
  ethers **51**, 36
  hydroxamic acid esters **51**, 142
  ketones, synthesis **51**, 371
  α-ketotrithioorthocarboxylic acid esters,
  with 1 extra C-atom **51**, 364
Carboxylic acid halides
- from
  carboxylic acids **51**, 207
- special s.
  acetyl chloride
  carboxylic acid fluorides
  ketocarboxylic acid halides
  phthalimidocarboxylic – –
- startg. m. f.
  alcohols, prim. **51**, 35
  2-azetidinones **51**, 402
  benzothiazoles **51**, 231
  β-diketones **51**, 403
  thiophenes **51**, 225
Carboxylic acid hydrazides
- special s.
  arylcarboxylic acid hydrazides
- startg. m. f.
  carboxylic acid esters **51**, 87
  – acids **51**, 87
Carboxylic acids
-, *o*-acylation with – **51**, 379
-, α-amination, asym. **51**, 182
- from
  alcohols, prim. **15**, 261s**51**
  carboxylic acid esters, asym. hydrolysis
  51, 6

– – hydrazides **51**, 87
α-diketones **51**, 399
1,3-dithiane 1-oxides, 2-acyl-,
  synthesis, asym. **51**, 399
α-halogenoketones **51**, 102
α-hydroxyketones, C-cleavage **51**, 105
α-hydroxymalonic acids, bis(decarbonylation) **51**, 106
-, resolution, kinetic by esterification,
  asym., lipase-catalyzed, review
  41, 175s**51**
-, solubilization as choline or 2-(2-
  methoxyethoxy)ethyl esters **51**, 5
- special s.
  acylaminocarboxylic acids
  aminocarboxylic –
  arylcarboxylic –
  ethylenecarboxylic –
  propionic acid
  ureidocarboxylic acids
- startg. m. f.
  α-acylaminocarboxylic acid amides
  51, 123
  alcohols, prim. **51**, 35
  carboxylic acid amides **51**, 187
  – – esters **51**, 78, 79
  – – halides **51**, 207
  – – thioamides **51**, 224
Carboxylic acid selenoamides
- as reagent **51**, 252
– – **thioamides**
- from
  carboxylic acids **51**, 224
  Grignard compds. **51**, 297
- special s.
  peptide thioamides
– – **vinyl esters** s. Enolesters
Carbozirconation
- special s.
  vinylzirconation
Carroll rearrangement
-, *o*-acetonylphenols via – **51**, 459
Cascade reactions (s.a. Ring closure,
  serial)
-, symposium reports **45**, 414s**51**
- –, **palladium-catalyzed**
-, review **46**, 679s**51**
Catalysis
- by acids, electrogenerated **51**, 54
Catalysis, asym. (s.a. Synthesis, asym.,
  catalyzed)
-, non-linear effects, review **47**, 646s**51**
-, **heterogeneous** s. under Supports,
  inorganic
-, **supramolecular 51**, 30
**Catalysts, heterobimetallic,
  multifunctional**
-, induction, asym. with – **51**, 309
Catalyst screening, high-throughput
- in parallel **51**, 476
Catecholalane
- as reagent **51**, 86
Catecholborane
- as reagent **51**, 22
Cation equivalents
- special s.
  vinyl cation equivalents
Cation exchangers
- special s.
  Dowex 50W
Cerium(IV) ammonium... s. Ammonium
  cerium(IV)...

Cerium(III) chloride 51, 265, 266
–**(III) dialkoxides, organo-, chiral**
- as reactant **51**, 266
–**(IV) periodates**
  tris[trinitratocerium(IV)] paraperiodate
  51, 2
Cesium fluoride 51, 369
Cesium hydroxide 51, 298
Chain transfer reaction, unimolecular
- via hydrogen atom transfer,
  intramolecular from silicon hydrides,
  organo- **51**, 323
Chloramine-T
- as reactant **51**, 132
B-Chlorocatecholborane
- as reagent **51**, 185
Chlorosilanes
- as reagent **51**, 469
4-Chromanone-3-acetic acid esters
  51, 342
Chromenes
-, epoxidation, asym. **46**, 106s**51**
3-Chromenes
- from
  α,β-ethyleneacetals **27**, 785s**51**
**Chromium α,β-acetylene(alkoxy)-
  carbene complexes**
- startg. m. f.
  5*H*-azepin-2(1*H*)-ones **51**, 425
Chromium alkoxycarbene complexes
- from
  dialkylzincs **43**, 860s**51**
- **[*o*-alkylideneamino)alkyl]-
  aminocarbene complexes**
- startg. m. f.
  4-imidazolidones, bridged **51**, 426
- **aminocarbene complexes**
- special s.
  chromium [*o*-(alkylideneamino)alkyl]
  aminocarbene complexes
–**(II) chloride/nickel(II) chloride/
  manganese
  51**, 288
- **pentacarbonyl migration** s.
  1,2-Pentacarbonylchromium group
  migration
–**(III) salen complex, chiral**
- as reagent **51**, 121
- **trioxide 51**, 134, 204
Cinnamaldehydes 51, 460
Cinnamyl alcohols 51, 71
Cinnamyl derivs.
-, carbothiation, asym. **51**, 299
Claisen rearrangement (s.a. CC∩OC,
  Ketal Claisen rearrangement)
- –, asym., chelation-controlled
  49, 703s**51**
o-Claisen –, double **2**, 621s**51**
Cobalt(II) acetoacetonate 51, 325
Cobalt(III) – 51, 76
Cobalt carbonyl 51, 324
– – **complexes, organo-**
- special s.
  acetylenedicobalt hexacarbonyl
  complexes
–**(II) chloride
  51**, 296, 404
- **complexes** (s.a. tetrakis(*tert*-butyl
  isocyanide)rhodiumtetracarbonylcobalt
  under Rhodium)

Cob – Cyc

Cobalt (II) Schiff base complex, chiral
(s.a. Sodium tetrahydridoborate/
cobalt(II) Schiff base complex, chiral)
–(II) **tetraphenylporphyrin 51,** 219
Coenzyme A derivs.
– special s.
S-acylcoenzyme A derivs.
Combinatorial synthesis (s.a. Compound
libraries and under Polymer-based ...)
–, reviews **50,** 555s**51**
– –, **liq.-phase**
– with fluorous systems **50,** 555s**51**
(review); **51,** 39
3-Component synthesis
– of acylamines **51,** 296
4-Component synthesis
– of 1,1'-iminodicarboxylic acid derivs.
with asym. induction **51,** 359
Compound libraries
– special s.
imidazole libraries
macrocyclic compd. –
oligosaccharide –
peptide –
1,4-phthalazinedione –
piperazine –
–, synthesis and screening, review
50, 555s**51**
Compounds, radio-labelled 41, 199s**51**
– from
stannanes, review **28,** 511s**51**
Conjugation
–, transmission of chirality through –
51, 20
**Copper/Amberlite IRA-400
tetrahydridoborate 51,** 216
Copper catalysis, asym.
–, review **48,** 772s**51**
Copper(II) acetate 51, 466
–(I) **cyanide 51,** 432, 434
–(I) **halides**
– chloride **51,** 270, 339, 403, 409, 493
– iodide **51,** 170, 228
–(II) **halides**
– bromide/alumina **51,** 199
– chloride **51,** 73
–(II) **pivalate 51,** 357
–(II) **sulfate 51,** 91
–(I) **thiophene-2-carboxylate 51,** 432
–(II) **p-toluenesulfinate**
– as reactant **51,** 228
–(I) **trifluoromethanesulfonate/bis(Δ²-
oxazolines), chiral**
– as reagent **51,** 72
–(II) **trifluoromethanesulfonate 51,** 388
**Coumarins, 4-aryl-
37,** 902s**51**
Cross-coupling
– of vinylzirconocene halides **51,** 410
Cross-coupling, asym.
–, diaryls by – **51,** 370
–, **Ni-catalyzed**
– of
aryl mesylates **51,** 381
ethylene-*prim*-iodides with zinc
compds., dialkyl- **51,** 412
–, **Pd- and/or Cu-mediated**
–, review **45,** 555s**51**
18-Crown-6 polyether
– as reagent **51,** 73, 298
– –, **dicyclohexyl-**
– as reagent **51,** 223

Cubanes
–, review **21,** 786s**51**
Cyanides s. Acylcyanides, Nitriles
α-**Cyanocarboxylic acid esters**
– startg. m. f.
carboxylic acid esters **51,** 43
Cyanohydrin carbonates 51, 287
Cyanohydrins
–, resolution, kinetic **22,** 693s**51**
α-**Cyanohydroximinohalides**
– from
1,1-nitroethylene derivs. **51,** 206
N-Cyanomethylation
–, activation of N-acylsulfonamides via –
51, 156
α-**Cyanomethylation, asym. 47,** 784s**51**
Cyanomethyl formate
–, N-formylation with – **51,** 139
Cyanosilanes
– special s.
trimethylsilyl cyanide
Cyanostannanes
– special s.
tri-*n*-butyltin cyanide
Cycloaddition (s.a. High-pressure
cycloaddition)
– with pyridinium betaines and ylids,
review **36,** 898s**51**
Cycloaddition, 1,3-dipolar
– to allenes, review **34,** 630s**51**
– –, **asym.**
– with nitrones and ethylene derivs.,
review **51,** 317s**51**
– –, –, **metal-catalyzed**
–, review **51,** 317s**51**
– –, –, **titanium(IV)-catalyzed 51,** 317
– –, **intramolecular**
– with nitrones, cyclic, *in situ*-generated
51, 336
– –, –, **stereospecific**
– with
enolates **51,** 350
N-(tetrahydropyran-2-yl)nitrones, *in
situ*-generated **51,** 352
– –, –, **stereospecific**
– with
carbonyl ylids, non-stabilized **51,** 443
N-benzylnitrones, *in situ*-generated
51, 352
N-(tetrahydropyran-2-yl)nitrones
51, 352
**Cycloaddition, transition metal-
mediated**
–, review **46,** 631s**51**
[2+2]-**Cycloaddition, intramolecular,
asym.**
– in aq. media **22,** 761s**51**
[3+2]-**Cycloaddition** (s.a. High-pressure
[3+2]-cycloaddition)
–, **intramolecular cation radical-
catalyzed 42,** 654s**51**
[4+1]-**Cycloaddition, carbonylative, Rh-
catalyzed 51,** 330
[4+4]-**Cycloaddition**
–, review **42,** 703s**51**
[6+2]-**Cycloaddition, asym. 46,** 631s**51**
3-Cycloalkenols
– from
[n+5]-oxabicyclo[n.2.1]alkenes **51,** 28
Cycloaromatization
– special s.
Myers-type cycloaromatization

Cyclobutane ring
– special s.
enecyclobutane ring
methylenecyclobutane –
– –, **siloxy- 47,** 694s**51**
Cyclobutanes
– special s.
methylenecyclobutanes
siloxycyclobutanes
vinylcyclobutanes
Cyclobutanols
– special s.
alkylidenecyclobutanols
Cyclobutanones
– startg. m. f.
cyclopentanones **51,** 408
Cyclobutenediones
– from
acetylene derivs. **34,** 811s**51**
Cyclobutene ring
– startg. m. f.
cyclopent-3-enol ring **51,** 69
Cyclobutenes
– special s.
alkylidenecyclobutenes
– startg. m. f.
(Z)-1,5-dienes **51,** 327
Cyclobutyl ketones, 2-arylthio-
–, reactions, review **49,** 499s**51**
Cyclocarbonylation, photochemical
–, γ-lactones by – **51,** 325
Cyclodec-3-ene-1,5-diynes
– from
1,10-dihalogeno-2,8-diynes **51,** 479
Cyclodextrins, modified
–, resolution by –, review **5,** 666s**51**
Cycloheptane ring
– special s.
methylenecycloheptane ring
1,4-Cyclohexadiene
– as reagent **51,** 312
**2,5-Cyclohexadienones, 4-peroxy-, 4-
subst.**
– as intermediates **51,** 472
2-Cyclohexenones
– startg. m. f.
o-iodophenols **51,** 197
–, **2-carbalkoxy- 46,** 721s**51**
Cycloisomerization, regiospecific
– via aminometalation-carbometalation,
intramolecular **51,** 340
1,5-Cyclooctadiene
– as reagent **51,** 60
Cyclooctane ring
– special s.
methylenecyclooctane ring
Cyclopentanes
– from
5-ethyleneselenides **51,** 427
– special s.
alkylidenecyclopent...
vinylcyclopentanes
Cyclopentanes, divinyl- 51, 327
Cyclopentanols
– startg. m. f.
piperidines **51,** 162
–, **3-α-aryltelluro- 51,** 355
–, **O-silyl-**
– from
acetyleneoxo compds. **51,** 315
Cyclopentanones
– from

cyclobutanones **51**, 408
Cyclopentanones, 2-α-iodo-
– from
  vinylcyclobutanes, 1-siloxy- **51**, 214
Cyclopent-3-enol ring
– from
  bicyclo[2.1.0]pentan-2-ol ring **51**, 69
  cyclobutene ring **51**, 69
2-Cyclopentenone ring
– from
  α,β-ethylenesulfones, asym. induction **51**, 454
  methylenecyclopentane ring, 3-sulfonyl- **51**, 454
2-Cyclopentenones
– special s.
  2-alkylidene-4-cyclopentenones
2-Cyclopentenones, 2-alkoxy- 51, 385
–, 5-β-hydroxy-
– from
  α-allyl-γ-lactones **51**, 344
3-Cyclopentenones
– special s.
  2-alkylidene-3-cyclopentenones
Cyclopent-1-enyl ketones
– from
  1,6-enynes **51**, 324
Cyclophanes
– special s.
  paracyclophanes
Cyclopropanation, asym.
– with iron carbene complexes **47**, 527s**51**
–, intramolecular, asym.
–, review **46**, 954s**51**
Cyclopropanecarbonitriles, 1-amino- 51, 367
Cyclopropanecarbonyl compds.
– special s.
  vinylcyclopropanecarbonyl compds.
Cyclopropanes
– from
  1,3-diol sulfates, cyclic **51**, 467
  glycol –, – **51**, 367
  methylene groups, active **51**, 367
– special s.
  siloxycyclopropanes
  vinylcyclopropanes
Cyclopropanes, alkoxy-
– from
  carboxylic acid esters **51**, 411
–, *tert*-amino-
– from
  carboxylic acid amides, N,N-disubst. **51**, 377
–, *cis*-trifluoromethanesulfonyl- **26**, 967s**51**
Cyclopropanols
– from
  carboxylic acid esters **51**, 378
  ethylene derivs., terminal **51**, 378
Cyclopropanols, condensed
– from
  ethylenecarboxylic acid esters **51**, 471
–, 2-β-hydroxy-
– from
  acoxy-3-ethylenes **51**, 316
N-Cyclopropylation, reductive 51, 143
– special s.
  N-dicyclopropylation, reductive
Cyclopropylcarbinols
– special s.

methylenecyclopropylcarbinols
Cyclopropyl ethers
– special s.
  cyclopropanes, alkoxy-
Cytosines
– from
  uracils **51**, 155

N-Deacylation (HN⇅C) **51**, 4
O-Deacylation (HO⇅C)
–, selective or preferential **51**, 4
S-Deacylation 51, 229
N-Debenzenesulfonylation, reductive 51, 13
O-De-*tert*-butoxy(diphenyl)silylation, preferential 46, 3s**51**
O-De-*tert*-butyldimethylsilylation, – 51, 3
–, selective **51**, 2, 3
N-Decarbalkoxylation (HN⇅C) **51**, 4
– special s.
  N-decarbo-*tert*-butoxylation
N-Decarbo-*tert*-butoxylation, catalytic 51, 15
–, partial and preferential
– at low pressure **51**, 17
Decarbonylation s. Bis(decarbonylation)
Dehydrogenation
– with vanadium oxide catalysts, review **3**, 708s**51**
3-Deoxy-4-ketosugars 51, 1
Deprotonation, asym.
– of
  N-benzyl-3-chloramines **51**, 478
  phosphines **51**, 357
  phosphine sulfides **51**, 357
Deprotonation, cathodic
– of acids, review **43**, 833s**51**
N-Depyrid-2-ylsulfonylation 51, 13
Deracemization (s.a. under Resolution, kinetic, dynamic)
– of alcohols, sec. **51**, 103, 499
O-Desilylation (HO⇅Rem)
– special s.
  O-de-*tert*-butoxy(diphenyl)silylation
  O-de-*tert*-butyldimethylsilylation
  O-detrimethylsilylation
–, preferential **51**, 3
–, review **46**, 3s**51**
N-Desulfonylation 51, 13
– special s.
  N-debenzenesulfonylation
  N-depyridylsulfonylation
  N-detosylation
Desymmetrization
– of
  dialdehydes **47**, 598s**51**
  diamines **51**, 146
  dicarboxylic acid anhydrides **51**, 56
  2-ene-1,4-diol carbamates, cyclic, N-protected **51**, 188
  – esters **37**, 527s**51**
  epoxides **48**, 767s**51** (review); **51**, 343
  α,β-ethyleneacylals **51**, 383
  glycols **51**, 83
–, enzymatic

–, review **48**, 767s**51**
N-Detosylation, reductive 51, 13
O-Detrimethylsilylation 51, 2
N-Detritylation, catalytic 51, 15
O-Detritylation, catalytic 51, 15
1-Deuterioalcohols 44, 42s**51**
α-Deuterio-α-aminocarboxylic acids, N-protected, chiral 31, 804s**51**
1-Deuteriohalides 47, 75s**51**
Diacoxy compds. s. Diol esters
1,1-Diacoxy compds. s. Acylals
1,2-Di(acylamines), chiral
– special s.
  (R,S)-1,2-bis[*o*-(diphenylphosphino)-benzoylamino]cyclohexane
Dialdehydes
–, desymmetrization **47**, 598s**51**
o-**Dialdehydes**
– startg. m. f.
  phthalimidines **51**, 144
Dialkyl phosphites
– startg. m. f.
  (E)-α,β-ethylenephosphonic acid esters **51**, 255
C-Diallylation 51, 303
Diamines
–, desymmetrization **51**, 146
Diamines, cyclic, chiral
– special s.
  (–)-sparteine
–, partially-protected
– special s.
  N-(carbo-*tert*-butoxy)diamines
  N-(carbofluorenylmethoxy)diamines
–, switching of protective groups **51**, 166
–, polymer-based
– as reagent **51**, 14
1,2-Diamines
– from
  amines (2 different molecules) **51**, 138
  glycol sulfates, cyclic, stereospecific conversion **51**, 138
  1,1-nitroethylene derivs. **51**, 149
– special s.
  ethylenediamine
1,1-Diarylalkanes
– special s.
  diarylmethanes
Diarylamines
– startg. m. f.
  triarylamines **51**, 170
Diaryl diselenides, chiral
– special s.
  bis[2,6-bis(1(R)-ethoxyethyl)phenyl] diselenide
1,1-Diarylethylenes
– from
  arenes **51**, 306
  arylacetylenes **51**, 306
Diaryliodonium salts
– from
  arenes **51**, 204
  iodides, ar. **51**, 204
Diarylmercury compds.
– startg. m. f.
  arenes **29**, 172s**51**
Diarylmethanes, sym.
– from
  benzyl ethers **51**, 463
Diaryls
– by
  cross-coupling, asym. **51**, 370

(Diaryls
- by)
  Suzuki coupling **51**, 416
- from
  arylboronic acids **51**, 453
  aryl mesylates **51**, 381, 453
  halides, ar. (chlorides) **51**, 453

Diaryls, sym.
- from
  aryl mesylates **51**, 381

Diaryl sulfones
- from
  arenesulfinic acids **51**, 228
  halides, ar. **51**, 228

Diaryl tellurides
–, Heck-type arylation with – **51**, 456

1,8-Diazabicyclo[5.4.0]undec-7-ene (s.a.
  tert-Butyl hydroperoxide/1,8-
  diazabicyclo[5.4.0]undec-7-ene)
- as reagent **51**, 247, 333, 431

**1,3,2-Diazaphospholidine 2-oxides,
  3-chloroallyl- chiral**
- as reactant **51**, 393

Diazides
- startg. m. f.
  azomethines, cyclic **51**, 163

α-Diazocarboxylic acid esters
- special s.
  ethyl diazoacetate

Diazo compds.
–, insertion, asym. into C-H bonds **51**, 476
- startg. m. f.
  ethylene derivs., trisubst. **51**, 391

α-Diazoketones
- special s.
  diazomethyl ketones
- startg. m. f.
  (E)-γ,δ-ethylene-α-hydroxyketones,
  asym. induction **51**, 390
  indolo[2,3-*a*]carbazole ring **51**, 389

Diazomethyl ketones
- startg. m. f.
  γ,δ-ethyleneketones, with 3 extra
  C-atoms **51**, 444
  methyl ketones **51**, 37

Diazo transfer
–, review **20**, 271s**51**

Dibenzyl diselenides, sym. 32, 578s**51**

Diboranes
- special s.
  oxydiboranes

Diboronic acid bis(pinacolate)
- as reactant **51**, 259

1,2-Di(boronic acid esters)
- from
  ethylene derivs. **51**, 237

(E)-1,4-Dibromobut-2-ene
- as reactant **51**, 397

1,3-Dibromo-5,5-dimethylhydantoin
- as reagent **51**, 213

1,2-Dibromoethane
- as reagent **51**, 248

Di-*n*-butyl(*tert*-butyl)tin hydride
- as reagent **51**, 93

Di-*tert*-butyl carbonate
  s. *tert*-Butoxyformic anhydride

**Di-*n*-butyl(chloro)stannane-
  hexamethylphosphoramide**
- as reagent **51**, 126

Di-*tert*-butyl hyponitrite
- as reagent **51**, 37

Di-*tert*-butyl peroxide

- as reagent **51**, 422

Di-*n*-butyltin oxide
- as reagent **51**, 352

Dicarbonic acid esters
- special s.
  tert-butoxyformic anhydride

β-Dicarbonyl compds. (s.a. β-Diketones,
  Malonic...)
–, α-alkylation with mesylates, with
  inversion **51**, 369
–, α-hydroxylation, Ni-catalyzed **51**, 52

α-Dicarboxylic... s. Succinic...

β-Dicarboxylic... s. Malonic...

Dicarboxylic acid amides
- from
  dicarboxylic acid imides **51**, 125

Dicarboxylic acid anhydrides
- startg. m. f.
  dicarboxylic acid imides **51**, 179
  – – monoesters, desymmetrization
    51, 56
  – – thioanhydrides **51**, 222
- acid derivs.
- special s.
  iminodicarboxylic acid derivs.
- acid imide ring
- startg. m. f.
  lactam ring **51**, 192
- acid imides
- from
  dicarboxylic acid anhydrides **51**, 179
  isocyanates **51**, 179
  isothiocyanates **51**, 179
- special s.
  N-acoxydicarboxylic acid imides
  N-hydroxydicarboxylic – –
- from
  dicarboxylic acid amides **51**, 125
- acid monoesters
- from
  dicarboxylic acid anhydrides,
  desymmetrization **51**, 56
- acids
- startg. m. f.
  dicarboxylic acid thioanhydrides
    51, 222
- acid thioanhydrides
- from
  dicarboxylic acid anhydrides **51**, 222
  – acids **51**, 222

2,3-Dichloro-5,6-dicyanoquinone
- as reagent **51**, 337, 461

2,6-Dichloropyridine N-oxide
- as reagent **51**, 108

2,2-Dichlorovinyl thioethers
- as intermediates **51**, 392

9,10-Dicyanoanthracene (s.a. Biphenyl/
  9,10-dicyanoanthracene)

Dicyanoketene ethylene acetal
- as reagent **51**, 61

Dicyclohexylborinyl triflate
- as reagent **51**, 275

**N-Dicyclopropylation, reductive
  51**, 143

Diels-Alder reaction (CC⇓CC; indexed
  under Diene synthesis in Vol. **1-50**)

Diels-Alder reaction, asym.
- with
  1-acylamino-1,3-dienes **49**, 622s**51**
  α,β-ethyleneketones, activated **51**, 307
  2-oxazolidones, 3-(α,β-ethyleneacyl)-
    46, 662s**51**

– –, **intramolecular** (CC∩CC)
- with eneboranes **34**, 693s**51**
– –, –, **ionic**
- with allyl cations, heteroatom-stabilized
    51, 474
– –, –, **regiostereospecific**
- using temporary vinylmetal alkoxide
  tethers **51**, 301
– –, **stereospecific**
- with 2-amino-1,3-dienes, review
  50, 155s**51**
– –, –, **heterogeneous, Lewis acid-
  catalyzed 51**, 311

Diels-Alder reactions
- in tandem, review **50**, 396s**51**

2,4-Dienals 51, 460

Dienes
–, ring closing metathesis, W-catalyzed
  51, 497

1,3-Dienes
- from
  acetylene derivs. (2 different molecules)
    51, 497
- special s.
  acylamino-1,3-dienes
  1-alkoxy-1,3-diene...
  amino-1,3-dienes
  polyfluoroalkyl-1,3-dienes
  siloxy-1,3-dienes
  silyl-1,3-dienes
  sulfinyl-1,3-dienes
  trifluoromethyl-1,3-dienes
- startg. m. f.
  4-aryl-2-ethylenesilanes **51**, 455
  3-ethylenealcohols, synthesis **51**, 319
  β,γ-ethylenecarboxylic acid amides,
    N,N-disubst. and esters, synthesis,
    regiospecific **51**, 319

1,3-Dienes, cyclic
- from
  bis(enestannanes) **51**, 493

1,4-Dienes, polysubst. 51, 302

1,5-Dienes
- from
  2-ethylenesilanes **51**, 437
  siloxy-2-ethylenes **51**, 437

(Z)-1,5-Dienes
- from
  cyclobutenes **51**, 327
  ethylene derivs., terminal **51**, 327

2,5-Dienols
–, synthesis, asym. **46**, 918s**51**

Diethylaminosulfur trifluoride
- as reagent **51**, 213

Diethylborinyl triflate
- as reagent **51**, 274

O,O-Diethyl dithiophosphoric acid
- as reagent **51**, 224

Diethyl methanephosphonate
- as reactant **51**, 365
- phosphorochloridate
- as reagent **51**, 365

1,1-Difluorides
- from
  1,3-dithiolanes **51**, 212

1,1-Difluoro-1,2-di(thioethers)
  51, 213

α,α-Difluoroketones
- from
  ketimines **51**, 202

1,1-Difluoro-2-ketothioethers
- from

α-hydroxytrithioorthocarboxylic acid
  esters **51**, 213
Difluoromethylene compds.
–, synthesis **32**, 867s**51** (review);
  47, 872s**51**
Dihalides
– startg. m. f.
  phosphinic acids, cyclic **51**, 254
1,1-Dihalides
– special s.
  1,1-difluorides
  ethylene-1,1-dihalides
1,2-Dihalides
– special s.
  1,2-dibromoethane
– startg. m. f.
  ethylene derivs. **51**, 480
1,3-Dihalides
– special s.
  polyfluoro-1,3-dihalides
o-**Dihalides**
– startg. m. f.
  benzene ring **51**, 409
2,2-Dihalogenalcohols
– from
  α,β-ethylenecarboxylic acids, with loss
    of 1 C-atom **51**, 215
3,3-Dihalogenalcohols
– from
  glycol sulfates, cyclic, with 1 extra
    C-atom **51**, 421
α,α-Dihalogenocarboxylic acid amides
  3, 722s**51**
**α,γ-Dihalogenocarboxylic acid esters,
  mixed**
  18, 776s**51**
1,10-Dihalogeno-2,8-diynes
– startg. m. f.
  cyclodec-3-ene-1,5-diynes **51**, 479
α,α-Dihalogenoketones
– special s.
  α,α-difluoroketones
1,1-Dihalogeno-2-ketothioethers
– special s.
  1,1-difluoro-2-ketothioethers
Dihalogenolactones
– startg. m. f.
  bicyclo[n.m.0]alkan-1-ols **51**, 483
**α,α-Dihalogeno-β-lactones,
  γ-functionalized, chiral**
  34, 811s**51**
Dihalogenomethylene compds.
– special s.
  difluoromethylene compds.
  β,β-dihalogenostyrenes
  diiodomethylene compds.
**(E)-Dihalogenomethylene compds.,
  mixed**
– from
  α,β-acetylenehalides **48**, 426s**51**
Dihalogenophosphines
– startg. m. f.
  phosphines, tert. **51**, 249
**β,β-Dihalogenostyrenes, *o*-function-
  alized**
–, ring closures **51**, 112
N,N-Dihalogenosulfonic acid amides
– startg. m. f.
  N'-sulfonylamidines **51**, 165
2,2-Dihalogenothioenolethers
– special s.
  2,2-dichlorovinyl thioethers

Dihydroxylation s.a. OC⇓CC
Dihydroxylation, asym.
–, update **47**, 114s**51**
Diiodomethylene compds.
  47, 872s**51**
(−)-Diisopinocampheylchloroborane
– as reagent **51**, 268
Diketene
– as reactant **51**, 459
α-Diketone monosiloximes
– startg. m. f.
  cis-2-aminoalcohols, reduction, asym.
    51, 8
Diketones
– startg. m. f.
  diols, reduction, asym. **29**, 36s**51**;
    51, 21
  ethylene derivs., macrocyclic **51**, 469
α-Diketones
– from
  acetylene derivs. **51**, 107
– startg. m. f.
  carboxylic acids **51**, 399
β-Diketones
–, α-alkylation **31**, 819s**51**
– from
  acetylene derivs., terminal **51**, 403
  α,β-acetyleneketones **51**, 403
  carboxylic acid halides **51**, 403
  enoxysilanes, catalytic conversion
    51, 449
γ-Diketones
– from
  α-aminonitriles, synthesis, asym.
    51, 395
  α,β-ethyleneketones, –, – **51**, 395
1,5-Dimethoxynaphthalene
– as electron donor **51**, 257
4-Dimethylaminopyridine
– as reagent **51**, 133, 459
Dimethylcarbamoyl chloride
– as reactant **51**, 319
Dimethyl carbonate
– as reactant, review **50**, 116s**51**
Dimethylchloroformiminium chloride
– as reagent **51**, 80
Dimethyldioxirane
– as reagent **51**, 52, 114
Dimethylformamide (s.a. Phosphorus
  oxide chloride/dimethylformamide)
**Dimethylketene methyl trimethylsilyl
  acetal**
– as reagent **51**, 223
N,N'-Dimethyl-N,N'-propyleneurea (s.a.
  Samarium diiodide/N,N'-dimethyl-N,N'-
  propyleneurea)
– as reagent **51**, 409
Dimethyl sulfoxide
– as reagent **51**, 105, 192
**(S)-4H-Dinaphth[2,1-c:1',2'-e]azepine-
  N-oxyl, 3,5-dihydro-3,3,5,5-
  tetramethyl-**
– as reagent **51**, 500
Dinucleoside phosphoramidates
  2, 314s**51**
1,2-Diol... s. Glycol...
1,4-Diol carbamates
– special s.
  2-ene-1,4-diol carbamates
1,3-Diol esters
– from
  aldehydes (2 molecules) **51**, 308
  enolesters **51**, 308

Diols
– from
  diketones, reduction, asym. **29**, 36s**51**;
    51, 21
Diols, differentially-protected
– from
  formals, cyclic **51**, 92
–, sym.
–, tetrahydropyran-2-ylation, partial **51**, 63
–, C$_2$-symmetric
  51, 21
1,2-Diols s. Glycols
1,3-Diols
– from
  aldehydes **51**, 298
  β-hydroxyacylsilanes **51**, 492
  1,2-oxasilacyclopentanes **51**, 298
  siliranes **51**, 298
– special s.
  ene-1,3-diols
  vinyl-1,3-diols
1,5-Diols
– special s.
  ene-1,5-diols
1,3-Diol sulfates, cyclic
– startg. m. f.
  cyclopropanes **51**, 467
1,1-Di(organometallics)
–, review **48**, 791s**51**
– special s.
  ene-1,1-di(organometallics)
**1,3,2-Dioxaborolan-2-ylmethylzinc
  iodide, 4,4,5,5-tetramethyl-**
– as reactant **51**, 434
1,3-Dioxane-4,6-diones, 5-alkylidene-
– startg. m. f.
  α,β-ethyleneselenolic acid esters
    51, 261
  α,β-ethylenethiolic – – **51**, 261
1,3-Dioxanes, 4-alkylidene-
– startg. m. f.
  pyrans, tetrahydro-, 4-hydroxy- **51**, 345
1,3,2-Dioxathiane 2,2-dioxides s. 1,3-
  Diol sulfates, cyclic
1,3,2-Dioxathiolane – s. Glycol sulfates,
  cyclic
1,2,4-Dioxathiolanes s. Monothioozonides
Dioxiranes
–, epoxidation, catalytic with – **51**, 66
– special s.
  bis(trifluoromethyl)dioxirane
  dimethyldioxirane
–, chiral, *in situ*-generated
– as reagent **51**, 65
Dioxo compds.
– startg. m. f.
  glycols, cyclic **51**, 286
1,3-Dioxolane-4,5-dimethanols, chiral
  (s.a. under Titanium(IV) alkoxides,
  chiral)
– as reagent **51**, 266
1,3-Dioxolanes
– from
  aldehydes (3 molecules) **51**, 373
– special s.
  glycol O,O-isopropylidene derivs.
1,3-Dioxolanes, 4-alkylidene-
– from
  1,3-dioxolan-4-ones **51**, 345
– startg. m. f.
  furans, tetrahydro-, 3-hydroxy- **51**, 345

Dio – Ene   262

1,3-Dioxolane-2-thiones
– startg. m. f.
  ethylene derivs. **49**, 938s**51**
1,3-Dioxolanium salts, 4-vinyl-
–, activation of α,β-ethyleneketones as –
  51, 307
1,3-Dioxolan-2-ones, 4-alk-1-ynyl-
– startg. m. f.
  furans, 2,5-dihydro-, 2-vinyl- **51**, 465
1,3-Dioxolan-4-ones
– startg. m. f.
  1,3-dioxolanes, 4-alkylidene- **51**, 345
2,2'-Dioxycarbenes
– as intermediates **51**, 468
Dipeptides, chiral 16, 384s**51**
–, cyclic, chiral
– as reagent **51**, 294
N-(Diphenylmethyl)aldimines
– startg. m. f.
  α-aminocarboxylic acids, conversion, asym. **51**, 294
α-(Diphenylmethylamino)nitriles, chiral
– as intermediates **51**, 294
Diphenylphosphinyl groups
– as stereodirecting group, review
  42, 595s**51**
(R,R,R)-[2-(4,5-Diphenyl-Δ²-oxazolin-2-yl)ferrocenyl]diphenylphosphine
– as reagent **51**, 24
Diphenylsilane
– as reagent **51**, 24
Diphenyl sulfoxide
– as reagent **51**, 49
Di(phosphines), chiral
– special s.
  (1R,1R')-2,6-bis[1-(diphenyl-phosphino)ethyl]pyridine
1,2-Di(phosphines)
– special s.
  1,2-bis(dicyclohexylphosphino)ethane
  1,2-bis(diphenylphosphino)ethane
–, C₂-symmetric
– from
  phosphines **51**, 357
Dipole stabilization
– of transition states, chiral **51**, 274
1,4-Diradicals
–, cascade cyclization via –, review
  45, 308s**51**
– special s.
  propargyl 1,4-diradicals
1,4-Diradicals, Myers-type
– as intermediates **51**, 312
Disaccharides
– by tail-to-tail coupling, review
  41, 173s**51**
Diselenides
– special s.
  diaryl diselenides
  dibenzyl –
Diselenides, sym.
– from
  aldehydes **51**, 243
  halides **51**, 252
Disilanes
– special s.
  allenyldisilanes
–, ar.
– startg. m. f.
  phenols **38**, 829s**51**
1,3-Di(silanes)
– special s.

  alkylidene-1,3-di(silanes)
1,2-Disilanyl group migration
  48, 674s**51**
Disilanyloxy-2-ethylenes
– startg. m. f.
  (E)-2-ethylenesilanes, 1,3-chirality transfer **51**, 262
Disilazanes
– from
  amines, prim. **51**, 118
  aminosilanes **51**, 118
Disodium 3,3'-thiodipropionate
– as reagent **51**, 64
Dispiroketals
–, review **50**, 400s**51**
Distannanes
– special s.
  hexabutyldistannane
  hexamethyldistannane
Distannoxanes
– special s.
  bis(tri-n-butyltin) oxide
Disulfides, sym.
– from
  mercaptans, disproportionation **51**, 216
  N-(organothio)phthalimides **51**, 217
  thiolic acid esters **51**, 216
S,S-Disulfones, sym.
– from
  sulfonic acid halides **22**, 591s**51**
1,2-Di(sulfones)
– from
  1,2-di(thioethers) **51**, 487
– startg. m. f.
  ethylene derivs. **51**, 487
1,3,2,4-Dithiadiphosphetane-2,4-disulfide, 2,4-bis(p-methoxyphenyl)-
– as reagent **51**, 225
1,3-Dithiane 1-oxides, 2-acyl-
–, C-alkylation, asym. **51**, 399
– startg. m. f.
  carboxylic acids, synthesis, asym.
    51, 399
1,3,2-Dithiaphospholane, 2-chloro-
– as reactant **51**, 247
1,3,2-Dithiaphospholane-2-thiones
– as intermediates **51**, 247
Dithioacetals s. Mercaptals
Dithioacetic acid esters
– startg. m. f.
  α-hydroxyketene mercaptals **51**, 396
Dithiocarbamic acid esters
– special s.
  vinyl dithiocarbamates
Dithiocarboxylic acid esters
– special s.
  dithioacetic acid esters
1,2-Di(thioethers)
– from
  α-acoxymercaptals **51**, 487
– special s.
  1,1-difluoro-1,2-di(thioethers)
  ene-1,2-di(thioethers)
– startg. m. f.
  1,2-di(sulfones) **51**, 487
1,3-Di(thioethers)
– special s.
  ene-1,3-di(thioethers)
1,3-Dithiolanes
– from
  oxo compds. **51**, 226
1,3-Dithiolan-2-ones 25, 167s**51**

Dithiolcarbonic acid esters
– special s.
  ethylenedithiolcarbonic acid esters
1,3-Dithiols
– special s.
  1,3-propanedithiol
Dithiophosphonic acid O-monoesters
– from
  Grignard compds. **51**, 247
Dithiophosphoric acid O,O-diesters
– special s.
  O,O-diethyl dithiophosphoric acid
Dithiothreitol
– as reagent **51**, 16
Diynes
– startg. m. f.
  benzene ring, condensed **26**, 725s**51**
1,3-Diynes, sym.
– from
  acetylene derivs., terminal **51**, 358
1,5-Diynes
– special s.
  ene-1,5-diynes
1,7-Diynes
– special s.
  1,10-dihalogeno-2,8-diynes
1,3-Di(zinc compds.)
– as intermediate **49**, 877s**51**
Domino reactions
–, review **45**, 414s**51**
Dowex 50W 51, 63

Electrolysis (s.a. Deprotonation, cathodic)
  51, 101, 127, 244
–, generation of acid catalysts under –
  51, 54
Emmons s. Horner
Enalanes
– startg. m. f.
  allylarenes **51**, 452
Enamines
– from
  2-ethyleneamines, asym. Rh(I)-catalysis, review **37**, 688s**51**
– special s.
  sulfonylenamines
Enazides
– from
  acetylene derivs. **51**, 130
– special s.
  β-azido-α,β-ethylene...
Eneboranes
–, Diels-Alder reaction, intramolecular with – **34**, 693s**51**
Enecyclobutane ring 22, 761s**51**
2-Ene-1,4-diol carbamates, cyclic, N-protected
–, desymmetrization **51**, 188
2-Ene-1,4-diol esters
–, desymmetrization **37**, 527s**51**
3-Ene-1,2-diol 2-monoethers
– from
  ketones, synthesis **51**, 385
anti-**3-Ene-1,2-diols, chiral**
  33, 865s**51**
5'-Ene-1,3-diols, cyclic

–, polymer-based synthesis **51**, 375
2-Ene-1,5-diols
– from
  3-ethyleneepoxides **51**, 279
  ketones **51**, 279
Ene-1,1-di(organometallics)
– as intermediates **51**, 302
Ene-1,2-di(thioethers)
– from
  α-hydroxymercaptals **51**, 487
– startg. m. f.
  acetylene derivs. **51**, 487
Ene-1,3-di(thioethers)
– special s.
  1,3-bis(arylthio)ethylenes
Enediynes
–, ring closure, double via 1,4-diradicals **51**, 312
3-Ene-1,5-diynes
17, 198s**51**; **29**, 973s**51**
–, chemistry, review **45**, 308s**51**
–, radical cascade cyclizations, review **45**, 308s**51**
3-Ene-1,5-diynes, cyclic
– special s.
  cyclodec-3-ene-1,5-diynes
Ene reaction (CC⇓OC, CC⇓CC)
– with selenonium ions **40**, 575s**51**
Ene reaction, asym.
– with carbonyl compds., review **44**, 568s**51**
– –, **intramolecular** (s.a. Knoevenagel condensation-intramolecular ene reaction)
– –, **stereospecific**
– with singlet oxygen, review **48**, 106s**51**
Ene-retroene reaction, intramolecular
–, isomerization of γ,δ-ethyleneketones by – **51**, 341
Eneselenides
– special s.
  halogeneneselenides
(E)-Eneselenides
– from
  acetylene derivs. **50**, 343s**51**
Enesilanes
– by radical cross-coupling **46**, 832s**51**
Enestannanes
– special s.
  alkoxyenestannanes
  bis(enestannanes)
Enetellurides
– startg. m. f.
  ethylene derivs., synthesis **37**, 868s**51**
Enethioethers s. Thioenolethers
Eniminophosphoranes
– from
  benzotriazoles, 1-α-azido- **51**, 117
Enisonitriles
– special s.
  1-isocyanocyclohexene
Enisothiocyanates
–, review **30**, 242s**51**
Enolates
–, cycloaddition, 1,3-dipolar, intramolecular, stereospecific with – **51**, 350
–, Michael addition, stereospecific, review **36**, 652s**51**
–, protonation, asym., review **50**, 389s**51**
– special s.
  alkali-metal enolates

aluminum –
boron[ester] –
lithium –
titanium ester –
zirconium –
(Z)-Enol borinates (s.a. Boron ester enolates)
– from
  enoxysilanes **30**, 621s**51**
Enol carbonates
– from
  enoxysilanes **51**, 100
– startg. m. f.
  α-formoxyketones **51**, 101
Enolesters
–, N-acylation with – **47**, 273s**51**
– from
  carboxylic acid fluorides **51**, 100
  enoxysilanes **51**, 100
– startg. m. f.
  1,3-diol esters **51**, 308
Enolethers
– from
  α-alkoxysulfones **51**, 420
  sulfones **51**, 420
–, Schmidt reaction with – **51**, 193
– special s.
  allyl enolethers
  aminoalkyl –
  vinyl ethers
– startg. m. f.
  α-azidoacetals **51**, 127
  4-ethylenealcohols via Claisen rearrangement **51**, 349
  monothioacetals **8**, 651s**51**
(E)-Enolethers
– from
  acetals, synthesis **51**, 366
Enolethers, cyclic
– from
  acoxyethylenes via metathesis **51**, 496
  1-(aziridin-1-ylimino)ethers **51**, 475
– startg. m. f.
  hydroxyalkoximes **51**, 122
  hydroxyhydrazones **51**, 122
Enol phosphites
–, S-phosphorylation with – **51**, 246
Enols, O-protected
– startg. m. f.
  α-hydroxyketones **51**, 99
Enol triflates
– startg. m. f.
  α,β-ethylenealdehydes **51**, 382
Enone s. α,β-Ethyleneketones
Enoxysilanes
–, aldol-type condensation, regiospecific with – **51**, 429
– special s.
  (arylseleno)enoxysilanes
  silylenoxysilanes
– startg. m. f.
  allyl enolethers **51**, 196
  O-allyl O-silyl α-halogenacetals **51**, 196
  β-diketones **51**, 449
  enol carbonates **51**, 100
  enolesters **51**, 100
  α,β-ethyleneketones, synthesis **51**, 428
  ketones, synthesis (of chiral compds.) **51**, 441
  α-(trifluoroacetylamino)ketones **51**, 178
Enoxysilanes, cyclic

– startg. m. f.
  lactams **51**, 191
  1,1-siloxyazides, cyclic **51**, 191
Envirocat EPZG
– as reagent **51**, 55
Enyne-allenes
–, chemistry, review **45**, 308s**51**
–, radical cascade cyclizations, review **45**, 308s**51**
Enynes
– special s.
  azaenynes
  hydroxylaminoenynes
1,3-Enynes
– from
  allenes **37**, 675s**51**
– startg. m. f.
  allenes, synthesis **51**, 334
  styrenes (from 2 molecules) **51**, 335
1,6-Enynes
– startg. m. f.
  cyclopent-1-enyl ketones **51**, 324
2,n-Enynones
–, ring closure, alkylative, regiostereo-specific **51**, 305
Enzymes (s.a. Synthesis, enzymatic)
  acylcholine esterases **51**, 5
  aldolase **48**, 607s**51** (review)
  chymotrypsin **51**, 145
  epoxide hydrolases **48**, 108s**51** (review)
  glycosidases **41**, 173s**51** (review)
  horseradish peroxidase **51**, 498
  hydantoinase **51**, 59
  laccase **51**, 74, 109
  lipase **41**, 175s**51** (review); **51**, 16, 94, 103
  –/silica **51**, 233
–, immobilized in asym. synthesis, review **49**, 305s**51**
  monoamine oxidase **51**, 181
  papain **51**, 5
  pig liver esterase **51**, 347
  subtilisin **51**, 146
Enzymes, cross-linked
  lipase, cross-linked **51**, 6
Epichlorhydrin
– as reactant **51**, 430
Epoxidation (s.a. OC⇓CC; Epoxides from ethylene derivs.)
Epoxidation, catalytic
– with dioxiranes **51**, 66
Epoxides (under Oxido compds. in Vol. 1-50)
–, desymmetrization **48**, 767s**51** (review); **51**, 343
– from
  ethylene derivs. **51**, 66
– –, asym. conversion (of *trans*-isomers) **51**, 65
– –, electron-deficient **51**, 62
  2-ethylenemonoperoxyketals, synthesis **51**, 406
–, reactions, catalytic with CO_2, review **23**, 139s**51**
–, resolution, kinetic **51**, 121
– special s.
  acetyleneepoxides
  ethyleneepoxides
  pyridinium salts, N-(2,3-epoxyalkyl)-
– startg. m. f.
  1,2-acoxyhalides **10**, 412s**51**
  aziridines **51**, 175

(Epoxides)
- startg. m. f.)
  glycol monoethers **51**, 54
  glycols **51**, 54
  – (enzymatic conversion), review
    48, 108s**51**
  1,3-oxathiolane-2-thiones **51**, 218
  2-siloxyazides, asym. ring opening
    51, 121
  β-siloxynitriles **49**, 589s**51**
Epoxides, fused
- special s.
  bicyclo[2.2.1]hept-2-ene-2,3-
    dicarboxylic acid esters, 5,6-*exo*-
    epoxy-
- startg. m. f.
  alcohols, bicyclic, desymmetrization
    51, 343
–, 1-metalated
- review **33**, 786s**51**
–, sym.
- from
  1,1-siloxyiodides **51**, 443
2,3-Epoxyalcohols 36, 670s**51**
- startg. m. f.
  β-hydroxyketones **51**, 1
2,3-Epoxyalcohols, cyclic
- startg. m. f.
  α,β-ethyleneketones, cyclic, with
    1,2-alkyl migration **51**, 468
Epoxy-*prim*-amines 20, 112s**51**
Epoxy-*tert*-amines 44, 117s**51**
2,3-Epoxyamines
- startg. m. f.
  2-aminoalcohols, 3-functionalized
    51, 120
α,β-Epoxycarboxylic acid esters s.
  Glycidic acid esters
2,3-Epoxyethers, fused
- startg. m. f.
  2-ethylenealcohols, cyclic, synthesis
    51, 363
2,3-Epoxyhalides
- special s.
  epichlorhydrin
β,γ-Epoxyketones
- from
  2-ethylenesilanes **51**, 225
- startg. m. f.
  thiophenes **51**, 225
α,β-Epoxy-δ-lactones 51, 62
1,2-Epoxysilanes 44, 117s**51**
2,3-Epoxysulfonylamines
- from
  aziridin-2-ylcarbinols, N-sulfonyl-
    51, 68
Estrane ring
- by Diels-Alder reaction, asym.
    36, 667s**51**
**Etherification, polymer-based,
  Rh-catalyzed 33**, 191s**51**
Ethers (s.a. Alkoxy..., Mitsunobu
  etherification)
- –, cleavage s. HOliC
- from
  carboxylic acid esters **51**, 36
  diazo compds., asym. induction
    33, 191s**51**
  thionocarboxylic acid esters **51**, 42
- special s.
  benzyl ethers
  enolethers

epoxyethers
ethyleneethers
phenolethers
- startg. m. f.
  ketones **51**, 115
Ethers, cyclic (s.a. O-Heterocyclics)
- from
  lactones **51**, 36
  thionolactones **51**, 42
- special s.
  halogenethers, cyclic
  sulfonyloxyethers, –
Ethyl diazoacetate
- as reactant **51**, 388, 462
Ethyldiisopropylamine
- as reagent **51**, 92, 148, 158, 172, 270,
    274, 285, 464
Ethylene
- as reactant **51**, 326
α,β-Ethyleneacylals
- startg. m. f.
  acoxy-2-ethylenes, synthesis, asym.
    51, 383
2-Ethylenealcohol O-derivs.
- –, hydrogenolysis, Pd-catalyzed, review
    34, 55s**51**
Ethylenealcohols
- startg. m. f.
  arylseleno-O-heterocyclics, asym.
    conversion **51**, 258
  2-ethyleneethers, cyclic **51**, 110
2-Ethylenealcohols
- –, C-α-allylation with – **22**, 782s**51**
- from
  3-ethylenealcohols, hydroxyl shift
    51, 71
- special s.
  allyl alcohol
  aryl-2-ethylenealcohols
  cinnamyl alcohols
  2-ene-1,5-diols
- startg. m. f.
  allyl enolethers **51**, 196
  4-ethylenealcohols via Claisen
    rearrangement **51**, 349
  2-ethylenedithiolcarbonic acid esters
    51, 221
  (E)-γ,δ-ethylene-α-hydroxyketones,
    asym. induction **51**, 390
  S-(2-ethylene)monothiolcarbonates
    51, 221
  2-ethylenestannanes **51**, 242
  γ-lactones, carbonylation **51**, 325
  2-pyrones, 5,6-dihydro-, 3-γ-keto-
    51, 329
**(E)-2-Ethylenealcohols, chiral
  49**, 81s**51**
2-Ethylenealcohols, cyclic
- from
  2,3-epoxyethers, fused, synthesis
    51, 363
3-Ethylenealcohols
- from
  aldehydes, synthesis, asym.,
    regiospecific **51**, 433
  –, –, asym. with 3 extra C-atoms
    51, 447, 450
  –, –, regiospecific **51**, 272, 281
  allyl phosphates, –, – **51**, 272
  1,3-dienes, – **51**, 319
  2-ethylenestannanes, –, asym.,
    regiospecific **51**, 433

2-ethylenetitanium(IV) alkoxides, –,
    regiostereospecific **51**, 281
  oxo compds., –, – **51**, 281
  – –, with 3 extra C-atoms **51**, 450
- startg. m. f.
  2-ethylenealcohols, hydroxyl shift
    51, 71
3-Ethylene-*tert*-alcohols
- from
  2-ethylenesilanes, synthesis, asym.
    51, 438
  ketones, –, – **51**, 438
3-Ethylenealcohols, cyclic
- special s.
  3-cycloalkenols
4-Ethylenealcohols
- from
  enolethers via Claisen rearrangement
    51, 349
Ethylenealdehydes
- startg. m. f.
  5'-ene-1,3-diols, cyclic **51**, 375
  isoxazolidine ring, N-unsubst. **51**, 352
α,β-Ethylenealdehydes (s.a. α,β-
  Ethyleneoxo compds.)
- from
  enol triflates **51**, 382
- special s.
  cinnamaldehydes
  2,4-dienals
(E)-α,β-Ethylenealdehydes
- from
  aldehydes, with 2 extra C-atoms **51**, 460
**β,γ-Ethylenealdehydes, chiral, α,α-
  disubst. 23**, 735s**51**
Ethyleneamines
- startg. m. f.
  halogeno-N-heterocyclics **48**, 434s**51**
Ethyleneamines, N-protected
- startg. m. f.
  2-ethyleneamines, cyclic, N-protected
    51, 110
2-Ethyleneamines
- –, cross-coupling, regiostereospecific with
    boronic acids or esters **51**, 451
- startg. m. f.
  enamines, asym. Rh(I)-catalysis, review
    37, 688s**51**
(E)-2-Ethylene-*tert*-amines
- from
  allenes **51**, 131
2-Ethyleneamines, cyclic, N-protected
- from
  ethyleneamines, N-protected **51**, 110
3-Ethyleneamines
- from
  aldimines, 1,3-asym. induction **51**, 295
  azomethines, synthesis, regiostereo-
    specific **51**, 292
  2-ethylenebarium halides, –, – **51**, 292
  2-ethylenecarbonic acid esters, 1,3-
    asym. induction **51**, 295
α,β-Ethyleneazomethines
- startg. m. f.
  5*H*-azepin-2(*1H*)-ones **51**, 425
2-Ethylenebarium halides
- startg. m. f.
  3-ethyleneamines, synthesis,
    regiostereospecific **51**, 292
(E)-α,β-Ethyleneboronic acid esters
- from
  acetylene derivs. **47**, 567s**51**

(Z)-α,β-Ethyleneboronic – –
  48, 48s51
α,β-Ethyleneboronic – –, in situ-
  generated
–, Suzuki coupling with – 51, 394
α,β-Ethylenecarbene complexes
– special s.
  titanocene α,β-ethylenecarbene
    complexes
2-Ethylenecarbonic acid esters
– startg. m. f.
  3-ethyleneamines, 1,3-asym. induction
    51, 295
α,β-Ethylenecarbonyl compds.
– startg. m. f.
  2-vinylcyclopropanecarbonyl compds.,
    functionalized, asym. synthesis
    51, 393
β,γ-Ethylenecarboxylic acid amides,
  N,N-disubst.
– from
  1,3-dienes, synthesis, regiospecific
    51, 319
α,β-Ethylenecarboxylic acid aryl esters
– special s.
  acrylic acid aryl esters
Ethylenecarboxylic acid esters
– startg. m. f.
  cyclopropanols, condensed 51, 471
  lactams 48, 305s51
(E)-Ethylenecarboxylic acid esters
– from
  ketones, cyclic 48, 751s51
α,β-Ethylenecarboxylic acid esters
– from
  β-ketocarboxylic acid esters 51, 44
– special s.
  azido-α,β-ethylenecarboxylic acid
    esters
(E)-α,β-Ethylenecarboxylic – –
– by Wittig synthesis 51, 439
– from
  α-allenecarboxylic acid esters,
    synthesis 50, 411s51
β,γ-Ethylenecarboxylic – –
– from
  1,3-dienes, synthesis, regiospecific
    51, 319
(E)-γ,δ-Ethylenecarboxylic – –
  36, 652s51
Ethylenecarboxylic acids
– from
  bicyclo[n.1.0]alkanes, 1-siloxy- 51, 97
– startg. m. f.
  (arylseleno)lactones, asym. conversion
    51, 258
  2-halogenolactones 51, 6
α,β-Ethylenecarboxylic acids
– startg. m. f.
  2,2-dihalogenalcohols, with loss of 1
    C-atom 51, 215
  α,β-ethylenehalides, – – – – – 51, 215
γ,δ-Ethylenecarboxylic acids
– startg. m. f.
  γ-lactones 51, 70
α,β-Ethylenecarboxylic acid thioamides
– by Horner synthesis 23, 879s51
Ethylene derivs. (s.a. under Allylation,
  Horner and Wittig)
–, C-β-alkylation, stereospecific of
  ethylene derivs., electron-deficient
  with – 51, 328

–, dihydroxylation s. OC⇓CC
– from
  α-acoxymercaptals 51, 487
  β-acoxysulfones 51, 489
  alkoxy-2-ethylenes, synthesis,
    regiostereospecific 51, 372
  1,2-dihalides 51, 480
  1,3-dioxolane-2-thiones 49, 938s51
  1,2-di(sulfones) 51, 487
  enetellurides, synthesis 37, 868s51
  glycol sulfates, cyclic 49, 938s51
  β-hydroxysulfones 51, 489
  lithium acetylides, synthesis,
    regiostereospecific 51, 302
  thiiranes 51, 490
–, hydroboration, regiospecific 51, 238
–, hydrogenation s. HC⇓CC
–, ozonolysis, improved workup 51, 64
– special s.
  acoxyethylenes
  alkoxyethylenes
  cinnamyl...
  diarylethylenes
  dien...
  disilanyloxyethylenes
  en...
  methylene compds.
  nitroethylene derivs.
  polyenes
  siloxyethylenes
  styrenes
  sulfonyloxyethylenes
  trien...
– startg. m. f.
  alcohols, via hydrozincation 51, 60
  amines, prim. 51, 129
  benzocyclobutenes 51, 414
  1,2-di(boronic acid esters) 51, 237
  epoxides (OC⇓CC) 51, 66
–, asym. conversion (of trans-isomers)
    51, 65
  2(5H)-furanones, 3-amino- 51, 111
  isoxazolidines, asym. conversion
    51, 317
  2-nitroacylamines 51, 128
  (E)-1,1-nitroethylene derivs. 51, 135
  α-nitroketones 51, 134
  thiiranes 51, 219
  cis-2-tosylaminoalcohols, asym.
    conversion 51, 132
(Z)-Ethylene derivs.
– from
  acetylene derivs. 16, 72s51
cis-Ethylene derivs.
– by Shapiro reaction, catalyzed 51, 475
Ethylene derivs., chiral
– by Shapiro reaction, catalyzed 51, 475
– –, cyclic (s.a. Alkylidenecyclo...)
–, ring opening metathesis 51, 327
– –, electron-deficient
–, C-β-alkylation, stereospecific 51, 328
–, Heck-type arylation with diaryl
  tellurides 51, 456
–, hydroacylation, intramolecular 51, 342
– startg. m. f.
  epoxides 51, 62
  furans, 2,5-dihydro-, 2-vinyl- 51, 465
– –, electron-rich
–, radical 1,2-addition of nitro compds. to
  – 51, 428
– –, H-labelled
– by Shapiro reaction 23, 935s51

– –, macrocyclic
– from
  diketones 51, 469
– –, polycyclic, anti-Bredt and hindered
–, synthesis, reviews, 43, 769s51
– –, terminal
– from
  aldehydes, with 1 extra C-atom 51, 434
– startg. m. f.
  alcohols, prim., synthesis, asym.
    51, 320
  cyclopropanols 51, 378
  (Z)-1,5-dienes 51, 327
– –, trisubst. 51, 489
– from
  diazo compds. 51, 391
  β-tert-hydroxyphosphine oxides
    51, 265
  β-ketophosphine –, synthesis,
    stereospecific 51, 265
  vinylmagnesium bromides 51, 391
Ethylenediamine (s.a. under Ruthenium
  complexes)
– as reagent 51, 14
α,β-Ethylenediazo compds.
– by Horner synthesis 39, 854s51
1,2-Ethylene-1,1-dihalides s.
  Dihalogenomethylene compds.
2,3-Ethylene-1,4-dihalides
– special s.
  1,4-dibromobut-2-ene
2-Ethylenedithiolcarbonic acid esters
– from
  2-ethylenealcohols 51, 221
3-Ethyleneepoxides
– special s.
  2-alkoxy-3-ethyleneepoxides
– startg. m. f.
  (E)-acoxy-3-ethylenes, synthesis,
    regiostereospecific 51, 290
  2-ene-1,5-diols 51, 279
cis-3-Ethyleneepoxides, chiral 51, 269
2-Ethyleneethers (s.a. Alkoxy-2-
  ethylenes)
– from
  allenyl ethers, synthesis 51, 334
2-Ethyleneethers, cyclic
– from
  ethylenealcohols 51, 110
Ethylenehalides
–, ring closure, radical (CC⋔Hal)
–, – – –, silyl radical-mediated with
  selenylation 51, 257
– special s.
  ethyleneiodides
α,β-Ethylenehalides
– from
  α,β-ethylenecarboxylic acids, with loss
    of 1 C-atom 51, 215
– special s.
  α,β-ethyleneiodides
– startg. m. f.
  vinylzirconocene halides 51, 410
β,γ-Ethylenehalides
– special s.
  allyl bromide
– startg. m. f.
  2-ethylenetosylamines, allyl shift
    51, 174
syn-3-Ethylene-2,1-halogenhydrins
– from
  aldehydes, synthesis, asym. 51, 269

Ethylene(halogeno)carboxylic acid esters
- startg. m. f.
  alcohols, tert., bicyclic **51**, 482
α,β-Ethylene-γ-hydroxyboronic – –
  18, 736s**51**
γ,δ-Ethylene-β-hydroxycarboxylic acid amides **39**, 592s**51**
γ,δ-Ethylene-α-hydroxyhydrazones, chiral **39**, 658s**51**
α,β-Ethylene(hydroxy)ketones
- special s.
  α-alkylidene-*o*-hydroxyketones
(E)-γ,δ-Ethylene-α-hydroxyketones
- from
  α-diazoketones, asym. induction **51**, 390
  2-ethylenealcohols, – – **51**, 390
α,β-Ethylene-γ'-hydroxyketones, cyclic **51**, 280
γ,δ-Ethylene-α-hydroxyphosphon-[amid]ates, chiral **39**, 658s**51**
(E)-α,β-Ethylene-γ-hydroxyphosphonic acid esters **38**, 858s**51**
(E)-α,β-Ethylene-γ-hydroxysulfoxides **38**, 858s**51**
Ethylene-*prim*-iodides
–, coupling, Ni-catalyzed with zinc compds., dialkyl- **51**, 412
(E)-α,β-Ethyleneiodides
- from
  acetylene derivs. **51**, 195
γ,δ-Ethylene-δ'-ketocarboxylic acid esters, cyclic
- special s.
  γ-alkylidene-δ-ketocarboxylic acid esters, medium-ring
Ethyleneketones
–, hydrogenation, selective, review **47**, 49s**51**
- startg. m. f.
  β,γ-ethyleneketones, cyclic **51**, 466
α,β-Ethyleneketones (s.a. α,β-Ethyleneoxo compds.)
–, activation as 1,3-dioxolanium salts, 4-vinyl- **51**, 307
–, 1,4-addition to – **51**, 304
–, aldol condensation, reductive, regiostereospecific with – **51**, 322
–, Diels-Alder reaction, asym. with – **51**, 307
- from
  aldehydes **51**, 448
  enoxysilanes, synthesis **51**, 428
  nitro compds., aliphatic **51**, 428
- special s.
  α-alkylideneketones
  bis(α,β-ethyleneketones)
  α,β-ethylene-γ-(organothio)ketones
  vinyl ketones
- startg. m. f.
  (E)-α-alkylideneketones, synthesis **51**, 322
  γ-diketones, synthesis, asym. **51**, 395
(Z)-α,β-Ethyleneketones
- by Stille coupling **39**, 887s**51**
- from
  α-diazoketones **37**, 946s**51**
α,β-Ethyleneketones, cyclic
- from

  2,3-epoxyalcohols, cyclic, with 1,2-alkyl migration **51**, 468
–, β-functionalization **51**, 280
- startg. m. f.
  3-ethylene-1-vinyllactones **51**, 397
  β'-stannyl-γ-vinyl-γ-lactols, bicyclic **51**, 397
β,γ-Ethyleneketones
- from
  carboxylic acid halides **10**, 625s**51**
β,γ-Ethyleneketones, cyclic
- from
  ethyleneketones **51**, 466
γ,δ-Ethyleneketones
- from
  diazomethyl ketones, with 3 extra C-atoms **51**, 444
–, isomerization **51**, 341
- special s.
  α-(2-alkylidenecyclopent-1-yl)ketones
  γ-aryl-γ,δ-ethyleneketones
δ,ε-Ethyleneketones
- special s.
  acoxy-δ,ε-ethyleneketones
α,β-Ethylenelactams
- special s.
  aryl-α,β-ethylenelactams
γ,δ-Ethylenelactones
- special s.
  α-allyllactones
α,β-Ethylenemercaptals, cyclic
- startg. m. f.
  vinylcyclopropanes **51**, 423
2-Ethylenemonoperoxyketals
- startg. m. f.
  epoxides, synthesis **51**, 406
(S)-(2-Ethylene)monothiolcarbonic acid esters **51**, 221
Ethylenenitrones
–, ring closures, selenylative **51**, 250
α,β-Ethylene-β-(organothio)aldehydes
- from
  acetylene derivs. **51**, 331
α,β-Ethylene-γ-(organothio)ketones
–, 1,2-addition, stereospecific to – **51**, 20
Ethyleneoxo compds.
- startg. m. f.
  alcohols, cyclic, with lactonization **51**, 313
  γ-lactones, bicyclic **51**, 315
(E)-α,β-Ethyleneoxo compds.
- from
  2-acetylenealcohols **51**, 33
α,β-Ethylenephosphonic acid esters
- special s.
  1,3-dienephosphonic acid esters
  α-methylenephosphonic – –
(E)-α,β-Ethylenephosphonic – –
- from
  dialkyl phosphites **51**, 255
  oxo compds., with 1 extra C-atom **51**, 365
  β-oxophosphonic acid esters **51**, 255
β,γ-Ethylenephosphonic – –
–, Michael addition, asym., regiospecific with – **49**, 630s**51**
2-Ethyleneselenides
- startg. m. f.
  2-ethylenetosylamines, allyl shift with asym. induction **51**, 174
5-Ethyleneselenides
- startg. m. f.

  cyclopentanes **51**, 427
α,β-Ethyleneselenolic acid esters
- from
  1,3-dioxane-4,6-diones, 5-alkylidene- **51**, 261
- special s.
  selenolcinnamic acid esters
2-Ethylenesilanes
–, 1,4-addition, intramolecular, asym. with – **38**, 911s**51**
- from
  2-ethylene-4-silylstannanes **33**, 883s**51**
- special s.
  acoxymethylethylenesilanes
  2-alkylidene-1,3-di(silanes)
  Si-allyl(hydroxy)silanes
  allyltrimethylsilane
  aryl-2-ethylenesilanes
  4-hydroxy-2-methylenesilanes
- startg. m. f.
  tert-alkoxy-3-ethylenes, synthesis, asym. **51**, 438
  1,5-dienes **51**, 437
  β,γ-epoxyketones **51**, 225
  3-ethylene-*tert*-alcohols, synthesis, asym. **51**, 438
  thiophenes **51**, 225
(E)-2-Ethylenesilanes
- from
  disilanyloxy-2-ethylenes, 1,3-chirality transfer **51**, 262
2-Ethylenesilanes, chiral **49**, 64s**51**
α,β-Ethylene-β-silylaldehydes, cyclic **51**, 332
2-Ethylenestannanes
- from
  2-ethylenealcohols **51**, 242
- special s.
  allyltri-*n*-butylstannane
- startg. m. f.
  3-ethylenealcohols, synthesis, asym., regiospecific **51**, 433
–, synthesis, asym. with – review **48**, 889s**51**
α,β-Ethylene-α-stannylphosphonic acid esters **42**, 725s**51**
α,β-Ethylenesulfones
- startg. m. f.
  2-cyclopentenone ring, asym. induction **51**, 454
  methylenecyclopentane ring, 3-sulfonyl-, asym. induction **51**, 454
β,γ-Ethylenesulfones
- from
  acoxy-2-ethylenes, asym. conversion **37**, 527s**51**
- special s.
  allyl sulfones
2-Ethylenesulfonylamines
- special s.
  2-ethylenetosylamines
α,β-Ethylene-N-sulfonylthioiminoesters
- from
  1-acetylene-1-thioethers **51**, 310
  N-sulfonylimines **51**, 310
(Z)-α,β-Ethylenesulfoxides
- by Horner synthesis **39**, 851s**51**
2-Ethylenetellurides
- as intermediates **51**, 174
α,β-Ethylenethioketones, *in situ*-generated

–, hetero-Diels-Alder reaction,
  transannular with – **51**, 353
α,β-Ethylenethiolic acid esters
– from
  1,3-dioxane-4,6-diones, 5-alkylidene-
  51, 261
2-Ethylenetitanium(IV) alkoxides
– startg. m. f.
  3-ethylenealcohols, synthesis,
  regiostereospecific **51**, 281
2-Ethylenetosylamines
– from
  β,γ-ethylenehalides, allyl shift **51**, 174
2-ethyleneselenides, – – with asym.
  induction **51**, 174
(Z)-2-Ethylene-1,1,1-trifluorides
  29, 854s**51**
1,2-Ethylene-1,1,2-trihalides
– special s.
  trichloroethylene
2-Ethyleneurethans
– special s.
  2-ene-1,4-diol carbamates
3-Ethylene-1-vinyllactones
– from
  α,β-ethyleneketones, cyclic **51**, 397
  β'-stannyl-γ-vinyl-γ-lactols, bicyclic
  51, 397
O-Ethylhydroxylamine
–, Michael addition with – **51**, 149

Ferrocenyl phosphine complexes, chiral
– special s.
  (R,R,R)-[2-(4,5-diphenyl-Δ²-oxazolin-
  2-yl)ferrocenyl]diphenylphosphine
Fluorenylmethyl chloroformate
– as reactant **51**, 166
Fluoride ion source
–, tetra-n-butylammonium (triphenylsilyl)-
  difluorosilicate as – **51**, 210
Fluorides
– from
  thioethers, ar. **51**, 212
  xanthates **51**, 205
– special s.
  carbanions, fluorinated
  carbenes, –
  heterocyclics, fluorine-containing
  ylids, fluorinated
Fluorination, selective
– with hypofluorites, review **50**, 271s**51**
Fluorine
– as reagent **48**, 210s**51** (review); **51**, 203
Fluoroalkyl radicals
–, review **41**, 463s**51**
Fluoroboric acid-dimethyl etherate
51, 307
N-Fluoro compds., electrophilic
–, review **39**, 458s**51**
N-Fluoroditriflimide
– as reagent **51**, 202
Fluoroperoxides
–, review **41**, 463s**51**
N-Fluorosulfonimides
– special s.
  N-fluoroditriflimide

Fluorous medium
–, combinatorial syntheses, liq.-phase in –,
  review **50**, 555s**51**
–, radical reactions in – **51**, 39
Fluorous two-phase medium
–, sensitized reactions with singlet oxygen
  in – **51**, 53
Formals, cyclic
–, ring opening, regiospecific, acylative
  51, 92
– startg. m. f.
  diols, differentially-protected **51**, 92
Formamidinesulfinic acid
  s. Thiourea dioxide
Formazans
– from
  aldehydes **8**, 421s**51**
Formic acid/triethylamine
– as H-donor **51**, 26, 29
Formic acid amides (s.a. N-Formylation)
Formic acid esters (s.a. O-Formylation)
– special s.
  cyanomethyl formate
α-Formoxyketones
– from
  enol carbonates **51**, 101
N-Formylamidrazones
– as intermediates **51**, 140
N-Formylation
– of α-aminocarboxylic acid esters
  51, 139
– with
  benzotriazole, 1-formyl- **51**, 157
  cyanomethyl formate **51**, 139
O-Formylation
– with benzotriazole, 1-formyl- **51**, 157
**Friedel-Crafts acylation, regiospecific,
  metal triflimide-catalyzed**
  51, 374
**Friedel-Crafts alkylation, Hf(IV)-
  catalyzed 50**, 466s**51**
o-Fries rearrangement (s.a. Thia-Fries
  rearrangement)
–, 1-indanones, 7-hydroxy- via – **51**, 346
– under microwave irradiation **51**, 407
– –, catalytic
  51, 407
**Furan-2-carboxylic acid esters,
  tetrahydro-, 3-hydroxy-**
– from
  β-benzyloxyoxo compds. **51**, 462
2(3H)-Furanones, 4,5-dihydro- s. γ-
  Lactones
2(5H)-Furanones, 5-acoxy-
–, hydrolysis, asym. **28**, 13s**51**
2(5H)-Furanones, 4-alk-1-ynyl-
– from
  acetylene derivs., terminal **51**, 333
  α,β-acetylene-γ-hydroxycarboxylic acid
  esters **51**, 333
–, 3-amino-
– from
  ethylene derivs. **51**, 111
  isoxazolidine-3-carboxylic acid esters
  51, 111
  nitrones **51**, 111
Furan ring, 2,5-dihydro-, bridged
– special s.
  [n+5]-oxabicyclo[n.2.1]alkenes
Furans, 3-α-acylamino- 51, 296
Furans, 2,5-dihydro-, meso-2,5-diacoxy-
– startg. m. f.

nucleosides, asym. conversion **51**, 153
–, –, 2-vinyl-
– from
  1,3-dioxolan-2-ones, 4-alk-1-ynyl-
  51, 465
–, tetrahydro-
– special s.
  poly(tetrahydrofurans)
–, –, 2-alkoxy- s. γ-Lactolides
–, –, chiral 20, 284s**51**
–, –, 2-α-chloro- 51, 209
–, –, 2,5-disubst. 51, 209
–, –, 2-α-hydroxy- 51, 113
–, –, 3-hydroxy-
– from
  1,3-dioxolanes, 4-alkylidene- **51**, 345
Furan-3-ylacetic acid esters
– from
  acetylene derivs., terminal **51**, 333
  α,β-acetylene-γ-hydroxycarboxylic acid
  esters **51**, 333

Gallium(III) iodide
  51, 273
Gas-solid reactions
– with nitrogen dioxide **51**, 96
Germanium hydrides, organo-
– special s.
  tris(trimethylsilyl)germane
Germylformylation 46, 672s**51**
Giese reaction s. CC⇓CC
Glycal epoxides 44, 117s**51**
Glycals
– startg. m. f.
  oligosaccharides, review **41**, 173s**51**
Glycidic acid esters
– from
  oxo compds. **51**, 388
Glycinamides
–, C-alkylation, asym. **51**, 398
– startg. m. f.
  α-aminocarboxylic acids, synthesis,
  asym. **51**, 398
Glycol ethers
– special s.
  pinacol ethers
Glycol O,O-isopropylidene derivs.
– startg. m. f.
  α-hydroxyketones **51**, 114
Glycol monoethers
– from
  epoxides **51**, 54
  ketones **50**, 531s**51**
– special s.
  ene-1,2-diol monoethers
Glycol monotosylates, chiral
  47, 32s**51**
Glycols (s.a. Dihydroxylation)
–, desymmetrization by O-acylation **51**, 83
– from
  aldehydes, with 1 extra C-atom **51**, 434
  epoxides **51**, 54
– (enzymatic conversion), review
  48, 108s**51**
– startg. m. f.
  α-hydroxyketones **51**, 114

Glycols, cyclic
– from
  dioxo compds. **51**, 286
***threo*-Glycols, sym.**
– from
  aldehydes **51**, 283
Glycol sulfates, cyclic
– special s.
  alkoxyglycol sulfates, cyclic
– startg. m. f.
  cyclopropanes **51**, 367
  1,2-diamines, stereospecific conversion **51**, 138
  3,3-dihalogenalcohols, with 1 extra C-atom **51**, 421
  ethylene derivs. **49**, 938s**51**
Glycoluril cage compds.
–, review **21**, 786s**51**
Glycosidation, enzymatic, polymer-based 50, 89s**51**
–, iterative **47**, 178s**51**
N-Glycosidation, intramolecular 48, 434s**51**
Glycosides
– from
  aldoses **44**, 166s**51**
  pyran-2-yl ethers, tetrahydro- **34**, 209s**51**
– special s.
  acyl glycosides
  tetrazolyl –
  thioglycosides
C-Glycosides
– special s.
  C-acylglycoside...
  C-allyl glycosides
  C-aryl –
Glycosyl fluorides
– from
  1H-tetrazol-5-yl glycosides **51**, 208
N-Glycosyl-N-heterocyclics (s.a. Nucleosides)
– from
  aldoses, unprotected **51**, 177
  2-pyridyl glycosides, 3-methoxy- **51**, 177
  siloxy-N-heterocyclics **51**, 177
β-Glycosyloxycarboxylic acids
– special s.
  amino-β-glycosyloxycarboxylic acids
Glyoxylic acid esters
–, hetero-Diels-Alder reaction, asym. with – **46**, 662s**51**
Gold complexes
  (triethylphosphine)gold(I) chloride **51**, 237
Grignard compds. (s.a. Magnesium halides, organo-)
– startg. m. f.
  carboxylic acid amides **51**, 297
  – – thioamides **51**, 297
  cyclopropanes, *tert*-amino- **51**, 377
  dithiophosphonic acid O-monoesters **51**, 247
  thiolic acid esters **51**, 297
  – acids **51**, 297
Guanidines
– from
  amines **51**, 158
Guanidinium azides
– special s.
  tetramethylguanidinium azide

Hafnium(IV) triflate 51, 379, 407
Halides (s.a. Halogenation)
– from
  alcohols **51**, 207
– special s.
  acetylenehalides
  benzyl halides
  deuteriohalides
  dihalides
  epoxyhalides
  ethylenehalides
  fluorides
  iodides
  siloxyhalides
– startg. m. f.
  alcohols **51**, 91
  aldehydes, with 1 extra C-atom **51**, 424
  diselenides, sym. **51**, 252
  hydroperoxides **51**, 93
  ketones **51**, 424
  mercaptans **51**, 229
  phosphonic acid benzyl esters **51**, 260
  tellurides (from 2 different molecules) **51**, 245
  thiolic acid esters **51**, 229
Halides, ar.
–, cross-coupling with aryl triflates **51**, 415
– startg. m. f.
  amines, ar. **51**, 171
  arylboronic acid esters **51**, 259
  4-aryl-2-ethylenesilanes **51**, 455
  arylstannanes **51**, 248
  benzocyclobutenes **51**, 414
  diaryls (from chlorides) **51**, 453
  diaryl sulfones **51**, 228
  styrenes **51**, 394
  triarylamines **51**, 170
α-Halogenacetals
– special s.
  O-allyl O-silyl α-halogenacetals
***o*-Halogenacylamines**
– startg. m. f.
  indoles, 1-acyl- **51**, 464
  indolines, 1-acyl-2-hydroxy- **51**, 464
Halogenalcohols s. Halogenhydrins
Halogenallenes
– special s.
  iodoallenes
3-Halogenamines
– special s.
  N-benzyl-3-halogenamines
Halogenation
– special s.
  bromination
  fluorination
Halogenation, ar.
– special s.
  iodination, ar.
–, –, oxidative **51**, 198
(E)-2-Halogeneneselenides 43, 625s**51**
1-α-Halogenethers, cyclic
–, ring expansion **51**, 77
2-Halogenethers, cyclic (s.a. Iodoetherification, intramolecular)
1,2-Halogenhydrins
– special s.
  2,2-dihalogenalcohols
  3-ethylene-2,1-halogenhydrins
1,3-Halogenhydrins

– special s.
  3,3-dihalogenalcohols
1,4-Halogenhydrins
– special s.
  methylene-1,4-halogenhydrins
Halogenimidium salts
– special s.
  halogenoformiminium halides
α-Halogenocarboxylic acid amides
– startg. m. f.
  2,4-oxazolidiones **51**, 169
Halogenocarboxylic acid esters
– special s.
  ethylene(halogeno)carboxylic acid esters
α-Halogenocarboxylic acid esters
– special s.
  α-halogenocarboxylic acid phenyl esters
– startg. m. f.
  β-aminocarboxylic acid esters, asym. induction **51**, 293
α-Halogenocarboxylic acid phenyl esters
– startg. m. f.
  α-halogeno-β-lactones **51**, 368
Halogenoformamidinium salts
– special s.
  tetramethylfluoroformamidinium ...
Halogenoformic acid esters
– special s.
  fluorenylmethyl chloroformate
Halogenoformiminium halides
– special s.
  dimethylchloroformiminium chloride
Halogeno-N-heterocyclics
– from
  ethyleneamines **48**, 434s**51**
α-Halogenoketones 37, 806s**51**
– special s.
  α,α-dihalogenoketones
– startg. m. f.
  carboxylic acids **51**, 102
  2-ketoselenides **51**, 251
1-Halogeno-2-ketothioethers
– special s.
  1,1-dihalogeno-2-ketothioethers
Halogenolactones
– special s.
  dihalogenolactones
2-Halogenolactones
– from
  ethylenecarboxylic acids **51**, 199
Halogeno-β-lactones
– special s.
  dihalogeno-β-lactones
α-Halogeno-β-lactones
– from
  α-halogenocarboxylic acid phenyl esters **51**, 368
Halogenolactonization (HalC↓H)
– special s.
  bromolactonization
  iodolactonization
Halogenophenols
– special s.
  iodophenols
Halogenophosphines
– special s.
  dihalogenophosphines
– startg. m. f.
  phosphines, tert. **51**, 249

Halogenosilanes
– special s.
   chlorosilanes
   trimethylsilyl halides
N-Halogeno-N-sodiosulfonic acid amides
– special s.
   Chloramine-T
Halogenostannanes
– special s.
   di-n-butyl(chloro)stannane
N-Halogenosulfonic acid amides
– special s.
   N,N-dihalogenosulfonic acid amides
   N-halogeno-N-sodiosulfonic – –
– – **imides**
– special s.
   N-fluorosulfonimides
2-Halogenothioenolethers
– special s.
   2,2-dihalogenothioenolethers
Heck arylation, intramolecular, polymer-based
–, macrocyclization by – **51**, 486
Heck reactions
– with palladacyclics as catalyst **51**, 416
Heck-type arylation
– with diaryl tellurides **51**, 456
p-**Hemiquinols**
– startg. m. f.
   o-acetonylphenols **51**, 459
Hetarylstannanes 51, 248
Heterocyclics, benzo-condensed 51, 417
–, **fluorine-containing**
–, synthesis, review **42**, 814s**51**
N-Heterocyclics
– from
   iminoesters, review **29**, 744s**51**
– special s.
   N-glycosyl-N-heterocyclics
   N-macroheterocyclics
   siloxy-N-heterocyclics
O-Heterocyclics
– special s.
   arylseleno-O-heterocyclics
   hydroxy-O-heterocyclics
N,O-Heterocyclics
– special s.
   N,O-macroheterocyclics
Hetero-Diels-Alder reaction
– with nitroso compds., review **41**, 305s**51**
– –, asym. **36**, 667s**51**
– with glyoxylic acid esters **46**, 662s**51**
– –, **transannular**
– with α,β-ethylenethioketones, *in situ* generated **51**, 353
Hexabutyldistannane
– as reagent **51**, 323, 355, 424
Hexadecyltrimethylammonium bromide
– as reagent **51**, 170, 222
1,1,1,3,3,3-Hexafluoro-2-propanol
– as reagent **51**, 7
Hexamethyldisilazane
– as reagent **51**, 254
Hexamethyldistannane
– as reagent **51**, 415
High-pressure coupling, Pd-catalyzed 51, 413
– cycloaddition, regiospecific, transition metal-catalyzed **51**, 457
– [3+2]-cycloaddition

– with trimethylenemethane equivalents **51**, 457
Hofmann elimination
–, regeneration of polymer support by – **51**, 172
Homoallyl... s.a. 3-Ethylene..., β,γ-Ethylene...
1,1-(Homoallyloxy)silanols
– from
   Si-allyl-1,1-hydroxysilanes **51**, 354
Homopeptides 2, 216s**51**
Horner synthesis (s.a. CC⇓Rem, 1,4-Addition-Horner synthesis)
– with resolution, kinetic **48**, 866s**51**
– –, **polymer-based 39**, 854s**51**
Horner-type synthesis, asym.
–, review **48**, 866s**51**
Hydantoins
– special s.
   1,3-dibromo-5,5-dimethylhydantoin
– startg. m. f.
   α-ureidocarboxylic acids, asym. hydrolysis **51**, 59
–, synthesis, polymer-based **31**, 452s**51**
Hydrazine
– as reagent **51**, 245
Hydrazines
–, reactions with α,β-unsatd. and β-dicarbonyl compds., review **2**, 368s**51**
– startg. m. f.
   hydroxyhydrazones **51**, 122
α-**Hydrazinocarboxylic acid amides 48**, 457s**51**
Hydrazones
–, cleavage **34**, 172s**51**
– special s.
   N-aziridin-1-ylimin...
   hydroxyhydrazones
   sulfonylhydrazones
   N-(thionocarbalkoxy)hydrazones
–, α-**functionalized, chiral 31**, 812s**51**
Hydroacylation, intramolecular
– of ethylene derivs., electron-deficient **51**, 342
Hydroalumination, stereospecific
– of acetylene derivs. **51**, 405
Hydroboration, regiospecific
– with pinacolborane **51**, 238
Hydrocarbons, hydrocarbon groups
–, aminomethylation, catalyzed, review **23**, 757s**51**
–, carboxylation, –, – **23**, 757s**51**
– special s.
   arenes
   diarylalkanes
– startg. m. f.
   ketones **51**, 73
–, trifluoroacetoxylation, regiospecific **51**, 75
Hydroformylation s. CC⇓CC
Hydrogenation s.a. HC⇓CC
Hydrogenation, asym., homogeneous
– in carbon dioxide, supercritical **51**, 31
– of 1-arylenacylamines **51**, 31
– –, **Noyori-type**
–, review **42**, 45s**51**
–, **catalytic**
– with retention of benzyl ethers **51**, 32
–, **homogeneous**
– of oxo compds., unsatd. **51**, 27

–, **ruthenium-catalyzed**
–, review **38**, 39s**51**
–, **selective**
– under supramolecular catalysis **51**, 30
Hydrogen atom transfer, intramolecular
– from silicon hydrides, organo-, chain transfer via – **51**, 323
Hydrogen bromide 51, 201
– –/**hydrogen peroxide 51**, 199
– **cyanide 51**, 294
– **fluoride/pyridine 51**, 208, 212
Hydrogenolysis, Pd-catalyzed
– of 2-acetylene- and 2-ethylene-alcohol O-derivs., review **34**, 55s**51**
Hydrogen peroxide 51, 107
Hydrolysis (s.a. HO⇊C)
Hydrolysis, asym.
– with enzymes, cross-linked **51**, 6
–, **asym.-Meinwald-type rearrangement 51**, 347
Hydroperoxides
– from
   dialkylzinc compds. **51**, 93
   halides **51**, 93
–, resolution, kinetic **51**, 498
– special s.
   tert-butyl hydroperoxide
α-**Hydroperoxycarboxylic acid esters**
–, resolution, kinetic **51**, 498
Hydrosilylation
– of
   ethylene derivs. s. RemC⇓CC
   lactones **51**, 25
Hydrosilylation, asym.
– of diketones **51**, 21
– with Pd-binap complexes, review **47**, 542s**51**
–, –, **iridium-catalyzed**
– of ketones **51**, 24
–, **catalytic**
– of carboxylic acid esters **51**, 36
Hydrostannylation
– of azomethines **51**, 126
Hydrovinylation, regiospecific 51, 326
Hydroxamic acid allyl esters
–, ring closures, selenylative **51**, 250
Hydroxamic acid esters
– as intermediates **51**, 371
– from
   carboxylic acid esters **51**, 142
– startg. m. f.
   acylamines **51**, 9
   isoxazoles **51**, 386
Hydroxamic acids
– startg. m. f.
   acylamines **51**, 9
– –, **O-acyl-** s. Acyl hydroxamates
Hydroximinohalides
– special s.
   azidohydroxyiminohalides
   cyanohydroxyiminohalides
β-**Hydroxyacylsilanes**
– startg. m. f.
   1,3-diols **51**, 492
β-**Hydroxyaldehydes**
– startg. m. f.
   2*H*-1,3-oxazines, 3,6-dihydro-, 3-tosyl- **51**, 173
Hydroxyalkoximes
– from
   alkoxylamines **51**, 122

enolethers, cyclic **51**, 122
Hydroxyamines s. Aminoalcohols
Hydroxyazomethines
- special s.
  (alkylideneamino)alcohols
β-Hydroxycarbonyl compds. (s.a. Aldol...)
- from
  acetals, cyclic, asym. synthesis
  40, 635s51
anti-**β-Hydroxycarboxylic acid esters, chiral**
  51, 285
syn-**β-Hydroxycarboxylic – –, –**
  51, 275
γ-Hydroxycarboxylic – –
- special s.
  acetylene-γ-hydroxycarboxylic acid esters
δ-Hydroxycarboxylic – –
- special s.
  acetylene-δ-hydroxycarboxylic acid esters
  allene-δ-hydroxycarboxylic – –
α-Hydroxycarboxylic acids
- special s.
  mandelic acids
– –, α-**subst.**
–, synthesis, asym., enzymatic, review **28**, 13s51
N-Hydroxydicarboxylic acid imides
- special s.
  N-hydroxyphthalimides
2-α-Hydroxy-1,3(4)-dienes 39, 887s51
trans-**4-Hydroxyenesilanes 36**, 879s51
2-α-Hydroxy-O-heterocyclics
- from
  alkoxyglycol sulfates, cyclic **51**, 113
Hydroxyhydrazones
- from
  enolethers, cyclic **51**, 122
α-Hydroxyketene mercaptals
- from
  dithioacetic acid esters **51**, 396
Hydroxyketones
- special s.
  ethylene(hydroxy)ketones
α-Hydroxyketones
- from
  aldehydes **51**, 418
  enols, O-protected **51**, 99
  glycol O,O-isopropylidene derivs.
  51, 114
  glycols **51**, 114
  Δ³-oxazolines, 2-alkoxy- **51**, 418
  1-sulfonyl-1-isonitriles **51**, 418
- special s.
  ethylene-α-hydroxyketones
- startg. m. f.
  carboxylic acids, C-cleavage **51**, 105
β-Hydroxyketones (s.a. Aldol...)
- from
  2,3-epoxyalcohols **51**, 1
  2-siloxy-1,3-dienes, synthesis **51**, 319
γ-Hydroxyketones
- from
  1,3-acoxyiodides **49**, 955s51
- special s.
  ethylene-γ-hydroxyketones
N-*o*-Hydroxylactams
- from
  azidoalcohols **51**, 159

ketones, cyclic **51**, 159
Hydroxylamines
- from
  nitro compds. **49**, 21s51
–, reactions with α,β-unsatd. and β-dicarbonyl compds., review **2**, 368s51
Hydroxylamines, bis-protected
–, review **44**, 7s51
–, **O-phosphinyl-** s. Phosphinyloxylamines
Hydroxylaminoenynes
- startg. m. f.
  oxazolidines, tricyclic **51**, 336
Hydroxylation, nickel-catalyzed
- of β-dicarbonyl compds. **51**, 52
α-Hydroxymalonic acids
- startg. m. f.
  carboxylic acids, bis(decarbonylation) **51**, 106
α-Hydroxymercaptals
- startg. m. f.
  acetylene derivs. **51**, 487
  ene-1,2-di(thioethers) **51**, 487
4-Hydroxy-2-methylenesilanes 40, 567s51
δ-Hydroxyoximes
- special s.
  5-hydroxypentanal oxime
5-Hydroxypentanal oxime
- as reactant **51**, 352
β-*tert*-Hydroxyphosphine oxides
- as intermediates **51**, 265
α-Hydroxyphosphonic acid diamides, chiral 49, 510s51
N-Hydroxyphthalimide
- as reagent **51**, 76
2-Hydroxyselenides
- startg. m. f.
  selenides, synthesis **51**, 441
threo-**2-Hydroxyselenides**
- from
  2-ketoselenides **9**, 61s51
1,1-Hydroxysilanes
- special s.
  allyl-1,1-hydroxysilanes
β-Hydroxysulfones
- from
  ketones **51**, 489
- startg. m. f.
  ethylene derivs. **51**, 489
o-**Hydroxysulfoxides**
- from
  sulfinic acid aryl esters **51**, 220
p-**Hydroxysulfoxides**
- startg. m. f.
  p-quinones **51**, 95
1,1-Hydroxythioethers
–, O-acylation, asym., with resolution, kinetic, dynamic **51**, 233
α-Hydroxytrithioorthocarboxylic acid esters
- startg. m. f.
  1,1-difluoro-2-ketothioethers **51**, 213
Hydrozincation
- of ethylene derivs. **51**, 60
–, **regiostereospecific**
- of acetylene derivs. **51**, 195
Hydrozirconation, regiospecific
- of amines, prim. via – **51**, 129
–, **stereospecific**
- of acetylene derivs. **51**, 405

Hypofluorites
–, fluorination, selective with –, review **50**, 271s51
Hypofluorous acid/acetonitrile 51, 99
Hypohalites s. Alkyl hypohalites, Hypofluorites
Hyponitrous acid esters
- special s.
  di-*tert*-butyl hyponitrite
Hypophosphorous acid
- as reactant **51**, 240

Imidazole libraries
–, synthesis **23**, 423s51
Imidazoles, 1-acyl-
- startg. m. f.
  acylsilanes **51**, 244
  alcohols, prim. **37**, 57s51
–, **4-acyl- 48**, 681s51
–, **1-carbalkoxy-**
- startg. m. f.
  enol carbonates **51**, 100
4-Imidazolidones, bridged
- from
  chromium [*o*-(alkylideneamino)alkyl]-aminocarbene complexes **51**, 426
–, **3-(α,β-ethyleneacyl)-**
–, 1,4-addition, asym. to – **37**, 657s51
–, **3-glycyl-**
–, C-alkylation, asym. **51**, 398
Imides s. Dicarboxylic acid imides
O-Imidocarbamates
–, prepn. **34**, 330s51
Imidoylstannanes
- from
  iminohalides **51**, 253
Imines (s.a. Azomethines)
- special s.
  silylimines
  sulfonylimines
1,1'-Iminodicarboxylic acid derivs.
–, 4-component synthesis, with asym. induction **51**, 359
Iminoesters
- from
  benzotriazoles, 1-imidoyl- **51**, 85
- special s.
  trichloroacetimidates
- startg. m. f.
  N-heterocycles, 5- and 6-membered, review **29**, 744s51
  tetrazoles **51**, 140
Iminoester salts, bicyclic
- as intermediates **51**, 159
- startg. m. f.
  lactams, functionalized **51**, 264
Iminohalide equivalents
–, benzotriazoles, 1-imidoyl- as – **51**, 85
Iminohalides
- startg. m. f.
  imidoylstannanes **51**, 253
Iminoiodinanes
- special s.
  sulfonyliminoiodinanes
Iminophosphoranes (s.a. Phosphine imines)

Iminyl radicals
– from
 N-(benzotriazol-1-yl)imines **51**, 189
 N-(thionocarbalkoxy)hydrazones
  51, 189
Immonium salts
– special s.
 azomethinium salts
 halogenimidium salts
1-Indanones, 7-hydroxy-51, 346
Indans
– from
 1,2,4,10-tetraen-6-ynes **51**, 312
– special s.
 polyindans
Indium 51, 481
– **(III) chloride 51**, 33
Indoles
– special s.
 biindoles
– startg. m. f.
 indolo[2,3-*a*]carbazoles (from 2 molecules) **51**, 389
– via styryl carbenes **51**, 112
Indoles, 1-acyl-
– from
 o-halogenacylamines **51**, 464
 indolines, 1-acyl-2-hydroxy- **51**, 464
–, **1-alkylthio-**
– as intermediates **51**, 144
–, **2-amino-**
– from
 N-arylcarboxylic acid hydrazides
  51, 473
Indolines, 1-acyl-2-hydroxy-
– as intermediates **51**, 464
Indolines, 7-aryl- 24, 839s51
–, **3-hydroxy- 46**, 956s51
–, **5-hydroxy-1-sulfonyl- 44**, 624s51
–, **3-imino-, chiral 35**, 496s51
Indolizines, hexahydro- 51, 340
Indolo[2,3-*a*]carbazole ring 51, 389
Initiator
–, di-*tert*-butyl hyponitrite as – **51**, 37
Interchange
– of ethylene derivs. s. under Metathesis
Iodides
– startg. m. f.
 alcohols **51**, 93
Iodides, ar.
– startg. m. f.
 diaryliodonium salts **51**, 204
Iodinanes
– special s.
 iminoiodinanes
Iodination, ar. 51, 203
***m*-Iodination 51**, 203
Iodine
– as reactant **51**, 214
– as reagent **51**, 18, 70, 197
Iodine compds., organo-, polyvalent
–, review **48**, 488s51
Iodine monochloride-pyridine 51, 203
Iodine(III) reagents
–, rearrangements, oxidative with –, review **27**, 162s51
Iodoallenes
– startg. m. f.
 erythro-3-acetylenealcohols, terminal
  51, 267
Iododifluorides
– special s.
 p-methyliodobenzene difluoride

Iodoetherification, intramolecular-lactonization, regiostereospecific
  51, 211
Iodolactonization, asym., polymer-based 39, 467s51
Iodonium salts
– special s.
 acetyleneiodonium salts
 diaryliodonium –
***o*-Iodophenols**
– from
 2-cyclohexenones **51**, 197
Iodosocarboxylates
– special s.
 phenyl iodosoacetate
N-Iodosuccinimide
– as reagent **51**, 196
Iridium complexes
 chloro(cyclooctadiene)iridium(I) dimer
  51, 24
Iridium trichloride 51, 119
Iron η³-(acoxymethyl)allyl complexes
–, syntheses via – **51**, 290
Iron complexes
 1,1'-bis(diphenylphosphino)ferrocene
  51, 10
**Isocarbostyril ring
 33**, 694s51
**Isocarbostyrils
 25**, 693s51
–, synthesis, polymer-based, multistep
  51, 486
Isocoumarins 35, 85s51; **48**, 839s51
Isocyanates
– from
 amines, prim. **51**, 133
 urethans **51**, 185
– startg. m. f.
 dicarboxylic acid imides **51**, 179
Isocyanides s. Isonitriles
1-Isocyanocyclohexene
– as reactant **51**, 380
Isocyclics (s.a. Arenes, Cyclo...)
– special s.
 spiroisocyclics
Isoindole ring
– from
 nitro compds., ar. **44**, 736s51
Isoindoles
–, chemistry, review **44**, 936s51
***1H*-Isoindol-1-ones, 2,3-dihydro-** s. Phthalimidines
Isomerization s. Cycloisomerization, Rearrangement
Isonitriles
– special s.
 enisonitriles
 sulfonylisonitriles
 1,1,3,3-tetramethylbutyl isocyanide
Isopropanol
– as H-donor **51**, 26
Isoquinolines, 1,2,3,4-tetrahydro-, 2-carbalkoxy-, 1-subst. 18, 744s51
1(2*H*)-Isoquinolones s. Isocarbostyrils
Isothiocyanates
– special s.
 enisothiocyanates
– startg. m. f.
 dicarboxylic acid imides **51**, 179
 thionourethans **49**, 105s51
Isoureas
–, chemistry, review **43**, 425s51

Isoxazoles
– from
 hydroxamic acid esters **51**, 386
 ketoximes **51**, 386
–, synthesis, polymer-based **2**, 368s51
Isoxazolidine-3-carboxylic acid esters
– startg. m. f.
 2(5*H*)-furanones, 3-amino- **51**, 111
Isoxazolidine ring, N-unsubst.
– from
 ethylenealdehydes **51**, 352
Isoxazolidines
– from
 ethylene derivs., asym. conversion
  51, 317
 nitrones, – – **51**, 317
– startg. m. f.
 1,3-oxazines, tetrahydro- **51**, 119
 2*H*-1,3-oxazin-2-ones – **51**, 119
Isoxazolidines, 4-arylseleno-
– from
 O-allyloximes **51**, 250
–, **chiral**
– by cycloaddition, review **51**, 317s51
Δ²-Isoxazolines
–, resolution **5**, 666s51

Ketal Claisen rearrangement, reductive, regiospecific
–, 4-ethylenealcohols via – **51**, 349
Ketals (s.a. Acetals, Peroxyketals)
– special s.
 azidoketals
 spiroketals
Ketene
–, review **47**, 623s51
Ketene acetals
–, review **47**, 623s51
– special s.
 dicyanoketene ethylene acetal
Ketene mercaptals
– special s.
 hydroxyketene mercaptals
 ketoketene –
Ketimines (s.a. Azomethines)
– startg. m. f.
 β-aminoketones **51**, 291
 α,α-difluoroketones **51**, 202
3-Ketoammonium salts, cyclic
– special s.
 piperidinium triflate, 1-dodecyl-1-methyl-4-oxo-
α-Ketocarboxylic acid esters
– from
 alkoxyacetylenes **51**, 67
β-Ketocarboxylic – –
– startg. m. f.
 α,β-ethylenecarboxylic acid esters
  51, 44
(β-Ketocarboxylic acid esters)
–, transesterification, heterogeneous
  51, 104
δ-Ketocarboxylic – –
– special s.
 ethylene-δ-ketocarboxylic acid esters

α-**Ketocarboxylic acid halides**
- startg. m. f.
  ketones, N-heterocyclic **51**, 458
α-**Ketoketene mercaptals**
- startg. m. f.
  ketones **51**, 41
  3-ketothioenolethers **51**, 41
  3-ketothioethers **51**, 41
Ketones (s.a. Carbonyl compds., Oxo compds.)
- from
  acetals, synthesis **51**, 366
  acetylene derivs. **38**, 111s**51**
  α-acoxyketones **51**, 34
  alcohols, sec. s. under Oxo compds. and OCfH
  aldehydes **51**, 448
  benzotriazoles, 1-α-alkoxy-, synthesis **51**, 384
  O-benzyl-1-sulfonyloximes **51**, 424
  carboxylic acid esters, synthesis **51**, 371
  enoxysilanes, – (of chiral compds.) **51**, 441
  ethers **51**, 115
  halides **51**, 424
  hydrocarbons **51**, 73
  2-hydroxythioethers, rearrangement **39**, 965s**51**
  α-ketoketene mercaptals **51**, 41
  oxo compds., synthesis, regiospecific **51**, 384
  tosylhydrazones **51**, 89
- special s.
  acetone
  acetyleneketones
  acoxyketones
  acylaminoketones
  aminoketones
  aryl ketones
  arylketones
  cyclopentenyl ketones
  diazoketones
  diketones
  epoxyketones
  ethyleneketones
  halogenoketones
  hydroxyketones
  methyl ketones
  nitroketones
  (organothio)ketones
  silylketones
- startg. m. f.
  tert-alkoxy-3-ethylenes, synthesis, asym. **51**, 438
  3-ene-1,2-diol 2-monoethers, – **51**, 385
  2-ene-1,5-diols **51**, 279
  3-ethylene-tert-alcohols, synthesis, asym. **51**, 438
  ethylene derivs., trisubst. **51**, 489
  β-hydroxysulfones **51**, 489
-, synthesis, polymer-based **37**, 806s**51**
Ketones, ar. s. Aryl ketones
-, **chiral (C₂-symmetric)**
- as reagent **51**, 65
-, **cyclic**
- from
  bis(organozinc halides) **51**, 404
-, ring expansion **51**, 384
-, – –, Barbier-type **51**, 481
- startg. m. f.
  alcohols, tert., cyclic, diastereoselectivity, review **37**, 623s**51**

(E)-ethylenecarboxylic acid esters **48**, 751s**51**
N-o-hydroxylactams **51**, 159
-, **N-heterocyclic**
- from
  stannanes, N-heterocyclic **51**, 458
-, **sym.**
- from
  zinc halides, organo- **51**, 404
-, **sym., functionalized 51**, 404
α-**Ketonitriles** s. Acylcyanides
β-**Ketophosphine oxides**
- startg. m. f.
  ethylene derivs., trisubst., synthesis, stereospecific **51**, 265
  β-tert-hydroxyphosphine oxides, –, – **51**, 265
Ketophosphinic acid esters
-, reduction, asym. **43**, 45s**51**
β-**Ketophosphonic – –, cyclic**
42, 176s**51**
δ-**Ketophosphonic – –, – 39**, 630s**51**
2-Ketoselenides
- from
  α-halogenoketones **51**, 251
4-Ketosugars
- special s.
  3-deoxy-4-ketosugars
β-**Ketosulfones**
- special s.
  acetylene-β-ketosulfones
β-**Ketosulfoxides**
- startg. m. f.
  allenes **51**, 488
3-Ketothioenolethers
- from
  α-ketoketene mercaptals **51**, 41
2-Ketothioethers
- special s.
  halogeno-2-ketothioethers
3-Ketothioethers
- from
  α-ketoketene mercaptals **51**, 41
4-Ketothioethers
- special s.
  γ-(organothio)ketones
β-**Ketothionophosphorus(V) compds.**
- from
  α-mercaptoketones **51**, 246
α-**Ketotrithioorthocarboxylic acid esters**
- from
  carboxylic acid esters, with 1 extra C-atom **51**, 364
Ketoximes (s.a. Oximes)
- startg. m. f.
  isoxazoles **51**, 386
Kharasch acoxylation, asym. 51, 72
Knoevenagel condensation-intramolecular ene reaction, polymer-based 51, 375
Kost reaction 51, 473
Kulinkovich reaction, intramolecular, stereospecific 51, 471
- – –, **modified**
  51, 378

Labelled compds.
- special s.
  compounds, radio-labelled

Lactam ring
- from
  dicarboxylic acid imide ring **51**, 192
Lactams
- from
  enoxysilanes, cyclic **51**, 191
  ethylenecarboxylic acid esters **48**, 305s**51**
  1,1-siloxyazides, cyclic **51**, 191
- special s.
  acyllactams
  ethylenelactams
  hydroxylactams
-, synthesis, asym. with Rh(II)-carboxamidates, chiral, review **47**, 955s**51**
Lactams, bicyclic
- by Schmidt reaction **51**, 193
-, **functionalized**
- from
  iminoester salts, bicyclic **51**, 264
-, **macrocyclic** s. Macrolactamization
β-**Lactams** s. 2-Azetidinones
γ-**Lactolides**
- special s.
  arylseleno-γ-lactolides
Lactols
- from
  lactones via hydrosilylation **51**, 25
γ-**Lactols**
- special s.
  vinyl-γ-lactols
Lactones
- special s.
  alkylidenelactones
  (arylseleno)lactones
  ethylenelactones
  halogenolacton...
  vinyllactones
- startg. m. f.
  ethers, cyclic **51**, 36
  lactols via hydrosilylation **51**, 25
-, synthesis, asym. **31**, 812s**51**
-, –, – with Rh(II)-carboxamidates, chiral, review **47**, 955s**51**
β-**Lactones**
- special s.
  halogeno-β-lactones
γ-**Lactones**
- from
  acoxy-3-ethylenes **51**, 470
  2-ethylenealcohols, carbonylation **51**, 325
  γ,δ-ethylenecarboxylic acids **51**, 70
- special s.
  alkylidene-γ-lactones
γ-**Lactones, bicyclic**
- from
  ethyleneoxo compds. **51**, 315
-, **γ,γ-disubst. 51**, 70
δ-**Lactones**
- special s.
  epoxy-δ-lactones
Lactonization s. Iodoetherification, intramolecular-lactonization
Lanthanum triisopropoxide/(R)-1,1'-bi-2-naphthol/potassium bis(trimethylsilyl)amide
51, 236

Lanthanum (III) trisodium tris[(R)-1,1'-bi-2-naphthoxide]
51, 309
Lead compds., organo-
– special s.
plumbanes
Lead tetraacetate
– as reagent 51, 397
Lewis acid catalysis
– in aq. media, review 49, 882s51
Lewis acids
bis[2-(o-hydroxyphenyl)-Δ^2-oxazolinato]-titanium and –zirconium bistriflate 33, 879s51
magnesium phenyl(tetraisopropoxy)titanate, bromo- 51, 282
metal triflimides 51, 78, 374
tin(IV) halides, organo- 51, 484
Lewis acids, strong
yttrium-zirconium-sulfur 51, 311
– –, super-
diisopropoxytitanium bis(triflimide) 51, 81
Lewis bases, internal
– as directing group 51, 372
Libraries of compds. s. Compound libraries
Lipopeptides 51, 5
Lithiation (s.a. Metalation)
–, asym., regiospecific
–, review 48, 794s51
–, reductive
– of
selenides 51, 427
selenoacetals 51, 427
– with arenide reagents, review 46, 811s51
Lithium/arenes
–, lithiation, reductive with –, review 46, 811s51
Lithium/4,4'-di-tert-butylbiphenyl
51, 362, 427, 467
– /naphthalene
51, 487
– acetylides
– startg. m. f.
ethylene derivs., synthesis, regiostereospecific 51, 302
– aluminum bis[(R,R)-1,1'-bi-2-naphthoxide] 51, 309
– amides
– bis(trimethylsilyl)amide 51, 44, 154 (reactant), 399, 479
– dicyclohexylamide 51, 269
– N-isopropylcyclohexylamide 51, 368
– –, chiral
– as reagent 51, 400
– bromide (s.a. under Potassium alkoxides) 51, 218
– chloride
51, 167, 398, 415, 434
– compds., organo- (s.a. under Carbolithiation, Lithiation, Metalation)
sec-butyllithium 51, 357, 478
–/potassium tert-butoxide 51, 366
tert-butyllithium 51, 112, 398
isopropyllithium 51, 343
– special s.
allyllithium compds.
aminoalkyllithium –
benzotriazoles, 1-α-lithiated
benzyllithium compds.

– enolates
– from
α-iodoketones 48, 616s51
– hydride (s.a. Nickel acetate/lithium hydride, Zinc iodide/– –)
– hydride complexes
–, reductions with – 51, 40
– hydridoborates, organo-
– hydridotri-sec-butylborate 51, 20
– hydridotriethylborate 51, 217
– N-(p-methoxyphenyl)acetimidide
– as base 51, 184
– perchlorate 51, 226,474

Macrocyclic compound libraries
51, 486
Macrocyclization
– by Heck arylation, intramolecular, polymer-based 51, 486
N- and N,O-Macroheterocyclics
–, Mannich synthesis, review 28, 765s51
Macrolactamization, regiospecific
– of tetraaminocarboxylic acid esters 51, 187
Magnesium 51, 248
–/ethanol 51, 489
–/methanol 51, 141
– alkoxides, organo-
– special s.
vinylmagnesium alkoxides
– bromide 51, 436
– halides, organo- (s.a. under Grignard compds. and Titanium(IV) alkoxides)
allylmagnesium bromide 51, 302
cyclohexylmagnesium chloride 51, 378
ethylmagnesium bromide 51, 480
isopropylmagnesium – 51, 316
– chloride 51, 6, 142
– special s.
vinylmagnesium bromides
– phenyl(tetraisopropoxy)titanate, bromo-
– as Lewis acid 51, 282
Malonic acid esters
– special s.
acetylenemalonic acid esters
acylaminomalonic – –
Malonic acids
– special s.
hydroxymalonic acids
Malononitriles
– startg. m. f.
nitriles 51, 43
Mandelic acids
– startg. m. f.
aryloxo compds. 23, 296s51
Manganese (s.a. under Chromium(II) chloride)
Manganese(II) acetate 51, 215
–(III) acetate
51, 466
–, radical ring closures, oxidative with –, review 46, 930s51
– complexes
acetyltetracarbonyl(triphenylphosphine)manganese 51, 36

nitridomanganese(V) salen complex 51, 178
β-oxoaldiminatomanganese(III) complexes, chiral 51, 48
–, oxidation, catalytic, biomimetic with –, review 48, 134s51
– dioxide 51, 95, 194
–(II) iodide 51, 303
–(III) salen complexes, chiral, polymer-based 46, 106s51
Mannich reaction
–, N-macroheterocyclics by –, review 28, 765s51
– –, asym., regiospecific
– via (Z)-2'-silylenoxysilanes 51, 440
Mannich-type reaction, regiostereospecific
– with
azomethinium salts 51, 291
ketimines 51, 291
Meinwald-type rearrangement
(s.a. Hydrolysis, asym.-Meinwald-type rearrangement)
Mercaptals
–, activation, chelation-controlled, review 48, 850s51
– from
acetals 51, 226
enolethers 8, 651s51
oxo compds. 51, 226
– special s.
acoxymercpatals
ethylenemercaptals
hydroxymercaptals
ketene mercaptals
–, synthesis, Ni-catalyzed with Grignards, review 48, 850s51
Mercaptans
–, S-acylation with N-acoxydicarboxylic acid imides 51, 223
– from
alcohols (with H$_2$S), review 44, 469s51
halides 51, 229
thiolic acid esters 51, 229
– special s.
dithiols
– startg. m. f.
1-acetylene-1-thioethers 51, 392
disulfides, sym., disproportionation 51, 216
silylethynyl thioethers 51, 392
Mercaptans, in situ-generated
–, Michael addition with – 51, 234
2-Mercaptoethanol
– as reagent 51, 144
α-Mercaptoketones
– startg. m. f.
β-ketothionophosphorus(V) compds. 51, 246
Mercury
– as sensitizer 51, 168
Mercury compds., organo-
–, review 33, 659s51
– special s.
diarylmercury compds.
Mesitylenesulfonyloxylamine
– as reactant 51, 129
1-α-Mesyloxyethers, cyclic
–, ring expansion 51, 77
Mesyloxy-2-ethylenes
– as intermediates
51, 242

Met – Nic   274

Metalation, asym. (s.a. under
  Deprotonation)
– special s.
  lithiation, asym.
Metalation, benzylic, asym.
– of benzyl ethers, tricarbonylchromium-
  complexed 51, 400
o-Metalation, asym.
– of aryl derivs., tricarbonylchromium-
  complexed 51, 400
Metal BINOL catalysts, chiral
–, synthesis, asym. with –, review
  42, 455s51
Metallocarbenoids
–, cascade processes, review 43, 943s51
Metal oxides, organo-
– as catalysts, review 50, 120s51
Metathesis (CC⇃C; CC⇃C; s.a.
  Interchange in Vol. 1-50; Ring closing
  metathesis; Ring opening –)
Methanesulfonic acid esters
–, α-alkylation of β-dicarbonyl compds.
  with inversion 51, 369
– special s.
  aryl mesylates
Methanesulfonyl chloride
– as reagent 51, 242, 487
(–)-B-Methoxydiisopinocampheyl-
  borane
– as reagent 51, 268
Methoxymethyl thioethers
– startg. m. f.
  sulfinic acid esters 51, 50
Methylamines, sec.
– from
  amines, prim. 28, 346s51
Methylarenes
– startg. m. f.
  aldehydes, ar. 51, 74
N-Methylation 51, 137
Methyl (S)-N-benzyl-N-methylalaninate
–, α-alkylation, asym. 51, 401
Methylene bromide
– as reactant 51, 421
Methylene chloride/n--butyllithium
– as reagent 51, 421
Methylene compds. s. Ethylene derivs.,
  terminal; Tebbe methylenation
Methylenecyclobutane ring, 2-γ-
  halogeno-
– startg. m. f.
  methylenecycloheptane ring 51, 485
Methylenecyclobutanes
–, chemistry, review 43, 989s51
Methylenecycloheptane ring
– from
  methylenecyclobutane ring, 2-γ-
  halogeno- 51, 485
Methylenecyclooctane ring
  51, 485
Methylenecyclopentane ring, 3-sulfonyl-
– from
  α,β-ethylenesulfones, asym. induction
  51, 454
– startg. m. f.
  2-cyclopentenone ring 51, 454
2-Methylenecyclopropylcarbinols
  42, 822s51
Methylene groups, active
– startg. m. f.
  cyclopropanes 51, 367
2-Methylene-1,4-halogenhydrins

36, 879s51
Methylene iodide
– as reactant 51, 408
α-Methylene-γ-lactones 51, 288
α-Methylenephosphonic acid esters
– from
  acetylene derivs., terminal 51, 239
3-Methylene-4-siloxyalcohols, chiral
  39, 658s51
Methyl iodide
– as reagent 51, 118
p-Methyliodobenzene difluoride
– as reagent 51, 205
Methyl ketones
– from
  diazomethyl ketones 51, 37
N-Methylmethoxylamine hydrochloride
– as reagent 51, 371
N-Methylmorpholine
– as reagent 51, 34
Methyl orthoformate
– as reagent 51, 147
Methyltriphenylphosphonium iodide
– as reactant 51, 430
Micellar media
–, reactions in –, review 41, 678s51
Michael addition
– with
  mercaptans, in situ-generated 51, 234
  C-nucleophiles s. CC⇃CC
Michael addition, asym.
– using catalysts, heterobimetallic,
  multifunctional 51, 309
– with oxazolidines, 3-acyl- 47, 664s51
– –, –, regiospecific
– with β,γ-ethylenephosphonic acid esters
  49, 630s51
– –, polymer-based
  51, 172
– –, stereospecific
– of enolates, review 36, 652s51
– addition-aldol condensation, asym.
  51, 309
– addition-alkylation, asym.
– with α-aminonitriles 51, 395
Michael-type addition
– with enoxysilanes s. CC⇃OC,
  CC⇃Rem
– –, Pd-catalyzed 51, 333
Microorganisms (s.a. Enzymes) 51, 499
Microwave irradiation
–, N-alkylation of prim. ar. amines 51, 151
–, bromination 47, 430s51
–, developments, review 44, 651s51
–, dicarboxylic acid imides from
  anhydrides 51, 179
–, o-Fries rearrangement 51, 407
–, radiolabelling 41, 199s51
–, Willgerodt reaction 2, 180s51
– –, solvent-less
–, acetalation 41, 172s51
–, dehydrohalogenation 3, 749s51
–, quinoxalines from α-diketones
  2, 378s51
– –, supported, solvent-less
–, benzyl iodides from alcohols
  39, 444s51
–, Knoevenagel condensation 46, 713s51
–, Wittig synthesis 51, 439
Microwave reactor, continuous 51, 84
Mitsunobu etherification, polymer-
  based 51, 82

Mitsunobu reaction (s.a. O-Acylation,
  asym., enzymatic-Mitsunobu reaction)
–, alternative 51, 80
–, review 46, 160s51
Molybdate, tetrathio- s. Benzyltriethylammonium tetrathiomolybdate
Molybdenum π-allyl complexes
– as intermediates 42, 750s51
Monoperoxyketals
– special s.
  ethylenemonoperoxyketals
Monothioacetals
– from
  acetals 51, 226
  enolethers 8, 651s51
– special s.
  monothioformals
Monothioacetals, cyclic
– special s.
  1,3-oxathiolanes
Monothioformals
– special s.
  methoxymethyl thioethers
Monothiolcarbonic acid esters
– special s.
  ethylenemonothiolcarbonic acid esters
Monothioozonides
–, sulfur atom transfer, transition metal-
  catalyzed with – 51, 219
Morpholines
– special s.
  N-methylmorpholine
Morpholines, chiral 22, 368s51
2-Morpholones 48, 955s51
Myers-type cycloaromatization (s.a.
  under 1,4-Diradicals) 51, 495

Naphthalenes
– special s.
  1,5-dimethoxynaphthalene
–, 1,2,3,4-tetrahydro-
  s. Tetral...
1-Naphthols
– from
  o-acetyleneketones 51, 338
1,4-Naphthoquinones
–, Diels-Alder reaction, stereospecific with
  – 51, 311
Nickel
  Raney nickel 51, 151
Nickel(II) acetate 51, 52
– –/lithium hydride/tert-butanol 51, 40
–(II) acetoacetonate
  51, 52, 60, 412, 451
– boride 51, 216
–(II) chloride (s.a. under Chromium(II)
  chloride)
– complexes
  bis[1,2-bis(diphenylphosphino)butane]-
  nickel(0) 51, 152
  bis(1,5-cyclooctadiene)nickel(0)
  51, 305
  bis(ethyl acetoacetato)nickel(II) 51, 270
  dichloro[1,2-bis(diphenylphosphino)-
  ethane]nickel(II) 51, 480

dichloro[1,1'-bis(diphenylphosphino)-
   ferrocene]nickel(II) **51**, 453
dichlorobis(triphenylphosphine)-
   nickel(II) **51**, 372, 381, 452
hexakis(acetonitrile)nickel(II)
   bis(tetrafluoroborate) **51**, 326
tetrakis(triphenylphosphine)nickel(0)
   51, 405
Nicotinamide, N-benzyl-1,4-dihydro-
– as reagent **51**, 12
N-Nitramines
– startg. m. f.
   amines **51**, 12
   aminyl radicals **51**, 12
Nitration, ar.
– of arenes, deactivated **51**, 136
N-Nitration-recyclization
–, nucleosides, ^{15}N-labelled by – **51**, 164
Nitric acid esters
–, resolution **5**, 666s**51**
– special s.
   (arylseleno)nitric acid esters
Nitric acid silyl esters
– special s.
   trimethylsilyl nitrate
Nitriles (s.a. Cyano...)
– from
   benzotriazoles, 1-cycloalkylidene-
      amino- with ring opening **51**, 189
   malononitriles **51**, 43
– special s.
   acoxynitriles
   alkoxynitriles
   aminonitriles
   cyclopropanecarbonitriles
   ketonitriles
   malononitriles
   siloxynitriles
– startg. m. f.
   aldehydes **51**, 86
   carboxylic acid amides **51**, 58
   2-nitroacylamines **51**, 128
Nitriles, ar.
– from
   aryl mesylates **51**, 381
2-Nitroacylamines
– from
   ethylene derivs. **51**, 128
   nitriles **51**, 128
α-**Nitroalkyl radicals**
– as intermediates **51**, 428
o-**Nitrobenzenesulfonyl chloride**
– as reagent **51**, 106
p-**Nitrobenzoic anhydride**
– as reagent **51**, 78
Nitro compds.
– from
   amines, prim. **21**, 125s**51**; **51**, 45
– startg. m. f.
   hydroxylamines **49**, 21s**51**
– –, aliphatic
–, 1,2-addition to electron-rich ethylene
   derivs. **51**, 428
– startg. m. f.
   α,β-ethyleneketones **51**, 428
– –, ar.
–, reactions, catalytic with CO, review
   47, 22s**51**
Nitroethylene derivs.
–, radical ring closure, double **51**, 428
1,1-Nitroethylene derivs.

–, [4+2]/[3+2]-cycloadditions with –,
   review **45**, 414s**51**
– startg. m. f.
   α-azidohydroximinohalides **51**, 206
   α-cyanohydroximinohalides **51**, 206
   1,2-diamines **51**, 149
(E)-**1,1-Nitroethylene derivs.**
– from
   ethylene derivs. **51**, 135
Nitrogen atom transfer
– from nitridomanganese(V) complexes
   51, 178
Nitrogen dioxide
–, gas-solid reactions with – **51**, 96
Nitrogen monoxide/alumina 51, 135
α-**Nitroketones**
– from
   ethylene derivs. **51**, 134
γ-**Nitroketones**
–, resolution, asym. **29**, 36s**51**
– startg. m. f.
   pyrroles **51**, 186
Nitrones
–, cycloaddition, 1,3-dipolar, asym. with –
   , review **51**, 317s**51**
–, –, –, intramolecular, stereospecific to
   enolates **51**, 350
– from
   amines, sec. **21**, 125s**51**
– special s.
   N-benzylnitrones
   ethylenenitrones
   N-(tetrahydropyran-2-yl)nitrones
– startg. m. f.
   2(5H)-furanones, 3-amino- **51**, 111
   isoxazolidines, asym. conversion
      51, 317
– –, **cyclic**, *in situ*-**generated**
–, cycloaddition, 1,3-dipolar,
   intramolecular with – **51**, 336
Nitroso compd. dimers s. Azo N,N'-
   dioxides
Nitroso compds.
–, hetero-Diels-Alder reactions, review
   41, 305s**51**
– startg. m. f.
   amines **30**, 10as**51**
o-**Nitrosulfonates**
– startg. m. f.
   amines, ar., prim. **51**, 10
Nitrosyl fluoroborate 51, 212, 463
– **hydrogen sulfate 51**, 201
Nitrous acid esters (s.a. Thionitrous acid
   esters)
Nitroxyls s. N-Oxide radicals
Nitryl iodide 51, 135
– **tetrakis(trifluoromethanesulfonato)-
   borate 51**, 136
Nozaki-Hiyama-Kishi reaction, catalytic
   51, 288
Nucleoside 5'-phosphates 48, 104s**51**
– –, **radio-P-labelled**
–, synthesis, review **26**, 97s**51**
– **phosphoramidates**
– special s.
   dinucleoside phosphoramidates
Nucleosides
   (s.a. under Vorbrüggen)
–, N-acylation, selective **43**, 295s**51**
– from
   furans, 2,5-dihydro-, *meso*-2,5-
      diacoxy-, asym. conversion **51**, 153

– special s.
   aminonucleosides
1,2-cis-Nucleosides 51, 177
Nucleosides, ^{15}N-**labelled 51**, 164
Nucleotides (s.a. OReml↑O; Nucleoside
   phosphates, Oligonucleotides)

Olefins s. Ethylene derivs.
Oligonucleotide synthesis
–, developments **17**, 169s**51**
Oligosaccharide libraries
   44, 211s**51**
Oligosaccharides
– from
   glycals, review **41**, 173s**51**
–, synthesis, polymer-based **44**, 211s**51**
–, –, review **41**, 173s**51**
–, – with glycosidases, review **41**, 173s**51**
Oppenauer-type oxidation 51, 448
β-**(Organothio)aldehydes**
– special s.
   ethylene-β-(organothio)aldehydes
β-**(Organothio)carbonyl compds.**
   51, 234
γ-**(Organothio)ketones**
– special s.
   ethylene-γ-(organothio)ketones
N-**(Organothio)phthalimides**
– startg. m. f.
   disulfides, sym. **51**, 217
α-**(Organothio)thiolic acid esters**
–, hydrolysis, enzymatic with dynamic
   kinetic resolution **51**, 94
Orthoformic acid esters
– special s.
   methyl orthoformate
Osmate s. Potassium osmate
[n+5]-Oxabicyclo[n.2.1]alkenes
– startg. m. f.
   3-cycloalkenols **51**, 28
1,2-Oxasilacyclopentanes
– from
   siliranes **51**, 298
– startg. m. f.
   1,3-diols **51**, 298
1,2,3-Oxathiazolidine 2,2-dioxides,
   3-acyl-
– as intermediates **51**, 230
1,3-Oxathiolanes
– from
   oxo compds. **51**, 227
– startg. m. f.
   oxo compds. **51**, 227
1,3-Oxathiolane-2-thiones
– from
   epoxides **51**, 218
1,2-Oxatitanacyclopentanes
– as intermediates **51**, 315
1,3,2-Oxazaborolidines, chiral
– as reagent **51**, 8, 21
–, –, **polymer-based**
– as reagent **43**, 45s**51**
1,3,2-Oxazaphospholidines, chiral
– as reagent **43**, 45s**51**
2H-1,2-Oxazine ring, 3,6-dihydro-
–, ring opening, reductive **51**, 9

1,2-Oxazines, tetrahydro-, 6-α-(arylseleno)- 51, 250
2H-1,3-Oxazines, 3,6-dihydro-, 3-tosyl-
– from
β-hydroxyaldehydes 51, 173
– startg. m. f.
3-(tosylamino)alcohols 51, 173
2H-1,3-Oxazines, tetrahydro-
– from
isoxazolidines 51, 119
2H-1,3-Oxazin-2-ones, tetrahydro-
– from
isoxazolidines 51, 119
2H-1,4-Oxazin-2-ones, 5,6-dihydro-
– from
Δ^2-oxazolines 51, 182
Oxaziridine, perfluoro-cis-2-butyl-3-propyl-
– as reagent 51, 115
Oxaziridines, polyfluoro-
–, review 49, 100s51
Oxazole-4-carboxylic acid esters
– from
β,γ-acetylene-α-acylaminomalonic acid esters 51, 116
Oxazoles, 2-tert-amino- 23, 441s51
Oxazolidines, 3-acyl-
–, Michael addition, asym. with –
47, 664s51
Oxazolidines, tricyclic
– from
hydroxylaminoenynes 51, 336
Oxazolidine-2-thiones
– startg. m. f.
thiazolidine-2-thiones 51, 232
–, 3-acyl-
–, aldol condensation, asym. with -
38, 632s51
2,4-Oxazolidiones
– from
α-halogenocarboxylic acid amides 51, 169
2-Oxazolidones
–, N-acylation 51, 167
2-Oxazolidones, 3-acyl-
–, N-acylation, preferential with – 51, 160
–, –, polymer-based
–, α-alkylation, asym. 44, 776s51
–, 3-(α,β-ethyleneacyl)-
–, Diels-Alder reaction, asym., with –
46, 662s51
–, radical 1,4-addition, asym., Lewis acid-catalyzed to – 51, 321
–, – –, intramolecular, regiostereospecific to – 51, 484
Δ^2-Oxazolines
– special s.
bis(Δ^2-oxazolin...)
– startg. m. f.
1,4-oxazin-2-ones, 5,6-dihydro- 51, 182
Δ^2-Oxazolines, chiral
– special s.
(R,R,R)-[2-(4,5-diphenyl-Δ^2-oxazolin-2-yl)...
–, 2-aryl-, tricarbonylchromium-complexed
–, 1,4-addition-alkylation, asym. 51, 400
–, 2-[o-(diarylphosphino)phenyl]-
– as reagent 51, 26
–, phosphino-
– as ligands, review 47, 646s51
Δ^3-Oxazolines, 2-alkoxy-

– from
1-sulfonyl-1-isonitriles 51, 418
– startg. m. f.
α-hydroxyketones 51, 418
2-Oxetanones s. β-Lactones
Oxidation, aerobic, N-hydroxyphthalimide-catalyzed 51, 76
–, catalytic, heterogeneous
– with molecular oxygen and sulfur, review 20, 647s51
Oxidation-reduction, microbial
–, deracemization of alcohols, sec. via –
51, 499
N-Oxidations
– with methylrhenium oxide/hydrogen peroxide 51, 45
N-Oxide radicals
– special s.
4H-dinaphth[2,1-c:1',2'-e]azepine-N-oxyl...
– –, stable
– as oxidant, review 39, 225s51
N-Oxides, cyclic
– special s.
2,6-dichloropyridine N-oxide
Oxides, organometallic s. Metal oxides, organo-
Oxide surfaces, well-defined
–, reactions at –, review 49, 372s51
Oxido compds. (from Volume 51 s. Epoxides)
Oximes
– special s.
hydroxyoximes
ketoximes
– startg. m. f.
amines 24, 23s51
Oximes, O-silyl- s. Siloximes
Oxo compds. (s.a. Aldehydes, Carbonyl compds., Ketones)
– from
alcohols (OCfl H)
–, (ketones), kinetic resolution 51, 500
anthraquinones, 1-alkoxy- 48, 255s51
1,3-oxathiolanes 51, 227
– special s.
alkoxyoxo compds.
aryloxo –
ethyleneoxo –
– startg. m. f.
2-acetylenealcohols 51, 273, 278
erythro-3-acetylenealcohols, terminal 51, 267
alcohols, reduction 51, 35
–, reduction, preferential and selective 51, 23
–, (sec. alcohols), reduction, asym. 51, 22, 24
–, (– –), synthesis 51, 284, 427
–, (– –), synthesis, asym. 51, 266
–, (– –), transfer-hydrogenation, asym. 51, 26
α-alkoxynitriles, asym. synthesis 51, 277
3-alkylidenecyclobutanols 51, 430
amines, sec. 51, 141
azomethines 51, 147
3-ethylenealcohols, synthesis, regiostereospecific 51, 281
–, with 3 extra C-atoms 51, 450
α,β-ethylene-γ-hydroxyketones, cyclic 51, 280

(E)-α,β-ethylenephosphonic acid esters, with 1 extra C-atom 51, 365
glycidic acid esters 51, 388
α-halogeno-β-lactones 51, 368
ketones, synthesis, regiospecific 51, 384
mercaptals 51, 226
1,3-oxathiolanes 51, 227
(E)-N-sulfonylimines 51, 154
Oxo compds., unsatd.
– startg. m. f.
alcohols, unsatd., hydrogenation, homogeneous 51, 27
Oxone s. Potassium peroxymonosulfate
β-Oxophosphonic acid esters
– startg. m. f.
(E)-α,β-ethylenephosphonic acid esters 51, 255
Oxyamination, asym. 51, 132
Oxydiboranes
– special s.
tetraoxydiboranes
Oxygen, singlet
–, ene reactions, stereospecific, review 48, 106s51
– from
azole-ozone complexes 51, 57
–, sensitized reactions in a fluorous 2-phase medium 51, 53
Oxygen atom transfer, catalytic
– from sulfoxides 51, 49
Oxyselenation, intramolecular, asym. 51, 258
Ozone 51, 454
Ozone-azole complexes
– as source of singlet oxygen 51, 57
Ozonides
– from
alkoximes 51, 88
Ozonolysis
– of ethylene derivs., improved workup 51, 64

Palladacyclics
– as catalyst 51, 416
Palladation s. Carbopalladation
Palladium(0)
–, addition, oxidative to P-H bonds 51, 239
Palladium/carbon 51, 149, 155
Palladium/carbon/ammonia 51, 32
–(II) acetate
51, 10, 110, 170, 229, 234, 333, 382, 394, 413, 447, 454, 464, 465, 486
–(II) acetoacetonate 51, 262
– π-allyl complexes
– as intermediates 51, 417
–, synthesis, asym. via –, review 48, 772s51
– π-allyl groups
–, substitution, nucleophilic 51, 358
–, catalysis, asym.
–, review 48, 772s51
– –, sequential
–, review 45, 555s51
–(II) chloride
51, 456

Palladium complexes
bis(η³-allylpalladium chloride) **51**, 457
bis(π-allylpalladium chloride) **51**, 383
bis(dibenzylideneacetone)palladium(0) **51**, 131, 358, 455
trans-di(μ-acetato)bis[*o*-(di-*o*-tolylphosphino)benzyl]dipalladium(II) **51**, 416
dichloro[1,1'-bis(diphenylphosphino)ferrocene]palladium(II) **51**, 259
dichlorobis(triphenylphosphine)palladium(II) **51**, 417
dichlorobis(tri-*o*-tolylphosphine)palladium(II) **51**, 171
dichloro[(S)-2-(dimethylamino)-1-(diphenylphosphino)-3-phenylpropane]palladium(II) **51**, 370
cis-dimethylbis(methyldiphenylphosphine)palladium **51**, 239
tetrakis(triphenylphosphine)palladium(0) **51**, 171, 190, 235, 289, 335, 410, 414, 415, 486
tris(dibenzylideneacetone)dipalladium **51**, 28, 153, 171, 188, 272, 334
Palladium complexes, organo-
– special s.
propargylpalladium complexes
– –, zwitterionic
– special s.
allenylpalladium complexes, zwitterionic
Paracyclophanes
51, 335
Pauson-Khand reaction, interrupted
51, 324
Payne rearrangement (s.a. Aza-Payne rearrangement)
1,2-Pentacarbonylchromium group migration 51, 425
1,2,2,6,6-Pentamethylpiperidine
– as reagent **51**, 270
Peptide aldehydes 45, 510s**51**
Peptide N-alkylamides
–, synthesis, polymer-based **46**, 317s**51**
Peptide libraries
–, synthesis and screening, review **50**, 555s**51**
Peptides
– from
α-aminocarboxylic acid choline esters **51**, 145
– – halides, review **46**, 350s**51**
– – silyl esters **51**, 176
– special s.
dipeptides
homopeptides
lipopeptides
Peptide synthesis (NC¹¹O)
– with
phosphorus oxide chloride/dimethylformamide **51**, 176
tetramethylfluoroformamidinium hexafluorophosphate **51**, 148
– –, enzymatic
– in aq. media **51**, 145
– –, solid-phase
–, update **19**, 33s**51**
Peptide thioamides
–, review **47**, 304s**51**
Peptidyl pentafluoroethyl ketones
37, 806s**51**
Perfluoro... s.a. Polyfluoro...

Perfluoroalkanes
–, photoamination, reductive, partial **51**, 168
Perfluoroalkylating agents
–, review **41**, 660s**51**
C-Perfluoroalkylation
– with perfluoroacyl peroxides **10**, 637s**51**
Peroxides
– special s.
cyclohexadienones, peroxy-di-*tert*-butyl peroxide
fluoroperoxides
Peroxides, cyclic
– special s.
thiophene endoperoxides
Peroxyketals (s.a. Monoperoxyketals)
Per-O-silylation
–, Vorbrüggen synthesis via – **51**, 177
Persulfate s. Ammonium persulfate, Tetra-*n*-butylammonium –
Peterson olefination
–, study of intermediates, review **13**, 831s**51**
2-Phase medium (s.a. Fluorous two-phase medium)
Phenolethers
–, *p*-acylation **51**, 374
– by Mitsunobu etherification, polymer-based **51**, 82
– special s.
aminophenolethers
– startg. m. f.
p-quinones **51**, 108
Phenols
–, *o*-acylation **51**, 379, 407
– from
arylboronic acids **51**, 98
disilanes, ar. **38**, 829s**51**
– special s.
acetonylphenols
aminophenols
halogenophenols
sulfinylphenols
Phenols, *p*-subst.
– startg. m. f.
2,5-cyclohexadienones, 4-peroxy-, 4-subst. **51**, 472
p-quinones, 2-subst. **51**, 472
Phenyl *tert*-butyldiphenylsilyl selenide
– as source of silyl radicals **51**, 257
Phenyl iodosoacetate
– as reagent **51**, 87, 97, 256
Phenylsilane
– as reagent **51**, 36, 313
Phenyl(tosylimino)iodination 51, 174
N-Phenyltriflimide
– as reactant **51**, 488
Phosphine imines
– special s.
benzotriazoles, 1-α-(phosphoranylideneamino)-
eniminophosphoranes
phosphoranylideneamino...
– startg. m. f.
aziridines **51**, 175
Phosphine oxides
– special s.
acylaminophosphine oxides
diphenylphosphinyl groups
hydroxyphosphine oxides
ketophosphine –

Phosphines
–, deprotonation, asym. **51**, 357
– special s.
acylaminophosphines
di(phosphines)
halogenophosphines
– startg. m. f.
1,2-di(phosphines), sym., asym. conversion **51**, 357
Phosphines, tert.
– from
dihalogenophosphines **51**, 249
diorganozinc compds. **51**, 249
halogenophosphines **51**, 249
– special s.
bis(dicyclohexylphosphino)...
bis(diphenylphosphino)...
tri-*n*-butylphosphine
triphenylphosphine
tris(2,6-dimethoxyphenyl)phosphine
Phosphines, tert., chiral
– as reagent **51**, 83
Phosphine sulfides
–, deprotonation, asym. **51**, 357
Phosphinic acids, cyclic
– from
dihalides **51**, 254
Phosphinyloxylamines
– startg. m. f.
amines, cyclic **51**, 183
Phosphites s. Dialkyl phosphites, Phosphorous acid esters
Phosphonic acid benzyl esters
– from
halides **51**, 260
Phosphonic acid esters
– special s.
aminophosphonic acid esters
diethyl methanephosphonate
ethylenephosphonic acid esters
oxophosphonic – –
phosphoryloxyphosphonic – –
Phosphonic acids
– special s.
aminophosphonic acids
Phosphonium salts
– special s.
methyltriphenylphosphonium iodide
Phosphonohalidates
– from
phosphonic acid monoesters **43**, 398s**51**
Phosphonous acids
– special s.
aminophosphonous acids
***o*-(Phosphoranylideneamino)thioethers**
– startg. m. f.
benzothiazoles **51**, 231
α-Phosphoranylidenecarboxylic acid esters
– special s.
(triphenylphosphoranylidene)acetic acid esters
Phosphoric acid esters
– special s.
allyl phosphates
– – –, cyclic
– special s.
2,4-bis[(4,6-dimethyl-1,3,2-dioxaphosphorinan-2-yl)oxy]pentane
– – –, mixed
– from
alcohols **51**, 51

Phosphoromonohalidates 43, 398s51
- special s.
  diethyl phosphorochloridate
Phosphorous acid diesters
- special s.
  dialkyl phosphites
Phosphorous acid esters
- special s.
  enol phosphites
  tribenzyl phosphite
  triethyl –
  triisopropyl –
  trimethyl –
Phosphorus(III) acid esters
- from
  phosphorus(III) acids 48, 104s51
–(III) acid halides
- startg. m. f.
  β-ketothionophosphorus(V) compds.
  51, 246
Phosphorus oxide chloride 51, 155, 473
– – –/dimethylformamide 51, 176
Phosphorus trichloride
- as reactant 51, 249
O-Phosphorylation, preferential
  51, 51
S-Phosphorylation
- with enol phosphites 51, 246
β-Phosphoryloxyphosphonic acid esters
- as intermediates 51, 365
Photoamination
–, heterocyclics by –, review 39, 292s51
–, reductive, partial
- of perfluoroalkanes 51, 168
1,4-Phthalazinedione library
–, synthesis 13, 472s51
Phthalazines
- special s.
  1,4-bis(9-O-dihydroquinine)phthalazine
Phthalides
- from
  benzocyclobutenones 43, 111s51
- special s.
  alkylidenephthalides
Phthalimides
–, cleavage 51, 14
- special s.
  N-hydroxyphthalimide
  N-(organothio)phthalimides
–, N-subst. 51, 179
–, tetrachloro-
–, cleavage, selective 51, 14
Phthalimidines
- from
  o-dialdehydes 51, 144
α-Phthalimidocarboxylic acid halides
–, Pictet-Spengler reaction, asym. with –
  51, 351
Pictet-Spengler reaction, asym. 51, 351
– –, polymer-based 8, 823s51
Pinacolborane
–, hydroboration, regiospecific with –
  51, 238
Pinacol ethers, sym. 34, 723s51
Pipecolic acid esters, 6-alkoxy-1-
  carbalkoxy- 46, 669s51
Piperazine library
–, synthesis 6, 465s51
Piperidines
- from
  cyclopentanols 51, 162

- special s.
  pentamethylpiperidine
  pipecolic...
Piperidinium acetate
- as reagent 51, 375
- triflate, 1-dodecyl-1-methyl-4-oxo-
- as reagent 51, 66
Pivalaldehyde
- as reagent 51, 48
Plumbanes
- special s.
  tetraethyllead
Polarity reversal
- of carbanions, nucleophilic, review
  36, 775s51
Polyenes
- special s.
  ubiquinones
Polyfluoro... s.a. Perfluoro...
Polyfluoroalkanes, [bi]cyclic
–, review 42, 865s51
(E,E)-1-Polyfluoroalkyl-1,3-dienes
  43, 650s51
Polyfluorocarbocations
–, review 42, 865s51
Polyfluoro-1,3-dihalides 29, 492s51
Polyfluoroethylene [bi]cyclic
–, review 42, 865s51
Poly(hydrogen fluoride) s. Hydrogen
  fluoride/pyridine
Polyindans, centro-
- by cyclodehydration, review 36, 430s51
Polymer-based reactants
- special s.
  N-acylsulfonic acid amides, polymer-
    based
  2-oxazolidones, 3-acyl-, –
  silanes, –
Polymer-based reactions (s.a. under
  Compound libraries)
–, review 50, 555s51
- special s.
  aza-Wittig synthesis, intramolecular,
    polymer-based
  Biginelli synthesis, –
  Bischler-Napieralski ring closure, –
  etherification, –
  Heck arylation, intramolecular, –
  Horner synthesis, –
  Knoevenagel condensation-
    intramolecular ene reaction, –
  Michael addition, –
  Mitsunobu etherification, –
  Pictet-Spengler reaction, –
  Ugi condensation, –
Polymer-based synthesis
- of
  α-prim-aminophosphonous acids
    49, 556s51
  1,4-benzodiazepine-2,5-diones 51, 184
  bicyclo[2.2.2]octan-2-ones 32, 641s51
  5'-ene-1,3-diols, cyclic 51, 375
  hydantoins 31, 452s51
  isoxazoles 2, 368s51
  ketones from hydroxamates 37, 806s51
  pyrazoles 2, 368s51
  pyrroles 30, 631s51
  4-thiazolidones 9, 672s51
Polymer-supports
–, regeneration by Hofmann elimination
  51, 172

Polymethylhydrosiloxane
- as reagent 51, 25
Polysilanes
- special s.
  (alkoxy)polysilanes
Poly(tetrahydrofurans) 51, 113
Porphyrins
- from
  pyrrole-2,5-dicarboxaldehydes 51, 461
  tripyrranedicarboxylic acid tert-butyl
    esters 51, 461
Porphyrins, sym. 51, 461
Potassium alkoxides
- tert-butoxide/lithium bromide 51, 429
Potassium amides
- bis(trimethylsilyl)amide 51, 338
- benzoate
- as reactant 51, 80
- bromide 51, 465
- cyanide 51, 381
- fluoride 51, 492
- –/alumina 51, 221
- hydride 51, 392
- osmate 51, 132
- permanganate 51, 67
- peroxymonosulfate 51, 66, 98
- –/chiral ketones 51, 65
- phosphate 51, 453
Pressure, high s. High pressure
L-Proline, trans-4-hydroxy-
- as chiral building block, review
  47, 646s51
Prolinates 51, 183
1,3-Propanedithiol
- as reagent 51, 34
Propargyl... s.a. 2-Acetylene...,
  β,γ-Acetylene...
N-Propargylanilines
- startg. m. f.
  quinolines, 1,2-dihydro- 51, 339
Propargyl 1,4-diradicals
–, 1,5-cyclization, review 45, 308s51
Propargylpalladium complexes
- as intermediates 51, 289
Propellanes 51, 318
Propionic acid
- as reagent 51, 341
Protection
- of amino groups as
  carbo-p-acetoxybenzoxyamines 51, 16
  carbo-p-azidobenzoxyamines 51, 16
  pent-4-enoylamines 51, 18
- of – –, prim. as
  1,3,5-dioxazines, dihydro- 46, 23s51
  tetrachlorophthalimides 51, 14
  triphenylphosphine imines 34, 271s51
- of 2-azetidinone nitrogen as
  N-bis(trimethylsilyl)methyl derivs.
    51, 402
- of carbonyl groups by
  aluminum complexation 51, 300
- of carboxyl groups as
  choline esters 51, 5
  2-hydroxyethyl esters 7, 246s51
  2-(2-methoxyethoxy)ethyl esters 51, 5
- of hydroxyl groups (alcohols) as
  tetrahydropyran-2-yl ethers 51, 61
  – – (partial protection) 51, 63
N-Protective groups, removal (HN!!S,
  HN!!C)
- of

bis(trimethylsilyl)methyl (oxidative removal) **51**, 15
carbo-*p*-acetoxybenzoxyl (enzymatic –) **51**, 16
carbo-*p*-azidobenzoxyl (reductive –) **51**, 16
pent-4-enoyl (selective –) **51**, 18
2-pyridylsulfonyl **51**, 13
tetrachlorophthalimide group (selective removal) **51**, 14
O-Protective groups, removal (HO↓↑S, HO↓↑Rem, HO↓↑C)
– of
choline ester groups (enzymatic removal) **51**, 5
2,4-dimethoxytrityl (selective removal) **51**, 7
4,4'-dimethoxytrityl **12**, 16s**51**; **51**, 7
2-hydroxyethyl **7**, 246s**51**
2-(2-methoxyethoxy)ethyl (enzymatic removal) **51**, 5
silyl s. O-Desilylation
tetrahydropyran-2-yl s. under Pyran-2-yl ethers, tetrahydro-
2,2,2-trichloroethyl **24**, 9s**51**
– –, solubilizing
– in peptide synthesis, enzymatic **51**, 145
Protective group strategies
–, review **46**, 3s**51**
Protonation, asym.
– of enolates and enols, review **50**, 389s**51**
Pummerer-type rearrangement
– of *p*-hydroxysulfoxides **51**, 95
– –, asym.
–, induction by O-silyl O-alkyl ketenacetals, review **50**, 252s**51**
2H-Pyran, 3,4-dihydro-
– as reactant **51**, 61, 63
2H-Pyrans, 3,4-dihydro, 2-alkoxy-, chiral 38, 688s**51**
Pyrans, tetrahydro- (s.a. Tetrahydropyran...)
–, –, 4-hydroxy-
– from
1,3-dioxanes, 4-alkylidene- **51**, 345
Pyran-2-yl ethers, tetrahydro- (s.a. O-Tetrahydropyran-2-ylation and under Protection)
–, cleavage **51**, 2
Pyrazoles
–, synthesis, polymer-based **2**, 368s**51**
Pyrazoles, 1-acyl-
– startg. m. f.
β-ketocarboxylic acid esters **48**, 808s**51**
Pyridinamides
– as chiral ligands, review **47**, 646s**51**
9H-Pyrid[3,4-b]indoles 40, 669s**51**
–, synthesis, review **46**, 921s**51**
–, 1,2,3,4-tetrahydro-, 2-acyl-, chiral **51**, 351
Pyridines
– from
pyridines, 1,4-dihydro-, with C-dealkylation **51**, 194
Pyridines, aryl- 51, 415
–, 4-aryl- **37**, 902s**51**
–, 1,4-dihydro-
– special s.
nicotinamide, N-benzyl-1,4-dihydro-
– startg. m. f.
pyridines, with C-dealkylation **51**, 194

–, 2-α-fluoro- **51**, 202
2-Pyridinethiones, 1-acoxy-
– startg. m. f.
azo N,N'-dioxides, sym. **51**, 180
Pyridinium betaines, oxido-
–, cycloaddition, review **36**, 898s**51**
– salts, N-(2,3-epoxyalkyl)-
– as intermediates **51**, 1
– tosylate
– as reagent **51**, 349
– ylids
– as nucleophile, review **36**, 898s**51**
–, cycloaddition, – **36**, 898s**51**
2-Pyridones, 3-cyano- 41, 636s**51**
2-Pyridylthio groups
–, cleavage **5**, 78s**51**
Pyrimidine-5-carboxaldehydes
– startg. m. f.
pyrimidines, 5-α-hydroxy-, automultiplication, asym., catalytic **51**, 271
Pyrimidines, acyl- 51, 458
–, 5-α-hydroxy-
– from
pyrimidine-5-carboxaldehydes, automultiplication, asym., catalytic **51**, 271
2(1H)-Pyrimidinones, 4-amino- s. Cytosines
2(1H)-Pyrimidinones, 3,4-dihydro-
–, synthesis, polymer-based **51**, 361
2-Pyrones
–, Diels-Alder reaction, asym. with – **36**, 667s**51**
2-Pyrones, 5,6-dihydro-, 3-γ-keto- 51, 329
Pyrrole-2-carboxylic acid esters 51, 186
Pyrrole-2,5-dicarboxaldehydes
– startg. m. f.
porphyrins **51**, 461
Pyrroles
– from
2-acetyleneamines **51**, 337
β-acylamino-α,β-ethyleneketones **47**, 940s**51**
α,β-ethylenesulfones **44**, 736s**51**
γ-nitroketones **51**, 186
Δ²-pyrrolines **51**, 337
thioiminoesters **44**, 736s**51**
–, reactions, photochemical, review **39**, 292s**51**
– special s.
tripyrr...
–, synthesis, polymer-based **30**, 631s**51**
Pyrroles, aryl- 37, 902s**51**
–, cyano- **51**, 337
Pyrrolidine, (S)-2-methoxymethyl-
– as chiral auxiliary, review **47**, 646s**51**
Pyrrolidines
– from
aldehydes **51**, 376
3-ethyleneamines **51**, 376
– special s.
prolin...
Pyrrolidines, 2,5-disubst.
–, synthesis, stereospecific, review **50**, 438s**51**
–, 2-aryl-, N-protected, chiral **51**, 478
–, N-condensed
– from
4-azadienes **51**, 340
–, 2,3-*trans*-disubst. **51**, 494

–, 3-methylene-4-vinyl-, chiral **49**, 699s**51**
–, 3-nitro- **44**, 897s**51**
Δ²-Pyrroline ring, N-condensed
– from
4-azaenynes **51**, 340
Δ¹-Pyrrolines 51, 189
Δ²-Pyrrolines
– from
2-acetyleneamines **51**, 337
– startg. m. f.
pyrroles **51**, 337
Pyrrolizines, tetrahydro- 51, 340
Pyrroloindoles
–, chemistry, review **49**, 366s**51**

2,4-Quinazolinediones
–, synthesis, polymer-based **18**, 544s**51**
Quinoline, 1,2-dihydro-, 2-ethoxy-N-ethoxycarbonyl-
– as reagent **51**, 79
Quinolines, 1,2-dihydro-
– from
N-propargylanilines **51**, 339
–, 1,2,3,4-tetrahydro-
25, 527s**51**; **38**, 760s**51**
–, synthesis, review **41**, 54s**51**
2(1H)-Quinolones s. Carbostyrils
Quinone methids
– as intermediates, review **40**, 384s**51**
Quinones, polycyclic
– from
arenes, polycyclic **51**, 108
o-**Quinones, electrogenerated 21**, 725s**51**
p-**Quinones**
– from
p-hydroxysulfoxides **51**, 95
phenolethers **51**, 108
– special s.
alkoxy-*p*-quinones
p-**Quinones, 2-subst.**
– from
2,5-cyclohexadienones, 4-peroxy-, 4-subst. **51**, 472
phenols, *p*-subst. **51**, 472

Racemates
–, conversion to one enantiomer s. Deracemization
Radical addition , asym., Lewis acid-catalyzed
– to 2-oxazolidones, 3-(α,β-ethyleneacyl)- **51**, 321
Radical 1,2-addition
– of nitro compds. to electron-rich ethylene derivs. **51**, 428
Radical 1,2-addition-allylation 51, 422
– 1,4-addition
– under chain transfer conditions, unimolecular **51**, 323

Radical 1,4-addition, intramolecular, regiostereospecific
- to 2-oxazolidones, 3-(α,β-ethyleneacyl)- **51**, 484
- **1,4-addition-allylation, asym., Lewis acid-catalyzed 51**, 435
- **C-allylation** (s.a. Radical 1,2-addition-allylation)
- with allyl sulfones **51**, 422
- **carbonylation**
-, review **50**, 357s**51**
- **cross-coupling, asym., Lewis acid-catalyzed 51**, 436
Radical-ion cyclization
-, review **48**, 818s**51**
Radical reactions (s.a. Chain transfer reactions)
- in tandem (with CO, RNC), review **48**, 822s**51**
- **–, oxidative, heterogeneous**
-, review **46**, 930s**51**
- **ring closure** (s.a. Acyl substitution, nucleophilic-radical ring closure)
- **– –, acylative 51**, 424
- **– –, double**
- of
  enediynes **51**, 312
  nitroethylene derivs. **51**, 428
- to benz[e]indenes, 2,3-dihydro- **51**, 312
- **– –, oxidative**
- with manganese(III) acetate, review **46**, 930s**51**
- **– –, –, regiospecific**
- of ethyleneketones **51**, 466
- **– –, regiostereospecific**
- of alkoxyl radicals, unsatd. **51**, 209
- **– –, –, silyl-mediated 51**, 257
- **– –, titanium(III)-mediated**
- of acetyleneepoxides **51**, 318
- **– – closure-transannular ring opening 51**, 485
- **– – expansion**
- of methylenecyclobutane ring, 2-γ-halogeno- **51**, 485
Radicals
- special s.
  acylaminoalkyl radicals
  alkoxyl –
  alkyl –
  aminoalkyl –
  aminyl –
  diradicals
  fluoroalkyl radicals
  iminyl –
  nitroalkyl –
  N-oxide –
  silyl –
  sulfenyl –
Radical substitution
- at heteroatoms, review **50**, 260s**51**
Radio-labelled compds. s. Compounds, radio-labelled
Rearrangement, [1.3]-sigmatropic, stereospecific
- of 1,3-diox[ol]anes, 4-alkylidene- **51**, 345
-, **[2.3]-sigmatropic**
-, 2-ethylenetosylamines via – **51**, 174
-, **[3.3]-sigmatropic**
-, 2-ethylenedithiolcarbonic acid esters via – **51**, 221
-, –, asym. review **50**, 433s**51**

-, [3.3]-sigmatropic-1,2-allyl group migration, asym.
51, 390
Reduction, microbial (s.a. Oxidation-reduction, microbial)
Reductions
- with lithium hydride complexes **51**, 40
Resolution (s. under Stereoisomers in Vol. **1-50**; s.a. Deracemization, and under **Res** section)
- by cyclodextrins, modified, review **5**, 666s**51**
Resolution, kinetic
- of
  alcohols, sec. **51**, 500
  aldehydes **48**, 866s**51**
  amines via N-acylation **51**, 160
- via N-carbalkoxylation **51**, 146
  carboxylic acids (enzymatically), review **41**, 175s**51**
  cyanohydrins **22**, 693s**51**
  epoxides **51**, 121
  hydroperoxides **51**, 498
  α-hydroperoxycarboxylic acid esters **51**, 498
- on ring closing metathesis **48**, 988s**51**
-, –, **dynamic**
- of
  1,1-hydroxythioethers **51**, 233
  α-(organothio)thiolic acid esters **51**, 94
-, **kinetic-Mitsunobu reaction**
-, deracemization via – **51**, 103
Retro-Diels-Alder scission
- of 1,2,4-triazoline-3,5-dione adducts **27**, 959s**51**
- **–, anionic 17**, 198s**51**
Rhenium complexes
  trichlorobis(triphenylphosphine)-oxorhenium(III) **51**, 49
Rhenium oxides, organo-
  methylrhenium oxide **51**, 388
- –/hydrogen peroxide **51**, 45, 107
Rhodium(II) acetate 51, 389, 390
–(II) **carboxamidates, chiral 47**, 955s**51** (review)
Rhodium-cobalt complexes
  tetrakis(tert-butyl isocyanide)rhodium tetracarbonylcobalt **51**, 332
Rhodium complexes
  bis(cyclooctadiene)rhodium(I) hexafluorophosphate **51**, 330
  carbonylhydridotris(triphenyl-phosphine)rhodium(I) **51**, 331
  chloro(1,5-cyclooctadiene)rhodium(I) dimer **51**, 119
  chlorotris(triphenylphosphine)-rhodium(I) **51**, 238
–(I) **hydride complexes, supramolecular 51**, 30
- **phosphine complexes, chiral**
- as reagent **51**, 21, 31
Ring closing metathesis (s. under Interchange, intramolecular in Vol. **1-50**; s.a. Tebbe methylenation-ring closing metathesis) **51**, 497
- **–, W-catalyzed 51**, 497
Ring closure
- special s.
  carbocyclization
  radical ring closure
  radical-ion – –

Ring closure, alkylative, regiostereospecific
- of 2,n-enynones **51**, 305
- **– –, double**
- special s.
  iodoetherification, intramolecular-lactonization
- **–, double, regiostereospecific**
- of ethylene(halogeno)carboxylic acid esters **51**, 482
- **–, serial, radical- or transition metal-catalyzed**
-, review **46**, 679s**51**
- **–, triple**
- via carbopalladation, intramolecular with allenes **51**, 417
Ring expansion (s.a. Radical ring expansion)
- of
  ketones, cyclic **51**, 384
  1-α-mesyloxyethers, – **51**, 77
- **–, Barbier-type**
- of ketones, cyclic **51**, 481
Ring opening metathesis
- with ethylene derivs., cyclic **51**, 327
Ritter-type reaction
-, 2-nitroacylamines via – **51**, 128
Rubidium fluoride 51, 223
Ruthenium(0)
-, addition, oxidative into C-H bonds, vinylic **51**, 328
Ruthenium π-allyl complexes
- as intermediates **51**, 71
- from
  ruthenium carbene complexes **51**, 391
- **carbene complexes**
- as reagent **51**, 327
- startg. m. f.
  ruthenium p-allyl complexes **51**, 391
- **complexes**
  (η⁶-arene)(N-arylsulfonyl-1,2-diamine)chlororuthenium(II) complexes, chiral **51**, 26, 29
  carbonyldihydrotris(triphenyl-phosphine)ruthenium(II) **51**, 328
  carbonyl(tetraarylporphyrinato)-ruthenium(VI) **51**, 108
  chloro(1,5-cyclooctadiene)-cyclopentadienylruthenium(II) **51**, 329
  chloro(η⁵-indenyl)bis(triphenyl-phosphine)ruthenium(II) **51**, 33
  dichlorotris(triphenylphosphine)-ruthenium(II) **51**, 26, 71, 472
  –/ethylenediamine/potassium hydroxide **51**, 27
- **–, low-valent**
- as redox catalysts, review **48**, 291s**51**

Sakurai-Hosomi reaction, Brönsted acid-catalyzed 51, 450
2-[N-(Salicylidene)amino]alcohols, chiral
- as reagent **51**, 46
Samarium/iodine/alcohols 51, 4
–/**mercury(II) chloride 51**, 443

Samarium complexes
 bis(η⁵-pentamethylcyclopentadienyl)-
   bis(trimethylsilyl)methylsam-
   arium(III) **51**, 340
 bis(η⁵-pentamethylcyclopentadienyl)-
   samarium **51**, 308
– diiodide
   51, 9, 11, 251, 308, 376, 408
–, sequencing reactions with –, review
   48, 818s**51**
– –/N,N'-dimethyl-N,N'-propyleneurea
   51, 13
– –/hexamethylphosphoramide
   51, 43, 261, 278, 279, 482, 483, 489
–(III) enolates
 – as intermediates **51**, 251
–(III) halides, organo-
 – special s.
   alkynylsamarium(III) iodides
Scandium(III) triflate 51, 277, 310, 407, 436
Scandium(III) triflimide
   51, 78
Schmidt reaction
 – of azides, aliphatic with alcohols
   51, 162
–, ring expansion by – **51**, 159
 – via 1,1-siloxyazides, cyclic **51**, 191
 – with
   azidoketals **51**, 193
   enolethers **51**, 193
Selenation (s.a. Aminoselenation, Oxyselenation)
Selenides (s.a. Arylseleno...)
 – from
   2-hydroxyselenides, synthesis **51**, 441
–, lithiation, reductive **51**, 427
 – special s.
   eneselenides
   ethyleneselenides
   hydroxyselenides
   ketoselenides
 – startg. m. f.
   alcohols (sec. alcohols), synthesis
   51, 427
Seleniranium ions
 – as intermediates **51**, 441
Selenium dioxide 51, 182
Selenoacetals
–, lithiation, reductive **51**, 427
Selenolcinnamic acid esters 51, 261
Selenolic acid esters
 – from
   aldehydes **51**, 256
 – special s.
   ethyleneselenolic acid esters
Selenonium ions
–, ene reaction with – **40**, 575s**51**
Selenylhalides
 – special s.
   benzeneselenyl bromide
Semicarbazones
–, cleavage **17**, 479s**51**; **34**, 172s**51**
Semicorrins, C₂-symmetric
 – as ligands, review **47**, 646s**51**
Sensitizers
 – special s.
   acetone
   biphenyl
   1,5-dimethoxynaphthalene
   5,10,15,20-tetrakis(heptafluoropropyl)porphyrin
Sensitizers, coupled 51, 445

Septanoses
–, synthesis, review **41**, 173s**51**
Shapiro reaction, catalyzed
 – with N-aziridin-1-ylimines **51**, 475
Shapiro reaction-Suzuki coupling
   51, 394
Sigmatropic rearrangement
   s. Rearrangement, sigmatropic
Silanes
–, oxidation, review **39**, 200s**51**
 – special s.
   acetylenesilanes
   acylsilanes
   allenesilanes
   di(silanes)
   enoxysilanes
   epoxysilanes
   ethylenesilanes
   halogenosilanes
   hydroxysilanes
   polysilanes
 – startg. m. f.
   alcohols, review **39**, 200s**51**
Silanes, polymer-based
–, cleavage **31**, 65s**51**
Silanols
 – special s.
   alkoxysilanols
Silica (s.a. lipase/silica under Enzymes)
   51, 17, 439
Silica surfaces
–, reaction on –, review **49**, 372s**51**
Silicon
–, reactivity in organic chemistry, review
   44, 65s**51**
Silicon hydrides, halogeno-
 – special s.
   trichlorosilane
– – –, organo-
–, chain transfer, unimolecular via
   hydrogen atom transfer, intramolecular from – **51**, 323
 – special s.
   phenylsilane
   polymethylhydrosiloxane
   trioctylsilane
Siliranes
 – startg. m. f.
   1,3-diols **51**, 298
   1,2-oxasilacyclopentanes **51**, 298
Siloximes
 – special s.
   diketone monosiloximes
4-Siloxyalcohols
 – special s.
   methylene-4-siloxyalcohols
1,1-Siloxyazides, cyclic
 – from
   enoxysilanes, cyclic **51**, 191
 – startg. m. f.
   lactams **51**, 191
2-Siloxyazides
 – from
   epoxides, asym. ring opening **51**, 121
Siloxycyclobutanes
 – special s.
   vinylcyclobutanes, 1-siloxy-
Siloxycyclopropane ring
 – special s.
   bicyclo[n.1.0]alkanes, 1-siloxy-
Siloxycyclopropanes, 1-alkoxy-
 – special s.
   trimethylsilyoxycyclopropane, 1-ethoxy-

2-Siloxy-1,3-dienes
 – startg. m. f.
   β-hydroxyketones, synthesis **51**, 319
Siloxy-2-ethylenes (s.a. Allyl silyl ethers)
 – startg. m. f.
   1,5-dienes **51**, 437
1,1-Siloxyhalides
 – special s.
   1,1-siloxyiodides
Siloxy-N-heterocyclics
 – startg. m. f.
   N-glycosyl-N-heterocyclics **51**, 177
1,1-Siloxyiodides
 – startg. m. f.
   carbonyl ylids, non-stabilized (from 2 molecules) **51**, 443
α-Siloxynitriles
 – from
   aldehydes **51**, 263
β-Siloxynitriles
 – from
   epoxides **49**, 589s**51**
1-Siloxy-1-silanes
 – from
   oxo compds. **34**, 591s**51**
1,1-Siloxysulfonium salts
 – as intermediates **36**, 879s**51**
Silver acetate 51, 456
– **fluoroborate 51**, 258
– **hexafluoroantimonate 51**, 476
– **trifluoromethanesulfonate 51**, 258
– –/(S)-2,2'-bis(diphenylphosphino)-1,1'-binaphthyl **51**, 433
β-Silylaldehydes
 – special s.
   ethylene-β-silylaldehydes
O-Silyl O-alkyl keteneacetals
–, induction of Pummerer-type rearrangement, asym. with –, review
   50, 252s**51**
 – special s.
   dimethylketene methyl trimethylsilyl acetal
2-Silyl-2-allenealcohols, chiral 51, 268
Silylating agents, super-
 – as reagent **51**, 276
Silylation (s.a. Carbonylation, silylative)
 – with silyl triflates, review **33**, 100s**51**
O-Silylation
 – special s.
   per-O-silylation
Silyl-1,3-dienes
 – in synthesis, review **42**, 209s**51**
Silyl enol ethers s. Enoxysilanes
(Z)-2'-Silylenoxysilanes
–, Mannich reaction, asym., regiospecific
   via – **51**, 440
Silyl ethers (s.a. Alkoxysilanes, Siloxy...)
 – special s.
   allyl silyl ethers
   benzyl – –
Silylethynyl thioethers
 – from
   mercaptans **51**, 392
Silylformylation, intramolecular, regiostereospecific
 – of acetylene derivs. **51**, 332
N-Silylimines
 – as intermediates **51**, 154
α-Silylketones
 – startg. m. f.
   β'-aminoketones, asym. induction
   51, 440

Sil – Sul

β-Silylketones
- from
  2-acetylenealcohols **15**, 494s**51**
Silyl radicals
–, generation from silyl selenides **51**, 257
Silyl selenides
– special s.
  phenyl *tert*-butyldiphenylsilyl selenide
Silylstannanes
– startg. m. f.
  4-aryl-2-ethylenesilanes **51**, 455
Silyl triflates
–, prepn. and silylation with –, review
  33, 100s**51**
1,1-Silylurethans
–, 1,4-addition, photochemical, co-
  sensitized with – **51**, 445
Smiles rearrangement
– of *o*-aminodiaryl ethers, review
  33, 315s**51**
Sodium/ammonia, liq. 51, 438
Sodium alkoxides
- *tert*-butoxide **51**, 171, 190
Sodium amides
– bis(trimethylsilyl)amide **51**, 177, 397
– **azide 51,** 127, 155, 256
– **bis(alkynyl)diethylaluminates**
– as intermediates **51**, 273
– **borate 51,** 431
– **hydridoaluminates, organo-**
– dihydridodiethylaluminate **51**, 273
– **hydrogen selenide/sodium selenide**
  51, 243
– **hypochlorite 51,** 500
– **iodide 51,** 193, 449
– **nitrite 51,** 128, 140
– **percarbonate 51,** 102
– **periodate 51,** 399
– **peroxide 51,** 198
– **sulfide 51,** 222
– **tetrahydridoborate 51,** 1, 162, 250
– **–/cobalt(II) Schiff base complex,**
  chiral 51, 19
– **trihydridocyanoborate 51,** 39, 143
Solid acids s.a. Acids, solid-supported
Solid acids, strong
– special s.
  zirconia, sulfated
Solid-phase synthesis s. under Polymer-
  based..., Supports, inorganic
(–)-Sparteine
– as reagent **51**, 299, 343, 357, 478
Spiro[4.5]dec-2-ene-1,6-diones 51, 344
Spiroisocyclics 51, 482
Spiroketals
– special s.
  dispiroketals
Stannanes (s.a. under Stille)
– from
  alkoxystannanes **51**, 241
  zirconium(IV) complexes, organo-
  51, 241
– special s.
  acetylenestannanes
  alkoxystannanes
  allenestannanes
  aminostannanes
  arylstannanes
  cyanostannanes
  enestannanes
  ethylenestannanes
  halogenostannanes
  hetarylstannanes
  imidoylstannanes
  silylstannanes
– startg. m. f.
  compds., radio-labelled, review
  28, 511s**51**
  sulfonic acid salts **47**, 512s**51**
Stannanes, N-heterocyclic
– startg. m. f.
  ketones, N-heterocyclic **51**, 458
Stannyllithium compds. s. Triorganostannyllithium compds.
β'-Stannyl-γ-vinyl-γ-lactols, bicyclic
– from
  α,β-ethyleneketones, cyclic **51**, 397
– startg. m. f.
  3-ethylene-1-vinyllactones **51**, 397
Stereoisomers
–, resolution s. under Resolution
Steroids
– special s.
  estrane...
Stetter reaction, intramolecular 51, 342
(E)-Stilbenes
– from
  diaryl tellurides **51**, 456
Stille coupling, Cu(I)-mediated 51, 432
in situ-**Stille coupling 51,** 415
Strecker synthesis, asym. 51, 294
Styrenes (s.a. *o*-Vinylation)
– from
  1,3-enynes (2 molecules) **51**, 335
  halides, ar. **51**, 394
  N-trisylhydrazones **51**, 394
  vinylzirconocene halides **51**, 410
– startg. m. f.
  allylarenes **51**, 326
  benzyllithiums **51**, 299
Styryl carbenes
–, ring closures via – **51**, 112
**Substitution, nucleophilic, intra-
  molecular**
– at nitrogen **51**, 183
– of styryl carbenes **51**, 112
Succinic acid esters
– special s.
  vinylsuccinic acid esters
– – –, **sym. 50,** 515s**51**
Sugars s. Carbohydrates
O-Sulfation, enzymatic 31, 81s**51**
Sulfenylation
– with benzothiazol-2-yl disulfides
  47, 470s**51**
Sulfenyl chlorides
– in synthesis, review **30**, 348s**51**
Sulfenyl radicals
–, generation from thionitrites **51**, 180
Sulfido compds. s. Thiiranes from Vol. 51
Sulfinic acid aryl esters
– startg. m. f.
  o-hydroxysulfoxides **51**, 220
Sulfinic acid esters
– from
  methoxymethyl thioethers **51**, 50
– – –, **cyclic** s. Sultines
Sulfinic acids
– special s.
  arenesulfinic acids
Sulfinic acid salts
– special s.
  copper(II) *p*-toluenesulfinate
**2-Sulfinyl-1,3-dienes, chiral
  39,** 887s**51**
p-**Sulfinylphenols** s. *p*-Hydroxysulfoxides

N-Sulfinyl-*p*-toluenesulfonamide
– as reactant **51**, 173
Sulfones
– special s.
  acetylenesulfones
  acoxysulfones
  alkoxysulfones
  (alkylthio)sulfones
  diaryl sulfones
  disulfones
  ethylenesulfones
  hydroxysulfones
  ketosulfones
– startg. m. f.
  enolethers **51**, 420
  thioenolethers **51**, 420
Sulfonic acid amides (s.a. N-Desulfonylation, Sulfonylamin...)
– from
  sulfonic acid azides **51**, 11, 163
– special s.
  N-acylsulfonic acid amides
  halogenosulfonic – –
  trifluoromethanesulfonic – –
Sulfonic acid azides
– startg. m. f.
  sulfonic acid amides **51**, 11, 163
– – **esters** (s.a. Sulfonyloxy...)
– special s.
  methanesulfonic acid esters
  nitrosulfonates
– – **halides**
– special s.
  arenesulfonic acid halides
– – **imides**
– special s.
  N-halogenosulfonic acid imides
– **acids**
– special s.
  trifluoromethanesulfonic acid
– **acid salts**
– from
  stannanes **47**, 512s**51**
Sulfonium salts
– special s.
  siloxysulfonium salts
1-Sulfonylalkoxines
– special s.
  O-benzyl-1-sulfonyloximes
N'-Sulfonylamidines
– from
  azomethines **51**, 165
  N,N-dihalogenosulfonamides **51**, 165
Sulfonylamines (s.a. Sulfonic acid
  amides)
– special s.
  epoxysulfonylamines
  ethylenesulfonylamines
o-**Sulfonylamino-α,β-acetyleneiodonium
  salts**
– startg. m. f.
  N-sulfonylenamines, bicyclic **51**, 477
Sulfonylaminoalcohols
– special s.
  tosylaminoalcohols
N-Sulfonylenamines, bicyclic
– from
  o-sulfonylamino-α,β-acetylene-
  iodonium salts **51**, 477
Sulfonylhydrazines
– special s.
  tosylhydrazines

Sulfonylhydrazones
- special s.
  tosylhydrazones
  trisylhydrazones
N-Sulfonylimines
- startg. m. f.
  α,β-ethylene-N-sulfonylthioiminoesters
  51, 310
(E)-N-Sulfonylimines
- from
  oxo compds. 51, 154
Sulfonyliminoiodinanes
- special s.
  phenyl(tosylimino)iodinane
1-Sulfonyl-1-isonitriles
- startg. m. f.
  α-hydroxyketones 51, 418
  Δ³-oxazolines, 2-alkoxy- 51, 418
1-α-Sulfonyloxyethers, cyclic
- special s.
  1-α-mesyloxyethers, cyclic
Sulfonyloxy-2-ethylenes
- special s.
  mesyloxy-2-ethylenes
Sulfonyloxylamines
- special s.
  mesitylenesulfonyloxylamine
N-Sulfonylthioiminoesters
- special s.
  ethylene-N-sulfonylthioiminoesters
Sulfoxides
- from
  thioethers 51, 47, 49
  –, asym. oxidation 51, 46
  –, – –, aerobic 51, 48
- special s.
  dimethyl sulfoxide
  diphenyl –
  hydroxysulfoxides
  ketosulfoxides
Sulfur atom transfer, transition metal-catalyzed
- with monothioozonides 51, 219
Sulfuric acid esters, cyclic
- special s.
  1,3-diol sulfates, cyclic
  glycol –, –
Sulfuric acid monoesters (s.a. O-Sulfation)
Sulfur trioxide 51, 204
Sultams
–, N-acylation 51, 167
Sultines
–, review 32, 974s51
Superacids
- special s.
  triflatoboric acid
Supercritical fluids
–, review 51, 31s51
Superoxide, electrochemically-generated
- as reagent 51, 169
Supersilylating agents s. Silylating agents, super-
Supports, inorganic
- in synthesis, review 49, 372s51
Supramolecular catalysis s. Catalysis, supramolecular
Suzuki coupling (s.a. CClRem, Shapiro reaction-Suzuki coupling)
–, catalysis with palladacyclics 51, 416
- with chlorides, ar. 51, 416
- –, nickel-catalyzed 51, 453

Synthesis, asym. (s.a. Catalysis, asym., Desymmetrization)
–, review 47, 646s51
- via palladium π-allyl complexes, review 48, 772s51
- with
  allenyl- and allyl-stannanes, chiral, review 48, 889s51
  auxiliaries, chiral, stoichiometric, reviews 47, 646s51
  Brintzinger-type complexes, review 50, 15s51
  lipases, immobilized, – 49, 305s51
  metal BINOL catalysts, chiral, – 42, 45s51
  reagents, chiral, – 47, 646s51
Synthesis, asym., absolute (s.a. Automultiplication, asym.)
–, –, Pd- and Cu-catalyzed
–, review 48, 772s51
Synthesis, enzymatic
- of
  α-aminocarboxylic acids, α-subst., review 28, 13s51
  2-amino-1-phenylethanols, – 28, 13s51
  α-hydroxycarboxylic acids, α-subst., – 28, 13s51

TADDOL... s. 1,3-Dioxolane-4,5-dimethanols and under Titanium
Tandem reactions s. under Specific reaction combinations)
–, review 45, 414s51
Tebbe methylenation 46, 870s51
- methylenation-ring closing metathesis 51, 496
Tellurides
- from
  halides (2 different molecules) 51, 245
- special s.
  ethylenetellurides
Tellurides, ar.
- special s.
  aryltelluro...
  diaryl tellurides
–, ar., unsatd.
–, radical ring closure with aryltelluro group migration 51, 355
–, sym.
- from
  halides 47, 563s51
Tellurium
- as reactant 51, 245
Tellurium compds., organo-
–, reactivity, review 43, 540s51
Tellurolic acid esters
- from
  aldehydes 51, 256
Tetraethyllead
- as reagent 51, 497
Tetraaminocarboxylic acid esters
–, macrolactamization, regiospecific 51, 187
Tetra-n-butylammonium bromide
- as reagent 51, 358, 416
- dihydrogentrifluoride
- as reagent 51, 213
- fluoride
- as reagent 51, 34, 100, 298, 465

- –/alumina
- as reagent 51, 25
- –/boron fluoride
- as reagent 51, 3
- hexafluorophosphate
- as reagent 51, 152
- persulfate
- as reagent 51, 47, 89
- (triphenylsilyl)difluorosilicate
- as source of fluoride ion 51, 210
1,2,4,10-Tetraen-6-ynes
- startg. m. f.
  indans 51, 312
Tetrahydridoborate, resin-supported
- special s.
  Amberlite tetrahydridoborate
O-Tetrahydropyran-2-ylation, partial
- of diols, sym. 51, 63
N-(Tetrahydropyran-2-yl)nitrones, in situ-generated
–, 1,3-dipolar cycloaddition, stereospecific via – 51, 352
5,10,15,20-Tetrakis(heptafluoropropyl)-porphyrin
- as sensitizer 51, 53
1-Tetralols, 2-amino-, chiral 51, 8
N,N,N',N'-Tetramethylazodicarbox-amide (s.a. Tri-n-butylphosphine/N,N,N',N'-tetramethylazodicarbox-amide)
1,1,3,3-Tetramethylbutyl isocyanide
- as reagent 51, 262
N,N,N',N'-Tetramethylethylenediamine
- as reagent 51, 344, 394, 411
Tetramethylfluoroformamidinium hexafluorophosphate
–, peptide synthesis with – 51, 148
Tetramethylguanidinium azide
- as reagent 51, 130
1,2,3,4-Tetraols, chiral, 2,3-O-protected
- special s.
  1,3-dioxolane-4,5-dimethanols, chiral
Tetraoxydiboranes
- special s.
  bis(catecholborane)
  diboronic acid bis(pinacolate)
Tetrazole ring, N-condensed
–, ring opening, reductive 51, 155
Tetrazoles
- from
  iminoesters 51, 140
Tetrazoles, 5-subst. 51, 140
Tetrazolo[1,5-c]pyrimidin-5(6H)-ones
- as intermediates 51, 155
1H-Tetrazol-5-yl glycosides
- startg. m. f.
  glycosyl fluorides 51, 208
Thallium(III) reagents
–, rearrangements, oxidative with –, review 27, 162s51
Thia-Fries rearrangement 51, 220
Thianaphthenes s. Benzo[b]thiophenes
Thiazoles, 5-amino- 6, 638s51
Thiazolidine-2-thiones
- from
  oxazolidine-2-thiones 51, 232
4-Thiazolidones
–, polymer-based synthesis 9, 672s51
Thiazolium chloride, 3-benzyl-5-(2-hydroxyethyl)-4-methyl-
- as reagent 51, 342

Thiiranes (s. under Sulfido compds. in Vol. 1-50)
- from
  ethylene derivs. **51**, 219
- startg. m. f.
  ethylene derivs. **51**, 490
Thioacetals s. Mercaptals, Monothioacetals
Thioamides s. Carboxylic acid thioamides
Thiocarbamic acid esters s. Thionourethans
Thiocarbonic acid esters s. Dithiolcarbonic acid esters, Monothiolcarbonic acid esters
Thiocarboxylic acid esters s. Dithiocarboxylic acid esters, Thiolic – –, Thionocarboxylic – –
Thioenolethers
- from
  α-(alkylthio)sulfones **51**, 420
  sulfones **51**, 420
- special s.
  ene-1,2-di(thioethers)
  α,β-ethylene-β-(organothio)...
  halogenothioenolethers
  ketothioenolethers
(E)-Thioenolethers
- from
  allenyl thioethers, synthesis **51**, 334
Thioethers (s.a. Alkylthio..., Organothio...)
- from
  alcohols (with H$_2$S), review **44**, 469s**51**
- special s.
  acetylenethioethers
  acoxythioethers
  (acylthio)thioethers
  disodium 3,3'-thiodipropionate di(thioethers)
  hydroxythioethers
  ketothioethers
  (phosphoranylideneamino)thioethers
- startg. m. f.
  sulfones **35**, 57s**51**
  sulfoxides **51**, 47, 49
  –, asym. oxidation **51**, 46
  –, – –, aerobic **51**, 48
Thioethers, ar. (s.a. Arylthio...)
- startg. m. f.
  fluorides **51**, 212
3-Thioglucosamine derivs. 51, 230
5-Thioglycosides 44, 211s**51**
Thiohydroxamic acids, O-acyl- s. Acyl thiohydroxamates
Thioiminoesters
- from
  benzotriazoles, 1-imidoyl- **51**, 85
- special s.
  sulfonylthioiminoesters
Thioketones
- special s.
  ethylenethioketones
Thiolactones s. Thionolactones
Thiolformylation, regiospecific 51, 331
Thiolic acid esters
–, alcoholysis, Pd-catalyzed **51**, 234
- from
  aldehydes **51**, 256
  Grignard compds. **51**, 297
  halides **51**, 229
- special s.
  ethylenethiolic acid esters

(organothio)thiolic – –
- startg. m. f.
  disulfides, sym. **51**, 216
  mercaptans **51**, 229
Thiolic acids
- from
  Grignard compds. **51**, 297
Thiols s. Mercaptans
Thionitrous acid esters
- special s.
  trityl thionitrite
N-(Thionocarbalkoxy)hydrazones
- startg. m. f.
  iminyl radicals **51**, 189
Thionocarboxylic acid esters
- startg. m. f.
  ethers **51**, 42
Thionolactones
- startg. m. f.
  ethers, cyclic **51**, 42
Thionophosphorus(V) compds.
- special s.
  ketothionphosphorus(V) compds.
Thionourethans
- from
  isothiocyanates **49**, 105s**51**
N-Thionylsulfonamides
- special s.
  N-sulfinyl-p-toluenesulfonamide
Thiophene endoperoxides
- as reagent **51**, 219
Thiophenes
- from
  carboxylic acid halides **51**, 225
  β,γ-epoxyketones **51**, 225
  2-ethylenesilanes **51**, 225
  hydrocarbons (and H$_2$S), review **28**, 543s**51**
Thiophosphonic acid esters s. Dithiophosphonic acid esters
Thiophosphorus(V) compds. s. Thionophosphorus(V) compds.
Thiourea dioxide
- as reagent **51**, 186
Thiourethans s. Thionourethans
Tin(IV) chloride 51, 462
Tin compds., organo-
- special s.
  distann...
  stann...
–(IV) enolates
- as intermediates **51**, 322
–(IV) halides, organo-
- as Lewis acid **51**, 484
- special s.
  tri-n-butyltin chloride
- **hydrides, organo-**
- special s.
  di-n-butyl(tert-butyl)tin hydride
  di-n-butyl(chloro)stannane
  tri-n-butyltin hydride
  triphenyltin hydride
  tris[2-(perfluorohexyl)ethyl]tin hydride
Tin-lithium compds., organo- (s.a. Triorganostannyllithium compds.)
Tin(IV) oxide, sulfated 51, 104
Tin oxides, organo-
- special s.
  bis(tri-n-butyltin) oxide
Tishchenko reaction 51, 256
Titanacyclopentanes (s.a. Aza- or Oxatitanacyclopentanes)

Titanacyclopentenes, vinyl-, fused
- as intermediates **51**, 315
Titanium, commercial
–, activation with chlorosilanes **51**, 469
Titanium, low-valent
–, review of reactions with – 30, 561s**51**
Titanium(IV) alkoxides
- tetraisopropoxide **51**, 351
- –/cyclohexylmagnesium chloride **51**, 378
- –/ethylmagnesium bromide **51**, 377
- –/isopropylmagnesium – **51**, 281, 316
- –/– chloride **51**, 295, 470
Titanium(IV) alkoxides, chiral
- tetraisopropoxide/1,1'-bi-2-naphthol/ diisopropyl D-tartrate **51**, 447
- –/1,3-dioxolane-4,5-dimethanols, chiral **51**, 22, 56
–(IV) –, disulfonato-, chiral
  ditosylatotitanium(IV) (R,R)-2,2-dimethyl-α,α,α',α'-tetraphenyl-1,3-dioxolane-4,5-dimethoxide **51**, 317
–(IV) –, halogeno-
  chlorotitanium triisopropoxide/n-butylmagnesium chloride **51**, 471
- –/isopropylmagnesium bromide **51**, 348
–(IV) –, organo-
- special s.
  2-ethylenetitanium(IV) alkoxides
–(III) η^3-allyl complexes
- as intermediates **51**, 319
–(IV) aroxides, chiral
  ethylene-1,2-bis(η^5-4,5,6,7-tetrahydro-1-indenyl)titanium-1,1'-binaphthyl-2,2'-diolate, chiral **50**, 15s**51** (review)
- **ate complexes, organotetraalkoxy-**
- special s.
  magnesium phenyl(tetraisopropoxy)-titanate, bromo-
- **complexes**
  bis(cyclopentadienyl)bis(trimethylphosphine)titanium(II) **51**, 315
  bis(cyclopentadienyl)titanium(III) chloride **51**, 283, 318
  diisopropoxytitanium bis(trifluoromethanesulfonimide) **51**, 81
  titanocene bis(p-chlorophenoxide) **51**, 25
- dichloride **51**, 160, 195, 423
- –/isopropylmagnesium chloride **51**, 319
–(IV) ester (Z)-enolates
- as intermediates **51**, 285
–(III) hydridoborates
  diisopropoxytitanium(III) tetrahydridoborate **51**, 35
  titanocene tetrahydridoborate **51**, 35
- **η^2-olefin complexes** (s.a. titanium tetraisopropoxide/alkylmagnesium halides under Titanium(IV) alkoxides)
–, addition, oxidative to C–O bonds **51**, 281
–, –, – to C–S bonds **51**, 423
- as intermediates **51**, 378, 470, 471
- **silicate 51**, 446
- **tetrachloride 51**, 206, 285, 472, 494
- –/zinc **51**, 411
- **trichloride/zinc/trimethylsilyl chloride 51**, 469
Titanocene α,β-ethylenecarbene complexes
- as intermediates **51**, 423

Titanocene equivalents
-, unusual reactions, review **49**, 939s**51**
Tolans
– from
halides, ar. **46**, 798s**51**
Tolans, sym. **46**, 798s**51**
p-Toluenesulfonic acid
– as reagent **51**, 225
cis-2-Tosylaminoalcohols
– from
ethylene derivs., asym. conversion
51, 132
3-Tosylaminoalcohols
– from
2*H*-1,3-oxazines, 3,6-dihydro-, 3-tosyl-
51, 173
(1S,2S)-N-Tosyl-1,2-diphenylethylene-
diamine
– as reagent **51**, 26, 29
Tosylhydrazines
– from
tosylhydrazones **25**, 37s**51**
Tosylhydrazones
– startg. m. f.
ketones **51**, 89
Transesterification, heterogeneous
– of β-ketocarboxylic acid esters **51**, 104
Transfer hydrogenation, asym.
– of
azomethines **51**, 29
ketones **51**, 26
Transition metal carbene complexes
(GroupVI)
– in synthesis, review **42**, 852s**51**
Pd→Zn-Transmetalation **51**, 289
Trialkylboranes
– startg. m. f.
alcohols, sec., synthesis **51**, 431
Trialkyl phosphites
s.a. Phosphorous acid esters
Trialkylsilyl tetrakis(triflyloxy)borates
– as reagent **51**, 276
Triarylamines
– from
diarylamines **51**, 170
halides, ar. **51**, 170
Triaryloboroxines
-, Suzuki coupling with – **37**, 902s**51**
1,2,4-Triazoline-3,5-dione [4+2]-adducts
-, cleavage **27**, 959s**51**
Tribenzyl phosphite
– as reactant **51**, 260
Tri-*n*-butylphosphine
– as reagent **51**, 161
-/N,N,N',N'-tetramethylazo-
dicarboxamide
– as reagent **51**, 82
2-(Tri-*n*-butylstannyl)vinyl ethyl ether
– as reactant **51**, 460
Tri-*n*-butyltin chloride
– as Lewis acid **51**, 484
– cyanide
– as reagent **51**, 287
– hydride
-, alternative **51**, 38, 39
– as reactant **51**, 397
– as reagent **51**, 12, 13, 37, 189, 286, 321, 322, 485, 490
Trichloroacetimidates
– from
alcohols **14**, 181s**51**
Trichloroacetonitrile
-, chemistry, review **38**, 470s**51**

Trichloroethylene
– as reactant **51**, 392
– in synthesis, review **42**, 771s**51**
Trichlorosilane
– as reagent **51**, 270
1,2,4-Trienes
– startg. m. f.
(E)-alkylidenecyclobutenes **51**, 356
2-alkylidene-3(4)-cyclopentenones
51, 330
1,2,5-Trienes **51**, 303
1,3(E),5-Trienes **29**, 973s**51**
1,3,6-Trienes, functionalized **51**, 303
1,4,7-Trienes **51**, 303
1(E),3(E),7-Trien-5-ols **36**, 879s**51**
1,2,4-Trien-6-ynes
– as intermediate **51**, 495
Triethylamine hydroiodide
– as reagent **51**, 131
Triethylammonium formate (s.a. Formic
acid/triethylamine)
– as H-donor **51**, 10
– hexafluorophosphate
– as reagent **51**, 33
Triethylamine tris(hydrogen fluoride)
-, review **46**, 540s**51**
Triethylborane
– as reagent **51**, 321, 406, 435, 436
Triethyl phosphite
– as reagent **51**, 423
Triethylsilane/dichloroacetic acid
– as reagent **51**, 7
Triflatoboric acid **51**, 136
Trifluoroacetic acid
– as reagent **51**, 193, 461, 474
– anhydride
– as reagent **51**, 75, 95, 178
Trifluoroacetoxylation, regiospecific
– of hydrocarbons **51**, 75
α-(Trifluoroacetylamino)ketones
– from
enoxysilanes **51**, 178
N-Trifluoroacetylation, partial and
preferential **36**, 340s**51**
Trifluoroacetyl nitrate
– as reagent **51**, 164
Trifluoromethanesulfonic acid
– as reagent **51**, 162, 250
– /boron triflate **51**, 136
– – amides, N,N-disubst.
– startg. m. f.
amines, sec., synthesis **51**, 419
– – esters
– special s.
aryl triflates
enol –
– anhydride
– as reagent **51**, 1, 297
(E,E)-1-Trifluoromethyl-1,3-dienes
23, 879s**51**
Trifluoromethyl groups
– in synthesis, review **44**, 577s**51**
Triisopropyl phosphite
– as reagent **51**, 454
Trimethyl borate
– as reagent **51**, 394
Trimethylenemethane equivalents
-, high-pressure [3+2]-cycloaddition with
– **51**, 457
Trimethyl phosphite
– as reactant **51**, 51
Trimethylsiloxycyclopropane, 1-ethoxy-
-, N-cyclopropylation, reductive with –

51, 143
Trimethylsilyl azide
– as reactant **51**, 121
– as reagent **51**, 191, 206
Trimethylsilyl chloride
– as reagent **51**, 161, 288, 304, 373, 387
Trimethylsilyl cyanide
– as reagent **51**, 121, 206, 263, 277
N-(Trimethylsilyl)diethylamine
– as reactant **51**, 118
2-[(Trimethylsilyl)methyl]allyl pivalate
– as reactant **51**, 454
Trimethylsilyl nitrate
– as reagent **51**, 134
Trimethylsilyl tetrakis(triflyloxy)borate
– as reagent **51**, 438
Trimethylsilyl triflate
– as reagent **51**, 120, 177, 227
Trioctylsilane
– as reagent **51**, 382
Triorganostannyllithium compds.
– as reactant **51**, 253
1,2,4-Trioxolanes (s.a. Ozonides)
Triphenylphosphine
– as reagent **51**, 131
–/N-bromosuccinimide
– as reagent **51**, 150, 207
–/diethyl azodicarboxylate
– as reagent **51**, 82, 103
(Triphenylphosphoranylidene)acetic
acid esters
-, alkylation with – **51**, 137
Triphenyltin hydride
– as reagent **51**, 42
Tripyrranedicarboxylic acid *tert*-butyl
esters
– startg. m. f.
porphyrins **51**, 461
Triquinanes **51**, 318, 482
Tris(2,6-dimethoxyphenyl)phosphine
– as reagent **51**, 333
Tris(methylthio)methane
– as reactant **51**, 364
Tris[2-(perfluorohexyl)ethyl]tin hydride
– as reagent **51**, 39
Tris(trimethylsilyl)germane
– as reagent **51**, 38
Tris(trimethylsilyl)silane
– as reagent **51**, 484
Trisylhydrazones
– special s.
aldehyde trisylhydrazones
– startg. m. f.
styrenes **51**, 394
Trithiocarbonic acid esters, cyclic
16, 671s**51**
Trithioorthocarboxylic acid esters
– special s.
hydroxytrithioorthocarboxylic acid
esters
ketotrithioorthocarboxylic – –
trithioorthoformic – –
Trithioorthoformic acid esters
– special s.
tris(methylthio)methane
Tri-*o*-tolylphosphine
– as reagent **51**, 171
Trityl ethers (s.a. under O-Detritylation)
Trityl thionitrite
– as reagent **51**, 180
Tungsten complexes
trans-dichlorobis(2,6-dibromo-
phenoxy)oxotungsten(VI) **51**, 497

Ubiquinones 51, 452
Ugi 4-component condensation
–, α-acylaminocarboxylic acids, esters,
   and N-unsubst. amides via – 51, 380
–, alternative 51, 123
– – –, polymer-based 51, 380
– in parallel 17, 809s51
Ugi condensation
–, α-(2-azetidinon-1-yl)carboxylic acid
   amides by – 51, 360
Umpolung s. Polarity reversal
Uracils
– startg. m. f.
   cytosines 51, 155
Ureas
– from
   urethans, polymer-based 13, 410s51
Ureas, cyclic
– special s.
   N,N'-dimethyl-N,N'-propyleneurea
α-Ureidocarboxylic acids
– from
   hydantoins, asym. hydrolysis 51, 59
Urethans (s.a. N-Carbalkoxylation)
– special s.
   carbo-*tert*-butoxy[di]amines
   carbofluorenylmethoxy[di]amines
   ethyleneurethans
   silylurethans
– startg. m. f.
   isocyanates 51, 185
Uronic acids
–, synthesis and reactions, review 18,
   299s51

Vanadium complexes
   tetracarbonyl(cyclopentadienyl)-
   vanadium/zinc 51, 373
Vanadium oxide catalysts
–, dehydrogenation with –, review
   3, 708s51
–(IV) salen complexes, chiral
– as reagent 51, 46
Vanadyl acetoacetonate
– as reagent 51, 46
Vilsmeier reagent s. Phosphorus oxide
   chloride/dimethylformamide
Vinyl... s.a. En..., α,β-Ethylene...
Vinylallenes s. 1,2,4-Trien...
Vinylaluminum alkoxides
– as tethers 51, 301
o-Vinylation
– with acetylene derivs. 49, 679s51
Vinylcarbenes
– special s.
   styryl carbenes
Vinyl cation equivalents
–, β-oxophosphonates as – 51, 255
Vinylcyclobutanes, 1-siloxy-
– startg. m. f.
   cyclopentanones, 2-α-iodo- 51, 214
Vinylcyclopentanes
– special s.
   cyclopentanes, divinyl-
2-Vinylcyclopropanecarbonyl compds.,
   functionalized
– from
   α,β-ethylenecarbonyl compds., asym.
   synthesis 51, 393

Vinylcyclopropanes
– from
   1,3-bis(arylthio)ethylenes 51, 423
   α,β-ethylenemercaptals, cyclic 51, 423
syn-2-Vinyl-1,3-diols, chiral 46, 918s51
(E)-Vinyl dithiocarbamates 23, 879s51
Vinyl ethers (s.a. Enolethers)
–, vinylzirconation, stereospecific with –
   51, 314
Vinyl halides s. α,β-Ethylenehalides
Vinyl ketones (s.a. α,β-Ethyleneketones)
   51, 384
γ-Vinyl-γ-lactols
– special s.
   stannyl-γ-vinyl-γ-lactols
1-Vinyllactones
– special s.
   ethylene-1-vinyllactones
γ-Vinyl-γ-lactones
– as intermediates 51, 397
Vinylmagnesium alkoxides
– as tethers 51, 301
– bromides
– startg. m. f.
   ethylene derivs., trisubst. 51, 391
Vinylsilanes s. Enesilanes
Vinylstannanes s. Enestannanes
α-Vinylsuccinic acid esters
   51, 344
Vinyl thioethers s. Thioenolethers
Vinylzinc compds., organo- (s.a. under
   Hydrozincation)
Vinylzirconation, stereospecific
– of acetylene derivs. 51, 314
Vinylzirconocene halides
– from
   α,β-ethylenehalides 51, 410
Vorbrüggen nucleoside synthesis
–, update 26, 446s51
– – –, modified 51, 177

Water
– as solvent, review 49, 882s51
Wittig synthesis
   (s.a. Aza-Wittig synthesis)
–, study of intermediates, review 13,
   831s51
– –, regiostereospecific
– with phosphine oxides, review
   42, 595s51
– –, stereospecific
– on inorganic supports 51, 439
– under microwave irradiation 51, 439

Xanthates
– startg. m. f.
   fluorides 51, 205
Xanthone
– as sensitizer 51, 325

Ylids, fluorinated
–, review 41, 463s51
P-Ylids s. Alkylidenephosphoranes
Ytterbium(III) triflate 51, 321
–(III) triflimide 51, 78
Yttrium-zirconium-sulfur
– as strong Lewis acid 51, 311

Zeolites 51, 55, 306
Zinc (s.a. under Vanadium complexes)
   51, 293, 404, 434
–/zinc chloride-N,N,N',N'-tetramethyl-
   ethylenediamine 51, 41
– acetate 51, 77
– bromide 51, 93, 284, 302, 375, 384,
   385, 394, 448
– chloride (s.a. Zinc/zinc chloride)
   51, 92, 267, 410, 437
– compds., dialkyl-
   diethylzinc 51, 60, 272, 289
– as reactant 51, 249, 271
–, cross-coupling, Ni-catalyzed with
   ethylene-*prim*-iodides 51, 412
– startg. m. f.
   hydroperoxides 51, 93
– –, diorgano- (s.a. Hydrozincation)
– special s.
   1,3-di(zinc compds.)
   vinylzinc compds., organo-
– – –, functionalized
–, synthesis, asym. with –, review
   42, 616s51
– halides, organo-
– special s.
   alkylzinc bromides
   1,3,2-dioxaborolan-2-ylmethylzinc
   iodide, tetramethyl-
– startg. m. f.
   ketones, sym. 51, 404
– iodide 51, 175, 449
– –/lithium hydride 51, 195
– mercaptides, chiral
–, review 47, 646s51
– triflate 51, 435
Zirconacyclopentadienes
– startg. m. f.
   benzene ring 51, 409
Zirconia, sulfated 51, 55
Zirconium complexes
   chlorobis(cyclopentadienyl)-
   hydridozirconium 51, 44, 129,
   238, 343s51 (review), 405
   di-*n*-butylzirconocene 51, 409, 410
   dichlorobis(1-neomenthylindenyl)-
   zirconium 51, 320
   diethylzirconocene 51, 314
– – , chiral
   ethylene-1,2-bis(η⁵-4,5,6,7-tetrahydro-
   1-indenyl)zirconium dichloride
   50, 15s51 (review)
–(IV) complexes, organo-
– startg. m. f.
   stannanes 51, 241
–, review 51, 446s51
– compds., organo- (s.a. Carbozirconation)
–(IV) enolates
– as intermediates 51, 44
Zirconocene
–, addition, oxidative to α,β-ethylene-
   halides 51, 410
– complexes, organo(chloro)-
–, synthesis with –, review 50, 343s51
– equivalents
–, unusual reactions, review 49, 939s51
– halides, organo-
– special s.
   vinylzirconocene halides
– startg. m. f.
   alcohols, sec. 51, 284
   ketones 51, 448

Supplementary References in Vol. 51

No.	Suppl.Ref. Vol., Page
Volume 1	
343	*51*, 71
419	*51*, 100
529	*51*, 142
591	*51*, 177
745	*51*, 222
790	*51*, 235
Volume 2	
147	*51*, 32
180	*51*, 41
216	*51*, 82
260	*51*, 51
288	*51*, 59
314	*51*, 64
368	*51*, 74, 238
378	*51*, 79
621	*51*, 164
Volume 3	
440	*51*, 97
657	*51*, 65
666	*51*, 184
708	*51*, 238
722	51, 73
749	*51*, 225
Volume 4	
48	*51*, 12
49	*51*, 12
154	*51*, 34
443	*51*, 87
667	*51*, 158
671	*51*, 164
676	*51*, 193
725	*51*, 177
Volume 5	
23	*51*, 7
78	*51*, 24
421	*51*, 95
444	*51*, 112
452	*51*, 111
666	*51*, 235, 236 (3), 238
Volume 6	
147	*51*, 27
202	*51*, 44
465	*51*, 84
638	*51*, 110
658	*51*, 112
Volume 7	
246	*51*, 47
407	*51*, 70
566	*51*, 98
836	*51*, 190 (2)
Volume 8	
48	*51*, 14
404	*51*, 68
421	*51*, 70
429	*51*, 75
651	*51*, 107
823	*51*, 172
Volume 9	
61	*51*, 12
74	*51*, 7
108	*51*, 23
174	*51*, 29
371	*51*, 60
431	*51*, 49
568	*51*, 91
672	*51*, 109
741	*51*, 129
816	*51*, 173
Volume 10	
182	*51*, 48
219	*51*, 57
255	*51*, 63 (2)
412	*51*, 95
573	*51*, 173
625	*51*, 191
637	*51*, 216
Volume 11	
95	*51*, 19
120	*51*, 20
235	*51*, 79
399	*51*, 9
489	*51*, 82
Volume 12	
15	*51*, 4
16	*51*, 4,
42	*51*, 8
229	*51*, 70
333	*51*, 59

Supplementary References

No.	Suppl.Ref. Vol., Page

Volume 12 continued

455	*51*, 76
867	*51*, 195
878	*51*, 219
952	*51*, 234

Volume 13

59	*51*, 12
410	*51*, 73
442	*51*, 77
472	*51*, 82
548	*51*, 94
831	*51*, 238
851	*51*, 41

Volume 14

181	*51*, 34
877	*51*, 203
888	*51*, 217
930	*51*, 182
968	*51*, 228
988	*51*, 234

Volume 15

19	*51*, 6
146	*51*, 4
190	*51*, 44
261	*51*, 59
494	*51*, 117
626	*51*, 85
634	*51*, 196

Volume 16

72	*51*, 18
140	*51*, 26
199	*51*, 38
203	*51*, 37
254	*51*, 49

384	*51*, 65
671	*51*, 111
722	*51*, 126
796	*51*, 173
853	*51*, 186
888	*51*, 143

Volume 17

169	*51*, 30
198	*51*, 232
393	*51*, 235
416	*51*, 76
431	*51*, 75
479	*51*, 49
503	*51*, 67
663	*51*, 109
733	*51*, 128
801	*51*, 170
809	*51*, 67, 172, 180
942	*51*, 222

Volume 18

109	*51*, 22
234	*51*, 46
299	*51*, 238
304	*51*, 44
369	*51*, 65
534	*51*, 90
544	*51*, 90
579	*51*, 98
636	*51*, 106
736	*51*, 129
744	*51*, 139
776	*51*, 155, 159

Volume 19

21	*51*, 8
33	*51*, 79
45	*51*, 11
260	*51*, 50
265	*51*, 238
734	*51*, 126
757	*51*, 140

764	*51*, 143
839	*51*, 177
844	*51*, 177
911	*51*, 216
915	*51*, 217

Volume 20

20	*51*, 2
55	*51*, 159
112	*51*, 37
271	*51*, 71, 238
284	*51*, 167
387	*51*, 91
456	*51*, 111
647	*51*, 238

Volume 21

125	*51*, 26
167	51, 34
312	51, 63
349	*51*, 72
426	*51*, 3
446	51, 79
504	*51*, 87
630	*51*, 106
725	*51*, 143
744	*51*, 161
786	*51*, 238
858	*51*, 40
937	*51*, 5

Volume 22

174	51, 43
355	*51*, 68
368	*51*, 74
408	*51*, 93
440	*51*, 82
591	51, 106
617	*51*, 216
652	51, 115
655	51, 115
693	*51*, 134
761	*51*, 170

No.	Suppl.Ref. Vol., Page

Volume 22 continued

782	51, 180
844	*51*, 191
864	*51*, 203

Volume 23

27	*51*, 8
51	*51*, 17, 18
139	*51*, 238
196	*51*, 28
296	*51*, 62
423	*51*, 77
427	*51*, 177
441	*51*, 83
572	*51*, 105
628	*51*, 113
735	*51*, 165
757	*51*, 125, 238
819	*51*, 184 (2), 185
871	*51*, 198
879	*51*, 202 (2)
927	*51*, 220
935	*51*, 223

Volume 24

9	51, 9
23	51, 6
91	*51*, 23
228	*51*, 53
236	*51*, 55
240	*51*, 45
312	*51*, 67
839	*51*, 191

Volume 25

15	*51*, 6
37	*51*, 17
46	*51*, 18
167	*51*, 52

205	*51*, 63
458	*51*, 114
470	*51*, 15
488	*51*, 128
527	*51*, 162
638	*51*, 216
693	*51*, 231
694	*51*, 235

Volume 26

97	*51*, 238
149	*51*, 40
446	*51*, 88 (2)
463	*51*, 59
564	*51*, 106
592	*51*, 108
723	*51*, 178
725	*51*, 155
792	*51*, 183
919	*51*, 216
967	*51*, 226
987	*51*, 233

Volume 27

15	*51*, 7
57	*51*, 17
162	*51*, 238
362	*51*, 72
485	*51*, 87
513	*51*, 91
558	*51*, 100
785	*51*, 175
790a	*51*, 177
833	*51*, 186
841	*51*, 186
843	*51*, 187
949	*51*, 223
959	*51*, 224

Volume 28

13	*51*, 3, 56, 238
17	*51*, 7
141	*51*, 45 (2)

144	*51*, 79
146	*51*, 45
243	*51*, 63
346	*51*, 75
481	*51*, 96
511	*51*, 238
543	*51*, 238
620	*51*, 139
648	*51*, 155, 163
683	*51*, 164
753	*51*, 78
765	*51*, 238
814	*51*, 191
851	*51*, 203
856	*51*, 203
916	*51*, 222

Volume 29

28	*51*, 5
36	*51*, 14
84	*51*, 29
107	*51*, 30
172	*51*, 25
492	*51*, 96
502	*51*, 99
547	*51*, 108
601	*51*, 117
744	*51*, 238
854	*51*, 198
868	*51*, 202
894	*51*, 128
910	*51*, 219
959	*51*, 128
964	*51*, 239
970	*51*, 123, 227
973	*51*, 230

Volume 30

10a	*51*, 5
66	*51*, 31
239	*51*, 71
242	*51*, 239
243	*51*, 72
247	*51*, 89

Supplementary References

No.	Suppl.Ref. Vol., Page

Volume 30 continued

348	*51*, 239
365	*51*, 102
463	*51*, 76
495	*51*, 147
561	*51*, 136, 239
621	*51*, 208
624	*51*, 213
631	*51*, 217

Volume 31

65	*51*, 25
81	*51*, 27
162	*51*, 42
163	*51*, 98
170	*51*, 47 (2)
336	*51*, 72
426	*51*, 86
452	*51*, 91
471	*51*, 95
501	*51*, 102
665	*51*, 161
719	*51*, 171
804	*51*, 187
812	*51*, 187
819	*51*, 190

Volume 32

47	*51*, 239
212	*51*, 60
523	*51*, 107
536	*51*, 113
578	*51*, 119
591	*51*, 119
614	*51*, 132
641	*51*, 145
715	*51*, 139
847	*51*, 198
867	*51*, 239
974	*51*, 239

985	*51*, 232

Volume 33

16	*51*, 4
43	*51*, 12
47	*51*, 17
71	*51*, 22
100	*51*, 239
138	*51*, 37
174	*51*, 47
191	*51*, 50
315	*51*, 239
477	*51*, 97
639	*51*, 145
659	*51*, 239
694	*51*, 167, 222
786	*51*, 239
854	*51*, 203
865	*51*, 204
879	*51*, 209, 211
883	*51*, 214

Volume 34

53	*51*, 6
55	*51*, 239
79	*51*, 28
172	*51*, 50
178	*51*, 50
190	*51*, 33
209	*51*, 57
222	*51*, 59
271	*51*, 64
330	*51*, 73
485	*51*, 98
541	*51*, 111
591	*51*, 120
610	*51*, 137
614	*51*, 129, 137 (2)
616	*51*, 133
630	*51*, 239
650	*51*, 149
693	*51*, 169, 170
723	*51*, 178
811	*51*, 192

825	*51*, 197
862	*51*, 215, 216 (2)

Volume 35

57	*51*, 27
85	*51*, 41
312	*51*, 92
327	*51*, 93
366	*51*, 105
383	*51*, 109
496	*51*, 167
501a	*51*, 169
634	*51*, 239

Volume 36

108	*51*, 33
117	*51*, 36
234	*51*, 58
235	*51*, 36
241	*51*, 100
340	*51*, 73
430	*51*, 239
561	*51*, 112
582	*51*, 116
584	*51*, 116
608	*51*, 124
652	*51*, 145, 239
667	*51*, 148, 153
670	*51*, 15
711	*51*, 167
740	*51*, 173
775	*51*, 239
879	*51*, 208
894	*51*, 215
898	*51*, 239

Volume 37

7	*51*, 5
57	*51*, 22
128	*51*, 38 (2)
182	*51*, 50
220	*51*, 58
264	*51*, 61

No.	Suppl.Ref. Vol., Page
Volume 37 continued	
415	*51*, 59
522	*51*, 109
527	*51*, 111
623	*51*, 126, 239
632	*51*, 137
646	*51*, 142
657	*51*, 145, 146
659	*51*, 191
669	*51*, 142
675	*51*, 158
688	*51*, 239
806	*51*, 184
828	*51*, 215
868	*51*, 204
871	*51*, 207
892	*51*, 201
902	*51*, 215
946	*51*, 224
994	*51*, 234
Volume 38	
39	*51*, 239
111	*51*, 35
221	*51*, 59
254	*51*, 63
307	*51*, 71
470	*51*, 239
488	*51*, 104
524	*51*, 111
584	*51*, 122
623	*51*, 129
632	*51*, 131
648	*51*, 155
661	*51*, 239
668	*51*, 150
669	*51*, 99
682	*51*, 148
688	*51*, 149
756	*51*, 177
760	*51*, 200
768	*51*, 180
793	*51*, 185
829	*51*, 54
858	*51*, 199
886	*51*, 207
890	*51*, 210
907	*51*, 203
911	*51*, 232
Volume 39	
59	*51*, 22
83	*51*, 28
107	*51*, 32, 33
109	*51*, 34
124	*51*, 39
128	*51*, 35, 37
189	*51*, 53
200	*51*, 54, 239
214	*51*, 55
225	*51*, 239
274	*51*, 65
292	*51*, 239
444	*51*, 101
458	*51*, 99, 240
467	*51*, 102
480	*51*, 59
576	*51*, 125
592	*51*, 128
593	*51*, 129, 240
630	*51*, 145
646	*51*, 154, 155
658	*51*, 164
798	*51*, 137
851	*51*, 203
854	*51*, 204
877	*51*, 55
887	*51*, 215
965	*51*, 229
966	*51*, 232
969	*51*, 232
993	*51*, 236
Volume 40	
81	*51*, 40
122	*51*, 53
152	*51*, 61
176	*51*, 65
266	*51*, 84
345	*51*, 98
384	*51*, 240
449	*51*, 137
475	*51*, 156 (2)
563	*51*, 191
567	*51*, 130, 131, 240
575	*51*, 194
633	*51*, 211
635	*51*, 207
669	*51*, 223
Volume 41	
54	*51*, 240
63	*51*, 24
115	*51*, 32
117	*51*, 33
160	*51*, 44
172	*51*, 45
173	*51*, 240
175	*51*, 240
199	*51*, 50
208	*51*, 52
221	*51*, 54
245	*51*, 60
305	*51*, 240
367	*51*, 81
427	*51*, 92
434	*51*, 102
463	*51*, 240
470	*51*, 98
500	*51*, 162
556	*51*, 115
564	*51*, 117
636	*51*, 173
660	*51*, 240
678	*51*, 240
723	*51*, 176
732	*51*, 184
736	*51*, 181
797	*51*, 192
810	*51*, 137

Supplementary References

No.	Suppl.Ref. Vol., Page

Volume 41 continued

812	*51*, 195
885	*51*, 165
892	*51*, 204
942	*51*, 224
993	*51*, 234
995	*51*, 234

Volume 42

3	*51*, 2
6	*51*, 3
24	*51*, 13
30	*51*, 17
45	*51*, 19, 240
58	*51*, 22
108	51, 32
125	*51*, 38
146	*51*, 44
176	*51*, 119
209	*51*, 240
220	*51*, 56
235	*51*, 59
290	*51*, 68
292	*51*, 143
311	*51*, 70
339	*51*, 73
448	*51*, 54
459	*51*, 96
469	*51*, 100
498	*51*, 28
506	*51*, 108
554	*51*, 116
562	*51*, 118
595	*51*, 240
616	*51*, 130, 240
621	*51*, 139
654	*51*, 170
703	*51*, 240
725	*51*, 174
750	*51*, 180
771	*51*, 240

814	*51*, 240
822	*51*, 193
826	*51*, 137
852	*51*, 240
865	*51*, 240
876	*51*, 207
908	*51*, 212
957	*51*, 228
960	*51*, 228

Volume 43

45	*51*, 13
49	*51*, 16
63	*51*, 20
83	*51*, 30
89	*51*, 31
111	*51*, 27, 37
121	*51*, 40
131	*51*, 41
137	*51*, 44
151	*51*, 240
162	*51*, 48
173	*51*, 101
200	*51*, 55
295	*51*, 76
385	*51*, 92
398	*51*, 95
425	*51*, 240
540	*51*, 241
555	*51*, 128
562	*51*, 130
571	*51*, 133
576	*51*, 133
584	*51*, 140
625	*51*, 97
650	*51*, 163
700	*51*, 241
703	*51*, 149, 191
749	*51*, 60
769	*51*, 241
833	*51*, 241
843	*51*, 204
847	*51*, 205
851	*51*, 211
860	*51*, 200

918	*51*, 90
943	*51*, 241
944	*51*, 170
954	*51*, 235
960	*51*, 228
972	*51*, 230
989	*51*, 241

Volume 44

7	*51*, 241
42	*51*, 16
65	*51*, 241
85	*51*, 26
114	*51*, 36
117	*51*, 36
118	*51*, 35
128	*51*, 117
139	*51*, 147
166	*51*, 47
211	*51*, 51, 57
214	*51*, 56
229	*51*, 59
312	*51*, 77
469	*51*, 241
531	*51*, 119
565	*51*, 131
568	*51*, 135, 241
577	*51*, 241
624	*51*, 148
651	*51*, 241
654	*51*, 165
663	*51*, 167
669	*51*, 169
736	*51*, 184, 199
776	*51*, 188
837	*51*, 241
889	*51*, 216
897	*51*, 216
929	*51*, 220
936	*51*, 241
955	*51*, 226

Volume 45

6	*51*, 241

No.	Suppl.Ref. Vol., Page
Volume 45 continued	
28	*51*, 21
41	*51*, 25
72	*51*, 39
106	*51*, 53
284	*51*, 97, 100
308	*51*, 241
377	*51*, 138
385	*51*, 137
414	*51*, 241
422	*51*, 145
439	*51*, 170
447	*51*, 181
462	*51*, 184, 218
476	*51*, 179
488	*51*, 192
510	*51*, 22
555	*51*, 241
Volume 46	
3	*51*, 2, 241
23	*51*, 11
60	*51*, 6
76	*51*, 23
84	*51*, 25
106	*51*, 29, 39
128	*51*, 39
160	*51*, 241
171	*51*, 55
197	*51*, 53
236	*51*, 43
316	*51*, 75
317	*51*, 76
350	*51*, 78, 79, 241
355	*51*, 85
377	*51*, 87
458	*51*, 225
484	*51*, 93, 241
490	*51*, 109
504	*51*, 225
540	*51*, 242
601	*51*, 129
605	*51*, 135
612	*51*, 140
631	*51*, 143, 242
656	*51*, 147
659	*51*, 155
662	*51*, 145, 147
669	*51*, 159
670	*51*, 159
672	*51*, 159, 160
679	*51*, 242
696	*51*, 145, 170
713	*51*, 178
720	*51*, 242
721	*51*, 176
798	*51*, 195
811	*51*, 242
832	*51*, 210
870	*51*, 210
918	*51*, 218
921	*51*, 242
930	*51*, 219, 242
954	*51*, 223, 242
956	*51*, 225
Volume 47	
18	*51*, 163
22	*51*, 7, 242
32	*51*, 13
46	*51*, 18
49	*51*, 242
62	*51*, 21
75	*51*, 24
80	*51*, 26
111	*51*, 39
114	*51*, 40
146	*51*, 48
178	*51*, 54
273	*51*, 73
304	*51*, 242
428	*51*, 98
430	*51*, 98
444	*51*, 242
450	*51*, 143
470	*51*, 105
487	*51*, 107
493	*51*, 109
512	*51*, 113
527	*51*, 201
542	*51*, 117, 242
563	*51*, 119, 121
567	*51*, 116
611	*51*, 167
616	*51*, 139
623	*51*, 242
646	*51*, 242
664	*51*, 145
694	*51*, 165
715	*51*, 171
750	*51*, 242
784	*51*, 188
797	*51*, 195
806	*51*, 192 (2)
829	*51*, 196
839	*51*, 196
872	*51*, 201
888	*51*, 207
901	*51*, 210
940	*51*, 221
954	*51*, 242
955	*51*, 224, 242
974	*51*, 229
Volume 48	
6	*51*, 2
8	*51*, 3
10	*51*, 2
31	*51*, 13
48	*51*, 18
61	*51*, 20
65	*51*, 48
104	*51*, 30
106	*51*, 243
108	*51*, 243
134	*51*, 243
135	*51*, 39
210	*51*, 243
255	*51*, 61
281	*51*, 68
291	*51*, 243

Supplementary References

No.	Suppl.Ref. Vol., Page
Volume 48 continued	
305	*51*, 74
426	*51*, 96
434	*51*, 98
439	*51*, 98
457	*51*, 44
488	*51*, 243
543	*51*, 116
588	*51*, 128
607	*51*, 134, 243
616	*51*, 127
625	*51*, 243
626	*51*, 140
640	*51*, 243
659	*51*, 150
661	*51*, 150
674	*51*, 154
681	*51*, 157
686	*51*, 159
692	*51*, 160
751	*51*, 177
767	*51*, 243
772	*51*, 181, 243
784	*51*, 185
791	*51*, 243
794	*51*, 243
808	*51*, 191
818	*51*, 133, 243
822	*51*, 243
831	*51*, 196
839	*51*, 196
850	*51*, 243
854	*51*, 199
866	*51*, 203, 243
889	*51*, 209, 243
955	*51*, 223
968	*51*, 228
976	*51*, 229
988	*51*, 233
995	*51*, 235

Volume 49	
16	*51*, 5
21	*51*, 7
45	*51*, 15
62	*51*, 20
64	*51*, 22
81	*51*, 25
100	*51*, 243
105	*51*, 34
108	*51*, 35
153	*51*, 45
158	*51*, 49
168	*51*, 51
277	*51*, 71
294	*51*, 78
305	*51*, 243
307	*51*, 77
366	*51*, 243
372	*51*, 243
407	*51*, 97
499	*51*, 244
510	*51*, 115
556	*51*, 115
560	*51*, 124
585	*51*, 130
589	*51*, 133
610	*51*, 140
622	*51*, 143
630	*51*, 144
655	*51*, 147
674	*51*, 156
679	*51*, 157 (2)
681	*51*, 158
697	*51*, 244
699	*51*, 163
703	*51*, 164
763	*51*, 178
877	*51*, 206
882	*51*, 244
898	*51*, 211 (2)
909	*51*, 210
934	*51*, 214
938	*51*, 220
939	*51*, 244
941	*51*, 179

955	*51*, 226
956	*51*, 227
985	*51*, 233 (2)
997	*51*, 237

Volume 50	
9	*51*, 6
15	*51*, 244
23	*51*, 18
43	*51*, 29
72	*51*, 40, 237
89	*51*, 46
97	*51*, 55
98	*51*, 224
116	*51*, 244
120	*51*, 244
123	*51*, 59
155	*51*, 244
157	*51*, 244
170	*51*, 79
252	*51*, 244
260	*51*, 244
271	*51*, 244
272	*51*, 100
307	*51*, 108
343	*51*, 116, 244
365	*51*, 129
368	*51*, 132
380	*51*, 140
383	*51*, 141
389	*51*, 244
396	*51*, 24
400	*51*, 244
411	*51*, 155
412	*51*, 154
415	*51*, 156
419	*51*, 244
433	*51*, 244
438	*51*, 244
444	*51*, 170
449	*51*, 171
466	*51*, 194
480	*51*, 185
500	*51*, 102
502	*51*, 190
515	*51*, 194

No.	Suppl.Ref. Vol., Page
Volume 50 continued	
531	*51*, 199
551	*51*, 209
555	*51*, 244
569	*51*, 221
579	*51*, 227
601	*51*, 237
Volume 51	
2	*51*, 2
21	*51*, 14
31	*51*, 245
131	*51*, 161
159	*51*, 126
171	*51*, 93
272	*51*, 138
289	*51*, 131
317	*51*, 245
345	*51*, 211
446	*51*, 245
453	*51*, 180

Derwent
Journal of Synthetic Methods

Every month this service covers everything new and important in synthetic organic chemistry . . .

Reactions selected from worldwide journal and patent literature

- Each reaction rigorously checked to ensure it only appears once
- Each reaction clearly illustrated with its own scheme
- Comprehensive indices, with an annual cumulation
- Retrospective retrieval through online access to over 80,000 reactions, including Theilheimer's 'Synthetic Methods'
- Structure-searchable either online on STN or in-house as REACCS-JSM.

Available in a variety of formats to suit your needs.

Scientific and Patent Information

For further details contact Annette Jones,
Derwent Information, 14 Great Queen Street, London WC2B 5DF UK
Telephone +44 171 344 2882 Fax +44 171 344 2901 Email eurinfo@derwent.co.uk